Rice: Genetics and Breeding

Rice: Genetics and Breeding

Edited by Abbie Chavez

MURPHY & MOORE
www.murphy-moorepublishing.com

Published by Murphy & Moore Publishing,
1 Rockefeller Plaza,
New York City, NY 10020, USA
www.murphy-moorepublishing.com

Rice: Genetics and Breeding
Edited by Abbie Chavez

International Standard Book Number: 978-1-63987-493-4 (Hardback)

Cataloging-in-Publication Data

Rice : genetics and breeding / edited by Abbie Chavez.
p. cm.
Includes bibliographical references and index.
ISBN 978-1-63987-493-4
1. Rice. 2. Rice--Genetics. 3. Rice--Breeding. I. Chavez, Abbie.
SB191.R5 R53 2022
641.331 8--dc23

TABLE OF CONTENTS

PREFACE

The branch of biology which deals with the study of genes, heredity and genetic variations in organisms is termed as genetics. Rice was one of the first sequenced crop genomes because it has relatively small genome size. The genome of rice can be genetically engineered by using various tools and techniques to increase the genetic potential of rice for yield. Genetic engineering can also be used to improve its biotic and abiotic stress tolerance. Rice breeding focuses on selecting superior overexpressing alleles of the same gene to enhance the grain quality and increase the grain length. The grain size of rice is a significant target for rice breeding because it affects the yield, appearance and eating quality. This book discusses the fundamentals as well as modern approaches of rice genetics and breeding. It provides significant information of this discipline to help develop a good understanding of various breeding methods and related fields. Researchers and students in this field will be assisted by this book.

This book is the end result of constructive efforts and intensive research done by experts in this field. The aim of this book is to enlighten the readers with recent information in this area of research. The information provided in this profound book would serve as a valuable reference to students and researchers in this field.

At the end, I would like to thank all the authors for devoting their precious time and providing their valuable contribution to this book. I would also like to express my gratitude to my fellow colleagues who encouraged me throughout the process.

Editor

1

Genome-Based Identification of Heterotic Patterns in Rice

Ulrike Beukert[1], Zuo Li[1], Guozheng Liu[1], Yusheng Zhao[1], Nadhigade Ramachandra[2], Vilson Mirdita[3], Fabiano Pita[4], Klaus Pillen[5] and Jochen Christoph Reif[1*]

Abstract

Background: Hybrid rice breeding facilitates to increase grain yield and yield stability. Long-term success of hybrid breeding depends on the recognition of high-yielding complementary heterotic patterns, which is lacking in crops like rice.

Result: The main goal of this study was to evaluate the potential and limits to use genomics for establishing heterotic patterns in rice. For this purpose, data of a commercial hybrid rice breeding program targeted to India was analyzed, including 1,960 phenotyped hybrids from three market segments and 262 genotyped parental lines. Our cross-validation study revealed that grain yield of all potential single-crosses can be accurately predicted. Based on the full matrix of hybrid performances, high-yielding heterotic patterns were identified. These heterotic patterns increased grain yield up to 9% compared to the currently employed groups. Heterotic groups of around 14 individuals reflect a good compromise between long-term and short-term selection response.

Conclusions: Our findings clearly underlined the benefits of a genome-based establishment of heterotic patterns in rice as a requirement for a sustainable long-term success of hybrid rice breeding.

Keywords: Heterotic Groups, Heterotic Pattern, Hybrid Rice, Genome-Wide Predictions

Background

Rice belongs to the three leading food crops providing the majority of calories to feed the world (International Rice Research Institute 2013). The demand for rice is raising steadily due to changes in customer priorities and population growth (Khush 2013). Thus, rice production has to be increased. Unfortunately, the potential of expanding the cultivated area is limited (Khush 2005) and, therefore, rice production has to be increased by enhancing yields per area (Khush 2013). Nevertheless, worldwide selection gain in rice is only 1.0% per year and much lower than the required increase in yield potential of 2.4% (Ray et al. 2013). Thus, there exits the strong need to develop and implement improved breeding tools (Phillips 2010; Zhu et al. 2016). A promising approach to boost yield per area consists in hybrid breeding, which has been successfully applied in rice improvement for some target regions such as China (Khush 2005; Khush 2013).

The advantages of hybrid versus line breeding are the exploitation of heterosis resulting in higher grain yield (Xu et al. 2014) and an enhanced yield stability (Longin et al. 2012). Furthermore, hybrid breeding simplifies to stack major dominant genes and results in substantial return on investments, which is required to refinance future breeding progress (Longin et al. 2012). Hybrid rice breeding programs were initiated in China in 1964 based on a cytoplasmic male sterility (CMS) system. The first hybrid rice variety was released in 1976 (Barclay 2010). In addition to CMS, photoperiod sensitive male sterility has been exploited to produce hybrid seeds on a large scale (Chen and Liu 2014; Li et al. 2007). Rice grain yield per area increased through hybrid breeding by approximately 40% during the past three decades (Zhu et al. 2016). As a consequence, hybrid varieties covered more than 50% of China's total rice growing area in 2003 (Cheng et al. 2007a). Moreover, the success of hybrid rice in China encouraged

* Correspondence: reif@ipk-gatersleben.de

[1]Department of Breeding Research, Leibniz Institute of Plant Genetics and Crop Plant Research (IPK) Gatersleben, Corrensstraße 3, 06466 Stadt Seeland, Germany

embarking on a national program for the development of hybrid rice for India in 1989 (Viraktamath 2010).

The optimum exploitation of heterosis requires that the available germplasm is structured into genetically diverse heterotic groups. A heterotic group refers to a collection of genotypes resulting in similar hybrid performance when crossed with individuals belonging to a complementary and genetically distinct germplasm group. A specific combination of two heterotic groups leading to high-yielding hybrids is defined as heterotic pattern (Melchinger and Gumber 1998). Hybrid breeding based on the concept of heterotic patterns leads to more pronounced variance of the breeding values in contrast to the variance of the dominance deviations, which enhances recurrent selection gain (Reif et al. 2007).

Heterotic patterns have been established in the past either considering hybrid seed production traits such as seed yield of the female lines and high pollination capability of the male lines (Reif et al. 2005) or they have been developed empirically by testing combining ability among available germplasm (Fischer et al. 2010). The latter approach is afflicted by the large number of possible hybrid combinations among available elite inbred lines. To solve this challenge, Zhao et al. (2015) developed a three-step approach to search for heterotic patterns: (1) The performance of all possible single-cross combinations is determined through genome-wide or metabolite-based hybrid predictions. (2) The predicted hybrid performances are used to identify superior heterotic patterns based on a simulated annealing algorithm. (3) The optimum size of the heterotic pattern is determined balancing the expected short- and long-term selection.

The genetically distant subgroups of *indica* and *japonica* have been suggested as a promising heterotic pattern for hybrid rice breeding (Cheng et al. 2007a). Nevertheless, fertility barriers hamper so far the exploitation of this heterotic pattern (Ikekashi and Araki 1984; Liu et al. 1996). Several approaches have been used to reduce the sterility occurring when crossing *indica* and *japonica* (Guo et al. 2016; Ouyang et al. 2010), but none of them has been successfully implemented into hybrid rice breeding programs. The search for heterotic groups within the major germplasm pools is limited to studies based on tropical *indica* rice (Wang et al. 2014; Xie et al. 2013). Moreover, previous studies used molecular marker-based genetic distances as proxy of the heterosis or hybrid performance, which is unreliable for unrelated parental inbred lines (Melchinger 1999; Xu et al. 2014; Xu et al. 2016). Thus, other approaches exploiting for example genome-wide hybrid predictions are needed to recognize promising heterotic groups for hybrid rice breeding.

Our study is based on genomic and phenotypic data of 1,960 rice hybrids adapted to the Indian sub-continent, which have been evaluated for grain yield in two to four locations. The hybrids were grouped based on grain size, shape and appearance into three market segments: Long grain (LS) segment with grain type of long slender, length of larger than 6 mm, length/breadth ratio of larger than 3; Medium (MM) grain segment with grain type of medium slender, length of less than 6 mm, length/breadth ratio of 2.5 to 3.0 mm; Short (SS) grain segment with grain type of short slender, length of less than 6 mm, length/breadth ratio of larger than 3. Our main goal was to evaluate the potential and limits of genome-based establishment of heterotic patterns in rice. In particular, the objectives were to (1) investigate the accuracy for genome-wide prediction of hybrid performance in rice, (2) evaluate the benefits of heterotic patterns identified with a simulated annealing algorithm, and (3) assess the optimal size of the heterotic patterns balancing short- and long-term selection gain.

Results

Variance due to specific combining ability effects played a prominent role in hybrid rice populations

We observed a wide range of Best Linear Unbiased Estimations leading to significant ($P < 0.001$) genetic variances within all three market segments (Fig. 1). In total, 19, 4, and 10% of the hybrids significantly ($P < 0.05$) outperformed leading commercial varieties for the market

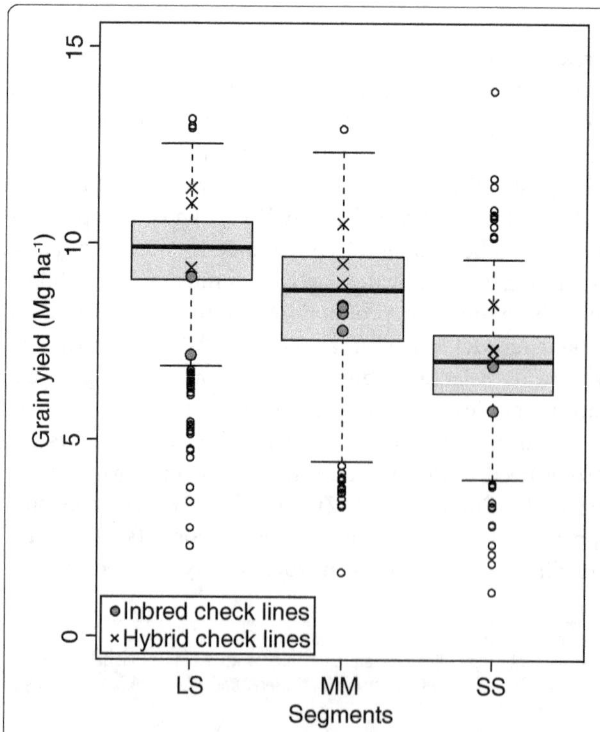

Fig. 1 Best linear unbiased estimations for grain yield performance of hybrids and checks of the market segments LS, MM, and SS

segments LS, MM, and SS, respectively. Genetic variance components involving general combining ability effects σ^2_{GCA} showed higher values for male than for female lines (Table 1). The variance of specific combining ability effects σ^2_{SCA} was for all segments significantly ($P < 0.05$) larger than zero and contributed on average to 42% of the total genetic variance. Heritability estimates ranged from 0.26 to 0.61.

Analyses of population structure revealed genetically distinct parental pools for market segments MM and SS but not for LS

The analyses of linkage disequilibrium (LD) revealed a fast decay of LD with increasing physical distance (Additional file 1: Figure S1). The sharp decay of LD underlines the potential of using the populations for high-resolution genome-wide association mapping. The population structure was examined by applying principal coordinate analyses. The analyses of the 262 genotyped lines revealed absence of a population structure among market segments (Additional file 1: Figure S2). Despite this, differences in quality requirements are large and parents are used only rarely across segments MM, SS, and LS. Thus, we analyzed in the following each market segment separately. In segment LS, male and female lines did not clustered separately (Fig. 2). This was further substantiated by the distributions of Rogers' distances with highest values between male lines as well as between male and female lines (Fig. 2b). In contrast, for segments MM and SS, principal coordinate analyses and distributions of pairwise Rogers' distances revealed genetically distinct parental pools (Fig. 2c, d, e and f).

Hybrid performance can be predicted with high accuracy

The predicted hybrid performance values of the full diallel matrix formed the basis to search for promising

Table 1 Second degree statistics for hybrid grain yield (Mg ha^{-1}) experiments in market segments LS, MM, and SS performed in three, two, and four environments, respectively

Source	LS	MM	SS
$\sigma^2_{Genotype}$	0.99***	0.53***	0.81***
$\sigma^2_{GCA\ male}$	0.30**	0.27	0.24***
$\sigma^2_{GCA\ female}$	0.24**	0.13	0.11***
σ^2_{SCA}	0.45***	0.13*	0.46***
$\sigma^2_{Genotype\ x\ Location}$	1.48***	2.10***	0.90***
$\sigma^2_{GCA\ male\ x\ Location}$	0.88***	1.39***	0.33***
$\sigma^2_{GCA\ female\ x\ Location}$	0.11***	0.26***	0.12***
$\sigma^2_{SCA\ x\ Location}$	0.49***	0.45**	0.45***
σ^2_{error}	0.59	0.97	1.34
h^2	0.59	0.26	0.61

Probability level: *** < 0.001, ** < 0.01, * < 0.05, not assigned not significant

heterotic patterns. We used a chess-board like cross-validation strategy to evaluate the prediction ability of the hybrid performance (Additional file 1: Figure S3). The hybrids were split into an estimation set and the test groups T2, T1, and T0 with decreasing relationship to the estimation set. The accuracy observed in the T2 scenario is relevant for compiling the full hybrid performance matrix, because the parents of all predicted hybrids were evaluated in other single-cross combinations. The prediction abilities ranged for the T2 scenario from 0.33 to 0.58 (Fig. 3). The prediction accuracies were estimated by standardizing the prediction abilities with the square root of heritabilities and amounted to 0.72 averaged across the three market segments.

Detected heterotic patterns are stable across varying group sizes

We used the predicted performances of all potential single-cross hybrids in order to identify high-yielding heterotic patterns with population sizes ranging from two to 20 lines using a previously developed simulated annealing algorithm. In this search, we assumed that lines can be clustered irrespective to their restoration ability. For all three segments, selection and clustering of lines into heterotic groups were stable across the full range of examined population sizes (Additional file 1: Table S1). Interestingly, only few female lines were selected to establish heterotic patterns: In market segment MM, no female line was selected. In the market segments LS and SS, 38 and 40% of the females were selected, respectively. The selected heterotic patterns substantially outperformed the overall mean with a maximum difference ranging from 14% for market segment SS to 28% for market segment MM (Fig. 4a, b and c).

The performances of the identified heterotic patterns were compared with the average hybrid performance of crosses between male and female lines. In order to make a fair comparison, we focused on a standardized population size of 256 hybrids. The hybrids of the identified heterotic patterns surpassed the superior female and male single crosses by 2, 9, and 7% for market segments LS, MM, and SS, respectively (Fig. 5).

Short- and long-term success of the identified heterotic patterns

We used the theoretical selection limit as a parameter to assess the long-term success of the detected heterotic groups. The selection limit increased with growing heterotic group size and plateaued at around 14 individuals across the different market segments (Fig. 4a, b and c). We further assessed the long-term success of the heterotic patterns by estimating genetic

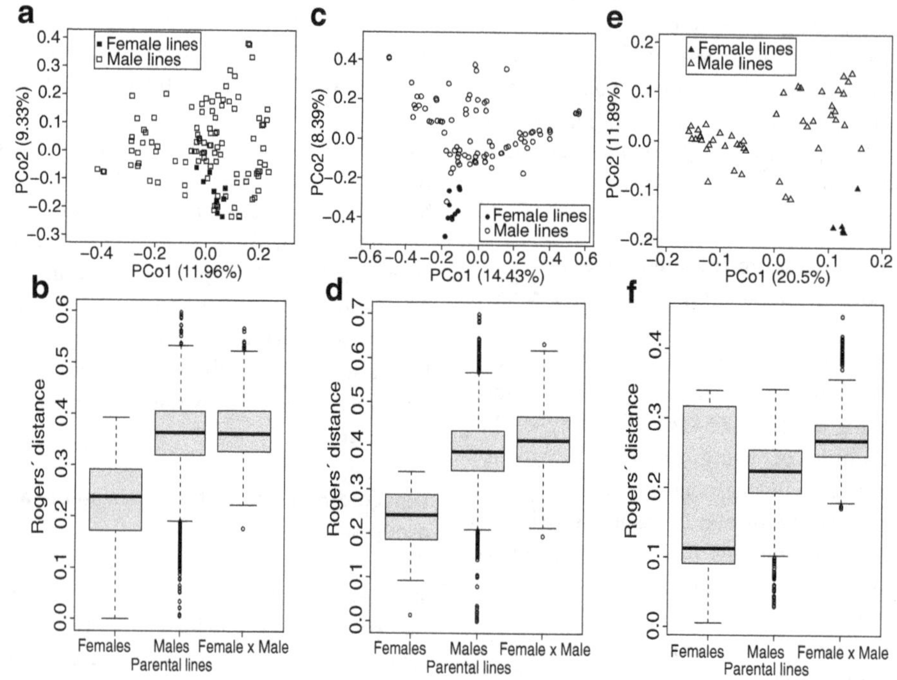

Fig. 2 Principal coordinate analyses of parental lines (**a**) for market segments LS, (**c**) MM, and (**e**) SS with relating distribution of Rogers' distances within and between parental pools (**b**) for market segments LS, (**d**) MM, and (**f**) SS

representativeness, which also plateaued at around 14 genotypes. In contrast, the short-term success assessed as the population mean was highest at small population sizes and decreased nearly linearly with increasing size of the heterotic patterns.

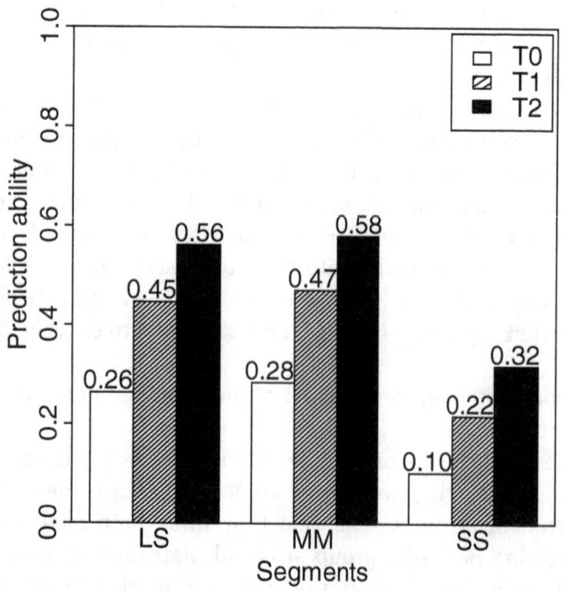

Fig. 3 Prediction ability of grain yield performance for different subgroups of market segments LS, MM, and SS

Discussion

More than 50% of the total cultivated rice area in China is grown with hybrid varieties, which was enabled by substantial investments in rice research and breeding (Cheng et al. 2007b). Encouraged by the success of hybrid rice in enhancing the rice production and productivity in China, India initiated a national program for the development and large scale adaption of hybrid rice in 1989 (Viraktamath 2010). Plant breeders in the government and private sectors have launched hybrid varieties for different states in India. These hybrids have a 10 to 44% higher yield compared to popular high-yielding line varieties (Wanjari et al. 2006). Thus, a good beginning has been made by ushering in to an era of hybrid rice breeding and production in India (Spielman et al. 2013). Long-term success of hybrid breeding depends on the establishment of complementary heterotic groups (Reif et al. 2007; Zhao et al. 2015). Nevertheless, heterotic groups in rice are not clearly defined (Xie et al. 2012). This encouraged us to assess the potential and limits of a recently developed three-step approach to search for high-yielding heterotic patterns using data of a commercial hybrid rice breeding program. The hybrid rice breeding program is centered on a CMS hybrid seed production strategy. Lines have been clustered based on their restoration ability into female and male groups. The molecular analyses revealed for two of the examined rice

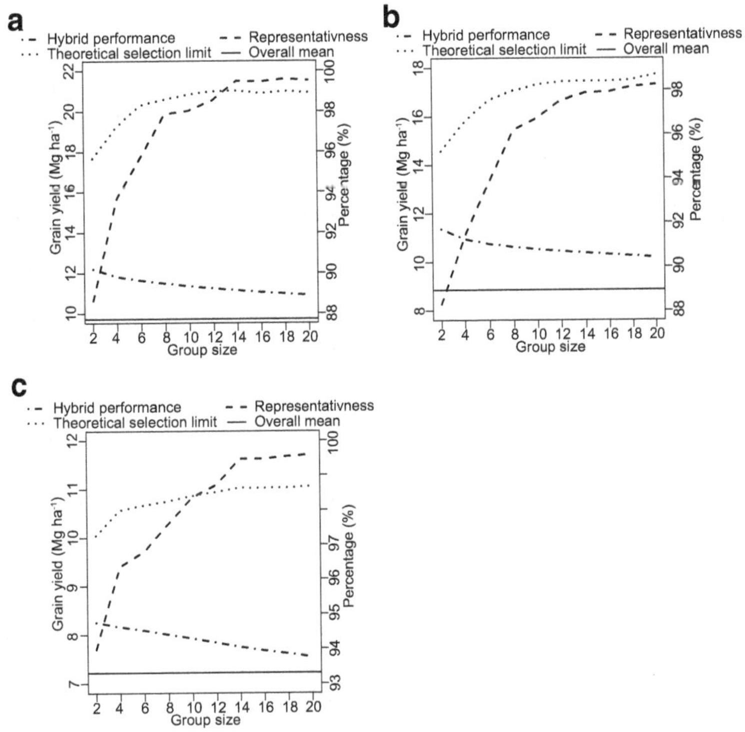

Fig. 4 Short- (Hybrid performance) and long-term success (Representativeness, Theoretical selection limit) in dependence of heterotic group size for market segments (**a**) LS, (**b**) MM, and (**c**) SS

market segments that male and female pools were genetically divergent (Fig. 2). Two major restorer genes *Rf3* and *Rf4* are known for the underlying CMS-WA cytoplasm (Zhang et al. 1997; Zhang et al. 2002). Thus, a conversion of female into male lines or vice versa can be realized but is sometimes hampered by modifier genes influencing the penetrance of the restorer genes. Despite this, we assumed in our study that the hybridization system is not restricting the grouping of lines.

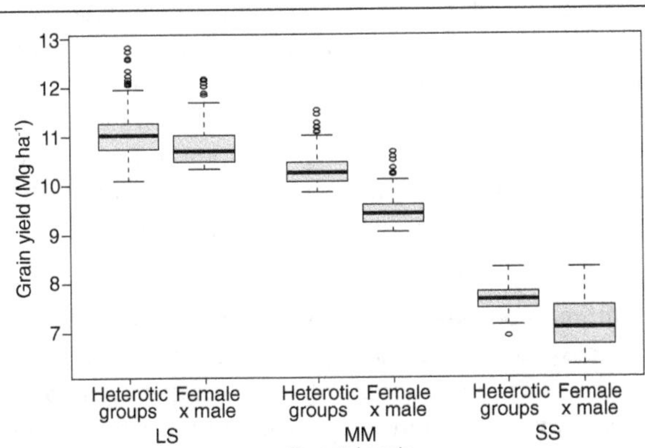

Fig. 5 Distribution of predicted hybrid performances for 256 best performing female x male crosses in comparison to heterotic groups of size 16

Hybrids significantly surpassed yield performance of commercial line varieties

The yield advantage of rice hybrids over released line varieties, which is often denoted as commercial heterosis, was reported to be in the range of 10 to 20% (Cheng et al. 2007a; Huang et al. 2016; Wang et al. 2014; Xie et al. 2013). We estimated the magnitude of the commercial heterosis by contrasting grain yield of the phenotyped hybrids with the performance of the highest-yielding line variety included in the trials. The commercial heterosis amounted to 25, 34, and 37% for the segments LS, MM, and SS, respectively (Fig. 1). The slightly higher values compared to the previously reported ones (Cheng et al. 2007a; Huang et al. 2016; Wang et al. 2014; Xie et al. 2013) can be explained by the moderate phenotyping intensity and the augmented design used in our study with unreplicated evaluation of hybrids but replicated evaluation of line varieties. The different allocation of resources for checks and hybrids was caused by restrictions in hybrid seed production. Summarizing, our findings underlined the yield advantage of hybrids over inbred lines and underpinned the potential to boost grain yield through hybrid rice breeding.

Pronounced variance of specific combining ability effects

A predominance of σ^2_{GCA} over σ^2_{SCA} leads to an accurate prediction of the hybrid performance based on GCA effects (Zhao et al. 2013). The experimental resources, which need to be installed to estimate GCA effects, are a linear function of the number of parents. This is in contrast to the resources required to estimate SCA effects, which are a quadratic function of the number of parents.

The incomplete factorial mating designs for the market segments LS, SS, and MM differed in the number of non-phenotyped hybrids (Additional file 1: Figure S4). Nevertheless, this should not bias the estimates of the variance components as we can assume missing at random (Little and Rubin 2002). A favorable ratio of σ^2_{GCA} versus σ^2_{SCA} is expected in crops with genetically divergent heterotic groups such as in hybrid maize breeding (Reif et al. 2007). In contrary to this, we observed that σ^2_{SCA} amounts to 25 to 57% of the total genetic variance in the hybrid populations for the three market segments (Table 1). The σ^2_{SCA} is higher compared to previous estimates in other selfing species such as wheat (Zhao et al. 2015) or barley (Philipp et al. 2016) and can be explained only partially by a lack of genetic divergence between the parental lines (Fig. 2). Another reason for the low estimates of σ^2_{GCA} is the reduced diversity in the pool of female lines, which is most likely due to the bottleneck in selecting suitable maintainer lines. The relevance of σ^2_{SCA} indicates that it is challenging to identify superior hybrid combinations based on GCA effects. Moreover, only slow recurrent selection gain is expected when interpreting σ^2_{GCA} as an estimate of the variance of the breeding values.

High accuracy observed for genome-wide hybrid prediction

We implemented ridge regression best linear unbiased prediction (RR-BLUP) of the hybrid performances considering additive and dominance effects (Zhao et al. 2014). The ability of prediction was evaluated applying a chess-board like cross-validation scenario (Additional file 1: Figure S3). The prediction ability averaged across the three market segments increased with increasing relatedness between the estimation and test sets from 0.21 for the T0 scenario to 0.49 for the T2 scenario (Fig. 3). These findings confirmed the high relevance of relatedness as the driving force of the prediction ability (Desta and Ortiz 2014; Zhao et al. 2015).

The prediction accuracy was estimated by standardizing the prediction ability with the square root of heritability estimates to compare our findings across market segments and also with other studies. Prediction accuracies were comparable for segment LS and MM but lower for segment SS (Fig. 3). This difference can be explained by a smaller population size (Additional file 1: Table S2) in combination with a more unbalanced factorial mating design (Additional file 1: Figure S4) for segment SS than LS and MM. Zhao et al. (2015) have shown using a resampling strategy based on 1,604 wheat hybrids that both the number of parents used in the factorial mating design and the balanced representation of parental lines used to generate hybrids are important factors driving the prediction accuracy.

Previous studies observed that the prediction accuracies were smaller for self-pollinating than for outcrossing crops (Philipp et al. 2016; Massman et al. 2012; Technow et al. 2014). Moreover, prediction accuracies were higher for crops with clearly defined heterotic groups versus crops with absence of heterotic groups (Philipp et al. 2016). In line with these expectations, we observed prediction accuracies, which were slightly smaller than reported previously for maize hybrids (Massman et al. 2012) and similar to those reported for wheat (Zhao et al. 2015) and rice hybrids (Xu et al. 2014; Xu et al. 2016). Summarizing, genome-wide prediction is a useful and appropriate method to predict hybrid performances in rice.

The identified heterotic patterns outyielded the currently used hybrid populations

The search for promising heterotic patterns, which maximizes the grain yield in the hybrid populations, was performed using a simulated annealing algorithm based on the predicted hybrid performance matrices. The detected heterotic patterns outyielded the population means by 14% for SS to 28% for LS and MM (Fig. 4). The lower advantage of SS compared with LS and MM can be explained by the limited genetic diversity of the SS segment (Fig. 2). Moreover, the detected heterotic patterns resulted in a yield improvement of 7 to 35% compared to hybrids between the existing male and female pools (Fig. 5). Interestingly, male lines were selected more frequently for the heterotic patterns than female lines. This finding is in line with those reported by Huang et al. (2016), who observed also a higher yield potential of male parents either pointing to more beneficial alleles carried by male than female lines or to a bottleneck in selecting stable maintainer lines.

Compromise between short- and long-term success was accomplished with heterotic groups each including 14 lines

The efficiency and success of heterotic patterns can be assessed by short- and long-term response to selection (Zhao et al. 2015). Short-term success is maximized by a high selection intensity choosing only the very best performing parental lines. In contrast, long-term success increases with increasing diversity and heterotic group size. Heterotic group sizes should reflect a compromise between short- and long-term successes.

Short-term selection response was assessed based on the hybrid performance of the selected heterotic pattern. We ignored a combination of hybrid performance and expected selection gain, i.e., the usefulness criterion, which was previously used (Zhao et al. 2015) because the usefulness criterion is mainly influenced by the hybrid performance. The assessment of long-term selection gain was based on the parameters of theoretical selection limit and genetic representativeness as suggested previously (Zhao et al. 2015). The genetic representativeness is an index estimating the proportion of specific genomes within the full population represented by the identified heterotic groups (Druet et al. 2014). The theoretical selection limit denotes the maximum hybrid performance realized by reciprocal recurrent selection based on the detected heterotic groups.

We observed that the parameters of long-term response plateaued at approximately 14 genotypes per heterotic group (Fig. 4). This group size was also related to an acceptable level of short-term selection gain; thus, representing a good compromise between short- and long-term selection response. Zhao et al. (2015) reported a similar heterotic group size of 16 individuals per group achieving a suitable balance between short-term and long-term selection gain for European hybrid wheat breeding. Summarizing, a beneficial hybrid breeding in rice can be realized with a heterotic group size of 14 parents, which is connected to a suitable level of short-term as well as long-term selection gain.

Conclusions

Hybrid breeding and the effective utilization of heterosis crucially depends on the identification of heterotic patterns. In this study, we have evaluated a three-step approach to detect heterotic patterns in rice using data of a commercial hybrid breeding program. Our findings revealed that hybrid rice breeding based on the identified heterotic patterns holds the potential to boost grain yield and represents an important step for the long-term success of hybrid rice breeding.

Methods

Plant material

The plant material included rice hybrids from the breeding program of Bayer Crop Science, adapted to the Indian sub-continent. The hybrids were grouped based on grain size, shape and appearance into three market segments: Long grain (LS) segment with grain type of long slender, length of larger than 6 mm, length/breadth ratio of larger than 3; Medium (MM) grain segment with grain type of medium slender, length of less than 6 mm, length/breadth ratio of 2.5 to 3.0 mm; Short (SS) grain segment with grain type of short slender, length of less than 6 mm, length/breadth ratio of larger than 3. All parental lines were derived from the indica rice pool with some introgressions from the japonica group. The lines were grouped into restorer (male) and maintainer (female) lines, because hybrids are produced based on a cytoplasmic male sterility (CMS) system. Most lines were used specifically for a certain market segment and only 17 lines were used as parents across different market segments (Additional file 1: Figure S5). The total number of unique lines amounted to 270, from which 262 were used for genotyping.

For the hybrid production in segment LS, 109 males and 13 females were used, segment MM consisted of 95 male and 16 female parents and in segment SS crosses were made between 47 male and 10 female lines (Additional file 1: Figure S4). The lower number of female than male lines is because it is easier to breed restorer than maintainer lines. The parental lines were crossed using an unbalanced factorial mating design. Phenotypic data comprised of 625, 935, and 400 hybrids for the LS, MM, and SS segment, respectively.

Field trials

The phenotypic data is based on grain yield trials conducted in the year 2014. The hybrids were evaluated in three, two, and four locations for LS, MM, and SS, respectively. The locations were Faizabad (FZB, 26.78°N; 82.13°E; loam, 97 m a.s.l.), Hyderabad (HYD, 17.44°N; 78.11°E, black loam, 569 m a.s.l.), Dhantori (DHN, 28.2° N; 75.45°E, loam, 257 m a.s.l.), Mysore (MYS, 12.5°N; 76.67°E, loam, 709 m a.s.l.), and Raipur (RPR, 21.19°N; 81.56°E, loam, 298 m a.s.l.). In all locations, genotypes were sown in nursery beds and 30 days aged seedlings were manually transplanted in the puddled main field conditions.

The experimental designs followed augmented designs including trials and blocks within replicated checks, but unreplicated entries. Restrictions in hybrid seed production and restricted budget were the main reason for choosing an augmented design with replicated checks, but unreplicated hybrids. The design of a field experiment is graphically illustrated for location HYD and segment MM (Additional file 1: Figure S6). The field experiments in the different locations were structured into trials. Every trial was further split into blocks including entries and five to seven checks. Plot size ranged from 3.6 m^2 to 4.2 m^2 and transplanting density was 33 seedlings m^{-2} using one seedling per hill.

Genomic marker data

In total 262 parental lines were genotyped with a 6 K SNP array based on an Illumina Infinium assay (Yu et al. 2014).

After excluding markers with minor allele frequency below 5%, 5,221 high-quality SNP markers were left for further analyses. For 612, 701, and 242 hybrids genotypic information was available for the LS, MM, and SS segment, respectively.

Statistical analyses of field trials

We analyzed every market segment independently. The variance components were estimated fitting the following model:

$$y_{jkmn} = \mu + m_j + g_j + l_k + t_{km} + b_{kmn} + (gl)_{jk} + \varepsilon_{jkmn},$$

where y_{jkmn} is the grain yield of jth genotype in n^{th} block, m^{th} trial and k^{th} location, μ is the overall mean, m_i is an effect for maturity of the observed genotype; g_j refers to the effect of the jth genotype, l_k to the effects of the kth location, t_{km} to the effect of the mth trial at the kth location, b_{kmn} to the effect of the nth block at the mth trial at the kth location, $(gl)_{jk}$ to the interaction effect between the jth genotype and the kth location and ε_{jkmn} to the residual. All effects except m_i were modeled as random. The estimated variance components were used to calculate the broad-sense heritability as:

$$h^2 = \frac{\sigma^2_{Genotype}}{\sigma^2_{Phenotype}} = \frac{\sigma^2_{Genotype}}{\sigma^2_{Genotype} + \frac{\sigma^2_{GxL}}{No.\ of\ location} + \frac{\sigma^2_{error}}{No.\ of\ location \times No.\ of\ replicates}}$$

The genotypic variance was further decomposed into the variance due to general (GCA) and specific combining ability effects (SCA). Significance of variance components were tested using a likelihood-ratio test. We used the above outlined model and assumed fixed genotype effects to obtain the best linear unbiased estimations (BLUEs) of every genotype (Additional file 2). All described statistical approaches were conducted using R (The R Core Team 2016) and linear mixed models were implemented with the package ASReml-R (Gilmour et al. 2009).

Analyses of population structure

The genomic data were used to estimate the allele frequency of every marker. Linkage disequilibrium (LD) was assessed by the LD measure r^2 (Weir 1996). Rogers' distances (Rogers 1972) among all pairs of parental lines were calculated. The matrix of Rogers' distances was used to perform principal coordinate analyses (Gower 1966).

Prediction of hybrid performance

The genome-wide prediction models make use of all hybrids for which genotypic and phenotypic information were available, i.e., 612, 701, and 242 hybrids for the LS, MM, and SS segment, respectively. For genome-wide prediction, we implemented ridge regression best linear unbiased prediction (RR-BLUP), modeling both additive and dominance effects as:

$$Y = 1_n\mu + Z_A a + Z_D d + e.$$

Y refers to BLUEs of the hybrids for grain yield. Vector 1_n includes only ones and its element number is equal to the number of n hybrids; μ refers to the overall mean. Design matrices Z_A and Z_D have a dimension of $n \times m$, where m is the number of markers and were defined using the F_∞ metric (Zhao et al. 2013). The vector e includes residual values for the single cross combinations. Estimations or predictions of a, d, and μ were done by implementing Henderson's mixed model equation (Henderson 1984). Normal distribution and constant variance for the additive and dominance effects were assumed (Zhao et al. 2013).

The accuracy of the prediction of grain yield was evaluated using cross-validations, which entails the division of the total population into estimation and test populations. Marker effects were estimated in the estimation population and used to predict the performance of hybrids in the test population. Since relatedness strongly influences prediction accuracy (Habier et al. 2007), a cross-validation strategy was used considering three test populations with varying degree of relatedness to the estimation population (Additional file 1: Figure S3): Test set T2 most closely related to the estimation set included only hybrids derived from the same parents as the hybrids that had been evaluated, while the less related test set T1 included hybrids sharing one parent with the hybrids in the estimation set and the least related test set T0 included only hybrids having no parents in common with the estimation set. Prediction ability was estimated as Pearson's correlation coefficient between the observed and the predicted hybrid performance.

Identification of heterotic groups

The search of heterotic patterns maximizing the hybrid performance was performed based on the predicted hybrid performances using a simulated annealing algorithm. The implementation of the simulated annealing algorithm is described in detail by Zhao et al. (2015). We assumed the same group size for two matching heterotic groups and performed the search for increasing group sizes from two to 20.

Assessing the short- and long-term success of heterotic groups

Beside the hybrid performance as a short-term success parameter, theoretical selection limit (Zhao et al. 2015) and genetic representativeness describing long-term success were determined to evaluate the performance of the identified heterotic groups. The genetic representativeness gives the proportion of specific genomes within the full population represented by the identified heterotic groups (Druet et al. 2014). Representativeness of two specific heterotic groups depends from the additive relationship within heterotic groups beside additive relationship between the full population and heterotic groups and the genome proportion of heterotic groups. Additive relationship was calculated as one minus the Rogers' distance values of parental pairs (Melchinger et al. 1991). The final representativeness for two matching heterotic groups was the result of the mean value of all parental lines included in the selected heterotic patterns (Zhao et al. 2015). Theoretical selection limit is closely linked to the concept of maximal long-term selection response (Walsh and Lynch 2014) and represents the maximum hybrid performance, which can be reached by reciprocal recurrent selection based on the identified heterotic groups performing an infinite number of selection cycles. We assumed absence of migration, mutation, and epistasis for estimating of theoretical selection limit based on the predicted marker effects (Zhao et al. 2015).

Additional files

Additional file 1: for Genome-based identification of heterotic. **Figure S1.** Decay of linkage disequilibrium measured using r^2 with physical map distance. **Figure S2.** Principal coordinate analysis of all parental lines for market segments LS, MM, and SS. **Figure S3.** Applied cross validation scenario exemplifying the selected fractions for market segment MM. **Figure S4.** Unbalanced factorial crossing designs for market segment (A) LS, (B) MM, and (C) SS. **Figure S5.** Venn diagram of overlapping parental lines of the hybrid evaluation trials in market segments LS, MM, and SS. **Figure S6.** Experimental design and distribution of checks for hybrid experiment of market segment MM at location HYD. Orange and black lines represent the size of trials and blocks, respectively. White plots were not phenotyped. **Table S1.** Composition of heterotic groups selected with the simulated annealing algorithm as well as the average hybrid performance between both heterotic groups (Inter) for the market segments LS, MM, and SS. **Table S2.** Composition of estimation and test populations for market segment LS, MM, and SS.

Additional file 2: Best linear unbiased estimations for hybrid grain yield experiments in market segments LS, MM, and SS.

Authors' contributions

JCR, KP, VM, and YZ conceived the design of this study. NG and FP coordinated the SNP genotyping and the experiments including the phenotypic trait measurements of the plant materials. UB, ZL, and GL conducted the analyses. UB and JCR wrote the manuscript. All authors have read and approved the final manuscript.

Competing interests

The authors declare that they have no competing interests.

Author details

[1]Department of Breeding Research, Leibniz Institute of Plant Genetics and Crop Plant Research (IPK) Gatersleben, Corrensstraße 3, 06466 Stadt Seeland, Germany. [2]Bayer Bioscience, 500081 Hyderabad, India. [3]European Wheat Breeding Center, Bayer Crop Science, Am Schwabenplan 8, 06466 Stadt Seeland, Germany. [4]Biometrics and Breeding Research US, Bayer Crop Science, 407 Davis Dr., Morrisville, NC, USA. [5]K. Pillen, Chair of Plant Breeding, Martin-Luther-University Halle-Wittenberg, Betty-Heimann-Str. 3, 06120 Halle/Saale, Germany.

References

Barclay A (2010) Hybridizing the world. Rice Today 9(4):32–5

Chen L, Liu Y-G (2014) Male Sterility and Fertility Restoration in Crops. Annu Rev Plant Biol 65:579–606. doi:10.1146/annurev-arplant-050213-040119

Cheng S-H, Zhuang J-Y, Fan Y-Y, Du J-H, Cao L-Y (2007a) Progress in Research and Development on Hybrid Rice: A Super-domesticate in China. Ann Bot 100:959–66. doi:10.1093/aob/mcm121

Cheng S-H, Cao L-Y, Zhuang J-Y, Chen S-G, Zhan X-D, Fan Y-Y, Zhu D-F, Min S-K (2007b) Super Hybrid Rice Breeding in China: Achievements and Prospects. J Integr Plant Biol 49:805–10. doi:10.1111/j.1672-9072.2007.00514.x

Desta ZA, Ortiz R (2014) Genomic selection: genome-wide prediction in plant improvement. Trends Plant Sci 19:592–601. doi:10.1016/j.tplants.2014.05.006

Druet T, Macleod IM, Hayes BJ (2014) Toward genomic prediction from whole-genome sequence data: impact of sequencing design on genotype imputation and accuracy of predictions. Heredity 112:39–47. doi:10.1038/hdy.2013.13

Fischer S, Maurer HP, Würschum T, Möhring J, Piepho H-P, Schön CC, Thiemt E-M, Dhillon BS, Weissmann EA, Melchinger AE, Reif JC (2010) Development of Heterotic Groups in Triticale. Crop Sci 50:584–90. doi:10.2135/cropsci2009.04.0225

Gilmour AR, Gogel BJ, Cullis BR, Thompson R (2009) ASReml User Guide Release 3.0. VSN International Ltd, Hemel Hempstead, HP1 1ES, UK. https://www.vsni.co.uk/downloads/asreml/release3/UserGuide.pdf. Accessed 25 May 2016.

Gower JC (1966) Some distance properties of latent root and vector methods used in multivariate analysis. Biometrika 53:325–38

Guo J, Xu X, Li W, Zhu W, Zhu H, Liu Z, Luan X, Dai Z, Liu G, Zhang Z, Zeng R, Guang T, Fu X, Wang S, Zhang G (2016) Overcoming inter-subspecific hybrid sterility in rice by developing indica-compatible japonica lines. Sci Rep 6: 26878. doi:10.1038/srep26878

Habier D, Fernando RL, Dekkers JCM (2007) The impact of genetic relationship information on genome-assisted breeding fs. Genetics 177:2389–97. doi:10.1534/genetics.107.081190

Henderson CR (1984) Applications of linear models in animal breeding, 3rd edn. University of Guelph, Ontario

Huang X, Yang S, Gong J, Zhao Q, Feng Q, Zhan Q, Zhao Y, Li W, Cheng B, Xia J, Chen N, Huang T, Zhang L, Fan D, Chen J, Zhou C, Lu Y, Weng Q, Han B (2016) Genomic architecture of heterosis for yield traits in rice. Nature 537: 629–33. doi:10.1038/nature19760

Ikekashi H, Araki H (1984) Varietal Screening of Compatibility Types Revealed in F1 Fertility of Distant Crosses in Rice. Jpn J Breed 34:304–13

International Rice Research Institute (2013) Rice as human food. In: Rice Almanac, 4th edn. Los Banos, Philippines, pp 10–4, http://books.irri.org/9789712203008_content.pdf. Accessed 25 April 2016

Khush GS (2005) What it will take to Feed 5.0 Billion Rice consumers in 2030. Plant Mol Biol 59:1–6. doi:10.1007/s11103-005-2159-5

Khush GS (2013) Strategies for increasing the yield potential of cereals: case of rice as an example. Plant Breed 132:433–6. doi:10.1111/pbr.1991

Li S, Yang D, Zhu Y (2007) Characterization and Use of Male Sterility in Hybrid Rice Breeding. J Integr Plant Biol 49:791–804. doi:10.1111/j.1672-9072.2007.00513.x

Little RJA, Rubin DB (2002) Statistical analysis with incomplete data, 2nd edn. John Wiley & Sons, New York

Liu KD, Zhou ZQ, Xu CG, Zhang Q, Saghai Maroof MA (1996) An analysis of hybrid sterility in rice using a diallel cross of 21 parents involving indica, japonica and wide compatibility varieties. Euphytica 90:275–80

Longin CFH, Mühleisen J, Maurer HP, Zhang H, Gowda M, Reif JC (2012) Hybrid breeding in autogamous cereals. Theor Appl Genet 125:1087–96. doi:10.1007/s00122-012-1967-7

Massman JM, Gordillo A, Lorenzana RE, Bernardo R (2012) Genomewide predictions from maize single-cross data. Theor Appl Genet 126:13–22. doi:10.1007/s00122-012-1955-y

Melchinger AE, Messmer MM, Lee M, Woodman WL, Lamkey KR (1991) Diversity and Relationships among U.S. Maize Inbreds Revealed by Restriction Fragment Length Polymorphisms. Crop Sci 31:669–78

Melchinger AE (1999) Genetic Diversity and Heterosis. In: Coors JG, Pandey S (eds) The Genetics and Exploitation of Heterosis in Crops, 1st edn. American Society of Agronomy, Crop Science of America, Madison, pp 99–118
Melchinger AE, Gumber RK (1998) Overview of Heterosis and Heterotic Groups in Agronomic Crops. In: Larnkey KR, Staub JE (eds) Concepts and Breeding of Heterosis in Crop Plants, 1st edn. Crop Science Society of America, Madison, pp 29–44
Ouyang Y, Liu Y-G, Zhang Q (2010) Hybrid sterility in plant: stories from rice. Curr Opin Plant Biol 13:186–92. doi:10.1016/j.pbi.2010.01.002
Philipp N, Liu G, Zhao Y, He S, Spiller M, Stiewe G, Pillen K, Reif JC, Li Z (2016) Genomic Prediction of Barley Hybrid Performance. Plant Genome 9:1–8. doi:10.3835/plantgenome2016.02.0016
Phillips RL (2010) Mobilizing Science to Break Yield Barriers. Crop Sci 50:99–108. doi:10.2135/cropsci2009.09.052
Ray DK, Mueller ND, West PC, Foley JA (2013) Yield trends are insufficient to double global crop production by 2050. PLoS One 8(6):e66428. doi:10.1371/journal.pone.0066428
Reif JC, Gumpert F-M, Fischer S, Melchinger AE (2007) Impact of Interpopulation Divergence on Additive and Dominance Variance in Hybrid Populations. Genetics 176:1931–4. doi:10.1534/genetics.107.074146
Reif JC, Hallauer AR, Melchinger AE (2005) Heterosis and heterotic patterns in maize. Maydica 50:215–23
Rogers JS (1972) Measures of genetic similarity and genetic distance. In: Wheeler MR (ed) Studies in Genetics VII, Univ, Texas Publ. 7213. University of Texas, Austin, pp 145–53
Spielman DJ, Kolady DE, Ward PS (2013) The prospects for hybrid rice in India. Food Sec. doi:10.1007/s12571-013-0291-7
Technow F, Schrag TA, Schipprack W, Bauer E, Simianer H, Melchinger AE (2014) Genome Properties and Prospects of Genomic Prediction of Hybrid Performance in a Breeding Program of Maize. Genetics 197:1343–55. doi:10.1534/genetics.114.165860
The R Core Team (2016) R: A Language and Environment for Statistical Computing. R Foundation for Statistical Computing, Vienna, https://cran.r-project.org/doc/manuals/fullrefman.pdf. Accessed 23 May 2016
Viraktamath BC (2010) Hybrid Rice in India - Current Status and Future Prospects. Directorate of Rice Research, Rajendranagar, Hyderabad, http://www.rkmp.co.in/sites/default/files/ris/research-themes/Hybrid%20rice%20in%20India.pdf. Accessed 25 Nov 2016
Walsh B, Lynch M (2014) Long-Term Response: Finite Population Size and Mutation. In: Evolution and Selection of Quantitative Traits., pp 181–202, http://nitro.biosci.arizona.edu/zbook/NewVolume_2/pdf/Chapter26.pdf. Accessed 23 Nov 2015
Wang K, Qiu F, Larazo W, Dela Paz MA, Xie F (2014) Heterotic groups of tropical indica rice germplasm. Theor Appl Genet 128:421–30. doi:10.1007/s00122-014-2441-5
Wanjari RH, Mandal KG, Ghosh PK, Adhikari T, Rao NH (2006) Rice in India: Present Status and Strategies to Boost Its Production Through Hybrids. J Sustain Agric 28:19–39. doi:10.1300/J064v28n01_04
Weir BS (1996) Genetic Data Analysis: Methods for Discrete Population Genetic Data, 2nd edn. Sinauer Associates, Sunderland
Xie F, Guo L, Ren G, Hu P, Wang F, Xu J, Li X, Qiu F, Dela Paz MA (2012) Genetic diversity and structure of indica rice varieties from two heterotic pools of southern China and IRRI. Plant Genet Res 10:186–93. doi:10.1017/S147926211200024X
Xie F, He Z, Esguerra MQ, Qiu F, Ramanathan V (2013) Determination of heterotic groups for tropical Indica hybrid rice germplasm. Theor Appl Genet 127:407–17. doi:10.1007/s00122-013-2227-1
Xu S, Xu Y, Gong L, Zhang Q (2016) Metabolomic prediction of yield in hybrid rice. Plant J. doi:10.1111/tpj.13242
Xu S, Zhu D, Zhang Q (2014) Predicting hybrid performance in rice using genomic best linear unbiased prediction. Proc Natl Acad Sci 111:12456–61. doi:10.1073/pnas.1413750111
Yu H, Xie W, Li J, Zhou F, Zhang Q (2014) A whole-genome SNP array (RICE6K) for genomic breeding in rice. Plant Biotechnol J 12:28–37. doi:10.1111/pbi.12113
Zhang G, Lu Y, Bharaj TS, Virmani SS, Huang N (1997) Mapping of the Rf-3 nuclear fertility-restoring gene for WA cytoplasmic male sterility in rice using RAPD and RFLP markers. Theor Appl Genet 94:27–33
Zhang QY, Liu YG, Zhang GQ, Mei MT (2002) Molecular mapping of the fertility restorer gene Rf-4 for WA cytoplasmic male sterility in rice. Acta Genet Sin 29:1001
Zhao Y, Li Z, Liu G, Jiang Y, Maurer HP, Würschum T, Mock H-P, Matros A, Ebmeyer E, Schachschneider R, Kazman E, Schacht J, Gowda M, Longin CFH, Reif JC (2015) Genome-based establishment of a high-yielding heterotic pattern for hybrid wheat breeding. Proc Natl Acad Sci 112:15624–9. doi:10.1073/pnas.1514547112
Zhao Y, Mette MF, Reif JC (2014) Genomic selection in hybrid breeding. Plant Breed 134:1–10. doi:10.1111/pbr.12231
Zhao Y, Zeng J, Fernando R, Reif JC (2013) Genomic Prediction of Hybrid Wheat Performance. Crop Sci 53:802–10. doi:10.2135/cropsci2012.08.0463
Zhu G, Peng S, Huang J, Cui K, Nie L, Wang F (2016) Genetic Improvements in Rice Yield and Concomitant Increases in Radiation- and Nitrogen-Use Efficiency in Middle Reaches of Yangtze River. Sci Report 6:21049. doi:10.1038/srep21049

2

Genetic and Root Phenotype Diversity in Sri Lankan Rice Landraces may be Related to Drought Resistance

Mayuri Munasinghe[1,2] and Adam H. Price[1*]

Abstract

Background: The development of relatively cheap and high throughput methods of genotyping and phenotyping plants offers the opportunity to explore local germplasm more thoroughly than before and should accelerate the identification of sources of genetic variation suitable for breeding. In this study, 135 Sri Lankan accessions, mostly identified as landraces, for which data was available at the International Rice Research Institute on drought scores were genotyped using a 384 SNP array and assessed for root depth using a newly developed buried herbicide method. Roots of 36 accessions were assessed using hydroponics and 12 using soil-filled rhizotrons to establish if variation in herbicide score could be attributed to root traits.

Results: Population structure based on the SNPs using STRUCTURE revealed six groups, being *tropical japonica*, *aus* and four *indica* subpopulations. Three of these *indica* subpopulations do not seem to be represented in the Rice Diversity Panel I (RDP1) of 372 global rice accessions and appear to represent genetic diversity so far poorly studied by the global scientific community. The herbicide score was highly discriminatory between landraces and correlated very strongly with hydroponic and rhizotron root traits. The mean herbicide score strongly differentiated between landraces according to the province and the latitude from which they were collected. It also differed between subpopulations, being high in *indica 2* and *tropical japonica* and low in *indica 1* and *aus*. Drought scores suggest that *indica 2* is more drought resistant than the other groups. Correlations indicate that those landraces with high herbicide scores are more drought resistant in the vegetative stage. The landrace Niyan Wee, whose name in Sinhalese means "drought rice" belongs to the *indica 2* subgroup, has high herbicide scores and deep roots.

Conclusions: Niyan Wee and other cultivars within the *indica 2* subgroup should be a valuable source of breeding for drought resistance at least partly because of their superior root traits, not normally associated with the *indica* rice cultivars.

Keywords: *Oryza sativa*, Rhizotron, SNP, Structure, Herbicide, Indica

Background

The rapid advance in genotyping technology including sequencing has led to an increased ability to accurately assess genetic variation across biological disciplines. In rice, the most important crop plant in terms of human consumption, this is exemplified by the development of SNP genotyping platforms (Zhao et al. 2011), low depth sequencing for genome wide association mapping (Huang et al. 2010) and now deeper sequencing of large collections (3K RGP 2014). The relative ease of genotyping promises to unlock some of the hitherto inaccessible diversity contained within local germplasm of our major crops (McCouch 2013).

In order to match genotype to phenotype, there has been a push to develop high throughput platforms that can provide relevant data on traits allowing genetic mapping to advance faster and more accurately. Araus and Cairns (2014) have reviewed advances in high throughput field-based phenotyping and highlight as an example "Shovelomics" for root traits where root numbers and their angle of growth can be assessed for a very large number of plants (Trachsel et al. 2011). One very simple

* Correspondence: a.price@abdn.ac.uk
[1]Institute of Biological and Environmental Sciences, University of Aberdeen, Aberdeen AB24 3UU, UK
Full list of author information is available at the end of the article

root phenotyping method has recently been described where a herbicide is buried in a box of soil and the speed of symptom development related to rooting depth (Al-Shugeairy et al. 2014).

Rice is the staple food in Sri Lanka, where it is cultivated as a wetland crop in all the districts of the island and occupies 34 % (0.77/million ha) of the total cultivated area (De Silva and Yamao 2009). It provides 45 % of the calories and 40 % of the protein for the population (De Silva and Yamao 2009). One-third of Sri Lanka's rice area are in rainfed lowlands or uplands, which are frequently subjected to water deficit of varying intensities and durations causing significant yield losses (De Costa 1998). During the Yala growing season (March to September: 310, 000 ha) a huge part of the island is drought-prone (De Costa 1998). According to the Climate Risk & Adaptation Country Profile of Sri Lanka 2011, (GFDRR 2011) the frequency of drought increased from 1974 to 2004.

Sri Lanka has more than 3,000 landraces of rice; the International Rice Research Institute have nearly 2,000 in their genebank while the Plant Genetic Resources Centre in Sri Lanka has about 3,000 (FAO 2010). Among these is a landrace called Niyan Wee which means "drought rice" in Sinhalese. The possibility exists that these landraces contain adaptation to specific climatic and edaphic conditions of Sri Lanka that might be usefully exploited for rice breeding in Sri Lanka and elsewhere. Furthermore, the possibility exists that the genetic diversity that is associated with this adaptation has not so far been significantly studied in projects assessing global variation in rice.

This paper describes a study on genetic diversity of 135 Sri Lankan accessions, almost all identified as landraces, including three accessions called Niyan Wee, with a local geographic focus using a marker system that allows comparison with global diversity panels. At the same time it assesses rooting depth using a new high throughput method that sheds light on potential adaptation mechanisms for drought that should prove useful in breeding programmes.

Results

Population structure and relation to drought scores

Based on analysis using STRUCTURE, it is most likely that the 129 Sri Lankan accessions analysed with 326 SNPs fall into six subpopulations since with K = 6 the Delta K value was over 16 while with K = 3–5 and 7–9 it was less than 2 (Additional file 1: Figure S1) (note 6 accessions were excluded from STRUCTURE analysis). The allocation of the cultivars to the K = 6 subpopulations is presented in Table 1. A more complete dataset including subpopulation, herbicide scores and data from the IRGCIS website is presented in Additional file 2: Table S1 while all the SNP data is presented in Additional file 3: Table S2.

A principal components analysis with 129 Sri Lankan accessions plus 18 accessions of the OryzaSNP is given in Fig. 1. It can be used to visualise the groupings identified by STRUCTURE. PC axis 1 (22.8 % loading) separates *Indica* (which includes OryzaSNP accession Aswina, Zhenshan 97, IR 64, Sanhuangzhan 2, Swarna and Minghui 63) from *Japonica* (which include OryzaSNP *tropical japonica* accessions Azucena and Moroberekan and *temperate japonica* M202, Nipponbare and Tainung 67) and PC axis 2 (9.2 %) separates *indica* from *aus* (which includes OryzaSNP accessions FR13A, Rayada, Dular and N22) (Fig. 1a). A combination of PC axis 3 (7.2 %) and PC axis 4 (4.2 %) clearly separate the *indica* subpopulations (Fig. 1b). This figure also clearly shows the relationship between the Sri Lankan accessions and the 18 OryzaSNP accession where the latter have no representation in *indica 2*, *3* or *4*.

A dendrogram of the 135 accessions with the 18 OryzaSNP accessions is given in Fig. 2 where each subpopulation is colour coded allowing the general pattern observed in the PCA to also be observed (note this tree is not bootstrapped so inferences must be treated with caution). A "nexml" format file of the NJ tree shown in Fig. 2 that can be visualised using Dendroscope is provided as Additional file 4: Online Resource 1. The OryzaSNP accessions are identifiable as cyan in the tree. OryzaSNP *aus* accession FR13A, Rayada, Dular and N22 clustered with, but were distinct from, 13 Sri Lankan accessions including Alnoliswee, while OryzaSNP *tropical japonica* accession Moroberekan and Azucena cluster within seven Sri Lankan accessions including Podiwee-1. *Temperate japonica* OryzaSNP accessions M202, Nipponbare and Tainung 67 cluster with, but are distinct from these *tropical japonica.* All the rest of the Sri Lankan accessions are considered to be *indica* because known *indica* OryzaSNP accessions Aswina, Zhenshan 97, IR 64, Sanhuangzhan 2, Swarna and Minghui 63 cluster with one group (called here *indica 1*) which contains 33 Sri Lankan accession. This *indica* 1 group includes Sri Lankan accessions BG12-1, BW100 and BW288-2, the only three of the accessions used here identified as breeding lines, not landraces. Three other groups recognised by STRUCTURE contain no representatives from the OryzaSNP set but appear to be *indica* since they are close to the *indica 1* group. The OryzaSNP cultivar Pokkali, a salt tolerant *indica* from India clusters close to the group called here *indica 2* which contains 16 Sri Lankan landraces including all three accessions named Niyan Wee. *Indica 3* contains 22 Sri Lankan landraces while *indica 4* contains 22. This leaves 22 Sri Lankan landraces considered as admix between these six subpopulations. This includes three landraces Suduru-Samba,

Table 1 Sri Lankan landraces used in this study with the subpopulation they are clusters in and herbicide score for deep rooting

Landrace Name	Sub-population	Herbicide score	Landrace Name	Sub-population	Herbicide score	Landrace Name	Sub-population	Herbicide score
AHAMBA	In 1	1.75	INGRISI WEE	In 3	1.87	PANDURU WEE	In 3	1.25
AKURAMBODA	In 1	1.25	KAHATA SAMBA1	In 3	1.25	PANNITI	Aus	2
AL WEE	Aus	0.75	KAHATA SAMBA2	In 3	1.75	PERIYA KARUPPAN	In 2	3.37
ALAGUSAMBA	Ad	1.75	KAHATAWEE	Ad	1.25	PERIYA SIVAPPU	In 2	2.5
ALAKU SAMBA	Ad	2.37	KAL VELLAI	In 2	4	PERIYAVELLAI	In 2	3
ALE WEE	In 1	1.87	KALU BALAWEE	Ad	2	PODI NIYAN WEE	In 2	2.75
ALEWEE	In 1	1.25	KALU GIRES	Ad	1.75	PODI WEE2	In 2	2.12
ANDIKULAN	Aus	0.5	KALUBALAWEE	Aus	3	PODIWEE1	TJ	5
ARNOLISWEE	Aus	0.25	KALUKUDA	In 4	2.37	POKKURU SAMBA	In 1	1.25
BADHIWEE	In 3	1.75	KALUNDAI	In 2	4	PUKURU SAMBA	In 1	2.25
BALA KAHARAMANA	In 4	2	KARUPPAN	In 2	3.37	PUSHPARAGA	Ad	2.25
BALAMAWEE	In 1	1.87	KARUPPU SEENADHI	In 4	2.75	RAMBUTTANWEE	In 3	2
BATA POLA EL	In 4	1.62	KATHARAMANA	In 3	1.83	RANGAMODAN	In 2	3.25
BATAPALA	In 1	1	KIRIBATHA	In 4	3.12	RANKIRI	In 1	2.75
BATAPOLAWEE	Ad	1.5	KIRINARAN	Ad	2	RATA THAVALU	In 4	2.12
BATHIRATAWEE	In 3	2.12	KOLANATHIWEE	In 3	1.37	RATATHAWALU	Ad	1.75
BATUDELLA	TJ	3.62	KOTHTHAMALLI SAMBA	In 1	1	RATAWEE1	In 1	0.5
BG12-1	In 1	1.12	KOTTAMALLI	In 3	1.62	RATAWEE2	Ad	2
BW100	In 1	1.25	KURUBALAWEE	Aus	1.12	RATH RANPODIWEE	Ad	1.87
BW288-2	In 1	0.5	KURULUWEE	Aus	1	RATHKARA	In 3	1.87
CHITHIRA KALI	In 1	2.25	KURUWEE	In 1	2.37	RATHU HEENATI	In 1	2.62
DAMPOTAWEE	In 4	2.37	LUMBINI1	In 3	1	RATHU WEE1	In 3	3
DIK WEE2	In 1	1.62	LUMBINI2	In 4	2.12	RATHUWEE2	In 3	2
DIKWEE1	In 1	2.5	MA GODA AL	In 1	2.17	RUWANRATHRAN	In 3	1
DURU WEE2	Aus	1	MADAEL	Ad	3	SARALI	In 2	3.5
DURUWEE1	Ad	1.87	MADAKALAPUWEE	Aus	3.5	SEDUKKAN SAMBA	In 1	1.5
GALPAWEE	Ad	2.37	MADOLUWA	In 3	1.75	SINNAVELLAI	In 2	4.33
GAMBADAWEE	In 4	0.75	MADURU SAMBA	In 4	1.75	SUDU HONDARAWALA	In 1	3.37
GAMBODA	In 4	2.87	MAHABATHAN	Ad	2.25	SUDU MADA THAVALU	In 4	1.87
GODAWEL	TJ	1	MAHAGODA AL	In 1	2.62	SUDU MURUNGA	In 1	1
GONABARU	TJ	1.5	MAHARATAWEE	Ad	1.75	SUDUBALAWEE	In 4	2.25
HAL AL1	In 4	2.37	MALWARIYA	TJ	2.75	SUDURU SAMBA	Ad	2.37
HAL AL2	In 4	1.75	MATARAWEE	In 3	1	SULAI	In 1	2
HAL ELIYA	In 4	1.87	MATHOLUWA	In 3	2.25	TAIWAN	In 1	1.25
HAM-ALE-WEE	In 1	1	MENIKWEE	In 4	2.5	THAHANALA	Aus	2.83
HEEN GODA WEE	TJ	3	MODDAI KARUPPAN	In 2	3	THATUWEL	TJ	3.25
HEENATI1	Aus	1.25	MOLLIGODA	In 3	2.37	THAVALU1	Ad	2.12
HEENATI2	Aus	3	MOOTHUKIRIEL	In 4	2.25	THAVALU2	In 4	1.87
HEENDIKWEE1	Ad	2	MOROGALLAWEE	Ad	2.12	TISSA WEE	In 3	1.75
HEENDIKWEE2	In 3	1	MUDUKIRIEL	In 1	2.25	VELLAI	In 2	3.62
HEENDIKWEE3	In 1	1.75	MUTHU SAMBA	In 1	2.5	VELLAI ILLANKALI	In 1	1.25
HONDARAWALA1	Ad	2.12	MUTHUSAMBA	In 1	3.25	WANNI DAHANALA	In 4	2

Table 1 Sri Lankan landraces used in this study with the subpopulation they are clusters in and herbicide score for deep rooting *(Continued)*

HONDARAWALA2	In 4	2.5	NIYAN WEE1	In 2	4	WEDA HEENATI1	In 3	2.62
HORANA MAWEE	In 4	2.5	NIYAN WEE2	In 2	4	WEDA HEENATI2	Aus	1.87
ILLUPAIPU SAMBA	Ad	1.5	NIYAN WEE3	In 2	3	YAKADAWEE	In 1	2.25

Further information including accession numbers, drought scores, maturity data and geography of their collection is provided in Additional file 2: Table S1

Thavalu 1 and Durwee 1 which clustered with the aromatic OryzaSNP accession Dom Sufid suggesting these three might be aromatic.

Examining the dendrogram produced when the Sri Lankan landraces were combined with the RDP1 (for 267 SNP markers) suggests that the *indica 2, 3* and *4* subpopulations are not represented in RDP1 while *indica 1* is (data not shown but nexml file supplied as Additional file 5: Online Resource 2). For example, Sri Lankan accessions Alewee 1, Alewee 2 and Ham-Ale-Wee (*indica 1*) cluster near RDP1 accession JM70, an *indica* from Mali, and Ratawee 1 (*indica* 1) clusters with RDP1 accession Jaya, an *indica* from India and quite close to the improved variety IR 8. A demonstration that this analysis is valid is provided by the fact that the *aus* landrace Kalubalawee used here clusters with the Sri Lankan *aus* RDP1 accession Kalubala vee. Another Sri Lankan accession in RDP1 (Rathuwee, an *indica*) clusters near the breeding line used here BW288_2 which is from *indica* 1 and not with the Sri Lankan landrace used here in *indica 3* that is also called Rathuwee. The other Sri Lankan accessions in RDP1 (LD 24, an *indica*) does not cluster with the Sri Lankan landraces used here.

Examining the distribution of the landraces of different subpopulations identified by the province of Sri Lanka from which they came shows they are not uniformly distributed (Fig. 3). *Tropical japonica* and *aus* landraces used in this study come from the south or the south and west of the county respectively while *indica 1* s are quite evenly spread. Most of the *indica 2* s come from the North but Niyan Wee is from Uva (South and East of the centre) and one comes from Sabaragamuwa (South and West of the centre). The *indica 3* subpopulation appears to come from the South and Southwest of the country while *indica 4* are present throughout the country except the two most northerly provinces.

One way analysis of variance (ANOVA) of date of maturity using the identified subpopulations was significant ($P = 0.001$) and when the outlying *indica 1* landrace Koththamalli Samba (224 days) was removed (160 days is the next highest maturity date) it was clear from a Tukey's test that *indica 2* are longer duration (146 days) compare to the other subpopulations all of around 130 days. ANOVA was also conducted on the drought score values. There were significant differences between subpopulations for drought score 1 (seedling vigour; $P = 0.05$), 3 (recovery after 2nd stress; $P < 0.001$), 4 (drought resistance at early vegetative stage; $P = 0.017$) and 5 (drought resistance at late vegetative stage; $P = 0.001$). Most striking were for drought 3 where 30 % of variation was explained by subpopulation and a Tukey's test revealed *indica 2* was more drought resistant than the other 5 subpopulations, but it is noteworthy

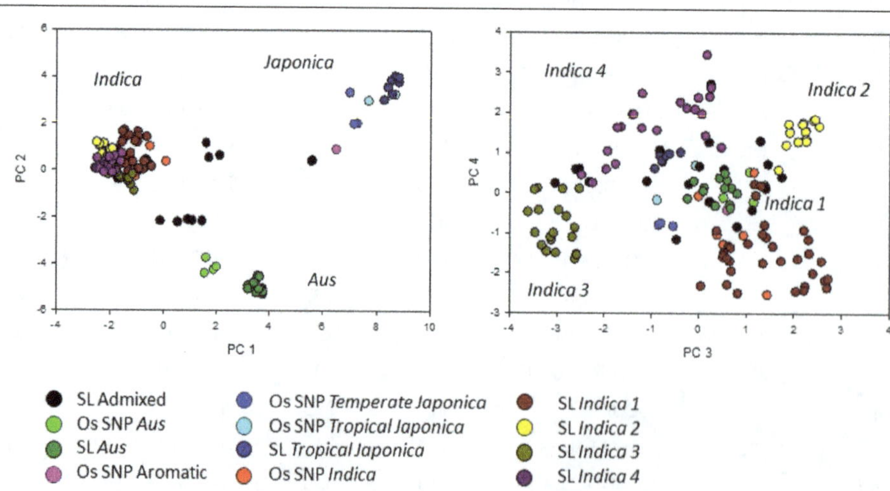

Fig. 1 Plots of the top four axis of principle components analysis where accessions have been colour coded to allow comparisons. SL means Sri Lankan landrace, Os SNP means OryzaSNP accession. Loadings are PCA 1 = 22.8 %, PCA 2 = 9.2 %, PCA 3 = 7.2 % and PCA 4 = 4.2 %

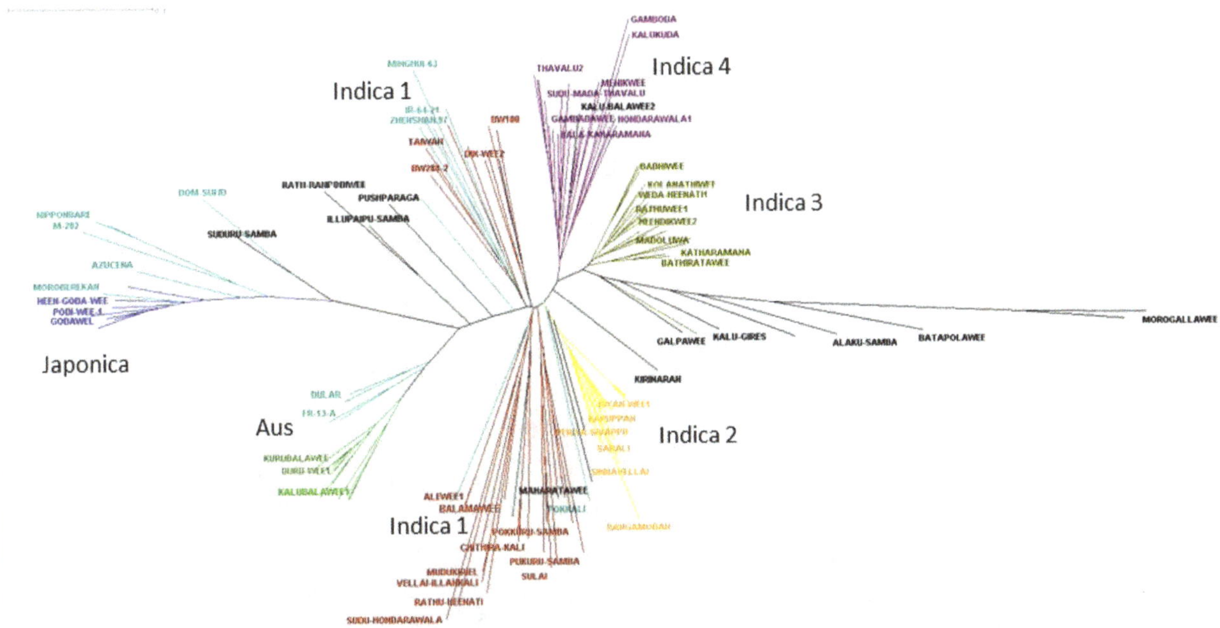

Fig. 2 Unrooted dendrogram of 135 Sri Lankan accessions and 18 OryzaSNP accession based on 326 SNP markers using colour codes of Fig. 1. This tree is available as a nexml file to open in Dendroscope as Additional file 4: Online Resource 1

that drought 4 and 5 also revealed *indica 2* are remarkable for possessing low scores indicative of high drought resistance (Fig. 4).

Root traits

Symptoms of herbicide-induced plant injury were first detected on Podiwee and Niyan Wee 2 on day 20 and by day 21 genotype explained 36 % of the variation in herbicide score (from one way ANOVA). The proportion of variation explained by genotype slowly increased to peak at 49 % on day 31, after which it decline very slowly as more and more plants reached a score of 5. All herbicide scores (at different days) correlated very strongly with each other, with herbicide scores on day 31 having $r = 0.935$ and $r = 0.872$ for correlations with day 37 and day 43 respectively and day 37 and day 43 with $r = 0.920$ (see Additional file 6: Table S3 for more correlations between herbicide scores on different dates). A histogram of mean herbicide scores at day 43 is presented in Additional file 7: Figure S2 when 43 % of variation was explained by cultivar. The herbicide score at day 43 was chosen as the most appropriate to examine relationships with genetic diversity and plant phenotype traits as it was roughly normally distributed while earlier scores are skewed by large numbers of landraces with scores of zero. As expected, check cultivar Black Gora had a high herbicide score throughout (over 3 by day 43) but was surpassed by Podiwee 1, Niyan Wee 1, Niyan Wee 2 and Sinnavellai. Poor-rooted check varieties Bala and IR 64 showed very little herbicide symptoms (below 1 by day 43). Some landraces showed equally low herbicide scores such as Arnoliswee and Ratawee 1. Herbicide scores correlated with drought scores 3, 4 and 5 only, and were strongest for drought 5 (late vegetative drought resistance) where herbicide score at day 31 and 43 gave correlation coefficients of $r = -0.335$ and -0.350 respectively ($p = 0.001$) indicating that a high herbicide score is associated with a low drought score (high drought resistance).

Herbicide score was significantly associated with subpopulation and a Tukey's test on herbicide score at day 43 revealed three groups, where *indica 2* had high scores, and *indica 1, 3* and *aus* had low ones ($P < 0.001$) (Fig. 5). The herbicide score was also very strongly associated with province and district, with these factors explaining a remarkable 28 and 35 % of the variation in one way ANOVA. The mean herbicide score according to province is presented on the map of Sri Lanka in Fig. 3 which shows that high herbicide scores are found in Northern, Central and Uva provinces. Herbicide score is also highly significantly related to the latitude from which the landraces were collected ($P < 0.001$; $R^2 = 20$ %) where those from the North (latitude 9°) had a higher score than others (from latitudes 5–8°) according to a Tukey's test.

Root and shoot traits measured on 36 hydroponically-grown rice genotypes are given in Additional file 8: Table S4. All traits were highly discriminatory between genotypes, (R^2 ranged from around 33 % for root/shoot dry weight ratio, root thickness and root mass to 50–60 % for maximum root and shoot length). A Tukey's test comparing the three chosen groups of Sri Lankan accessions (those

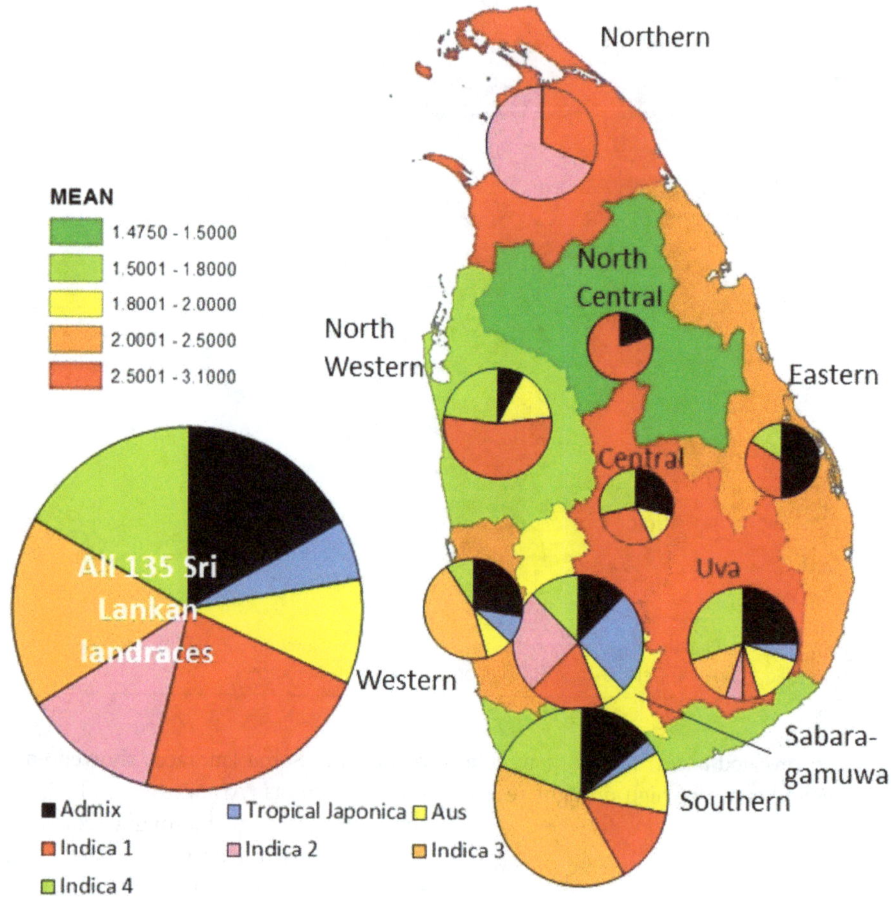

Fig. 3 Provinces of Sri Lanka colour coded by herbicide score with pie graphs indicating group of landraces by province (area proportional to number of accessions)

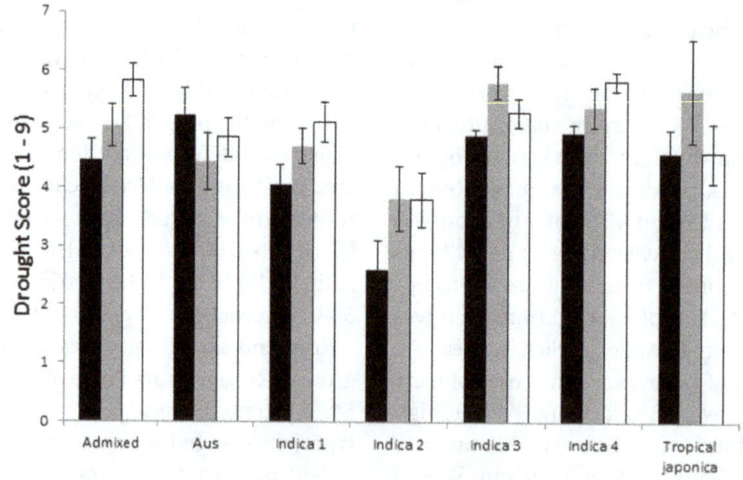

Fig. 4 Mean drought scores obtained from IRGIS database for Sri Lankan landraces grouped according to identified subpopulation. Black is drought score 3, grey is 4 and open is 5. In all cases, low numbers represent high drought resistance. Bar is standard error

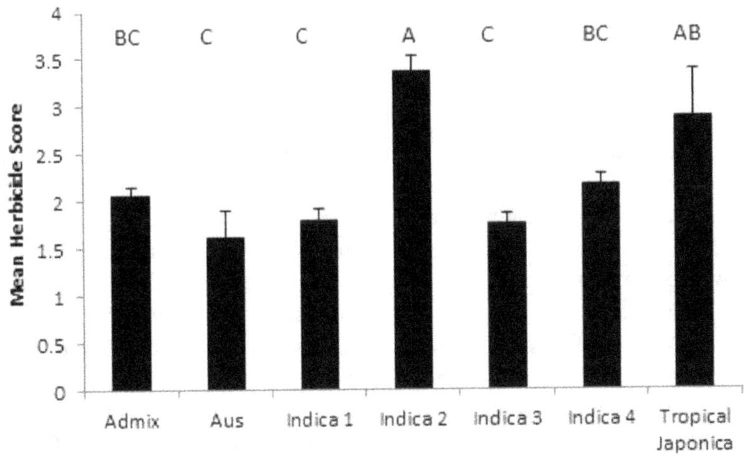

Fig. 5 Mean herbicide score at day 43 for Sri Lankan landraces grouped according to identified subpopulation. Bar is standard error. Columns sharing the same letter are not different

with high, average and low herbicide scores) showed that root length and root weight distinguished the high herbicide accession from the rest, but root thickness and % root mass (mass of root over mass of plant) did not. Most traits measured in hydroponics correlated with herbicide score, but the correlations were strongest with root traits (see Additional file 9: Table S6 for complete correlation data). The strongest correlation of all was with maximum root length at day 28 where $r = 0.750$ (Fig. 6) whereas the strongest correlation for shoot traits was with shoot length at day 35 ($r = 0.586$). Figure 6 shows that check varieties behaved as expected (as shown in Shrestha et al. 2014), with Black Gora very long rooted, Azucena reasonably long rooted, IR 64 reasonable short rooted and Bala very short rooted. There were Sri Lankan landraces across the spectrum of these check varieties including some very long and very short rooted.

All of the 27 traits measured on the 12 genotypes grown in rhizotrons (presented in Additional file 10: Table S5) were highly discriminatory between genotypes ($P < 0.005$) with R^2 values from one way ANOVA ranging from 27 % (number of roots past 50 cm depth at 35 days) to 83 % (shoot length at 42 days). For maximum visible root length at day 35 and day 42 R^2 was 62 and 60 % respectively. Almost all traits also correlated with herbicide score at day 43 (see Additional file 11: Table S7 for complete correlation data) with the two strongest being root length at day 21 ($r = 0.881$) and root angle at 42 days

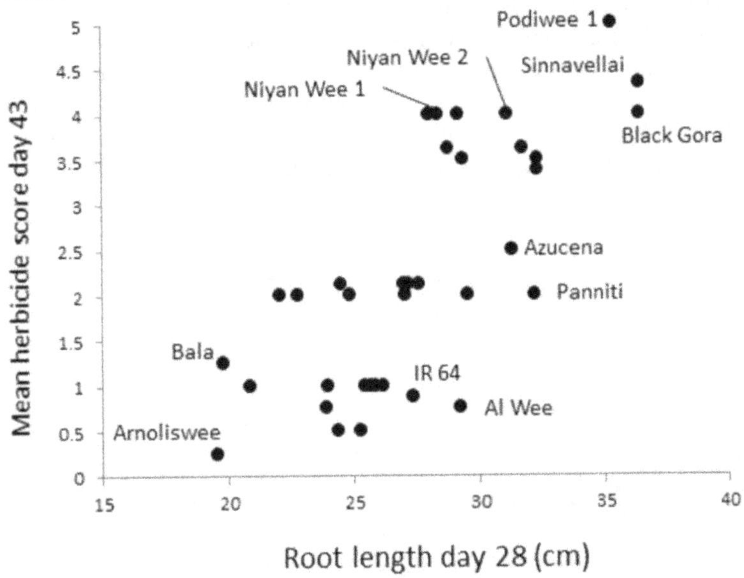

Fig. 6 Plot of hydroponic root length at day 28 vs. mean herbicide score at day 43

($r = -0.901$). These remarkable correlations are presented in Fig. 7. The lowest correlation was with shoot length at day 14 ($r = 0.603$).

Discussion

Sri Lankan landraces are genetically distinct

The rapid reduction in the cost of genotyping should mean that for world crops a better appreciation of local diversity within the context of the worldwide genetic variation of the species will become available. As yet, however, this process is only just beginning and is very limited for Sri Lankan rice. The Rice Diversity Panel 1 which has been genotyped with 44,000 SNPs (Zhao et al. 2011) has only three Sri Lankan accession, a recent expansion of that set to 1571 accessions of the Rice Diversity Panel 2 (RDP2) with 700,000 markers has 48 (McCouch et al. 2016) while the recent sequencing of 3,000 rice accessions (3K RGP 2014) includes only 47 Sri Lankan accessions. Of the 48 Sri Lankan accessions in RDP2, only four have the same name as used here and none have the same IRGC id (that is KURUWEE, MADAEL, RATHUWEE and WEDA HEENATI). Of the 47 accessions of the 3 K rice genomes project only four are the same as used here (8944 ALAGUSAMBA, 8901 MATHOLUWAA, 15485 MODDAI KARUPPAN and 11938 PODIWEE) while three have the same name but different accession numbers to this study (HEENDIKWEE, HONDARAWALA and KURULUWEE). A study using AFLPs to examine specifically Sri Lankan rice assessed 74 accessions plus six wild rices (Rajkumar et al. 2011), applying 772 molecular markers to characterise local diversity but since the marker system does not transfer to other studies, placing those results into the wider context of global diversity is difficult. In the study reported here, by using SNP markers that have been characterised on the OryzaSNP set (McNally et al. 2009) and most of which have been tested on the RDP1, the global context is more easily evaluated.

Structure analysis reveals six subpopulations within 129 of the accessions, which by comparative analysis with 18 OryzaSNP cultivars can be identified as *aus*, *tropical japonica* and four which appear to be *indica*. This grouping is clearly demonstrated in PCA plots of Fig. 1. This grouping is not reflected in the two large clusters of Sri Lankan accessions identified by Rajkumar et al. (2011); 14 of the accessions have common names revealing that their “cluster 2” contains 10 landrace used here from *indica 1, 3, 4, tropical japonica* and admix while their “cluster 3” has four landraces used here from *indica 1* and *2*, *aus* and admix. The splitting of *indicas* into distinct subpopulations is not common in rice diversity studies but has been reported in China (Huang et al. 2010) where evidence showed latitude played a role in differentiating three *indica* subpopulations. Analysis of the 3,000 sequenced rice genomes presented in Alexandrov et al. (2015) also suggests the presence of three *indica* subpopulations but the details of the structure analysis method are lacking and it is not possible to reliably compare results with those presented here.

While the four *indica* subpopulations reported here were roughly equally represented in the landraces as a whole, and the *indica 1* subpopulation was represented in almost all of Sri Lanka’s 9 provinces (not Western Province), the *indica 2* and *3* subpopulations were found in only three provinces while the *indica 4* subpopulation was absent in the two most northerly provinces (Fig. 3). Clearly, therefore, these subpopulations of *indica* do not appear to be randomly distributed throughout Sri Lanka.

What seems to be relatively clear is that the *indica 1* subpopulation is typical of *indica* present in global collections but this does not appear to be true for the other three. This is suggested because the OryzaSNP *indicas* cluster in *indica 1* (Figs. 1b and 2) and when the RDP1 data of 267 SNPs are analysed with the Sri Lankan cultivars, *indica 1* landraces cluster with RDP1 *indica* but *indica 2, 3* and *4* landraces do not. This implies that the cultivars from *indica 2, 3* and *4* represent germplasm and genes that may have so far been poorly studied by the global rice community.

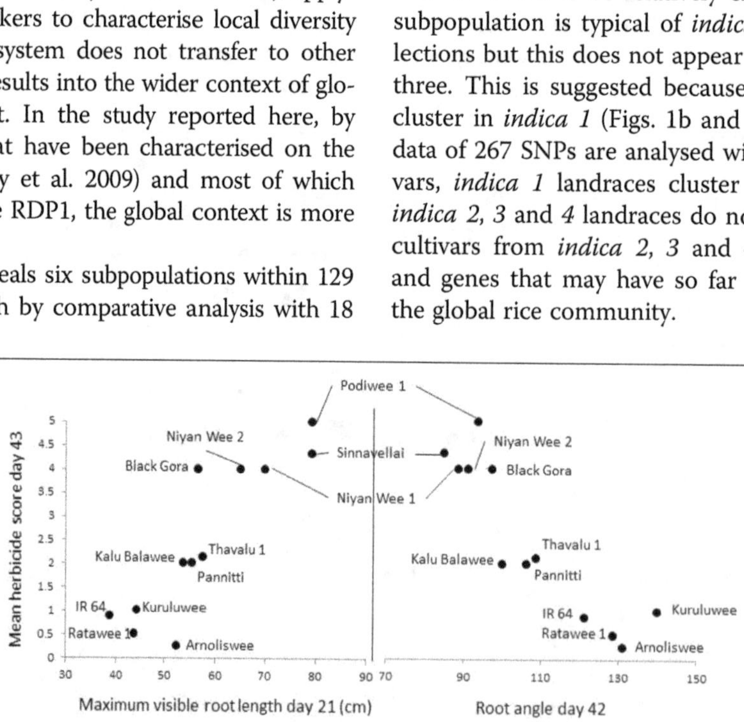

Fig. 7 Scatter plots of rhizotron root length (left) and root angle (right) against mean herbicide score

The herbicide screen reveals differences in root traits

The herbicide screening system which has been recently developed to screen rice (Al-Shugeairy et al. 2014) proved highly discriminatory when used on the 135 Sri Lankan accession. This revealed some landraces having higher herbicide scores than Black Gora, the deep rooted check. When developing the high throughput method it was observed that there are differences in sensitivity of accessions to the herbicide (Al-Shugeairy et al. 2014). The authors commented that this was small compared to difference observed in the timing of symptoms when the herbicide is buried, but they nevertheless cautioned that the screen is a positive screen for deep roots since a plant that develops symptoms rapidly must be deep rooted while a plant that develops symptoms slowly or not at all might be shallow rooted or be insensitive to the herbicide. For this reason, the reliability of the herbicide screen to deduce that accessions differ in rooting depth was confirmed by a hydroponic screen on 36 genotypes and a rhizotron screen on 12 where Sri Lankan landraces from high, low and average herbicide scores plus check accessions were used. In both cases very high correlations were revealed between herbicide scores and root growth measured directly giving great confidence that these do reflect genuine differences in rooting behaviour. The fact that all traits measured in these root screens, including shoot traits, correlate with herbicide score indicates that an underlying driving factor for herbicide score is seedling vigour. However, the fact that the most strongly associated traits are root traits, not shoot traits and that in rhizotrons root angle is strongly correlated to herbicide score emphasises that the development of symptoms is most strongly related to the direction and length of root growth rather than any other trait. This confirms the observations in Al-Shugeairy et al. (2014) and should strengthen confidence about using this method for assessing large numbers of genotypes so long as the caveats detailed in that publication are understood. Confidence in this method should also be taken from the fact that the herbicide score is correlated negatively with drought scores 3, 4 and 5.

Diversity in Sri Lankan rice is associated with rooting depth and drought resistance

The herbicide score discriminates between subpopulations of Sri Lankan landraces such that *indica 2* is high and *indica 1, 3* and *aus* are low. In general it is reported that rooting depth of rice accessions differs between subpopulations. For example Shrestha et al. (2014) report that rhizotron screens rank subgroups in the order *aus* > *tropical japonica* > *indica* ≈ *temperate japonica* for maximum root length which was in agreement with an earlier report of Lafitte et al. (2001). This order has been confirmed testing the RDP1 with the herbicide screen (unpublished results) where *aus* has the highest average scores, followed by *tropical japonica* with *indica* and *temperate japonica* both similarly low. In this context, two surprising observations are made in the current study. First is that the Sri Lankan *aus* are low, and may reflect the fact that these landraces are somewhat genetically distinct from global *aus* cultivars. That this is the case is suggested by the observation that they do not cluster exactly with the OryzaSNP *aus* accessions (Figs. 1a and 2) and examining the dendrogram produced when the Sri Lanka accessions are combined with the RDP1 data (Additional file 5: Online resource 2), when they form a distinct cluster within the *aus* which only has RDP1 cultivars Kalubala Vee (a Sri Lankan aus) and Khao Pahk Maw (an aus from Thailand). The second, highly noteworthy observation is that the *indica 2* subpopulation has high herbicide scores (Fig. 5) which when considered in the context of the genetic data discussed above suggests that *indica 2* represents a distinct type of *indica* that has not previously been studied and which is deep rooted. This group includes Sinnavellai that was the cultivar with the 2nd longest roots in rhizotrons (Fig. 7) and all three cultivars called Niyan Wee, two of which were tested in the rhizotrons and had root traits surpassing the deep rooted check Black Gora (note they did not surpass Black Gora in hydroponics while Sinnavellai matched it).

The subpopulations are also differentiated in the distribution of their drought scores such that the *indica 2* subpopulation appears to be more drought resistant, at least in terms of drought in the vegetative stage and recovery from stress (from drought score 3, 4 and 5) (Fig. 4). This emphasises the unique and potentially valuable nature of the landraces in *indica 2* and, taken together with the paragraph above implies that these cultivars are have a drought resistance that is linked to root growth.

Conclusions

The study characterises 135 Sri Lankan accession, almost all landraces, and suggests the existence of genetic diversity within *indica* that has not been studied until now. It further shows that *indica 2* group has deep roots (unlike most *indica*) and high drought resistance. Since Niyan Wee is in this group, and its name means "drought rice" it seems highly plausible that accessions with this name are drought resistant and they certainly have deep roots when tested in herbicide screening and rhizotrons. These landraces and their partners in this subpopulation should be more thoroughly studied in order to determine the genetic and physiological background of their drought resistance and rooting traits.

Methods

Plant material

A total of 135 Sri Lankan rice accessions were supplied by the International Rice Research Institute (IRRI) chosen based on information on the International Rice Genebank Collection Information System database (http://www.irgcis.irri.org:81/grc/IRGCISHome.html). Only those identified on IRGCIS as landraces were considered except three breeding lines BG12-1, BW100 and BW288-2. From more than 2,000 Sri Lankan accessions listed, the landraces were further selected to represent most of the geo-climatic regions of the wet zone and dry zone of the country. The geographical location of each landrace including the province and district with longitude and latitude were obtained by searching detailed germplasm information of IRGCIS database for collecting location data (http://www.irgcis.irri.org:81/grc/SearchData.htm). Landraces were also chosen if there was available drought scores in the IRGCIS database (searching detailed germplasm information for reaction to physio-chemical stress) to ensure a range of scores. This data gives nine categories of drought score, Drought 1–9 but for only the first 5 are there sufficient data points to be valuable (>41). That is, in order; 1 = Seedling vigour (data for 117 accessions); 2 = Rate of recovery after the 1st stress (data for 107 accessions); 3 = Rate of recovery after the 2nd stress (data for 90 accessions); 4 = Drought resistance score at early vegetative stage (50–60 days) (data for 110 accessions); 5 = Drought resistance score at late vegetative stage (80–100 days) (data for 88 accessions). In all cases, drought scores are given on a scale of 1 to 9 where a high value indicates highly sensitive to drought.

A list of the accessions is given in Table 1 and full detail of the IRGC accessions is provided in Additional file 2: Table S1. It includes some accessions which share the same name. Where this occurs, they have been given a number to distinguish them given by ascending order of accession number. Thus there are three accessions called Niyan Wee, which are named here Niyan Wee 1, 2 and 3 for accession numbers 66525, 67646 and 67647 respectively.

In addition to the Sri Lankan material, four check cultivars Azucena (*tropical japonica*), Bala (*indica*), Black Gora (*aus*) and IR 64 (*indica*) were used for root screening, using seeds from plants grown in Aberdeen but originally supplied from IRRI.

Genotyping using a 384 SNP array and genetic analysis

DNA was extracted from fresh leaves using the Invitrogen ChargeSwitch gDNA plant kit (Invitrogen, Paisley, UK) using the manufacturer's instructions. The DNA was analysed on a custom designed 384 SNP Illumina GoldenGare array using the Illumina Beadxpress platform at the Molecular Marker Laboratory at the James Hutton Institute, Invergowrie, Dundee, DD2 5DA, Scotland, UK. The design of the 384 SNP array is described in Travis et al. (2015) but briefly it use Veracodes listed on the Rice Diversity web site (http://www.ricediversity.org/data/index.cfm) selected to include evenly-spaced SNPs with a mixture picked from the RiceOPA1.0 (quality control), RiceOPA2.1 (*Indica/Indica*), RiceOPA3.1 and RiceOPA7.0 (*Indica/Japonica*) and RiceOPA4.0 (*Japonica/Japonica*) SNP sets.

From the 384 SNPs, data of 58 markers were removed because of recognisable variation between plates of 96 samples. All SNP data generated from 326 markers (given in Additional file 3: Table S2) were combined with data for the same markers for 18 accessions from diverse set of 20 rice cultivars and landraces in the OryzaSNP panel (McNally et al. 2009) in order to use them as check accessions of known subpopulation origin in identifying different groups in population structure analysis.

Cluster analysis of SNP data were conducted using TASSEL (Trait Analysis by aSSociation, Evolution and Linkage) Version 4 (Bradbury et al. 2007). This software produced a data matrix showing relative closeness of accessions based on SNP data. Neighbour joining clustering method was used in this analysis and TASSEL produced a dendrogram (tree plot) which was visualized in detail in software Dendroscope (Huson and Scornavacca 2012). For 267 of the SNPs used here, data for the 372 accessions of the Rice Diversity Panel 1 (Zhao et al. 2011) are available from the Rice Diversity web site. TASSEL was used to produce a dendrogram of these accessions with the 135 Sri Lankan landraces. Both of the trees produced this way are available as online resources that can be opened in Dendroscope.

A total of six landraces (Alaku-samba, Batapolawee, Heendikwee-1, Kahatawee, Madael and Morogallawee) which produced unusually long branches in the dendrogram were removed from the list before further analysis to determine the number of populations within the Sri Lankan accessions. Population structure in 129 Sri Lankan accessions (without the OryzaSNP cultivars) was analysed in STRUCTURE Version 2.3 (Pritchard et al. 2000) software by implementing a model - based clustering method for inferring population structure. The model consisted of 10000 burn-in period and 50000 MCMC replicates. The program was run with number of populations assumed (K) from 1 to 10 each with 10 replicates. The true number of subpopulations (K) was determined by plotting a graph of K versus Delta K using mean log likelihood for each K (Ln P (D)) as described in the tutorial of the STRUCTURE software (Additional file 1: Figure S1). The quantitative genetics program TASSEL (Bradbury et al. 2007) was used to perform Principal Component Analysis (PCA) of the SNP data including the OryzaSNP set of 18 accessions, after filtering and imputation of missing SNPs using default parameters.

Assessing rooting depth using buried herbicide

The buried herbicide method for assessing rooting depth in rice has been described in full (Al-Shugeairy et al. 2014). Briefly, a box of dimensions 386 cm long x 81.5 cm wide and 40 cm deep was made in a greenhouse using plywood covered in plastic sheet. In the bottom, a 5 cm deep layer of sieved loam topsoil (Rolawn, UK) was watered with 11.75 l of Yoshida's nutrient solution (Yoshida et al. 1976). This was covered in filter paper soaked in TRIK (active ingredients Diuron (46.4 %), Aminotriazole (26.6 %) 2,4-D (11.2 %) herbicide (no longer available)) at a dose of 100 mg of herbicide per plant. Two soil moisture meters (Theta Probe, Delta-T, Cambridge) were placed above the filter paper and a depth of 15 cm more soil added before a further two moisture probes were added. A further 15 cm of soil was added (meaning the herbicide layer was 30 cm below the soil surface) and the whole box watered with 65 l of nutrient solution. The 135 Sri Lankan accessions plus five check cultivars (Azucena, Bala, Black Gora, IR 64 and Nipponbare) were sown 5 cm apart in four replicate blocks in a randomised complete block design on the 19th July 2010. The greenhouse conditions maintained a minimum temperature of 25 °C. While there was no supplemental lighting, at this time of year in Aberdeen (57° latitude) the day length is very long (going from 17.75 h to 15.5 h during experiment) so the plants received a relatively high amount of light (but it was not monitored). The box was watered to maintain a volumetric water content between 24 and 25 % with tap water. After 19 days plants started to show symptoms of herbicide-induced damage and were scored every day using a visual "Herbicide Score" of 0–5 where 0 is no symptoms, 1 is given when 5–15 % of leaf area shows visible symptoms (leaf yellowing), 2 when 15–50 % of leaf area shows visible symptoms, 3 when more than 50 % of the leaf area shows symptoms, 4 when 15–50 % of the leaf area is dead and 5 when more than 50 % of the leaf area is dead. Note this score has been shown to be essentially parametric by correlation with % affected leaf area score in subsequent screens (Al-Shugeairy et al. 2014).

Assessing root traits using hydroponics

A total of 32 Sri Lankan landraces plus check cultivars Azucena, Bala, Black Gora and IR 64 were assessed in a hydroponic system previously described (Price et al. 1997, Shrestha et al. 2014). The Sri Lanka landraces were chosen to represent 11 with the highest herbicide scores, 11 with lowest scores and 10 with near average scores. Four completely randomised replicate boxes containing 38 L of nutrient were used. Seeds were washed in tap water, surface sterilised in 1 % sodium hyperchlorite then pre-germinated in Petri dishes at 37 °C for two days. Germinated seedlings were planted in plastic trays fitting in to covered plastic tubs with aerated Yoshida's nutrient solution in a controlled environment growth room under a 12 h light regime with a light intensity of approximately 400 μmol m^{-2} s^{-1} PAR for 12 h per day with 25 °C at night and 28 °C in the day. The nutrient solution was replaced every week while maintaining a constant pH of 5.5 daily. The maximum root and shoot length of each plant were recorded weekly for 5 weeks. After 35 days the plants were harvested, roots and shoots were separated, dried and weighed. The root thickness of each plant was measured with a Leitz stereo dissecting microscope using 1 cm long pieces of three thick nodal (adventitious) roots sampled near the base of the plant which were kept in water at 4 °C for a maximum of one week.

Assessing root traits using rhizotrons

A total of 12 genotypes were assessed for root traits using the rhizotron system that is described in detail in Price et al. (2012) which has been widely used to assess rice (Price et al. 2002; Shrestha et al. 2014). Four replicates were arranged in completely randomised blocks. The genotypes were the four highest, three lowest and three average herbicide scoring Sri Lankan landraces plus check cultivars Black Gora and IR 64. Briefly, the rhizotrons were made by taping together two sheets of 4 mm thick glass (1.2 x 0.3 m width x length) 15 mm apart with duct tape and filling the space with topsoil (as used in herbicide box). Approximately 7 kg of moist (20 % volumetric water content) soil were used to give a dry bulk density of about 1.1 kg l^{-1}. The chambers had 5 mm diameter drainage holes at the bottom of each side. Before sowing, each chamber was saturated with 500 ml of Yoshida's full strength nutrient solution at pH 5.5 (Yoshida et al. 1976). Two seeds were sown in each chamber on 20th July 2011. After seedling emergence, the seedlings were thinned to one. Plants were grown for 6 weeks in a greenhouse (minimum temperature 25 °C) under daylight conditions (no supplementary lighting). Each rhizotron was connected to a drip irrigation system and supplied with 250 ml of Yoshida's nutrient solution three times a week for the first 3 weeks, and 5 times a week for the final three weeks. In the final week rhizotrons received 125 ml of water a day in addition to the 250 ml nutrient solution. Plants received their last watering on the evening of day 39 after sowing and received no more water until harvested on day 42.

On a weekly basis shoot growth was monitored as shoot length (length from the soil to the tip of the longest leaf) while the length of the longest visible root and the number of roots passed 25 cm at 21 days, 50 cm at

35 days, 75 cm at 42 days were recorded. On day 21 and 42 the angle of spread of the root system (the maximum angle between the least vertical main root axes on either side of the plants base) was measured with a protractor. The weight of each rhizotron was measured every day from day 39 until day 42 to assess water use in the final three days of the experiment. On the evening of day 39, a digital image was taken of each rhizotron using a 12 MPixel camera. On day 42 shoots were removed in a single day, dried at 80 °C for two days and subsequently weighed. Over a 1-week period, each chamber was opened. Short sections of three of the thickest nodal (adventitious) roots were removed from each root system at the base of the shoot, placed in water and stored at 4 °C, before being used to assess root thickness under a Leitz stereo dissecting microscope. The entire root system was divided into 3 sections, 0–40 cm (the top), 40–80 cm (middle) and 80–120 cm (bottom), washed, dried and subsequently weighed.

Additional files

Additional file 1: Figure S1. Optimum K value of 6 revealed in the K vs delta K plot from STRUCTURE.

Additional file 2: Table S1. Accessions, subpopulation, drought scores and mean herbicide score. (XLSX 49 kb)

Additional file 3: Table S2. Genotype of 326 SNPs for 135 Sri Lankan landraces plus 18 OryzaSNP accessions (the last 18 entries). (XLSX 198 kb)

Additional file 4: Online Resource 1 (NEXML tree of SL and OryzaSNP accessions 152 kb)

Additional file 5: Online Resource 2 (NEXML tree of SL and RDP1 accessions 127 kb)

Additional file 6: Table S3. Correlations between herbicide scores (HS) at different days.

Additional file 7: Figure S2. Frequency distribution of mean herbicide score at 43 days for 135 Sri Lankan landraces.

Additional file 8: Table S4. Mean of traits measured in hydroponics with summary test statistics for one way ANOVA at the bottom.

Additional file 9: Table S6. Correlations between herbicide score day 43 (HS43) and hydroponic traits.

Additional file 10: Table S5. Mean rhizotron data for 12 accessions.

Additional file 11: Table S7. Correlations between herbicide scores day 43 (HS43) and rhizotron traits.

Competing interest
Both authors declare that they have no competing interests.

Authors' contributions
Project conception and design (MM, AHP). All experimental work and data analysis (MM). Manuscript preparation (MM, AHP). Both authors read and approved the final manuscript.

Acknowledgements
Mayuri Munasinghe was supported by a Commonwealth Scholarship (ref no. LKCS-2009-384). The development and use of the SNP chip was funded by a BBSRC grant BB/J003336/1. The authors thank Owen Price (University of Wollongong, Australia) for producing the coloured province map of Sri Lanka, Gareth Norton (Aberdeen) for merging the RDP1 SNP data with the Sri Lankan data and Tony Travis (Aberdeen) for help with PCA.

Author details
[1]Institute of Biological and Environmental Sciences, University of Aberdeen, Aberdeen AB24 3UU, UK. [2]Current address: University of Sri Jayewardenepura, Nugegoda, Sri Lanka.

References
3K RGP (2014): The 3,000 rice genomes project. GigaScience3:7
Alexandrov N, Tai S, Wang W, Mansueto L, Palis K, Fuentes RR, Ulat VJ, Chebotarov D, Zhang G, Li Z et al (2015) SNP-Seek database of SNPs derived from 3000 rice genomes. Nucleic Acids Res 43:D1023–D1027
Al-Shugeairy Z, Islam MS, Shrestha R, Al-Ogaidi F, Norton GJ, Price AH (2014) High throughput screening of rooting depth in rice using buried herbicide. Annals of Applied Biology 165:96–107
Araus L, Cairns JE (2014) Field high-throughput phenotyping: the new crop breeding frontier. Trends in Plant Science 19:52–61
Bradbury PJ, Zhang Z, Kroon DE, Casstevens TM, Ramdoss Y, Buckler ES (2007) TASSEL: Software for association mapping of complex traits in diverse samples. Bioinformatics 23:2633–2635
De Costa WAJM (1998) Prediction of the effects of soil water stress on grain yield and radiation use efficiency of rice (*Oryza sativa* L.) using a simulation model. Journal of the National Science Council of Sri Lanka 26(2):101–123
De Silva DAM, Yamao M (2009) Rice pinch to war thrown nation: an overview of the rice supply chain of Sri Lanka and the consumer attitudes on government rice risk management. Journal of Agricultural Sciences 4:77–96.
FAO (2010) The Second Report on the State of the World's Plant Genetic Resources for Food and Agriculture- Country Report Sri Lanka http://www.fao.org/docrep/013/i1500e/SriLanka.pdf.
GFDRR (2011) Vulnerability, Risk Reduction, and Adaptation to Climate - Sri Lanka (http://sdwebx.worldbank.org/climateportalb/doc/GFDRRCountryProfiles/wb_gfdrr_climate_change_country_profile_for_LKA.pdf) Accessed Apr 2016.
Huang X, Wei X, Sang T, Zhao Q, Feng Q et al (2010) Genome-wide association studies of 14 agronomic traits in rice landraces. Nat Genet 42:961–967
Huson DH, Scornavacca C (2012) Dendroscope 3: An Interactive Tool for Rooted Phylogenetic Trees and Networks. Systematic Biology 61:1061–1067
Lafitte HR, Champoux MC, McLaren G, O'Toole JC (2001) Rice root morphological traits are related to isozyme group and adaptation. Field Crops Research 71:57–70
McCouch SR (2013) Feeding the future. Nature 499:23–24
McCouch SR, Wright MH, Tung CW, Maron LG, McNally KL, Fitzgerald M, Singh N, DeClerck G, Agosto-Perez F, Korniliev P, Greenberg AJ, Naredo MEB, Mercado SMQ, Harrington SE, Shi Y, Branchini DA, Kuser-Falcao PR, Leung H, Ebana K, Yano M, Eizenga G, McClung A, Mezey J (2016) Open access resources for genome-wide association mapping in rice. Nature Communications 7:10532. doi:10.1038
McNally KL, Childs KL, Bohnert R, Davidson RM, Zhao K, Ulat VJ, Zeller G, Clark RM, Hoen DR, Bureau TE, Stokowski R, Ballinger DG, Frazer KA, Cox DR, Padhukasahasram B, Bustamante CD, Weigel D, Mackill DJ, Bruskiewich RM, Rätsch G, Buell CR, Leung H, Leach JE (2009) Genomewide SNP variation reveals relationships among landraces and modern varieties of rice. Proceedings of the National Academy of Science USA 106:12273–12278
Price AH, Shrestha R, Piffanelli P, Lupotto E, Casella L (2012) Soil-filled glass rhizotrons for visualizing roots. In: Shashidhar HE, Henry A, Hardy B (eds) Methodologies for root drought studies in rice. International Rice Research Institute, Los Baños (Philippines), pp 9–14
Price AH, Steele KA, Gorham J, Bridges JM, Moore BJ, Evans JL, Richardson P, Jones RGW (2002) Upland rice grown in soil-filled chambers and exposed to contrasting water-deficit regimes: I. Root distribution, water use and plant water status. Field Crops Research 76:11–24
Price AH, Tomos AD, Virk DS (1997) Genetic dissection of root growth in rice (Oryza sativa L.) I: A hydroponic screen. Theor Appl Genet 95:132–142.
Pritchard JK, Stephens M, Donnelly P (2000) Influence of population structure using multilocus genotype data. Genetics 155:945–959
Rajkumar G, Weerasena J, Fernando K, Liyanage A (2011) Assessment of genetic diversity among Sri Lankan rice varieties by AFLP markers. Plant Genetic Resources: Characterization and Utilization 9:224–228
Shrestha R, Al-Shugeairy Z, Al-Ogaidi F, Munasinghe M, Radermacher M, Vandenhirtz J, Price AH (2014) Comparing simple root phenotyping methods on a core set of rice genotypes. Plant Biology 16:632–642

Trachsel S, Kaeppler SM, Brown KM, Lynch JP (2011) Shovelomics: high throughput phenotyping of maize (*Zea mays* L.) root architecture in the field. Plant Soil 341:75–87

Travis AJ, Norton GJ, Datta S, Sarma R, Dasgupta T, Savio FL, Macaulay M, Hedley PE, McNally KL, Sumon MH, Islam MR and Price AH (2015) Assessing the genetic diversity of rice originating from Bangladesh, Assam and West Bengal. Rice (in press).

Yoshida S, Forno DA, Lock JH, Gomez KA (1976) A laboratory manual for the physiological studies of rice. IRRI, Manila, pp 61–66

Zhao K, Tung CW, Eizenga GC, Wright MH, Ali ML et al (2011) Genome-wide association mapping reveals a rich genetic architecture of complex traits in *Oryza sativa*. Nat Comm 2:467

3

Detection of Novel QTLs Regulating Grain Size in Extra-Large Grain Rice (*Oryza sativa* L.) Lines

Shuhei Segami[1,2], Tatsuya Yamamoto[1], Katsuyuki Oki[1], Tomonori Noda[3], Hiroyuki Kanamori[4], Harumi Sasaki[4], Satomi Mori[4], Motoyuki Ashikari[3], Hidemi Kitano[3], Yuichi Katayose[4], Yukimoto Iwasaki[1*] and Kotaro Miura[1*]

Abstract

Background: Grain size is an important trait that affects rice yield. Although many genes that contribute to grain size have been cloned from mutants or by quantitative trait locus (QTL) analysis based on bi-parental mapping, the molecular mechanisms underlying grain-size determination remain poorly understood. In this study, we identified the lines with the largest grain size and detected novel QTLs affecting the grain size.

Results: We screened the National Institute for Agrobiological Sciences Genebank database and identified two rice lines, BG23 with the widest grain and LG10 with the longest grain. Using these two lines, we performed QTL analysis for grain size. Eight QTLs were detected during the QTL analyses using F_2 populations derived from crosses between the large-grain lines BG23 or LG10 and the middle-size grain cultivars Nipponbare and Kasalath. Both BG23 and LG10 possessed large-grain alleles of four major QTLs: *GW2*, *GS3*, *qSW5/GW5*, and *GW8*. Other three minor QTLs were derived from BG23. However, these QTLs did not explain the differences in grain size between these two lines. Additionally, four QTLs for grain length or width were detected in an F_2 population derived from a cross between BG23 and LG10; this population lacked the strong effects of the four major QTLs shared by both parent plants. Of these newly detected QTLs, the effects of two QTLs, *GL3b* and *GL6*, were confirmed by progeny testing. Comparison of the length of inner epidermal cells in plants homozygous for BG23 and LG10 alleles indicated that *GL3b* and *GL6* genes regulate cell elongation and cell division, respectively.

Conclusions: In this study, we detected 12 loci including 14 QTLs regulating grain size from two lines with largest grains available in Japanese stock. Of these loci, we confirmed the effect of two gene loci and mapped their candidate region. Identification of novel genes regulating grain size will contribute to our understanding of the molecular mechanisms controlling grain size.

Keywords: QTL, Grain size, Rice (*Oryza sativa* L.)

Abbreviations: BS, Brassinosteroid; CAPS, Cleaved amplified polymorphic sequence; Chr6L, Long arm of Chr6; Chr6S, Short arm of Chr6; QTL, Quantitative trait locus; SEM, Scanning electron microscopy; SNP, Single nucleotide polymorphism; SSR, Simple sequence repeat; TRM, TONNEAU1 recruiting motif

* Correspondence: iwasaki@fpu.ac.jp; miura-k@fpu.ac.jp
[1]Fukui Prefectural University, Faculty of Biotechnology, 4-1-1 Kenjojima, Matsuoka, Eiheiji-cho, Yoshida-gun, Fukui 910-1195, Japan
Full list of author information is available at the end of the article

Background

Grain size is an important trait in breeding, serving as a factor determining rice yield. Generally, genes regulating grain size are classified into two groups, those controlling cell elongation and those controlling cell division. A short-grain phenotype due to shortened cell length in the loss-of-function mutants of brassinosteroid (BR)-related gene *D61* implies that these genes control cell elongation (Segami et al. 2012). Furthermore, *SRS3* and *SRS5* are also involved in cell elongation. *SRS3* and *SRS5* encode a kinesin-13 protein, and an α-tubulin, respectively (Kitagawa et al. 2010, Segami et al. 2012). These *srs* mutants commonly show short-grain phenotypes due to shortened cell lengths, similar to the BR-related mutants. However, the *Srs5* mutant is seemingly not involved in BR signal transduction. *D1* and *TUD1* are involved in cell division (Izawa et al. 2010; Hu et al. 2013). The mutants *d1* and *tud1* commonly show a short-grain and dwarf phenotype owing to reduced cell numbers. *D1* and *TUD1* encode the α-subunit of a heterotrimeric G-protein and U-box ubiquitin ligase, respectively, and *d1* is epistatic to the *tud1* mutation (Ashikari et al. 1999; Fujisawa et al. 1999; Hu et al. 2013). Of these mutants, we succeeded in classifying *srs3, Srs5, d1,* and *d61* into cell elongation- or division-type mutants by comparing the cell length and cell number of the inner-epidermal cells of lemma (Izawa et al. 2010; Kitagawa et al. 2010; Segami et al. 2012). Quantitative trait locus (QTL) analysis also revealed other genes involved in grain size regulation. *GS3* controls grain length and weight (Fan et al. 2006; Mao et al. 2010; Takano-Kai et al. 2009); *GW2* and *qSW5/GW5* control grain width and weight (Shomura et al. 2008; Song et al. 2007; Weng et al. 2008); *GS5* controls grain width, filling, and weight (Li et al. 2011); *GW8* controls grain width (Wang et al. 2012); *GL3.1* (also named *OsPPKL1*) controls grain length (Qi et al. 2012; Zhang et al. 2012); *TGW6* and *GW6a* control grain weight (Ishimaru et al. 2013; Song et al. 2015); and *GL7/GW7* controls grain length and width (Wang et al. 2015a, b). Interestingly, among these genes cloned by QTL analysis, only *GL7/GW7* was involved in cell elongation, whereas the others were involved in cell division. It is noteworthy that these genes encode a variety of different proteins involved in various signal transduction pathways. *GS3* encodes the γ-subunit 3 of a heterotrimeric G-protein (Li et al. 2012); *GW2* encodes a RING-type domain with E3 ubiquitin ligase activity (Song et al. 2007); *qSW5/GW5* encodes a novel nucleoprotein that interacts with polyubiquitin in a yeast two-hybrid experiment (Weng et al. 2008); *GS5* encodes a putative serine carboxypeptidase and acts as a positive regulatory factor in the cell cycle (Li et al. 2011); *GW8* encodes the transcription factor Squamosa promoter-binding protein-like 16 containing the *miR156*-targeted site (Wang et al. 2012); *GL3.1* encodes a Ser/Thr phosphatase of the protein phosphatase of the kelch-like family (Qi et al. 2012; Zhang et al. 2012); *TGW6* encodes a protein with indole-3-acetic acid-glucose hydrolase activity (Ishimaru et al. 2013); *GW6a* encodes a functional GCN5-related *N*-acetyl-transferase-like protein that harbors intrinsic histone acetyl-transferase activity (Song et al. 2015); and *GL7/GW7* encodes a homolog of the *Arabidopsis thaliana* TONNEAU1 recruiting motif (TRM) protein (Wang et al. 2015a, b). Although many genes were cloned and identified, the molecular mechanisms underlying grain size determination are not well understood, because most of these genes are not involved in the same signal transduction pathways. To clarify these mechanisms, it is necessary to identify and classify additional genes that contribute to grain size, whether through cell elongation or cell division, by unified methodology. Therefore, the aim of this study was to identify, from the National Institute for Agrobiological Sciences (NIAS) Genebank database (http://www.gene.affrc.go.jp), the rice lines with the largest grain, and using these lines, detect novel QTLs and elucidate how these QTLs control grain size.

Results

BG23 and LG10 Display the Extra-Large Grain Phenotype

The parental lines used in QTL analysis for grain size were selected by screening for the largest-grain rice lines in the public database NIAS Genebank. The grain width and length of screened lines ranged from 1.7 to 4.6 mm and 3.4 to 14.3 mm, respectively (Additional file 1: Figure S1A and B). Of these screened lines in NIAS Genebank, BG23 (grain width 4.6 mm) and LG10 (grain length 14.3 mm), which produced seeds with the largest width and length, respectively, were selected as parental lines for QTL analysis. Because the grain length of Nipponbare and Kasalath was 6.97 ± 0.28 mm and 7.99 ± 0.33 mm, respectively, and near to the median value of the NIAS Genebank database, we used these lines as middle-size lines for grain length and width comparison (Fig. 1, Table 1). The grain length of BG23 and LG10 was 11.24 ± 0.42 mm and 14.25 ± 0.46 mm, respectively (Fig. 1, Table 1). The grain width of BG23 and LG10 was 4.43 ± 0.22 mm and 4.06 ± 0.36 mm, respectively, whereas

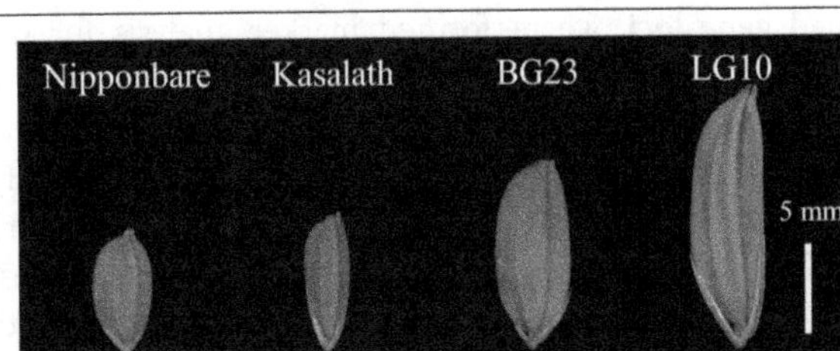

Fig. 1 Grain size and morphology of middle-size grain cultivars Nipponbare and Kasalath and extra-large grain lines BG23 and LG10. Bar = 5 mm

Table 1 Comparisons of grain length and width in rice lines Nipponbare, Kasalath, BG23, and LG10

Traits	Nipponbare	Kasalath	BG23	LG10
Grain length (mm)	6.97 ± 0.28	7.99 ± 0.33	11.24 ± 0.42	14.25 ± 0.46
Grain width (mm)	3.27 ± 0.27	2.55 ± 0.21	4.43 ± 0.22	4.06 ± 0.36

Student's t-test was used to generate the *P* values. An asterisk indicates a statistically significant difference from the neighboring value ($P < 0.01$). Significant difference of grain length was detected between Nipponbare and BG23, Kasalath and BG23, Nipponbare and LG10, and BG23 and LG10 ($P < 0.01$). Significant difference of grain width was detected between Nipponbare and BG23, Kasalath and BG23, and Nipponbare and LG10 ($P < 0.01$). The difference of grain width detected between BG23 and LG10 was not significant ($P > 0.05$)

that of Nipponbare and Kasalath was 3.27 ± 0.27 mm and 2.55 ± 0.21 mm, respectively (Fig. 1, Table 1). Because the grains of BG23 and LG10 were significantly larger compared to those of the middle-size grain cultivars, these lines were expected to possess different composition of genes regulating grain size. Clarifying the compositions and effects of each gene will provide information about epistasis and effect of interactions of these genes. Therefore, this analysis was expected to provide novel and useful genes with the potential to increase grain yield by controlling grain size.

BG23 and LG10 Shared Four Major QTLs for Grain Size

To identify the genes regulating grain length and width in BG23 and LG10, four combinations of F_2 populations were derived from the following crosses: Kasalath × BG23, Kasalath × LG10, Nipponbare × BG23, and Nipponbare × LG10. A plot of grain length and width of the F_2 individuals in each population showed a continuous distribution (Fig. 2a–h), indicating that the differences in grain traits between the extra-large grain lines and middle-size grain lines were regulated by QTLs. QTL analysis for grain length and width detected two loci on Chr2 and Chr3 that were shared by all four F_2 populations (Fig. 2i–l, Table 2, Additional file 2: Figure S2A–H) and one locus on Chr5 shared by two F_2 populations derived from Kasalath crosses (Fig. 2i and j, Table 2). Unshared QTLs were detected on Chr1, Chr4, Chr10, and Chr8 (Fig. 2i–k, Table 2). In total, we detected eight QTLs including seven loci in this analysis (Fig. 2i–k, Table 2).

As several QTLs were detected around previously cloned gene loci, we performed marker analysis for the genes *GW2, GS3, qSW5/GW5*, and *GW8*. Both BG23 and LG10 possessed all of the alleles responsible for large grain size in all four genes (Additional file 3: Figure S3A–H). The QTLs for grain width and length, detected on Chr2 and Chr3, respectively, in all tested populations correspond to *GW2* and *GS3*, whereas the QTLs for grain width detected on Chr5 and Chr8 correspond to *qSW5/GW5* and *GW8*.

The QTLs for grain length on Chr4 and Chr10 detected from the Kasalath × LG10 cross and Chr1 detected from the Nipponbare × BG23 cross include *qGL1, qGL4b*, and *qGL10*; loci have been previously reported (Kato et al. 2011; Wang et al. 2012; Yan et al. 2014) (Fig. 2j and k). In total, we detected seven loci including four cloned and three non-cloned QTLs for grain size. However, as *GW8* was detected only in the F_2 population derived from the Kasalath × BG23 cross, these minor QTLs appeared to be masked by the effects of the major QTLs. Here, we examined whether the QTL analysis for grain length using F_2 population derived from BG23 and LG10 cross is possible to detect the same loci on Chr1, Chr4, and Chr10 detected in other combinations of F_2 populations and whether these loci can fully explain the differences between BG23 and LG10. Hence, the QTLs for grain length and width were investigated using the F_2 population derived from the BG23 × LG10 cross, thus avoiding the strong effects of the four QTLs shared by the parents.

Novel QTLs Detected from the BG23 × LG10 Cross

Screening for simple sequence repeat (SSR) markers for QTL analysis that could indicate the differences in genotypes between BG23 and LG10 was unsuccessful, because these two parental lines exhibited a highly similar banding pattern on the gel. To construct the linkage map, we performed re-sequencing of the parents and designed BeadArray markers using detected single nucleotide polymorphisms (SNPs).

The histograms of grain length and width in F_2 individuals showed a continuous distribution, indicating that the differences between BG23 and LG10 with respect to these parameters were likely regulated by multiple QTLs (Fig. 3a and b). QTL analysis of the F_2 population from the BG23 × LG10 cross revealed four QTLs for grain length on Chr1, Chr3, the short arm of Chr6 (Chr6S), and the long arm of Chr6 (Chr6L) and two QTLs for grain width on Chr1 and Chr6S (Fig. 3c, Table 3, and Additional file 3: Figure S3I and J). Additionally, the QTL for grain length on Chr6L included *GW6a* and *GW6b* (Song et al. 2015) (Fig. 3c). Overall, we detected six QTLs including five loci (Fig. 3c, Table 3). The QTL for grain width on Chr1 and that for grain length on Chr3 and Chr6S represent novel loci regulating grain size. A two-dimensional scan revealed no significant epistatic interactions among these QTLs in this population (Additional file 4: Figure S4).

Phenotypic Evaluation of Novel QTLs

To confirm the effect of detected QTLs on grain length, we screened F_3 lines including one segregating locus and three other fixed loci. Of the four loci, we screened two segregation lines for QTLs detected on Chr3 and Chr6L (Additional file 5: Figure S5) and from these segregation lines selected plants homozygous for BG23 and LG10 alleles on Chr3 (Chr3-BG23 and Chr3-LG10) and Chr6L

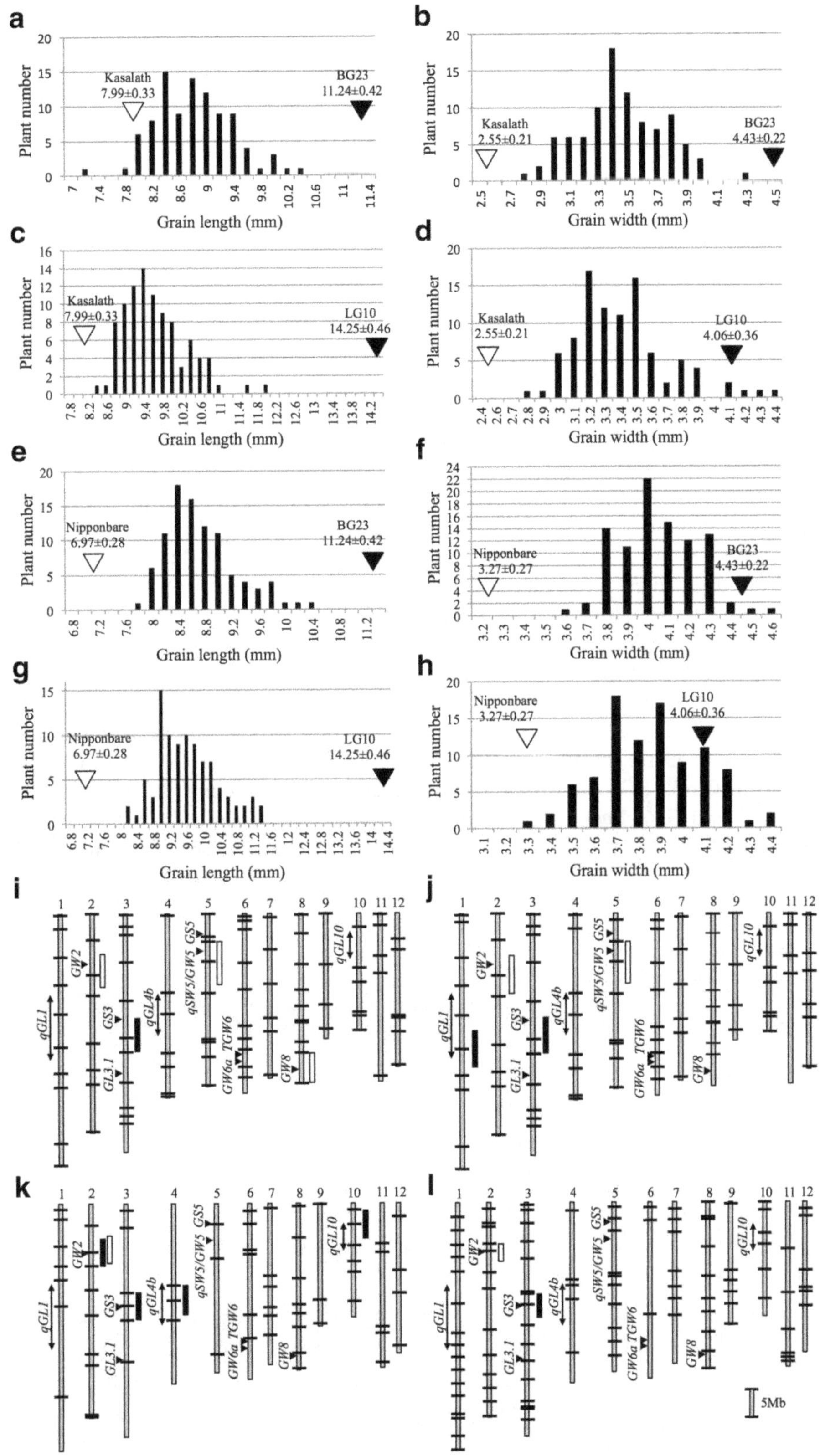

Fig. 2 (See legend on next page.)

(See figure on previous page.)
Fig. 2 Frequency distribution of grain length and width and quantitative trait loci (QTLs) detected in F_2 populations. Frequency distribution of grain length is shown in **a**, **c**, **e**, and **g**. Frequency distribution of grain width is shown in **b**, **d**, **f**, and **h**. F_2 populations derived from Kasalath × BG23 (**a** and **b**), Kasalath × LG10 (**c** and **d**), Nipponbare × BG23 (**e** and **f**), and Nipponbare × LG10 (**g** and **h**) crosses. Vertical white and black arrowheads indicate the means of the middle-size grain cultivars and extra-large grain lines, respectively. Genetic linkage maps of F_2 populations derived from Kasalath × BG23 (**i**), Kasalath × LG10 (**j**), Nipponbare × BG23 (**k**), and Nipponbare × LG10 (**l**). Boxes indicate positions of regions harboring QTLs with the logarithm of the odds > 3.0 for grain length (black) and width (white). Horizontal black arrowheads indicate the position of previously cloned QTLs for grain size. Double-headed arrows indicate the position of previously detected QTLs for grain size. Bars indicate marker positions

(Chr6L-BG23 and Chr6L-LG10). The grain length of Chr3-LG10 and Chr6L-LG10 were significantly longer compared to those of Chr3-BG23 and Chr6L-BG23, respectively (Fig. 4a, c). These results clearly indicate that these two loci include genes responsible for grain length. LG10 alleles on Chr3 and Chr6L have a positive effect on grain length. We named these two genes *GL3b* and *GL6*. To determine whether these grain-size regulating genes are involved in cell division or cell elongation, we compared the length of the inner epidermal cell in lemma by using scanning electron microscopy (SEM) (Fig. 4b). The length of the inner epidermal cells was significantly longer in Chr3-LG10 than in Chr3-BG23, whereas no significant difference was observed between Chr6L-BG23 and Chr6L-LG10 (Fig. 4d). In contrast, the estimated number of cells in Chr6L-LG10 was significantly higher than that in Chr6L-BG23 (Fig. 4e). These results indicate that *GL3b* and *GL6* affect cell elongation and cell division, respectively.

Genetic Mapping of *GL3b* and *GL6*

To narrow down the candidate region of *GL3b* and *GL6*, we performed genetic mapping using recombinant lines of each locus. Recombinant screening of these two loci using F_3 population of segregating lines resulted in four and three recombinants in *GL3b* and *GL6* regions, respectively (Fig. 5a and b). Because the F_3 plants possess heterozygous chromosomes, we selected F_4 plants homozygous for BG23 or LG10 and recombinant chromosome of each *GL3b* and *GL6* region to compare the phenotype of grain length. Using the data on grain length in F_5 seeds obtained from selected F_4 plants from each line, we performed mapping analysis for *GL3b* and *GL6* gene loci.

At the *GL3b* locus, we screened recombinant plants with a recombination breakpoint between the PCR markers RM7000 and BGLG_3-37.2_Hinf and obtained four F_3 recombinant plants (Fig. 5a). These plants were grown to obtain F_4 seeds, which were then sawn and the

Table 2 Quantitative trait loci for grain length and width detected in F_2 populations derived from the crosses Kasalath × BG23, Kasalath × LG10, Nipponbare × BG23, and Nipponbare × LG10

Populations	Traits	Chr[a]	Source	Marker interval	LOD[b]	Additive	Dominance	PVE[c]
Kasalath × BG23 F_2	Grain length	3	BG23	RM5551 -RM5488	5.22	0.44	−0.46	20.28
	Grain width	2	BG23	RM7581 -RM6639	7.43	0.25	0.00	28.27
		5	BG23	TG93 -d1-8	4.02	0.18	−0.06	17.88
		8	BG23	RM3452 -RM3155	5.2	0.15	−0.16	18.54
Kasalath × LG10 F_2	Grain length	1	LG10	B1075D06 -RM3143	3.93	0.44	0.03	17.41
		3	LG10	RM5551 -RM5488	6.39	0.60	−0.59	27.36
	Grain width	2	LG10	RM7581 -RM6639	11.12	0.30	−0.10	41.15
		5	LG10	TG93 -d1-8	6.47	0.23	−0.06	27.16
Nipponbare × BG23 F_2	Grain length	2	BG23	RM5654 -RM1313	5.71	0.39	0.06	24.38
		3	BG23	RM3033 -GS3-PstI	16.82	0.58	−0.44	60.39
		4	BG23	RM2565 -RM5714	3.55	0.33	0.07	15.87
		10	BG23	RM5271 -RM5147	3.34	0.18	0.36	15.07
	Grain width	2	BG23	RM5654 -GW2-HinfI	14.49	0.20	−0.04	46.39
Nipponbare × LG10 F_2	Grain length	3	LG10	RM3033 -RM2346	14.99	0.71	−0.42	51.99
	Grain width	2	LG10	RM5654 -RM1313	25.59	0.27	−0.04	71.48

[a]Chromosome number. [b]log10 of the odds ratio. [c]Phenotypic variance explained

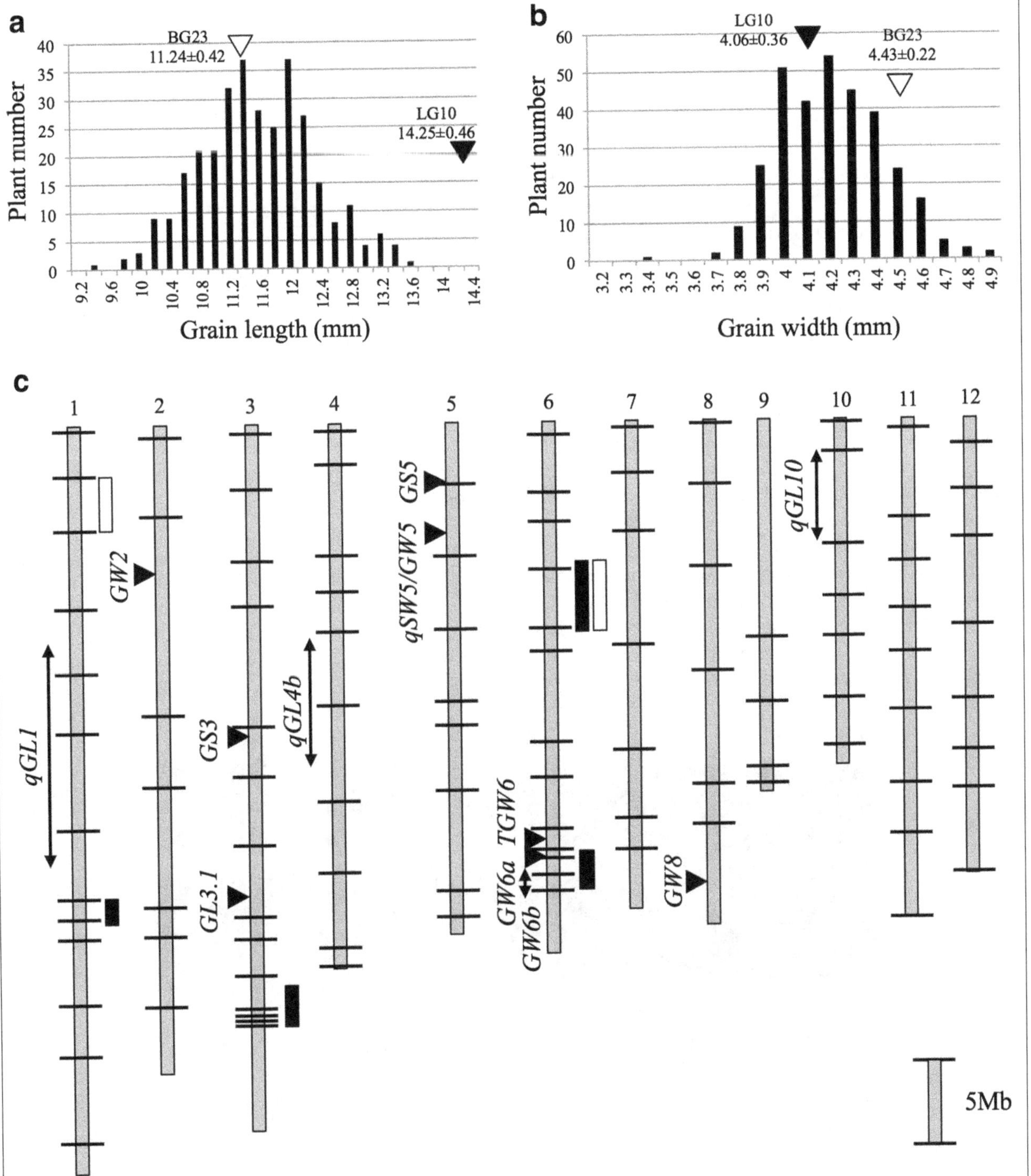

Fig. 3 Frequency distributions of grain length and width and quantitative trait loci (QTLs) detected in F_2 populations derived from the BG23 × LG10 cross. Frequency distributions of grain length (**a**) and width (**b**). Vertical white and black arrowheads indicate the means for BG23 and LG10, respectively. (**c**) Boxes indicate the positions of regions harboring QTLs with the logarithm of the odds > 3.0 for grain length (black) and width (white). Horizontal black arrowheads indicate the positions of previously cloned QTLs for grain size. Double-headed arrows indicate the positions of previously detected QTLs for grain size. Bars indicate marker position

resulting plants were screened homozygous for BG23 or LG10 chromosome as well as for recombinant chromosomes and compared the grain length of the resulting F_5 seeds. The recombinant plant selected from F_4 line no. 1 with a recombination event between 35.20 Mb and 35.35 Mb showed a significantly longer grain compared

Table 3 Quantitative trait loci for grain length and width detected in F_2 population derived from the BG23 × LG10 cross

Populations	Traits	Chr[a]	Source	Marker interval	LOD[b]	Additive	Dominance	PVE[c]
BG23 × LG10 F_2	Grain length	1	LG10	aa01009984 -BGLG_1-29.92_Hinf	4.16	0.26	−0.07	5.59
		3	LG10	BGLG_3-36.1_Pst -aa03002747	9.52	0.36	−0.18	12.85
		6	LG10	ac06000397 -ab06001036	5.40	0.27	−0.23	7.44
		6	LG10	bglg6_27 -BGLG_6-28.12_Hae	20.04	0.51	0.06	24.73
	Grain width	1	LG10	aa01004844 -ab01000509	5.80	0.10	−0.01	7.77
		6	LG10	ac06000397 -ab06001036	11.74	0.12	−0.10	15.13

[a]Chromosome number. [b]log10 of the odds ratio. [c]Phenotypic variance explained

to that of sibling plants homozygous for BG23, indicating that the causal gene of *GL3b* is located on the right from 35.20 Mb (Fig. 5a). Similarly, the F_4 line no. 7 also indicates that *GL3b* is located to the right from 35.20 Mb because grain length of the recombinant plant is the same as that in plants homozygous for BG23. Similarly, the right border of *GL3b* was indicated at 36.37 Mb according to F_4 lines 16 and 18 (Fig. 5a). Taken together, we concluded that *GL3b* locus is located in a region 1.2 Mb long, between 35.20 Mb and 36.37 Mb on Chromosome 3.

At the *GL6* locus, we screened three recombinant plants possessing recombination breakpoint between the PCR markers BGLG_6-26.79_Hinf and BGLG_6-28.12_Hae (Fig. 5b). Because the recombinant plant selected from the F_4 line no. 1 with a recombination event between 26.79 Mb and 27.09 Mb shows same overlap in grain length with sibling plants homozygous for BG23, we conclude that the causal gene *GL6* locates to the right from 26.79 Mb (Fig. 5b). Similarly, the right border of *GL6* was indicated as 28.12 Mb by F_4 lines 36 and 20 (Fig. 5b). Taken together, we concluded that *GL6* locus locates in a 1.3-Mb long region, between 26.79 Mb and 28.12 Mb on Chromosome 6.

Discussion

In this study, we detected eight QTLs on seven loci from four populations derived from crosses between large-grain lines and middle-size grain lines. Furthermore, we newly detected six QTLs on five loci from the population

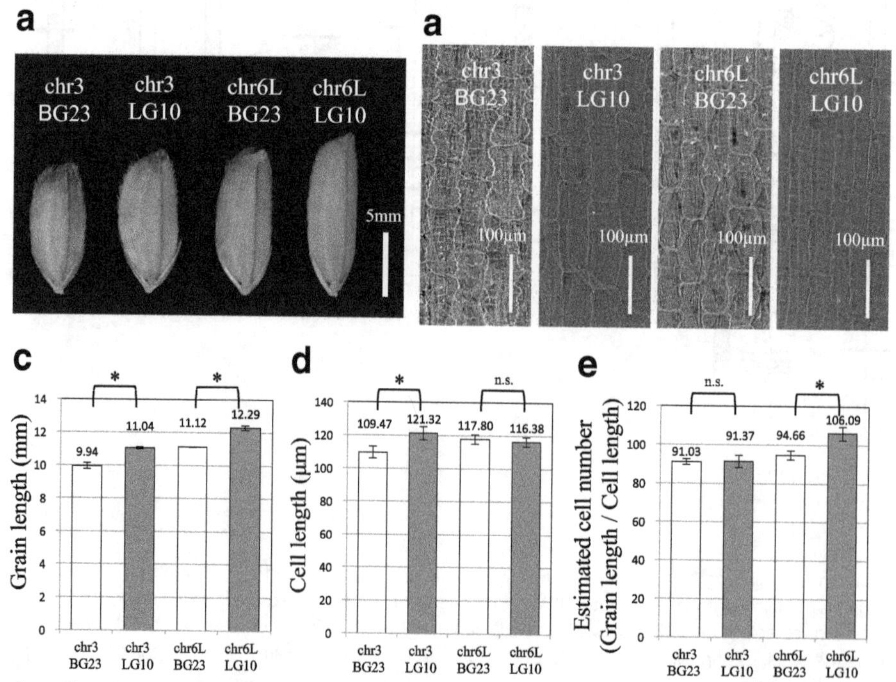

Fig. 4 Comparison of grain length and inner-epidermal cell length in segregation lines. **a** Grain morphology of Chr3-BG23, Chr3-LG10, Chr6-BG23, and Chr6-LG10. Bar = 5 mm. **b** Inner-epidermal cells in the lemma of Chr3-BG23, Chr3-LG10, Chr6-BG23, and Chr6-LG10. Bars = 100 μm. **c** Grain length of Chr3-BG23, Chr3-LG10, Chr6-BG23, and Chr6-LG10. **d** Cell length of Chr3-BG23, Chr3-LG10, Chr6-BG23, and Chr6-LG10. **e** Estimated cell number of Chr3-BG23, Chr3-LG10, Chr6-BG23, and Chr6-LG10. Numbers in **c**–**d** indicate the averages. Student's t-test was used to generate the *P* values. * indicates statistical significance from the neighboring value ($P < 0.05$). n.s. indicates no significance ($P > 0.05$)

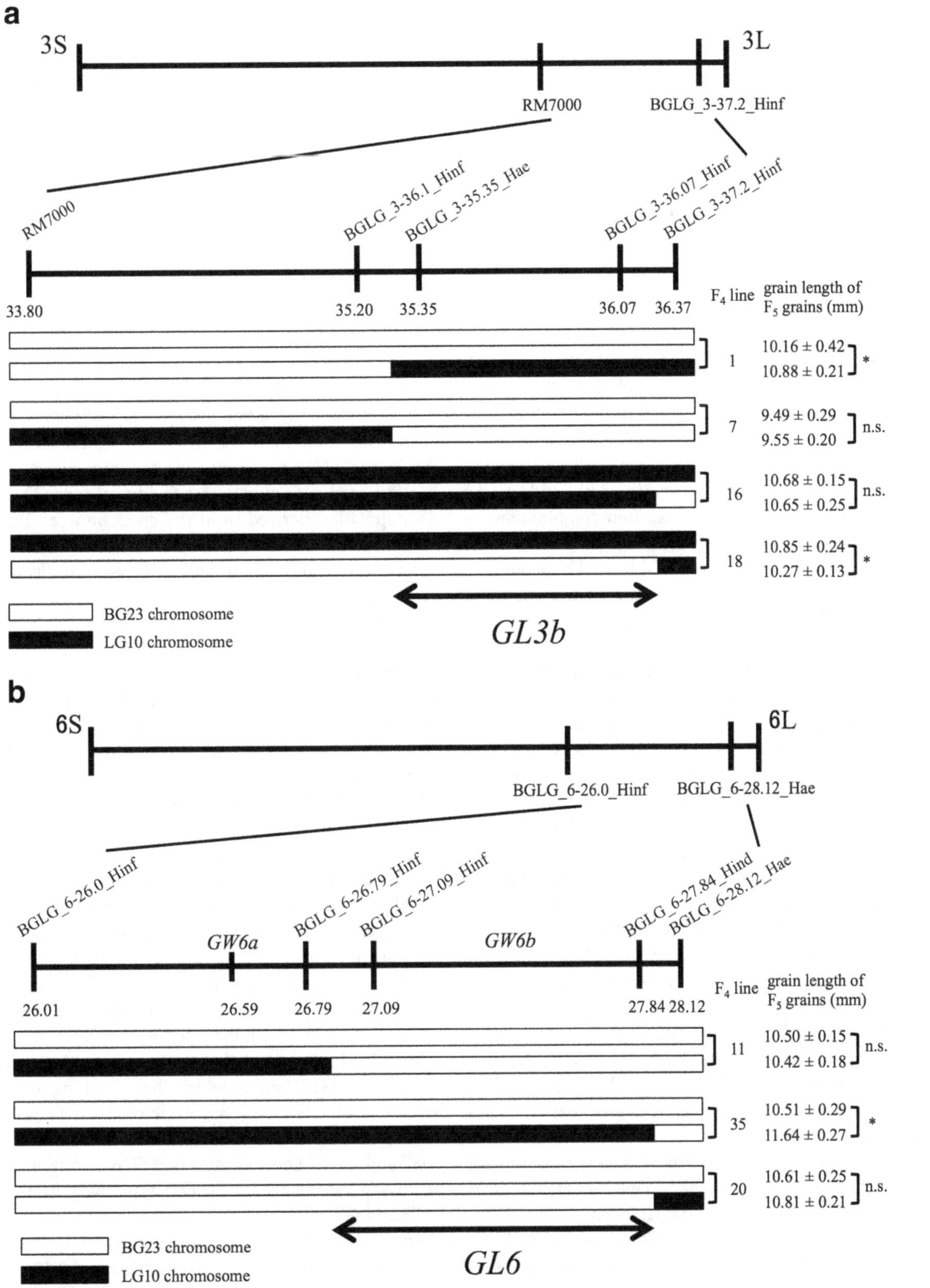

Fig. 5 (See legend on next page.)

(See figure on previous page.)

Fig. 5 Genetic mapping of *GL3b* and *GL6*. **a** Recombinant plants possessing breakpoint at the Chr3 locus were screened from the F_3 population of segregation lines by using PCR markers RM7000 and BGLG_3-37.2_Hinf. We found four F_3 recombinant plants. Because the F_3 plants were heterozygous, we selected F_4 plants homozygous for BG23 or LG10 chromosome and recombinant chromosome to compare the phenotype for grain length. At least three homozygous plants were selected for each genotype. The grain length of these homozygous F_4 plants was measured by using 50 F_5 seeds. Two types of genotype selected among homozygous plants from each line are indicated to the left of the F_2 line number. The grain length of each line is indicated to the right of the F_2 line number. The results of the *t*-test for grain length between the plants with or without recombination are indicated to the right of the grain length of F_5 grains. *GL3b* gene locus was mapped into a 1.2-Mb region between BGLG_3-36.1_Hinf and BGLG_3-37.2_Hinf. **b** Recombinant plants possessing breakpoint at the Chr6L locus were screened from the F_3 population of segregation lines by using PCR markers BGLG_6-26.0_Hinf and BGLG_6-28.12_Hae. We found three F_3 recombinant plants. The grain length of homozygous F_4 plants was measured by using 50 F_5 seeds. *GL6* gene locus was mapped into a 1.3-Mb region between BGLG_6-26.79_Hinf and BGLG_6-28.12_Hae. White and black boxes indicate BG23 and LG10 chromosomes, respectively. Student's *t*-test was used to generate the *P* values. * indicates statistical significance from the neighboring value ($P < 0.05$). n.s. indicates no significance ($P > 0.05$). The numbers under the marker name indicate marker's position (Mb)

derived from the cross between large grain lines. In total, we detected 12 loci including four cloned QTL genes. Both BG23 and LG10 contained four major grain-size regulating genes: *GW2, GS3, qSW5/GW5,* and *GW8* (Fig. 2i-l). The results presented herein indicate that the additive effect of *GW2* locus for grain width, and of *GS3* locus for grain length, detected from four combinations of F_2 population were distributed from 0.2 to 0.3 and from 0.44 to 0.7, respectively (Fig. 2i-l, Table 2). These additive effects are consistent with previously reported effects of *GW2* and *GS3* (Fan et al. 2006; Song et al. 2007). The dominant *GS3* locus detected in our four populations was distributed from −0.42 to −0.59. This indicates that plants heterozygous for *GS3* locus show equal phenotype as plants homozygous for functional *GS3* gene. Thus, the *GS3* locus detected in our study behaves in a recessive manner, which is consistent with previous report (Fan et al. 2006). The locus *qSW5/GW5* was detected in the F_2 population derived from Kasalath × BG23 and Kasalath × LG10 crosses (Fig. 2i and j, Table 2). Originally, *qSW5* was detected from descendants derived from the Kasalath × Nipponbare cross (Shomura et al. 2008), with the Nipponbare allele of *qSW5* being recessive. Our results corroborate this feature of *qSW5*. *GW8* was detected only in the F_2 population derived from the cross between BG23 × Kasalath, whereas LG10 also contained large grain allele (Fig. 2i, Additional file 3: Figure S3D). The *GW8* locus detected in the BG23 × Kasalath cross showed additive effect of 0.15 and dominance effect of −0.16 (Table 2). This result indicates that the large grain allele of *GW8* locus behaves in a recessive manner, although previously *GW8* was reported as semi-dominant gene (Wang et al. 2012). Because the effects of *GW8* locus for grain size regulation are comparatively weaker compared to the effects of the other three genes and not detected in the BG23× Nipponbare cross, the effects of *GW8* seemed to be masked or distorted by other three major QTLs. Therefore, the effects of minor QTLs may not be detectable. Although it is difficult to compare the effects of QTLs detected from different populations because the genetic background affects the effect of each QTL, we assumed that there should be some alternative QTLs explain the differences between BG23 and LG10. Thus, we tried to detect QTLs for grain size between BG23 and LG10 while avoiding the four major QTLs. As we expected, we could detect additional four QTLs in the F_2 population derived from the cross between BG23 and LG10. Of these four QTLs, two loci located on Chr3 and Chr6L and named *GL3b* and *GL6* were confirmed to be involved in cell elongation and cell division, respectively (Fig. 4a-e). Observation of the inner epidermal cells in lemma revealed that *GL3b* promotes grain length by promoting cell elongation (Fig. 4b-e). Using the same method, the genes responsible for cell elongation had been identified from a BR-related mutant (*d61*) and microtubule-related mutants (*srs3* and *Srs5*) (Kitagawa et al. 2010; Segami et al. 2012). TRM-containing protein (*GL7/GW7*) cloned in the QTL analysis also regulates cell elongation in rice (Wang et al. 2015b). Because these studies cloned only those genes involved in cell elongation, genetic diversity of the grain size in rice was considered to have developed mainly owing to the differences in the genes responsible for cell division. Therefore, clarifying the molecular functions of *GL3b* via cloning and genetic analysis of these genes will be important in elucidating the complex genetic mechanism underlying grain size regulation. *GL6* is involved in cell division (Fig. 4b-e). Since it was detected in the F_2 population derived from the crossing of BG23 and LG10, both of which share the long-grain alleles of *GS3* and *GW8* (Fig. 3c, Additional file 3: Figure S3), *GL6* promotes grain length in an additive manner with *GS3* and *GW8*. This indicates that *GL6* belongs to a genetically separated signal transduction pathway from those of *GS3* and *GW8*.

In contrast, the grain of BG23 is wider than the grain of LG10 (Fig. 1, Table 1). Nevertheless, LG10 provided positive additive effects for all detected QTLs for grain width in the F_2 population derived from the crossing of BG23 and LG10. This is explained by the grain length of LG10, which is too long in the longitudinal direction to fill grain hull. Therefore, starch input is not sufficient to fill the volume of the grain hull. This starch shortage leads to unfilled and thinner grains than predicted by the genetic potential of LG10.

Conclusion

In this study, we detected 12 loci that are involved in the regulation of grain size, using rice lines with largest grain size in Japanese stock (Figs. 2i-l, 3c). Among these loci, we identified two loci *GL3b* and *GL6* that promote grain length through increased cell length and cell division, respectively (Fig. 4b-e). Although we confirmed the effects of the two loci detected in this study, the effects of six loci remain unknown (Fig. 2i-l, and Fig. 3c). The results indicate that the molecular mechanisms underlying grain size determination are complex. To clarify the mechanisms and genetic architecture, we need to isolate the genes and test the genetic interactions of each gene in future studies.

Methods

Plant Materials

Two rice (*Oryza sativa* L.) cultivars, Nipponbare (*japonica*) and Kasalath (*indica*), and two extra-large grain rice lines, BG23 (*japonica*) and LG10 (*japonica*), were used in this study. BG23 (accession number: 8966) and LG10 (accession number: 84410) were obtained from NIAS Genebank after screening for lines with the widest and the longest grain, omitting the data with input errors (containing wrong trait annotations, low reliability of the reads, etc.).

For the QTL analysis, five F_2 populations were produced from the following parental crosses: Kasalath × BG23, Kasalath × LG10, Nipponbare × BG23, Nipponbare × LG10, and BG23 × LG10. Phenotypic evaluations for QTL analysis using these F_2 populations were performed in the 2012 growing season. Phenotypic evaluations for F_2, F_3, and F_4 populations derived from BG23 × LG10 cross were performed in the 2013, 2014, and 2015 growing seasons, respectively. Rice seeds were sown at the beginning of April, and at the beginning of May the seedlings were transplanted into a paddy field at the Research Center for Bioresources Development, Fukui Prefectural University, Fukui, Japan, and grown under natural conditions.

Phenotypic Evaluation

Harvested rice seeds were air-dried and stored in a seed dryer for at least 2 weeks. Fully filled grains were then evaluated for grain length and width. Fifty grains were randomly selected from each plant and scanned with a scanner (GT8300, EPSON, Nagano, Japan) to obtain images that were then used to measure each size using Smart Grain software (Tanabata et al. 2012).

The inner-epidermal cells of lemmas in mature grain hull were observed by SEM (TM3030, Hitachi, Tokyo, Japan). We measured cell length of the central cell layer of the hull ($n > 20$).

QTL Analysis

Genomic DNA was extracted from each F_2 progeny. The number of individuals and markers used from each F_2 population for each genotype were as follows: 94 individuals and 92 markers in Kasalath × BG23 F_2 population, 94 individuals and 90 markers in Kasalath × LG10 F_2 population, 94 individuals and 66 markers in Nipponbare × BG23 F_2 population, 94 individuals and 105 markers in Nipponbare × LG10 F_2 population, and 318 individuals and 106 markers in BG23 × LG10 F_2 population (Additional file 6: Table S1 and Additional file 7: Table S2). PCR-based markers, including SSR markers (McCouch et al. 2002; Ware et al. 2002; http://www.gramene.org/) and cleaved amplified polymorphic sequence (CAPS) markers (Konieczny and Ausubel 1993), were used to construct linkage maps. An Illumina GoldenGate Assay with VeraCode (San Diego, CA, USA) was used according to the manufacturer's instructions.

QTL analysis was performed using the R/qtl software (http://www.rqtl.org; Broman et al. 2003). Putative QTLs were detected by using the interval mapping function. Two QTLs of the same chromosome were confirmed by composite interval mapping function. The QTLs were assumed present when the logarithm of the odds score was 3.0 or higher. The additive and dominant effects and the phenotypic variance explained by each QTL at the maximum LOD score were estimated by using the sim.geno, makeqtl, fitqtl, and effectplot functions in the R/qtl software (Broman et al. 2003).

Marker Analysis

GW2-specific CAPS markers were amplified using the primers GW2-Hinf1-F and GW2-Hinf1-R. *GS3* marker analysis was carried out according to the protocol described by Fan et al. (2009) using primers SF28-U and SF28-L. *qSW5/GW5* marker analysis was conducted according to the method presented in Weng et al. (2008) using primers Indel2-F and Indel2-R, and *GW8* marker analysis was carried out using indel markers to detect a 10-bp deletion in the promoter region (Additional file 8: Table S3).

Resequencing Analysis of the Extra-Large Grain Lines

Construction of the shotgun library and sequencing were conducted with a Genome Analyzer-IIx (GA-IIx; Illumina, USA). Rice genomic DNA was fragmented by Covaris (Woburn, MA, USA). The size-selected DNA (average length 300 bp) was purified with an AmpureXP (Beckman Coulter, Inc., Fullerton, CA, USA) and gel extraction. The standard short-read library was built using a TruSeq DNA Sample Prep Kit v2-SetA (Illumina, USA) and TruSeq SBS v5 (Illumina, USA) according to manufacturer's instructions for sequencing runs at 2 × 100 bp total. After sequencing, GA-IIx real time analysis 1.13.48.0 and CASAVA 1.8.2 (Illumina, USA) were utilized for base calling.

The rice genome IRGSP sequence (IRGSP Build5 Pseudomolecules of the Rice Genome, http://rgp.dna.affrc.go.jp/E/IRGSP/Build5/build5.html) was used for reference mapping. The pair-ended reads were used for reference mapping with BWA ver.0.6.1 (Li and Durbin 2010) and SAMtools ver. 0.1.18 (Li et al. 2009), converting the mapped data to a SAM file and then generating a pileup of the sequence data. SNP, deletions, and insertions were detected using CLC Genomics Workbench v6.5.1 (CLCbio, Qiagen, Aarhus, Denmark). After selecting the nucleotide variations with frequency equal to 100.00, we identified 252,343 SNPs and INDELs between BG23 and Nipponbare and 326,351 SNPs and INDELs between LG10 and Nipponbare. SNPs included in the center of single-copied 60-bp sequence were selected to design a marker by running a BLAST search of the Rice Annotation Project Database (http://rapdb.dna.affrc.go.jp/).

Additional files

Additional file 1: Figure S1. Frequency distribution of grain length and width in rice public dataset at NIAS Genebank. Frequency distribution of grain length (A) and grain width (B). Arrowheads indicate the trait data of BG23 and LG10 with measured place and year.

Additional file 2: Figure S2. Logarithm of odds (LOD) curves of five quantitative trait locus (QTL) analyses. LOD curves of QTL analyses for grain length are shown in A, C, E, G, and I. LOD curves of QTL analyses for grain length are shown in B, D, F, H, and J. F_2 populations derived from Kasalath × BG23 (A and B), Kasalath × LG10 (C and D), Nipponbare × BG23 (E and F), Nipponbare × LG10 (G and H), and BG23 × LG10 (I and J) crosses. Dashed lines indicate LOD = 3.0 as threshold value.

Additional file 3: Figure S3. Marker analysis of *GW2*, *GS3*, *qSW5/GW5*, and *GW8* alleles in Nipponbare, Kasalath, BG23, and LG10 lines. The results of marker analysis of *GW2* (A), *GS3* (B), *qSW5/GW5* (C), and *GW8* (D) alleles. Arrows indicate the larger grain (L) and smaller grain (S) alleles.

Additional file 4: Figure S4. Heat map for a two-dimensional genome scan with two-quantitative trait locus (QTL) models in F_2 population derived from the BG23 × LG10 cross. The heat map of the maximum logarithm of odds (LOD) score. Upper left triangle: interaction (Full–Add) model. Lower right triangle: full model. Color-coded scales indicate the values on the left for the interaction model (LOD threshold = 5.35) and on the right for the full model (LOD threshold = 8.13).

Additional file 5: Figure S5. Selection of segregation lines to evaluate the effect of quantitative trait loci (QTLs) detected on Chr3 and Chr6L. F_3 lines with one segregating locus for Chr3 or Chr6L and three other fixed loci were selected. From these two lines, F_4 plants homozygous for BG23 or LG10 were selected to evaluate the effect of QTLs on Chr3 and Chr6L. White and black boxes indicate BG23 and LG10 chromosomes, respectively.

Additional file 6: Table S1. Primer sequences used for the quantitative trait locus (QTL) analysis of four populations.

Additional file 7: Table S2. Primer sequences employed in the quantitative trait locus (QTL) analysis and genetic mapping using the population from the BG23 × LG10 cross.

Additional file 8: Table S3. Primer sequences used in marker analysis.

Acknowledgments
We thank Drs. K. Takeda, I. Takamure, and T. Takita for informative suggestions about large grain rice lines.
This study was supported by grants from the Ministry of Agriculture, Forestry, and Fisheries of Japan (Genomics-based Technology for Agricultural Improvement, RBS-2001, IPG-0002) and JSPS KAKENHI Grant Number 25•10703.

Authors' contribution
SS carried out the QTL analysis and drafted the manuscript. TY and KO conducted the QTL analysis. TN and MA performed the GoldenGate Assay. HK, HS, SM, and YK performed re-sequencing. HK developed all of the F_2 populations. YI and KM conceived and participated in the design and coordination of the study and helped to draft the manuscript. All authors read and approved the final version of the manuscript.

Competing interests
The authors declare that they have no competing interests.

Author details
[1]Fukui Prefectural University, Faculty of Biotechnology, 4-1-1 Kenjojima, Matsuoka, Eiheiji-cho, Yoshida-gun, Fukui 910-1195, Japan. [2]Japan Society for the Promotion of Science, Chiyoda-ku, Tokyo 102-8472, Japan. [3]Bioscience and Biotechnology Center, Nagoya University, Furo-cho, Chikusa, Nagoya, Aichi 464-8601, Japan. [4]Agrogenomics Research Center, National Institute of Agrobiological Sciences, 2-1-2 Kannondai, Tsukuba, Ibaraki 305-8602, Japan.

References

Ashikari M, Wu J, Yano M, Sasaki T, Yoshimura A (1999) Rice gibberellin-insensitive dwarf mutant gene Dwarf 1 encodes the alpha-subunit of GTP-binding protein. Proc Natl Acad Sci U S A 96:10284–10289

Broman KW, Wu H, Sen Ś, Churchill GA (2003) R/qtl: QTL mapping in experimental crosses. Bioinformatics 19:889–890

Fan C, Xing Y, Mao H, Lu T, Han B, Xu C, Li X, Zhang Q (2006) *GS3*, a major QTL for grain length and weight and minor QTL for grain width and thickness in rice, encodes a putative transmembrane protein. Theor Appl Genet 112:1164–1171

Fan C, Yu S, Wang C, Xing Y (2009) A causal C-A mutation in the second exon of *GS3* highly associated with rice grain length and validated as a functional marker. Theor Appl Genet 118:465–472

Fujisawa Y, Kato T, Ohki S, Ishikawa A, Kitano H, Sasaki T, Asahi T, Iwasaki Y (1999) Suppression of the heterotrimeric G protein causes abnormal morphology, including dwarfism, in rice. Proc Natl Acad Sci U S A 96:7575–7580

Hu X, Qian Q, Xu T, Zhang Y, Dong G, Gao T, Xie Q, Xue Y (2013) The U-Box E3 Ubiquitin Ligase TUD1 Functions with a Heterotrimeric G α Subunit to Regulate Brassinosteroid-Mediated Growth in Rice. PLoS Genet 9:e1003391

Ishimaru K, Hirotsu N, Madoka Y, Murakami N, Hara N, Onodera H, Kashiwagi T, Ujiie K, Shimizu B, Onishi A et al (2013) Loss of function of the IAA-glucose hydrolase gene *TGW6* enhances rice grain weight and increases yield. Nat Genet 45:707–711

Izawa Y, Takayanagi Y, Inaba N, Abe Y, Minami M, Fujisawa Y (2010) Function and Expression Pattern of the α Subunit of the Heterotrimeric G Protein in Rice. Plant Cell Physiol 51:271–281

Kato T, Segami S, Toriyama M, Kono I, Ando T, Yano M, Kitano H, Miura K, Iwasaki Y (2011) Detection of QTLs for grain length from large grain rice (*Oryza sativa* L.). Breed Sci 61:269–274

Kitagawa K, Kurinami S, Oki K, Abe Y, Ando T, Kono I, Yano M, Kitano H, Iwasaki Y (2010) A novel kinesin 13 protein regulating rice seed length. Plant Cell Physiol 51:1315–1329

Konieczny A, Ausubel F (1993) A procedure for mapping *Arabidopsis* mutations using co-dominant ecotype-specific PCR-based markers. Plant J 4:403–410

Li H, Durbin R (2010) Fast and accurate long-read alignment with Burrows-Wheeler transform. Bioinformatics 26:589–595

Li H, Handsaker B, Wysoker A, Fennell T, Ruan J, Homer N, Marth G, Abecasis G, Durbin R (2009) The Sequence Alignment/Map format and SAMtools. Bioinformatics 25:2078–2079

Li Y, Fan C, Xing Y, Jiang Y, Luo L, Sun L, Shao D, Xu C, Li X, Xiao J et al (2011) Natural variation in *GS5* plays an important role in regulating grain size and yield in rice. Nat Genet 43:1266–1269

Li S, Liu W, Zhang X, Liu Y, Li N, Li Y (2012) Roles of the Arabidopsis G protein γ subunit AGG3 and its rice homologs GS3 and DEP1 in seed and organ size control. Plant Signal Behav 7:1357–1359

Mao H, Sun S, Yao J, Wang C, Yu S, Xu C, Li X, Zhang Q (2010) Linking differential domain functions of the GS3 protein to natural variation of grain size in rice. Proc Natl Acad Sci U S A 107:19579–19584

McCouch SR, Teytelman L, Xu Y, Lobos KB, Clare K, Walton M, Fu B, Maghirang R, Li Z, Xing Y et al (2002) Development and mapping of 2240 new SSR markers for rice (*Oryza sativa* L.). DNA Res 9:199–207

Qi P, Lin Y, Song X, Shen J, Huang W, Shan J, Zhu M, Jiang L, Gao J, Lin H (2012) The novel quantitative trait locus *GL3.1* controls rice grain size and yield by regulating Cyclin-T1;3. Cell Res 22:1666–1680

Segami S, Kono I, Ando T, Yano M, Kitano H, Miura K, Iwasaki Y (2012) *Small and round seed 5* gene encodes alpha-tubulin regulating seed cell elongation in rice. Rice 5:4

Shomura A, Izawa T, Ebana K, Ebitani T, Kanegae H, Konishi S, Yano M (2008) Deletion in a gene associated with grain size increased yields during rice domestication. Nat Genet 40:1023–1028

Song XJ, Huang W, Shi M, Zhu M, Lin H (2007) A QTL for rice grain width and weight encodes a previously unknown RING-type E3 ubiquitin ligase. Nat Genet 39:623–630

Song XJ, Kuroha T, Ayano M, Furuta T, Nagai K, Komeda N, Segami S, Miura K, Ogawa D, Kamura T et al (2015) Rare allele of a previously unidentified histone H4 acetyltransferase enhances grain weight, yield, and plant biomass in rice. Proc Natl Acad Sci U S A 112:76–81

Takano-Kai N, Jiang H, Kubo T, Sweeney M, Matsumoto T, Kanamori H, Padhukasahasram B, Bustamante C, Yoshimura A, Doi K et al (2009) Evolutionary history of GS3, a gene conferring grain length in rice. Genet 182:1323–1334

Tanabata T, Shibaya T, Hori K, Ebana K, Yano M (2012) SmartGrain: High-throughput phenotyping software for measuring seed shape through image analysis. Plant Physiol 160:1871–1880

Wang S, Wu K, Yuan Q, Liu X, Liu Z, Lin X, Zeng R, Zhu H, Dong G, Qian Q et al (2012) Control of grain size, shape and quality by *OsSPL16* in rice. Nat Genet 44:950–954

Wang S, Li S, Liu Q, Wu K, Zhang J, Wang S, Wang Y, Chen X, Zhang Y, Gao C, Wang F, Huang H, Fu X (2015a) The *OsSPL16-GW7* regulatory module determines grain shape and simultaneously improves rice yield and grain quality. Nat Genet 47:949–954

Wang Y, Xiong G, Hu J, Jiang L, Yu H, Xu J, Fang Y, Zeng L, Xu E, Xu J, Ye W, Meng X, Liu R, Chen H, Wang Y, Zhu X, Li J, Qian Q (2015b) Copy number variation at the *GL7* locus contributes to grain size diversity in rice. Nat Genet 47:944–948

Ware DH, Jaiswal P, Ni J, Yap IV, Pan X, Clark KY, Teytelman L, Schmidt SC, Zhao W, Chang K et al (2002) Gramene, a tool for grass genomics. Plant Physiol 130:1606–1613

Weng J, Gu S, Wan X, Gao H, Guo T, Su N, Lei C, Zhang X, Cheng Z, Guo X et al (2008) Isolation and initial characterization of *GW5*, a major QTL associated with rice grain width and weight. Cell Res 18:1199–1209

Yan B, Liu R, Li Y, Wang Y, Gao G, Zhang Q, Liu X, Jiang G, He Y (2014) QTL analysis on rice grain appearance quality as exemplifying the typical events of transgenic or backcrossing breeding. Breed Sci 64:231–239

Zhang X, Wang J, Huang J, Lan H, Wang C, Yin C, Wu Y, Tang H, Qian Q, Li J et al (2012) Rare allele of *OsPPKL1* associated with grain length causes extra-large grain and a significant yield increase in rice. Proc Natl Acad Sci U S A 109:21534–21539

4

Analysis of Allelic Imbalance in Rice Hybrids Under Water Stress and Association of Asymmetrically Expressed Genes with Drought-Response QTLs

Nelzo C. Ereful[1,4*], Li-Yu Liu[2], Eric Tsai[2], Shu-Min Kao[2], Shalabh Dixit[1], Ramil Mauleon[1], Katrina Malabanan[1,5], Michael Thomson[1,6], Antonio Laurena[3], David Lee[4], Ian Mackay[4], Andy Greenland[4], Wayne Powell[7] and Hei Leung[1]

Abstract

Background: Information on the effect of stress on the allele-specific expression (ASE) profile of rice hybrids is limited. More so, the association of allelically imbalanced genes to important traits is yet to be understood. Here we assessed allelic imbalance (AI) in the heterozygote state of rice under non- and water-stress treatments and determined association of asymmetrically expressed genes with grain yield (GY) under drought stress by *in-silico* co-localization analysis and selective genotyping. The genotypes IR64, Apo and their F1 hybrid (IR64 × Apo) were grown under normal and water-limiting conditions. We sequenced the total RNA transcripts for all genotypes then reconstructed the two chromosomes in the heterozygote.

Results: We are able to estimate the transcript abundance of and the differential expression (DE) between the two parent-specific alleles in the rice hybrids. The magnitude and direction of AI are classified into two categories: (1) symmetrical or biallelic and (2) asymmetrical. The latter can be further classified as either IR64- or Apo-favoring gene. Analysis showed that in the hybrids grown under non-stress conditions, 179 and 183 favor Apo- and IR64-specific alleles, respectively. Hence, the number of IR64- and Apo-favoring genes is relatively equal. Under water-stress conditions, 179 and 255 favor Apo- and IR64-specific alleles, respectively, indicating that the number of allelically imbalanced genes is skewed towards IR64. This is nearly 40–60 % preference for Apo and IR64 alleles, respectively, to the hybrid transcriptome. We also observed genes which exhibit allele preference switching when exposed to water-stress conditions. Results of *in-silico* co-localization procedure and selective genotyping of Apo/IR64 $F_{3:5}$ progenies revealed significant association of several asymmetrically expressed genes with GY under drought stress conditions.

Conclusion: Our data suggest that water stress skews AI on a genome-wide scale towards the IR64 allele, the cross-specific maternal allele. Several asymmetrically expressed genes are strongly associated with GY under drought stress which may shed hints that genes associated with important traits are allelically imbalanced. Our approach of integrating hybrid expression analysis and QTL mapping analysis may be an efficient strategy for shortlisting candidate genes for gene discovery.

Keywords: Rice (*Oryza sativa* L.), Allele-specific expression (ASE), Allelic imbalance (AI), Drought, RNA-seq, Selective genotyping, Co-localization analysis, Quantitative trait loci (QTL)

* Correspondence: nelzo.ereful@niab.com
[1]Genetics and Biotechnology Division, International Rice Research Institute (IRRI), Los Baños, Laguna, Philippines
[4]The John Bingham Laboratory, National Institute of Agricultural Botany (NIAB), Huntingdon Road, Cambridge CB3 0LE, UK
Full list of author information is available at the end of the article

Background

Studies reveal that heritable variations do not reside in protein-coding DNA alone but also in the regulatory sequences. It is now increasingly clear that DNA segments coding for proteins account for only one aspect of heritable variations.

Non-coding variants, gene imprinting, epigenetic factors, *trans*-acting elements, among others were revealed to play important roles in variations and diversity. Even an alteration in one or more nucleotides in the *cis*-acting regulatory elements residing in non-coding regions of the DNA contributes to phenotypic variations. It can alter the binding of a transcription factor (TF) affecting transcription initiation, rate and stability which consequently brings about asymmetric expression of alleles residing in one genome, a phenomenon called allele-specific expression (ASE) imbalance or simply allelic imbalance (AI) (Syvänen et al. 2007; Heap et al. 2010; Bell and Beck 2009).

Preferential expression of one allele over another can be detected by differences in transcript number. ASE is ascribed to *cis*-acting polymorphism, *cis*-acting epigenetic effectors, or a combination of both genetic and epigenetic effectors as *trans*-acting factors have an equal opportunity to affect both alleles within a nucleus (Bell and Beck 2009; Guo et al. 2015).

AI has been the subject of several studies in humans (some are described by Syvänen et al. 2007; Tsuchiya et al. 2009; Voutsinas et al. 2010) and mouse hybrids (Crowley et al. 2015). A review of the research in AI and its regulatory mechanism was published by Gaur et al. (2013). It has been likewise reported to occur in several crop hybrids such as maize (Springer and Stupar, 2007a), barley (von Korff et al., 2009), and rice (Guo et al. 2015). A recent report showed that ASE exists in the natural populations of *Cirsium arvense*, an invasive Compositae weed (Bell et al. 2013). In maize, a review showed the potential contributions of allelic variation to heterosis (Springer and Stupar, 2007b) and the prevalence of *cis*-acting regulatory variations which contribute to biased allelic expression (Springer and Stupar, 2007a).

Only a limited number of studies on AI in rice has been conducted and hence it is still poorly understood. He et al. (2010) found that gene activity was correlated with DNA methylation and both active and repressive histone modifications in the transcribed region. Using SNPs as markers, they observed a high correlation of allelic bias of epigenetic modifications or gene expression in reciprocal hybrids with differences in the parental lines. The study concluded that transcriptional and epigenetic trends in reciprocal rice hybrids contribute to heterosis.

A similar study by Wang et al. (2011) indicated that drought-induced genome-wide DNA methylation changes accounted for ~12.1 % of the total site-specific methylation differences in the rice genome. These epigenetic changes can be considered as an important regulatory mechanism for rice plants to adapt to drought and other environmental stresses.

A recent study on epigenomic analysis using ChIP-seq on *indica-indica* hybrids concluded that histone modifications regulate AI (Guo et al. 2015). This study confirms the contribution of epigenetic mechanism in AI. Earlier studies in rice also suggested that *trans* effects mediate the majority of the transcriptional differences in hybrid offspring and to a lesser extent, intergenerational epimutations (Chodavarapu et al. 2012). These studies in rice showed the important role of *trans*-acting factors on asymmetric expression of parent-specific alleles. However, little is known about ASE in rice under contrasting environmental conditions.

Here we identified genes exhibiting AI under two different water regimes and demonstrated their association with drought-yield QTLs. We reasoned that by exposing the hybrids to two contrasting conditions, we can identify genes that exhibit AI and condition-mediated changes in their ASE profiles. We hypothesized that such genes can be highly associated with drought-tolerance response and can be linked with previously identified drought-response QTLs in the same genetic background.

Specifically, we measured AI in the F1 hybrids and determined changes in ASE profiles of genes between non- and water-stress treatments in the *indica-indica* genetic background using the two varieties IR64 and Apo (IR55423-01). IR64 is a high-yielding variety but was previously shown to be moderately susceptible to drought conditions. Apo, on the other hand, is a drought-tolerant indica variety.

Our study is limited to one-way hybrid cross (IR64 × Apo) exposed to two contrasting water regimes to provide cross-specific information on the effect of stress on allelic imbalance then assess the association of asymmetrically expressed genes with previously identified QTLs known to be involved in drought response. We employed RNA-seq platform to: (1) identify genes asymmetrically expressed between the two genotype-specific alleles in the hybrid in non- and water-stress conditions, (2) assess the level of expression and parental allele preference (magnitude and direction, respectively) of AI on a genome-wide scale, (3) compare the ASE profile of genes under non- and water-stress conditions and then assess changes in expression patterns as effected by the two treatments, and (4) assess the participation of AI genes on drought response.

Results and Discussion

Using RNA-seq, AI was assessed in one-way hybrid cross in rice after exposure to non- and water-stress conditions. Water-stress treatment was initiated at the early stage of flowering after which collection of leaf samples was performed. This is the stage when rice is most sensitive to drought consequently affecting grain yield (Boonjung and Fukai, 1996). Genes identified to exhibit asymmetric expression using AI assay were further tested for their association with drought-yield QTLs by *in-silico* colocalization procedure and selective genotyping.

To generate the hybrids, two genotypes with contrasting response to water-limiting conditions were crossed: (i) IR64, a popular high-yielding drought-susceptible *indica*, and (ii) Apo (IR55423-01), an upland drought-tolerant *indica*. Both materials have been used in previous studies on selection for yield (Venuprasad et al. 2007; Venuprasad et al. 2008) and QTL mapping (Venuprasad et al. 2009) under drought stress. We sequenced the RNA of twelve samples which includes the two parental genotypes and the hybrids exposed to two contrasting treatments (3 genotypes × 2 treatments × 2 replicates) using Illumina platform (described in the Materials and Methods) and generated a pseudo-molecule based on the *O. sativa* MSU7 cDNA (Nipponbare, a *japonica*). A pseudo-reference sequence was created for the subsequent data analysis because of the sequence variations between *japonica* and *indica* (i.e. reference genome versus our parental *indica* genotypes, respectively).

Construction of a Pseudo-Reference Sequence

To create a pseudo-reference sequence, SNPs and InDels were identified between *japonica* and *indica*. After mapping the reads from the parental genotypes Apo and IR64 (4 samples each) to the MSU7 cDNA sequences, we found 106,884 sites (SNPs) which are common to both Apo and IR64 but different from those from the reference, and 8340 sites (InDels) which are common insertions or deletions in Apo and IR64 but different from the reference. These variations imply that there are about 0.92 SNPs and 0.07 InDels per kilobase. The identified SNPs and InDels in this step were used to replace the corresponding nucleotides in the reference MSU7 cDNA sequences to construct the pseudo-reference sequence. The alignment rates of the reads generated from the 12 samples which mapped against the pseudo-reference range from 67.21 to 85.31 % while the alignment rates range from 66.11 to 84.19 % if the reads were mapped against the original MSU7 cDNA sequences. These results show a 0.7–1.6 % improvement rates (Table 1).

We also mapped the reads of the *indica* rice variety 93-11 from Beijing Genomics Institute (BGI) against the original MSU7 cDNA sequences and the pseudo-reference. Their alignments yielded 70.06 % and 71.15 % mapping rates, respectively. These figures indicate an improvement of the mapping rate by 1.09 % which is approximately the same as that of the observed values from our 12 samples. The increase in the alignment rates shows that extra information was incorporated in the pseudo-molecule sequences by modifying the SNP and InDel sites.

Identification of SNPs to Identify the IR64 and Apo Alleles

In this study, the SNPs were used as molecular markers to identify which alleles in the hybrids belong to which parental genotypes. The reads from the eight samples of Apo and IR64 were mapped against the pseudo-reference by bowtie2 (Langmead and Salzberg, 2012). We found 48,175 sites which showed polymorphism between Apo and IR64. This modest level of polymorphism may be possibly due to the similarity of the two genomes since both are *indica*. Low level polymorphism had been previously reported between parents belonging to the same ecotypes (Ali et al. 2000).

Using these 48,175 SNPs, we were able to identify which SNP alleles in the hybrids belong to which parental genotypes. We refer our strategy of finding parent-specific alleles in the hybrids as 'read-wise' which is described in the Materials and Methods. Under normal conditions, 2.56 % and 5.41 % of the F1 reads from reps 1 and 2, respectively, can be identified as either IR64- or Apo-specific alleles. Under water-stress conditions, 1.52 % and 5.23 % of the F1 reads from reps 1 and 2, respectively, can be identified (Table 2). The F1 reads with known parental origins were further analyzed for AI.

A modest percentage level of F1 reads containing information on their parental origins was obtained due to the low densities of parental SNPs (48175/66338 = 0.72 SNPs per gene on average), i.e. between Apo and IR64 transcriptomes. Most of the sequence reads do not show SNPs between the parents and therefore were discarded for further analysis on AI.

Genome-Wide Analysis of ASE in the F1

The magnitude and direction of ASE of genes in the rice hybrids were estimated and classified into two categories depending on the fold change (FC) values between Apo and IR64 (Apo/IR64) the cross-specific paternal and maternal alleles, respectively. These categories include: (1) symmetrically or biallelically expressed, and (2) asymmetrically expressed or imbalanced genes. Symmetrical expression happens when genes have relatively equal expression between the two genotype-specific alleles in the hybrids at fold expression ratio lower than 1.25×. Because both alleles are expressed at a relatively equal magnitude, the expression is biallelic. The minimum fold value of 1.25× has been used previously (Bell et al. 2013; McManus et al. 2010) and will be widely used in this paper. Asymmetrical or imbalanced, on the other hand, happens when the allelic expression fold

Table 1 Comparison of the mapping rates using the reads generated from the samples as query sequence and the original MSU7 cDNA reference and the pseudo-reference as the subject sequences

Treatments		Rep 1			Rep 2		
		Original MSU7 (%)	Pseudo-reference (%)	Improvement (%)	Original MSU7 (%)	Pseudo-reference (%)	Improvement (%)
Control	Apo	77.42	78.43	1.01	84.19	85.31	1.12
	IR64	77.92	79.03	1.11	78.07	79.68	1.61
	F1	76.65	77.77	1.12	83.78	85.23	1.45
Stress	Apo	78.8	79.91	1.11	83.53	84.7	1.17
	IR64	66.11	67.21	1.1	70.28	71.75	1.47
	F1	78.34	79.11	0.77	81.01	82.6	1.59

difference between the two alleles is equal to or more than 1.25×. This can be further classified into two subcategories: IR64- and Apo-favoring genes. The former type happens when IR64-specific allele of the same locus supersedes the expression of the other by at least 1.25 fold; the latter when the reverse happens.

This classification is similar to a previous report (Song et al. 2013). However, the maternal and paternal allelic contributions were effectively estimated in the present study while the former report could not distinguish the parental alleles in the hybrids mainly due to technological limitations at that time. Furthermore, monoallelic expression was not significantly observed in our study. This happens when only one of the two parent-specific alleles is expressed.

In this study, some genes have been identified to have more than one splice variant as a consequence of alternative splicing. Thus, genes will be represented in terms of gene models or transcript variants as described in MSU Rice Genome Annotation located at http://rice.plantbiology.msu.edu/.

AI at 1.25× Fold Expression

Using our pipeline, we found 2337 and 2092 genes with SNPs between the two genotype-specific alleles in the hybrid under non- and water-stress conditions, respectively (Additional file 1: Table S1 and Additional file 2: Table S2). The modest number of genes is due to the limited variations between the two parental genomes in the heterozygote. Of these, only 560 genes exhibited significant DE between the two alleles under normal conditions at $P < 0.05$ (Additional file 3: Table S3). This corresponds to 24.0 % of the total number of genes detected (560/2337). On the other hand, under water-stress conditions, 676 genes showed significant DE between the two genomes in the heterozygous state which corresponds to 32.2 % of the total number of genes (676/2092) (Additional file 4: Table S4). These results indicate that a higher number of genes significantly differentially expressed between the two alleles are induced in response to water-stress conditions.

Under non-stress treatment, 179 and 183 genes favorably expressed Apo- and IR64-specific alleles, respectively (Table 3) using a 1.25× fold expression ratio as threshold. These values show that under normal conditions, the number of genes favoring each parental allele in the heterozygote was about equal. This suggests that AI is inherent to the organism, not necessarily induced by stress. There are 198 genes biallelically expressed under normal conditions. This means that both parental alleles at the same locus are expressed in a relatively equal copy number. Additional file 5: Figure S1A was generated to show genes in the F1 hybrids exhibiting ASE under non-stress conditions.

On the other hand, 179 and 255 genes favored the expression of Apo- and IR64-specific alleles, respectively, under water deficit (Table 3; Additional file 6: Figure S1B). These figures correspond to 41 %-59 % (nearly 40-60)

Table 2 Number of reads which can be assigned as IR64- or Apo-specific allele based on parental SNPs

	Rep 1			Rep 2		
	F1 Mapped Read Counts			F1 Mapped Read Counts		
Control	21,860,178	From Apo	276,827	51,364,894	From Apo	1,378,264
		From IR64	282,901		From IR64	1,403,493
		Unspecified	21,300,450		Unspecified	48,583,137
Stress	36,853,236	From Apo	275,132	54,525,736	From Apo	1,399,485
		From IR64	288,589		From IR64	1,454,190
		Unspecified	36,289,515		Unspecified	51,672,061

Table 3 Number of genes exhibiting preferential expression for Apo- and IR64-specific alleles in the hybrids exposed to two contrasting conditions at a minimum fold differences of 1.25× and 2.0× (at $P < 0.05$)

Description	For at least 1.25-fold		For at least 2.0-fold	
	Non-stress	Water-stress	Non-stress	Water-stress
No. of genes which prefer Apo-specific allele	179	179	77	69
No. of genes which prefer IR64-specific allele	183	255	80	95
Total no. of genes exhibiting AI	362	434	157	164
% genes exhibiting AI (over total number of genes significantly DE at $P < 0.05$)	64.6	64.2	28.0	24.3

preferential expression for Apo and IR64 alleles (at 1.25× fold-expression threshold and $P < 0.05$), respectively, for nuclear genome. These figures do not account for the contribution of organellar genomes (i.e. mitochondria and plastids). There appears to be a bias expression of IR64- (maternal) over Apo-specific allele (paternal) for this specific hybrid cross. These data showed that water-stress conditions skewed preferential expression towards IR64-specific allele, the cross-specific maternal allele. There are 242 genes biallelically expressed under water-stress conditions. Notably, the number of genes (179) favorably expressing Apo-specific allele under non- and water-stress conditions remained unchanged; only the number of genes preferentially expressing IR64-specific allele increased from non- to water-stress conditions. Genes in the F1 hybrids exhibiting ASE under water-stress conditions are shown in Additional file 6: Figure S1B.

In total, 362 and 434 genes showed AI under normal and stress conditions, respectively, demonstrating a higher number of genes asymmetrically expressed under water-stress conditions. However, if we compute the number of genes exhibiting AI over the total number of transcripts significantly differentially expressed, 64.6 % are imbalanced under normal condition (Table 3). This ratio does not significantly deviate from the ratio obtained under stress conditions (64.2 %). This suggests that the stress did not change the ratio of genes exhibiting asymmetric expression over the total number of genes identified to have significant DE between the two alleles. The stress only enhances gene expression and skews the number of AI genes toward the IR64 allele, the cross-specific maternal allele.

A summary of results on AI is shown in Table 3. Heat maps were also generated to show the allele-specific preference of each F1 gene under non- and water-stress conditions (Additional file 5: Figure S1A and Additional file 6: Figure S1B).

DE at 2.0× Fold Expression

If we are to impose a minimum FC of 2.0×, 77 and 80 showed preferential expression for Apo- and IR64-specific alleles, respectively, under non-stress conditions. A 2.0× fold minimum expression difference was used in previous studies (Song et al. 2013; Paschold et al. 2012). These data show that 157 genes (28.0 %) are asymmetrically expressed under normal conditions. There are 403 and 512 biallelically expressed genes under normal and water-stress conditions, respectively. Under water-stress conditions, of the 676 genes, 69 preferentially expressed for the Apo-specific allele, two of which are monoallelically expressed, and 95 preferentially expressed the IR64-specific allele. This further affirms that water-stress skews the number of genes in favor of the IR64-specific allele, a bias expression towards the cross-specific maternal allele.

If we consider 2.0× fold expression ratios, these ratios (28.0 % and 24.3 %, under non- and water-stress, respectively) are lower than the values reported in the literature. However, if the cut-off is set at 1.25× fold, these ratios are more consistent with those reported previously. A study in barley showed that 63 % of genes (19/30) tested showed allelic expression imbalances (von Korff et al. 2009). In maize, 43-53 % of the 316 analyzed genes (depending on the cross and tissues) showed unequal allelic expression (Springer and Stupar, 2007b). A similar study on genome-wide ASE analysis using massively parallel signature sequencing showed that 60 % of the genes in the maize hybrid meristems exhibited differential allelic expression (Guo et al. 2015). Additionally, different approaches to measure ASE may influence these variations (von Korff et al. 2009).

In summary, the hybrid expresses higher number of genes significantly differentially expressed between the two alleles (at $P < 0.05$) during water-stress conditions as compared to during normal conditions indicating that water stress enhances gene expression. Our results further suggest that genes exhibiting AI is inherent to the organism regardless of the conditions and is not a consequence of stress. On a genome-wide scale, the number of genes favoring each of the parental alleles approaches a normal distribution curve (using 2.0× fold minimum level of expression; $P < 0.05$) (Fig. 1). However, at water-limiting conditions, the distribution curve is skewed towards IR64-specific allele – a global distortion of preferential expression. This indicates that more genes favor the maternal over the paternal allele under stress

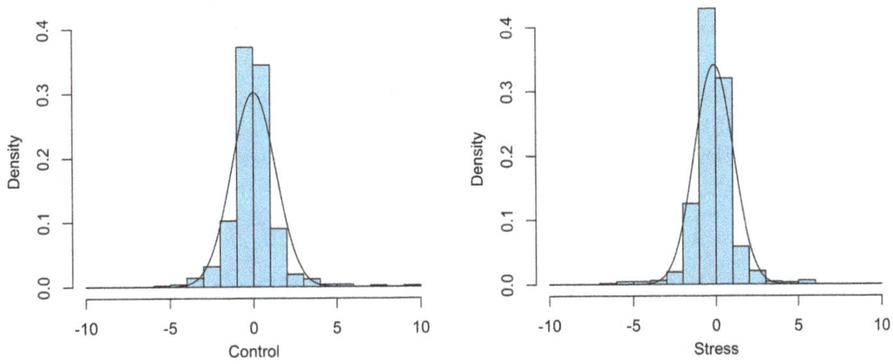

Fig. 1 Genome-wide analysis of ASE in F1 at (left) non- and (right) water-stress conditions (for 2x fold minimum expression). The right side of both graphs starting from the origin 0 (positive) shows FC values when Apo-specific allele is preferentially expressed over IR64-specific allele; the left side (negative values) when IR64-specific allele is preferentially expressed over Apo-specific allele

conditions in this particular hybrid cross. It is surprising to note that while Apo is the drought-tolerant genotype, IR64-specific allele is preferentially expressed in the hybrid. This is contrary to our initial speculation that the tolerant genotype (Apo) should exhibit preferential expression in the heterozygote.

We speculate that this genome-wide change is a mechanism to respond to water-limiting conditions and that *trans*-sensory effectors may have contributed in the global distortion favoring IR64 allele. This was earlier reported by Tirosh et al. (2009) in which *trans* effects were condition-dependent and that expression difference in *trans* reflected a differential response to the environment. F1 hybrids exposed under two contrasting treatments have theoretically the same structural sequences including *cis*-acting elements, thus these regulatory regions may not contribute to the expression differences between the samples. While *trans*-sensory effectors may have contributed on these changes, epigenetic factors cannot be entirely excluded as it was earlier reported (Wang et al. 2011).

Whether the genome-wide preferential expression as effected by water-limiting conditions is maternal parent-of-origin or genotype-dependent is yet to be concluded. Further study using reciprocal hybrid cross is recommended to address this question.

Variations in the *cis*-acting regulatory sequences cause asymmetric expression of the two genotype-specific alleles in the hybrids because both alleles are exposed to the same pool of *trans* factors within a nuclear environment. However, changes in environmental cues may have affected *trans*-sensory effectors subsequently leading to changes in the genome-wide expression profile. The stress which acted as a "pressure" distorted the normal ASE profile toward IR64 allele. This supports previous studies that such regulatory, as well as epigenetic factors, may have mediated the interactions between the environment and the genome (Tirosh et al. 2009; Herceg 2016). Furthermore, it was earlier reported that changes in regulatory regions, *cis* and *trans*, may contribute to evolutionary changes (Wittkopp et al. 2004). However, further studies are needed to elucidate the effect of environmental pressures to evolutionary changes vis-à-vis regulatory factors and the epigenetic landscape.

Relative Transcript Abundance

The relative transcript abundance of the two parent-specific alleles in the F1 can now be estimated and the relative expression ratios can be derived owing to the advent of sequencing platforms and computational methods. Our data show varying expression ratios between the two transcripts of the same locus in the F1. In some genes, two alleles of the same locus are expressed symmetrically (biallelic) in which both parent-specific transcripts contributed nearly equally to the hybrid transcriptome. For several genes (listed in Additional file 3: Table S3 and Additional file 4: Table S4), one of the two alleles shows higher level of expression (dominant) over the other one with lower expression level. For example, in the gene LOC_Os01g27360.1 (glutathione S-transferase), IR64- and Apo-specific alleles contributed 201.08 and 4.75 transcript copies (after normalization and read count adjustment), respectively, to the hybrid transcriptome under non-stress conditions. This is consistent with a recent finding (Gonzàlez-Porta et al. 2013), in which most protein-coding genes in humans have one major transcript expressed at significantly higher level than others. This spectrum of ratios between the parent-specific transcripts indicates the varying possible interactions that exist between the two genomes in the hybrid.

Furthermore, the ability of the hybrids to "skew" towards a preferred allele during harsh conditions suggests that the presence of two genomes in the hybrids allow flexibility in response to stressful environments. The expression of two diverse genomes, which is demonstrated by biallelic genes, may subsequently result in a more diverse peptide population when translated. Previous studies in humans suggested that not all the transcripts that contribute to transcriptome diversity are equally likely to contribute to protein diversity (Gonzàlez-Porta et al. 2013). Jovanovic et al. (2015) cited studies indicating a considerable discrepancy between RNA and protein levels. Little is known about genotype-specific alleles in hybrids of crop species.

Genes with Extreme Cases of AI

An allelic expression level difference of 582.6-fold was the highest observed FC which is exhibited by LOC_Os10g24004.1 (unknown), an Apo-favoring gene (Additional file 3: Table S3). Consistently, this gene was observed to exhibit the highest expression fold difference of 63.1-fold during water-stress condition, also favoring the Apo-specific allele.

A gene encoding for SCP-like extracellular protein (LOC_Os01g28450.1) exhibited the second highest expression fold difference of 173.2× during non-stress condition. The same gene also exhibited a relatively high expression difference (47.3× fold) under water-stress condition favoring the Apo allele. On the other hand, genes encoding for glutathione S-transferase (LOC_Os01g27360.1), regulator of ribonuclease (LOC_Os02g52430.1), metallothionein-like protein (LOC_Os05g11320.1) and thaumatin (LOC_Os 12g43450.1) showed the highest levels of FC (42.3-, 27.5-, 23.5- and 14.9-fold, respectively) which favor the IR64-specific allele under non-stress conditions. The gene thaumatin was previously shown to confer tolerance against fungal infection and abiotic stresses (Rajam et al. 2007).

Under water-stress condition, four splice variants of the gene encoding for jacalin-like lectin proteins (LOC_Os01g24710) were expressed favoring the Apo-specific allele. Two of these variants exhibited monoallelic expression, an extreme case of AI in which the IR64-specific allele is completely silenced. Studies in wheat suggested that these proteins are component of the salicylic acid- and jasmonic acid-dependent defense signaling pathways (Xiang et al. 2011). The other two variants of jacalin-like proteins showed the second and third highest expression difference of 50.2- and 51.8-fold favoring the Apo-specific allele under water-stress condition.

As in the non-stress conditions, the gene encoding for regulator of ribonuclease showed the highest level of FC (67.2-fold) which favor the IR64-specific allele under water-stress condition (Additional file 4: Table S4). Additionally, another thaumatin gene was found to have high expression fold difference of 22.5 × .

Notably, genes with extreme fold differences between the two alleles in the heterozygote either confer tolerance or are important components of defense systems under stress conditions. So far, we have established that the number of genes favoring Apo-specific allele is relatively equal to the number of genes preferring IR64-specific under non-stress conditions. However, in terms of the magnitude of expression, several genes strongly express the Apo-specific allele, the paternal allele for this specific hybrid cross, as shown by the high fold-differences between the two parental alleles. This difference in quantitative expression may account for the phenotypic expression of tolerance to water stress in Apo.

ASE Profiles of Genes Between the two Conditions

We performed pairwise comparison of the ASE profiles of each gene with differentially expressed alleles between the two contrasting conditions to characterize any condition-mediated expression changes. Our method of pairwise comparison is described in the Materials and Method. We found 976 genes, of which 885 are commonly expressed in both conditions if we use a lenient approach (i.e. $P < 0.05$ in one or both conditions); 57 and 34 genes uniquely expressed under non- and water-stress conditions, respectively. Additional file 7: Table S5 shows the result of this pairwise comparison. Heatmaps were created (Additional file 8: Figure S2A, Additional file 9: Figure S2B, Additional file 10: Figure S2C, Additional file 11: Figure S2D and Additional file 12: Figure S2E) to visualize these patterns.

By comparing the ASE profiles of genes expressed at 1.25× fold difference between the two conditions (at $P < 0.05$ in one or both conditions), we are able to identify: (1) transcript Presence/Absence Variation or tPAV, (2) bi-directional and (3) unidirectional expression behaviors.

tPAV

There are 91 genes (9.3 % of the 976 genes) that exhibit tPAV. This happens when a gene is expressed in one condition but not in the other condition regardless of its specific allele preference. Of these, 57 are expressed only during normal conditions but are repressed during the stress conditions (tPAVnormal in Additional file 7: Table S5; Additional file 8: Figure S2A). On the other hand, 34 are induced only during stress conditions but are repressed during normal condition (tPAVstress in Additional file 7: Table S5; Additional file 9: Figure S2B).

Bidirectional

Only 50 genes (5.1 %) exhibited bidirectional expression. This happens when the allele preference of a gene changes from one condition to the other contrasting condition, hence a change in direction. The gene exhibits the ability to switch from one to the other allele as effected by environmental changes, hence, condition-dependent expression deviations. We speculate that *trans*-sensory and epigenetic factors may have influenced these changes.

Of these, 18 showed IR64- to Apo-specific allele change from non- to water-stress conditions, respectively. Some of these genes encode for 6-phosphofructokinase (LOC_Os01g09570), Ser/Thr protein phosphatase family protein (LOC_Os11g05400), and ultraviolet-B-repressible protein (LOC_Os03g22370.1). These changes in expression profile are illustrated in Additional file 10: Figure S2C (green cells in the normal to red cells in the water-stress conditions).

On the other hand, 32 genes showed Apo- to IR64-specific allele change, from normal to stress conditions, respectively (illustrated in Additional file 10: Figure S2C; red cells in the normal to green in stress conditions). These genes include transmembrane amino acid transporter (LOC_Os01g41420), membrane protein (LOC_Os03g51650), and RNA recognition motif (LOC_Os03g17010.1). Another splice variant of the last-named gene (LOC_Os03g17010.2) was found to exhibit tPAV(normal). Additionally, LOC_Os12g18650 has two variants with varying ASE behaviors: versions 6 and 7 exhibit unidirectional (described below) and bidirectional, respectively. These results show that splice variants of a particular gene have varying ASE behaviors with respect to changes in environmental conditions.

The identification of genes exhibiting bidirectional expression behavior may potentially demonstrate that preferential expression (and possibly dominance) for some genes is condition-mediated as a consequence of gene-environment interactions.

Unidirectional

If there are genes exhibiting bidirectional expression behavior, there are also genes that the organism consistently prefers in both conditions. We call this behavior "unidirectional expression." The change in ASE profile is in the magnitude alone (expression level), not on the direction (allele preference) since the gene prefers a particular allele in both conditions. Unlike bidirectional, unidirectional genes are minimally influenced by changes in environmental cues. There are 350 (35.9 %) genes which exhibit this type of behavior. Of these, 179 consistently favored IR64-specific allele in both conditions (Additional file 11: Figure S2D); 171 favored Apo-specific allele (Additional file 12: Figure S2E).

To summarize, the changes in ASE profiles are ascribed to the presence of two different genomes in the F1 hybrid which allows dynamic expression of either or both alleles under any given condition. Two contrasting genomes in the hybrid provide a wider dynamic response to changes in environmental cues and allow greater potential adaptive advantage. The heterozygosity between the genomes lead to structural and functional diversity of peptides (depending on the functional consequences of translation) inside the cell or tissues which may allow the organism to respond to stress in a way that may not be possible otherwise. Tolerance in this manner is attributed to the ability of the organism to respond with environmental cues vis-à-vis condition-mediated changes in ASE behaviors. These mechanisms provide the organism to adapt to harsh conditions. Parent-specific variations in hybrids allow dynamic interactions and changes in the genome that may endow hybrids with greater ability to adapt to environmental changes or stresses. These information may potentially contribute to the genomic and epigenetic insights explaining heterosis.

In-Silico Association of AI Genes with Drought-Response QTL

To initially test the involvement with grain yield (GY), AI genes found to respond in two contrasting water regimes (listed in Additional file 7: Tables S5) were aligned *in-silico* with genetic markers identified to have significant effects on GY under drought stress (Venuprasad et al. 2009). These molecular markers were monitored in the same parental backgrounds (Apo/IR64) as our study using $F_{2:3}$ lines. Using Gramene (*O. sativa indica* genome), this approach of co-localization entails *in silico* anchoring the positions of the SSR markers and the AI genes to estimate their physical-genetic linkage.

This procedure identified a number of AI genes closely linked with the QTL markers known to participate in drought stress in the same parental background. Results of this co-localization analysis are summarized in Table 4. Genes that co-localize with QTL markers are distributed across six chromosomes – 1, 3, 6, 8, 9 and 12. The physical distance (in kb) between the AI genes and the co-localizing markers are estimated based on Gramene (*O. sativa indica* genome), most of which are less than 300 kb.

Markers coinciding with AI genes encode for important stress-response factors involved in transcription regulation and binding. Zinc finger (ZF), a known regulatory factor coincides with the marker RM6703 (139 cM) in chromosome 1. A tetratricopeptide repeat, a binding protein is closely linked with RM71 at chromosome 2 while a stress responsive alpha/beta barrel protein is adjacent to the QTL marker RM3387 at chromosome 3 (0 cM).

The DNA-binding proteins histone H3 and core histone H2A/H2B/H3/H4 are closely linked with the markers RM510 on chromosome 6 (21 cM) and RM256 at chromosome 8, respectively. The former exhibits bidirectional bias expression (Additional file 7: Table S5).

The AI gene encoding for no-apical meristem (NAM) transcription factor (LOC_Os12g29330) is closely linked to the marker RM511, with a physical distance of 9 kb. Interestingly, this is a major gene in the large-effect QTL $qDTY_{12.1}$ region identified to be a drought-yield

Table 4 A summary of the AI genes co-localizing with SSR markers previously identified to be associated with GY under drought stress in the study by Venuprasad et al. (2009) using the same genetic backgrounds (Apo/IR64). Markers are based on Gramene (*O. sativa indica* genome)

Chr #	SSR Name	AI genes co-localizing with the QTL marker	Description/ Function	Estimated distance between the marker and the genes (kb) in the *indica* genome
1	RM6703	LOC_Os01g59980	zinc finger	209
2	RM71	LOC_Os02g15520	transposon	67
		LOC_Os02g15660	tetratricopeptide repeat	44
3	RM3387	LOC_Os03g02020	stress responsive A/B barrel domain containing proteins	20
		LOC_Os03g53710	aldose-1-epimerase	110
	RM520	LOC_Os03g54980	expressed protein	330
		LOC_Os03g55030	UPD-gucosyl transferase	360
6	RM510	LOC_Os06g04030	Histone H3	210
8	RM256	LOC_Os08g38300	core histone H2A/H2B/H3/H4	11
		LOC_Os08g39150	expressed protein	470
		LOC_Os08g39160	formyl transferase	
9	RM201	LOC_Os09g35940	cytochrome P450	520
12	RM511	LOC_Os12g29330	No apical meristem	9

QTL using Vandana/Way Rarem cross (Bernier et al. 2007; Dixit et al. 2015). It is a mainstay of the intra-QTL region and plays roles on spikelet fertility and root proliferation under drought. This was found to exhibit unidirectional expression behavior which consistently prefers IR64 allele in both conditions (Additional file 7: Table S5).

In summary, Table 4 provides a number of AI genes involved in drought confirmed solely by *in silico* co-localization procedure using markers previously known to co-segregate with QTL regions. Genes identified particularly those lying closely to QTL markers are promising candidates for drought-tolerance response. The involvement of these loci in this important abiotic stress is of practical interest for breeding.

Association of AI Genes in Drought Response by Selective Genotyping

To further validate the involvement of several AI genes with GY during drought stress, selective genotyping was performed using Apo/IR64 $F_{3:5}$ recombinant inbred lines (RILs). These lines were phenotyped to confirm the involvement of the identified AI genes with GY under water-limiting conditions. Both tail ends which constitute 25 % of the whole population were genotyped.

QTL regions validated to participate in drought response in the co-localization procedure were further tested by adding markers contiguous to their locations. These were supplemented with additional SSR markers distributed across the genome which coincide with the AI genes. The results of the initial polymorphism survey between the parental genotypes showed a modest level of polymorphism. Of the 153 SSR markers identified to reside closely with these genes, 22 (14 %) exhibited clear polymorphism between the two parents. The parental genotypes exhibited a low level polymorphism since both lines belong to the same subspecies (*indica*).

Results of selective genotyping are shown in Table 5. Six of the 22 polymorphic markers showed significant association with GY (at varying P values of 0.05 and 0.01) under water-stress. AI genes co-localizing with the markers associated with GY were distributed across four chromosomes 1, 2, 3, and 8.

The marker region RM6333 in chromosome 1 showed a highly significant association with the GY under drought stress. It coincides with the AI genes encoding for DNA polymerase I and receptor-like protein kinase. The contiguous marker RM11943 likewise exhibited significant association with GY under drought stress. A cluster of AI genes was found to coincide with this marker.

The markers RM5789 and RM3688 in chromosome 2 were found to coincide with the AI genes encoding for photosystem II 5 kDa and protein-binding protein. On the other hand, RM3387 at chromosome 3, which was found to align with the AI gene encoding for stress responsive A/B barrel protein (LOC_Os03g02020.2) in the co-localization analysis, is further shown here to have significant association with GY under drought stress. These findings reinforce the results obtained from the co-localization procedure.

The marker RM80 on chromosome 8 exhibited significant association with GY under drought stress. This is contiguous to the marker region RM256 (Venuprasad et

Table 5 AI genes which co-localize with markers confirmed to respond during drought stress by selective genotyping. These genes are potential candidates for gene discovery

Chr#	Marker	P	R^2 (%)	AI genes colocalizing with the markers	Estimated distance between markers and genes (kb) in the *indica* genome
1	RM6333	**	21.12	LOC_Os01g65400 (DNA polymerase I)	52
				LOC_Os01g65650 (receptor-like protein kinase HAIKU2 precursor)	113
	RM11943	*	19.11	LOC_Os01g64960 (chlorophyll A-B binding protein)	188
				LOC_Os01g65130 (peptide transporter)	48
				LOC_Os01g65220 (XRN 5'-3' exonuclease N-terminus domain)	0.2
				LOC_Os01g65169 (proton-dependent oligopeptide transport)	25
				LOC_Os01g64790 (AP2)	306
2	RM5789	**	24.48	LOC_Os02g37030 (protein binding protein)	0.5
	RM3688	**	24.33	LOC_Os02g37060 (photosystem II kDa protein)	10
3	RM3387	*	14.33	LOC_Os03g02020 (stress responsive A/B barrel domain-containing protein)	23
				LOC_Os03g01360 (expressed protein)	92
8	RM80	*	14.18	LOC_Os08g38300 (core histone H2A/H2B/H3/H4)	220
				LOC_Os08g39150 (expressed protein)	248
				LOC_Os08g39160 (formyl transferase)	251
				LOC_Os08g39300 (aminotransferase)	356

**, * Significant at 0.01 and 0.05 probability levels, respectively

al. 2009) described in the co-localization procedure. AI genes coinciding with these markers are strong candidates for further drought-yield QTL analysis.

Co-Localization Procedure and Selective Genotyping

Integrating the co-localization procedure and selective genotyping, allelically imbalanced genes are distributed across seven rice chromosomes – 1, 2, 3, 6, 8, 9, and 12. Two regions were confirmed using both *in silico* co-localization procedure and selective genotyping and therefore are the strongest candidates for further study on drought response. These include the regions RM3387 and RM80/RM256 in chromosomes 3 and 8, respectively. The marker region RM3387 coincides with stress responsive A/B barrel (LOC_Os03g02020); while RM80 with core histone (LOC_Os08g38300), expressed protein (LOC_Os08g39150) and formyl transferase (LOC_Os08g39160).

In summary, the ASE pipeline assayed in two contrasting conditions yielded a list of promising genes for further study. Validation of these AI genes by co-localization procedure and selective genotyping shows a strong association of these genes with drought-yield QTLs. This demonstrates that ASE method assayed on two contrasting conditions revealed the involvement of AI genes under water-stress and provided potential candidate genes for functional validation.

Our approach and findings share similarities with eQTL analysis which represents a straightforward approach in identifying candidate genes (reviewed by Druka et al. 2010). However, eQTL approach utilizes more advanced lines (e.g. recombinant inbred lines, F_3) or double haploid whereas AI is assayed on F1 hybrids. We showed that integrating hybrid expression analysis with QTL mapping studies provides convergent evidence for potential candidate genes involved in drought response. This study opens up two avenues for further investigation: (1) test the participation of the promising candidate genes in drought response and (2) further analyze the regulatory sequences of these genes.

This paper may shed insights in assisting breeding design for hybrid crops to improve adaptation and yield to changing climate conditions. Studies using reciprocal hybrid cross is recommended for future investigation.

Conclusions

This study reveals that AI was found inherent to rice hybrids regardless of the condition. On a genome-wide scale, water-stress conditions induce gene expression and skew the number of genes exhibiting asymmetric expression in favor of IR64-specific allele. Furthermore, changes in ASE profiles as effected by the two contrasting conditions were revealed, the most interesting of which is the bidirectional expression behavior. This happens when a gene reverses its allele preference from one condition to the other. This gives information that preferential expression (and possibly dominance) for some genes is condition mediated as a consequence of gene-environment interactions.

The integration of two inbred genomes in the F1 may bring polymorphism in the *cis*-acting regulatory elements that causes AI since both alleles are exposed to the same pool of *trans* factors. The two inbred genomes allow the hybrid to dynamically "select" either one or both alleles, allowing more "options" to express the superior alleles which may provide the hybrids a wider expression and subsequently phenotypic advantages.

We then validated the involvement of these AI genes using *in-silico* co-localization analysis and selective genotyping. Our results showed a strong association of the genes asymmetrically expressed with previously published drought-QTL markers and markers selected in this study for selective genotyping. This suggests that important trait-associated genes are asymmetrically expressed in the F1 and therefore AI assayed in the hybrid may be an efficient approach to determine candidate genes associated with the phenotype.

Methods

Dry-Down Experiment

In this study, rice (*Oryza sativa* L.) F1 hybrids were obtained by crossing IR64 and Apo to study AI under normal and drought conditions. Seeds from the parents and the hybrids were grown in petri plates and were transferred to small pots after germination. Leaf samples from the seedlings were collected to test the hybrid status of the plants. Genomic DNA was extracted using the modified CTAB method as described by Murray and Thompson (1980). Hybrids were confirmed using SSR markers RM 269, RM 511, and RM 80. After hybrid confirmation, seedlings were transplanted in large pots. The parental and hybrid plants were adequately fertilized and grown under controlled conditions in the phytotron, at IRRI. For water-limiting condition, the fraction of transpirable soil water (FTSW) dry-down approach was used as previously described (Cal et al. 2013; Serraj et al. 2014; Sinclair and Ludlow 1986). Water-limiting condition was imposed on stress-treated plants by initiating a soil dry-down protocol starting 10 days before heading until the plants reached 0.5 FTSW.

All pots were weighed daily to calculate the amount of water lost. All genotypes were replicated four times (IR64, Apo and their F1), for each of the two treatments (control and 0.5 FTSW), giving a total of 24 samples, randomly arranged in the phytotron.

RNA Extraction

Leaf samples from each plant were collected at the end of the dry-down treatment. RNA was extracted using the TRIzol method according to the instructions provided by the supplier (Invitrogen, San Diego, Calif., USA). RNA samples were treated with DNAse to remove DNA contamination. RNA concentrations were estimated using a NanoDrop 1000 Spectrophotometer (NanoDrop Technologies, Wilmington, Del., USA).

RNA-seq libraries were made as described in Illumina's standard protocol for RNA-seq using the parental (IR64 and Apo) and F1 RNA samples from each treatment (normal and water-stress). Libraries were sequenced on Illumina GAIIx, generating 38- and 90-base paired end (PE) reads, for the first and second sequencing protocols, respectively. Two biological replicates were sent for sequencing to satisfy minimum requirements for RNA-seq as stipulated in the Standards, Guidelines and Best Practices of the ENCODE Consortium (http://genome.ucsc.edu/). Raw reads were received in fastq format.

Data Processing and Analysis

Quality Checking and Trimming

Quality filtering of PE reads was performed using FASTQ Quality Filter. Filtered reads were trimmed off by FASTQ Quality Trimmer from FASTX-Toolkit (http://hannonlab.cshl.edu/). We checked the 75 % percentile of the sequencing qualities for each base of the PE reads in each replicate. Bases with 75 % percentile of the sequencing qualities < 28 were removed for all reads in the replicate. All commands are described in the Additional file 13: S1.

Mapping and Generating Consensus Pseudo-Reference

After read trimming, a pseudo-reference was generated by mapping Apo and IR64 reads separately against the indexed reference sequence *O. sativa* ssp. *japonica* (cv. Nipponbare) MSU v7 cDNA (http://rice.plantbiology.msu.edu/) using bowtie2. Because of the variations between *indica* and *japonica* (Nipponbare reference genome), alignment procedure was set at lenient parameters to allow more significant mapping using the 'score function'.

Post-Processing

After mapping, we used samtools sort and mpileup commands (Li and Durbin 2009) to parse the mapping results. PERL scripts were written to call SNPs and InDels common to Apo and IR64 sequences but different from the Nipponbare reference (see Additional file 14: Figure S3 for schematic diagram). SNPs and InDels were called between the *indica* genotypes and the Nipponbare reference genome, a *japonica*. SNP reads must have at least 5× coverage and SNP proportion must be more than 0.8.

For InDels, read coverage must also be at least 5× but InDel proportion should be at least 0.5.

The consensus pseudo-reference was generated by (1) replacing the SNPs common to Apo and IR64 but not in the MSU v7 cDNA, and (2) incorporating the common InDels found in both Apo and IR64 (Additional file 14: Figure S3).

The pseudo-reference was mapped against the 93-11 *indica* RNAseq reads downloaded from BGI (http://rise2.genomics.org.cn/page/rice/index.jsp). We mapped the 93-11 reads (*indica*) against the Nipponbare. This is to assess any increase in alignment among the reads. *Indica-indica* alignment (i.e., 93-11 and our pseudo-reference) should be higher than *indica-japonica* (i.e., 93-11: Nipponbare reference genome).

SNP Calling

Bowtie2 was used to map Apo, IR64 and F1 to the pseudo-reference. Samtools sort and mpileup commands were then used to parse the mapping results. A PERL script was used to find the SNPs between Apo and IR64 sequences. SNP reads must have at least 3× read coverage and SNP proportion must be more than 0.8 (Additional file 15 Figure S4).

Identifying the F1 Reads by Read-Wise Approach

To identify F1 reads, two approaches are possible: the SNP- and Read-wise approaches (Additional file 16 Figure S5). SNP-wise approach entails the counting of the number of SNPs from each parent-specific allele. A schematic diagram is shown in Additional file 16 Figure S5 to illustrate this strategy. The number of SNPs corresponding to Apo were 10; IR64, 6. Apparently, SNP-wise approach tends to overestimate the read counts.

Read-wise approach, on the other hand, uses SNPs between the parents to guide us in identifying which genotype-specific reads belongs to which parental genotype in the hybrid. The number of reads corresponding to Apo were 8; IR64, 4. We preferred the Read- over SNP-wise approach as it gives more accurate number of read counts. This approach uses SNPs between the two parent-specific alleles as copy-specific tags. These variants allow us to distinguish and quantify the two SNP alleles in the heterozygote. A PERL script was used to carry out this strategy.

Parental Reads Extraction and Gene Expression Estimation

Using the mapping results of Apo and IR64 for the pseudo-reference sequence (bam files) and SNP information, we extracted the parental reads containing SNP by a PERL script. We used eXpress (Roberts and Pachter 2013) to perform gene expression analysis. We then counted the reads to estimate gene expression level from mapping results of F1 to the pseudo-reference. We identified genotype-specific reads guided by the parent-specific SNPs in the F1.

Differences in Gene Expression Analysis Using DESeq

Allele-specific expression levels were estimated in the hybrids between the two treatments in both replicates using DESeq. Before testing, we normalized samples by dividing the corresponding size factor to minimize the difference between samples. We tested the significance of differential expression in the hybrids.

AI test

Binomial test was employed to test significance of AI between the two alleles in the hybrid. We used DESeq package in R (Anders and Huber 2010) to calculate the size factors for normalization, then we normalized the expression values by dividing the corresponding size factors. The normalized expression of the two replicates were combined and tested for significance of imbalance using binomial and chi square tests.

To satisfy sample size requirement of normal approximation by chi-square test, genes with expected values less than 5 were removed. Gramene (www.gramene.org) was used to find the annotation and biological function of genes.

Calculating FC and Determining Condition-Mediated ASE Profiles of Genes

After normalization and read count adjustments, expression ratios and $\log_2$FC were calculated. Expression ratios were obtained by dividing the transcript expression of Apo over IR64 (Apo/IR64) then were log-transformed. To infer condition-mediated ASE changes of genes, we pairwise-compared their $\log_2$FCs under non- and water-stress conditions (Additional file 1: Table S1 and Additional file 2: Table S2). We only considered genes with P values < 0.05 in one or both conditions. Using this lenient strategy, 976 genes were shortlisted for this analysis (Additional file 7: Table S5). If we impose a more stringent strategy (i.e. $P < 0.05$ in both conditions except for tPAV genes which are expressed in only one condition), only 351 genes are shortlisted (data not shown). We preferred the lenient approach to obtain more information on expression profiles and include a wider number of genes. Asymmetrically expressed genes (biallelic) have expression values lower than the cutoff of 1.25× fold. Their condition-mediated ASE behaviors cannot be ascertained. Both alleles are expressed relatively equal and therefore cannot be classified as Apo- or IR64-specific allele preferring.

Association Analysis of AI Genes with Drought-Response QTL

In-Silico Association

To initially confirm putative involvement with grain yield (GY) under drought condition, asymmetrically expressed genes were aligned with the markers previously identified (Venuprasad et al. 2009). These molecular marker loci showed highly significant allele frequency differences between stress-selected (SS) and non-stress selected (NS) sets of $F_{2:3}$ lines derived from Apo/IR64 cross. This approach entails anchoring the positions of the SSR markers and AI genes to estimate their physico-genetic linkage on Gramene using *O. sativa indica* (www.gramene.org).

Selective Genotyping

After the alignment procedure, the involvement of the AI genes with drought tolerance was tested by selective genotyping. Forty (20 highest and 20 lowest yielding lines from each of the tails of the whole population) of 160 Apo × IR64 $F_{3:5}$ RILs were genotyped using 153 SSR markers residing closely to or co-localizing with the genes.

Phenotypic Data Collection

The RILs were screened under lowland drought stress in dry season in 2009 at the experiment station of the International Rice Research Institute (IRRI), Los Baños, Laguna, Philippines.

Trial management, population generation and phenotypic data collection of progenies comprising of 160 lines were performed using the protocol outlined under lowland conditions. All the trials were laid out as alpha lattice designs, with plot length of 5.25 m, and spacing between rows was 0.20 m under drought stress. Twenty-one-day-old seedlings were transplanted then stress was applied after 28 days. The number of rows per plot was two; two replications were used in all trials.

From the upland (stress) trial of DS 2009, phenotypic data on grain yield were collected. Least square (LS) means was computed between two replicates of each line for all phenotypic data.

Genotyping

All molecular marker work was conducted at the Molecular Genetics Laboratory, Genetics and Biotechnology (GB) Division, IRRI. Leaf samples of the 160 lines were collected from Field 202, IRRI experiment station. Samples were placed in -20 °C freezer until extraction. Miniprep scale DNA extraction was performed using a modified CTAB protocol by Sambrook and Russell (2001). The concentration of DNA was analyzed using NanoDrop® ND-1000 Spectrophotometer and adjusted to ~50 ηg/μL.

A parental polymorphism survey was carried out using 153 SSR markers between Apo and IR64 parents. Only 22 (14 %) of the markers exhibited clear polymorphism and were used to genotype the lines identified to belong to the two tail ends.

PCR amplification of all markers was performed in 20 μL reactions containing 2 μL 10 × buffer, 1.5 μL 25 mM $MgCl_2$, 1.6 μL 2 mM dNTPs, 1 μL 5 μM forward and reverse primers and 0.5 μL of Taq DNA polymerase (4U/ μL) and 10-50 ηg μL-1 gDNA using 96-well thermal cycler. After initial denaturation for 5 min at 94 °C, each cycle comprised 1 min denaturation at 94 °C, 2 min annealing at 50-65 °C (depending on the primer pair), and 2 min extension at 72 °C with a final extension for 5 min at the end of 35 cycles. PCR products were mixed with bromophenol blue loading dye and were analyzed by PAGE (8 % gel solution) for 1.5 h, 100 V, 100 mA. Gels were stained with SYBR Safe and viewed using GelDoc System.

The two tails of the RILs were scored according to the parental banding pattern as IR64 homozygotes (1), Apo homozygotes (2) or heterozygotes (3). When no banding pattern or non-parental bands were observed, they were treated as missing values (0).

Association analysis between the genotypic and phenotypic data was performed using Single Marker Regression Analysis. All rice microsatellite (RM) markers described in this paper were taken from Gramene (http://www.gramene.org/) as described by Temnykh et al. (2001).

Additional Files

Additional file 1: Table S1. List of genes in the hybrids with SNPs between the two parent-specific alleles under normal conditions.

Additional file 2: Table S2. List of genes in the hybrids with SNPs between the two parent-specific alleles under water-stress conditions.

Additional file 3: Table S3. List of genes in the hybrids exhibiting significant DE between the two parent-specific alleles at $P < 0.05$ under normal conditions.

Additional file 4: Table S4. List of genes in the hybrids exhibiting significant DE between the two parent-specific alleles at $P < 0.05$ under water-stress conditions.

Additional file 5: Figure S1A. Heatmaps showing genes in the F1 hybrid exhibiting ASE under normal conditions. Green cells represent values when IR64 allele is preferred; red, when Apo is preferred in the F1 hybrid. The black cells represent a symmetrical (or equal) expression between the two genotype-specific alleles. Color-scale bar is also shown.

Additional file 6: Figure S1B. Heatmaps showing genes in the F1 hybrid exhibiting ASE under water-stress conditions. Description is similar to Additional file 5: Figure S1A.

Additional file 7: Table S5. Pairwise comparison of the expression preference of the hybrid genes exposed under normal and water-stress conditions. It shows the condition-mediated changes in expression preference of each gene.

Additional file 8: Figure S2A. Heatmaps of genes between non- (Column 1) and water-stress (Column 2) conditions exhibiting tPAV. Genes are induced during non-stress conditions (tPAVnormal). The first column represents the ASE preference of a gene under normal conditions (Control) and the second under stress conditions (Stress). Descriptions of color and color scale bar are similar to Additional file 5: Figure S1A.

Additional file 9: Figure S2B. Heatmaps of genes between non- (Column 1) and water-stress (Column 2) conditions exhibiting tPAV. Genes are induced during water-stress conditions (tPAVstress). Descriptions of color and color scale bar are similar to Additional file 5: Figure S1A and Additional file 8: Figure S2A.

Additional file 10: Figure S2C. Heatmaps of genes between non- (Column 1) and water-stress conditions (Column 2) exhibiting bidirectional expression behavior. Green (Column 1; Control) to red cells (column 2; Stress) indicate genes with preferential expression changes from IR64- to Apo-specific allele. Red (Column 1; Control) to green cells (column 2; Stress) indicate genes with preferential expression changes from Apo- to IR64-specific allele. Description of colors and color scale bar is similar to Additional file 5: Figure S1A and Additional file 8: Figure S2A.

Additional file 11: Figure S2D. Heatmaps of genes between non- (Column 1) and water-stress conditions (Column 2) exhibiting unidirectional expression behavior (IR64-favoring genes). Green (Column 1; Control) to green cells (column 2; Stress) indicate genes which consistently prefer the IR64-specific allele regardless of the conditions. Description of colors and color scale bar is similar to Additional file 5: Figure S1A and Additional file 8: Figure S2A.

Additional file 12: Figure S2E. Heatmaps of genes between non- (Column 1) and water-stress conditions (Column 2) exhibiting unidirectional expression behaviour (Apo-favor genes). Red (Column 1; Control) to red cells (column 2; Stress) indicate genes which consistently prefer the Apo-specific allele regardless of the conditions. Description of colors and color scale bar is similar to Additional file 5: Figure S1A and Additional file 8: Figure S2A.

Additional file 13: S1. Commands used for AI analysis. (DOCX 15 kb)

Additional file 14: Figure S3. Schematic diagram illustrating our method of generating a pseudo-reference sequence. Bases common to Apo and IR64 but different from the Nipponbare reference genome were called. These variations were used to generate the pseudo-reference.

Additional file 15: Figure S4. A schematic representation to call SNPs. By using the SNPs between the parents, we can use the information to determine the expression of alleles in the F1 hybrids. In the figure, Apo and IR64 have two SNP loci. In the first locus the read coverage is more than 3 for Apo and most of the genotypes are T (more than 80 %) so this locus is defined as 'T' SNP. Compared to Apo, IR64 has a read coverage of more than 3 and most of the genotypes are C (more than 80 %) so this locus is defined as 'C' SNP for IR64.

Additional file 16: Figure S5. Schematic diagram illustrating Read- and SNP-wise approaches to identify genotype-specific alleles in F1. Using SNPs between the parents as copy-sequence tags, genotype-specific alleles in the hybrid can be identified.

Abbreviations

AI: Allelic imbalance; ASE: Allele-specific expression; DE: Differential expression; FTSW: Fraction of transpirable soil water; GY: Grain yield; NAM: No apical meristem; QTL: Quantitative trait loci; SNP: Single nucleotide polymorphism; TF: Transcription factors

Acknowledgement

The work is supported in part by a grant from Biotechnology and Biological Sciences Research Council (BBSRC: BB/F004265/1), National Institute of Agricultural Botany (NIAB), and the International Rice Research Institute (IRRI). We also acknowledge Dr. Konstantia Stamati for the interim supervision of the project; Alice Bordeos and Marietta Baraoidan for managing the generation of hybrids; Dr. David Parry of the University of Cambridge for reviewing and editing the manuscript.

Authors' Contributions

NCE undertook analysis of the NGS results, genotyping, *in silico* association analysis, writing of the manuscript; LL performed statistical analysis, project management and coordination between IRRI and NTU; ET and SK undertook bioinformatics analysis and PERL scripting; SD performed the phenotyping and field data collection and management; RM for drafting the bioinformatics analysis pipeline and data handling; KM provided laboratory assistance and data analysis; AL managed laboratory activity and did manuscript review; MT undertook manuscript review, consultation, project coordination; DL undertook critical review of the manuscript, laboratory supervision, consultation and project coordination between IRRI and NIAB; IM, reviewed manuscript and provided statistical analysis; AG coordinated the work and acquired financial support; WP wrote the original proposal, sought project funding and managed the initial part of the study; and HL coordinated the work among IRRI, NIAB and NTU and undertook project management. All authors read and approved the final manuscript.

Competing Interests

The authors declare that they have no competing interests.

Author details

[1]Genetics and Biotechnology Division, International Rice Research Institute (IRRI), Los Baños, Laguna, Philippines. [2]Department of Agronomy, National Taiwan University (NTU), Taipei City 100, Taiwan. [3]Institute of Plant Breeding, University of the Philippines, Los Baños, Laguna, Philippines. [4]The John Bingham Laboratory, National Institute of Agricultural Botany (NIAB), Huntingdon Road, Cambridge CB3 0LE, UK. [5]Crop Science Cluster, College of Agriculture, University of the Philippines, Los Baños, Laguna 4031, Philippines. [6]Texas A &M, Department of Soil and Crop Sciences 2474 TAMU, College Station, TX 77843-2474, USA. [7]SRUC, Peter Wilson Building, West Mains Road, Edinburgh EH9 3JG, UK.

References

Ali ML, Pathan MS, Zhang J, Bai G, Sarkarung S, Nguyen HT (2000) Mapping QTLs for root traits in a recombinant inbred population from two *indica* ecotypes in rice. Theor Appl Genet 101:756–766

Anders S, Huber W (2010) Differential expression analysis for sequence count data. Genome Biol doi:10.1186/gb-2010-11-10-r106

Bell CG, Beck S (2009) Advances in the identification and analysis of allele-specific expression. Genome Med 1(5):56.1–56.5

Bell CG, Graeme DM, Nolan CK, Rieseberg LH, Adams KL (2013) RNA-Seq analysis of allele-specific expression, hybrid effects, and regulatory divergence in hybrids compared with their parents from natural populations. Genome Biol Evol 5(7):1309–1323. doi:10.1093/gbe/evt072

Bernier J, Kumar A, Venuprasad R, Spaner D, Atlin GN (2007) A large-effect QTL for grain yield under reproductive stage drought stress in upland rice. Crop Sci 47:505–516

Boonjung S, Fukai H (1996) Effects of soil water deficit at different growth stages on rice growth and yield under upland conditions. 2. Phenology, biomass production and yield. Field Crop Res 48:47–55

Cal AJ, Liu D, Mauleon R, Hsing YC, Serraj R (2013) Transcriptome profiling of leaf elongation zone under drought in contrasting rice cultivars. PLoS 8(1). doi:10.1371/journal.pone.0054537

Chodavarapu RK, Feng S, Ding B, Simon SA, Lopez D, Jia Y, Wang GL, Meyers BC, Jacobsen SE, Pellegrini M (2012) Transcriptome and methylome interactions in rice hybrids. Proc Natl Acad Sci 109(30):12040–12045. doi:10.1073/pnas.1209297109

Crowley JJ et al (2015) Analyses of allele-specific gene expression in highly divergent mouse crosses identifies pervasive allelic imbalance. Nat Gen 47(4):353–360

Dixit S et al. (2015) Action of multiple intra-QTL genes concerted around a co-localized transcription factor underpins a large effect QTL. Nature. doi:10.1038/srep15183

Druka A, Potokina E, Luo Z, Jiang N, Chen X, Kearsey M, Waugh R (2010) Expression quantitative trait loci analysis in plants. Plant Biotechnol J 8(1):10–27. doi:10.1111/j.1467-7652.2009.00460.x

Gaur U, Li K, Shuqi M, Liu G (2013) Research progress in allele-specific expression and its regulatory mechanisms. J Appl Gen 54:271–283

Gonzàlez-Porta M, Frankish A, Rung J, Harrow J, Brazma A (2013) A transcriptome analysis of human tissues and cell lines reveals one dominant transcript per gene. Genome Biol 14:R7. doi:10.1186/gb-2013-14-7-r70

Guo Z, Song G, Liu Z, Qu X, Chen R, Jiang D, Sun Y, Liu C, Zhu Y, Yang D (2015) Global epigenomic analysis indicates that epialleles contribute to allele-specific expression via allele-specific histone modifications in hybrid rice. BMC Genomics 16:232. doi:10.1186/s12864-015-1454-z

Heap GA, Yang JH, Downes K, Healy BC, Hunt KA, Bockett N, Franke L, Dubois PC, Mein CA, Dobson RJ, Albert TJ, Rodesch MJ, Clayton DG, Todd JA, van Heel DA, Plagnol V (2010) Genome-wide analysis of allelic expression imbalance in human primary cells by high-throughput transcriptome resequencing. Hum Mol Genet 19(1):122–134

He G, Zhu X, Elling AA, Chen L, Wang X, Guo L, Liang M, He H, Zhang H, Chen F, Qi Y, Chen R, Deng XW (2010) Global epigenetic and transcriptional trends among two rice subspecies and their reciprocal hybrids. Plant Cell 22:17–33

Herceg Z (2016) Epigenetic Mechanisms as an Interface Between the Environment and Genome. In: RoachRC, Hackett PH, Wagner PD (ed) Hypoxia. Advances in Experimental Medicine and Biology, vol 903. Springer, New York, pp 3–15

Jovanovic M et al. (2015) Dynamic profiling of the protein life cycle in response to pathogens. Science 347 (6226). doi: 10.1126/science.1259038

Langmead B, Salzberg SL (2012) Fast gapped-read alignment with Bowtie 2. Nat Methods 9:357–359. doi:10.1038/nmeth.1923

Li H, Durbin R (2009) Fast and accurate short read alignment with Burrows-Wheeler transform. Bioinformatics 25:1754–1760

McManus CJ, Coolon JD, Duff MO, Mains JE, Graveley BR, Wittkopp PJ (2010) Regulatory divergence in *Drosophila* revealed by mRNA-seq. Genome Res 20: 816–825. doi:10.1101/gr.102491.109

Murray MG, Thompson WF (1980) Rapid isolation of high molecular weight plant DNA. Nucleic Acids Res 8(19):4321–4325

Paschold A, Jia Y, Marcon C, Lund S, Larson NB, Yeh CT, Ossowski S, Lanz C, Nettleton D, Schnable PS, Hochholdinger F (2012) Complementation contributes to transcriptome complexity in maize (*Zea mays* L.) hybrids relative to their inbred parents. Genome Res 22(12):2445–2454. doi:10.1101/gr.138461.112

Rajam MV, Chandola N, Saiprasad Goud P, Singh D, Kashyap V, Choudhary ML, Sihachakr D (2007) Thaumatin gene confers resistance to fungal pathogens as well as tolerance to abiotic stresses in transgenic tobacco plants. Biol Plant 51(1):135–141

Roberts A, Pachter L (2013) Streaming fragment assignment for real-time analysis of sequencing experiments. Nat Methods 10(1):71–73. doi:10.1038/nmeth.2251

Sambrook J, Russell DW (ed) (2001) Molecular cloning - a laboratory manual, 3rd edn, Cold Spring Harbor Laboratory Press, New York

Serraj R, Dongcheng L, Hong H, Sellamuthu R, Impa S, Cairns J, Dimayuga G, Torres R (2014) Novel approaches for integration of physiology, genomics and breeding for drought resistance improvement in rice., http://www.intlcss.org/

Sinclair T, Ludlow M (1986) Influence of soil water supply on the plant water balance of four tropical grain legumes. Aust J Plant Physiol 13:329. doi:10.1071/pp9860329

Song G, Guo Z, Liu Z, Cheng Q, Qu X, Chen R, Jiang D, Liu C, Wang W, Sun Y, Zhang L, Zhu Y, Yang D (2013) Global RNA sequencing reveals that genotype-dependent allele-specific expression contributes to differential expression in rice F1 hybrids. BMC Plant Bio 13:221. doi:10.1186/1471-2229-13-221

Springer NM, Stupar RM (2007a) Allele-Specific Expression Patterns Reveal Biases and Embryo-Specific Parent-of-Origin Effects in Hybrid Maize. Plant Cell 19(8): 2391-2402

Springer NM, Stupar RM (2007b) Allelic variation and heterosis in maize: How do two halves make more than a whole? Genome Research 17(3):264-275

Syvänen AC, Milani L, Gupta M, Dhar S, Fryknäs M, Isaksson A, Larsson R (2007) Allelic imbalance in gene expression as a guide to *cis*-acting regulatory single nucleotide polymorphisms in cancer cells. Nucleic Acids Res 35(5):1–10

Temnykh S, Declerck G, Lukashova A, Lipovich L, Cartinhour S, Mccouch S (2001) Computational and experimental analysis of microsatellites in rice (Oryza sativa L.): frequency, length variation, transposon associations, and genetic marker potential. Genome Res 11(8):1441–1452

Tirosh I, Reikhav S, Levy AA, Barkai N (2009) A yeast hybrid provides insight into the evolution of gene expression regulation. Science 324:659–662

Tsuchiya N, Narita S, Kumazawa T, Inoue T, Ma Z, Tsuruta H, Saito M, Horikawa Y, Yuasa T, Satoh S, Ogawa O, Habuchi T (2009) Clinical significance of a single nucleotide polymorphism and allelic imbalance of matrix metalloproteinase-1 promoter region in prostate cancer. Oncol Rep 22:493–499

Venuprasad R, Bool ME, Dalid C, Bernier J, Kumar A, Atlin GN (2009) Genetic loci responding to two cycles of divergent selection for grain yield under drought stress in a rice breeding population. Euphytica 167:261–269

Venuprasad R, Lafitte R, Atlin GN (2007) Response to direct selection for grain yield under drought stress in rice. Crop Sci 47(1):285–293. doi:10.2135/cropsci2006.03.0181

Venuprasad R, Sta Cruz MT, Amante M, Magbanua R, Kumar A, Atlin GN (2008) Response to two cycles of divergent selection for grain yield under drought stress in four rice breeding populations. Field Crops Res 107(3):232–244. doi:10.1016/j.fcr.2008.02.004

von Korff M, Radovic S, Choumane W, Stamati K, Udupa SM, Grando S, Ceccarelli S, Mackay I, Powell W, Baum M, Morgante M (2009) Asymmetric allele-specific expression in relation to developmental variation and drought stress in barley hybrids. Plant J 59(1):14–26

Voutsinas GE, Stavrou EF, Karousos F, Dasoula A, Papachatzopoulou A, Syrrou M, Verkerk AJ, Van Der Spek P, Patrinos GP, Stöger R, Athanassiadou A (2010) Allelic imbalance of expression and epigenetic regulation within the alpha-synuclein wild-type and P.Ala53Thr alleles in Parkinson disease. Hum Mutat 31(6):685–691

Wang WS, Pan YJ, Zhao XQ, Dwivedi D, Zhu LH, Ali J, Fu BY, Li ZK (2011) Drought-induced site-specific DNA methylation and its association with drought tolerance in rice (*Oryza sativa* L.). J Exp Bot 62(6):1951–1960

Wittkopp PJ, Haerum BK, Clark AG (2004) Evolutionary changes in *cis* and *trans* gene regulation. Nature 430:85–88. doi:10.1038/nature02698

Xiang Y, Song M, Wei Z, Tong J, Zhang L, Xiao L, Ma Z, Wang Y (2011) A jacalin-related lectin-like gene in wheat is a component of the plant defence system. J Exp Bot 62(15):5471–83

5

Genome-wide Association Mapping of Cold Tolerance Genes at the Seedling Stage in Rice

Dan Wang[1†], Jinling Liu[1†], Chengang Li[2†], Houxiang Kang[3], Yue Wang[1], Xinqiu Tan[2], Minghao Liu[1], Yufei Deng[1], Zhilong Wang[1], Yong Liu[2], Deyong Zhang[2], Yinghui Xiao[1*] and Guo-Liang Wang[1,4*]

Abstract

Background: Rice is a temperature-sensitive crop and its production is severely affected by low temperature in temperate and sub-tropical regions. To understand the genetic basis of cold tolerance in rice, we evaluated the cold tolerance at the seedling stage (CTS) of 295 rice cultivars in the rice diversity panel 1 (RDP1), these cultivars were collected from 82 countries.

Results: The evaluations revealed that both temperate and tropical japonica rice cultivars are more tolerant to cold stress than indica and AUS cultivars. Using the cold tolerance phenotypes and 44 K SNP chip dataset of RDP1, we performed genome-wide association mapping of quantitative trait loci (QTLs) for CTS. The analysis identified 67 QTLs for CTS that are located on 11 chromosomes. Fifty-six of these QTLs are located in regions without known cold tolerance-related QTLs.

Conclusion: Our study has provided new information on the genetic architecture of rice cold tolerance and has also identified highly cold tolerant cultivars and CTS-associated SNP markers that will be useful rice improvement.

Keywords: Oryza sativa, Cold tolerance, Quantitative trait locus (QTL), GWAS

Background

Temperature adaptability is a critical factor for rice domestication and production. The differentiation of two major domesticated rice subspecies (indica and japonica) is associated with temperature. Japonica is tolerant to low temperature and is mainly planted in the temperate, sub-temperate and high altitude areas of the subtropics; indica, in contrast, is sensitive to low temperature and is mainly grown in tropical and sub-tropical areas (Zhao et al. 2013). Cold stress is often a major limiting factor for stable rice production in temperate areas and sub-tropic areas with high altitude. Cold stress restricts the growth and development of rice at all the growth stages and results in low germination and seedling vigor, delayed seedling growth with leaf wilting and browning, prolonged duration of cultivation, pollen sterility, poor grain filling, and reduced yields (Suh et al. 2010; Pan et al. 2015).

Rice cold tolerance is genetically controlled by multiple quantitative trait loci (QTLs). Traditional QTL mapping using bi-parental or multiple cross populations identified more than 250 QTLs on all 12 chromosomes for rice cold tolerance at different growth and development stages (Yang et al. 2015; Xiao et al. 2015; Mao et al. 2015; Zhu et al. 2015). Among these QTLs, several genes have been fine mapped, including *Ctb1* (Saito et al. 2004), *qCT8* (Kuroki et al. 2007), *qCTB7* (Zhou et al. 2010), *qCTB3* (Shirasawa et al. 2012), and *qCT-3-2* (Zhu et al. 2015) for cold tolerance at the booting stage, *qCTS12* (Andaya and Tai, 2006), *qCTS4* (Andaya and Tai, 2007), *qCtss11* (Koseki et al. 2010), *qSCT1* and *qSCT11* (Kim et al. 2014), *qLOP2* and *qPSR2-1* (Xiao et al. 2015) for CTS, *qLTG3-1* for germination cold tolerance

* Correspondence: xiaoyh@hunau.edu.cn; wang.620@osu.edu
†Equal contributors
1Southern Regional Collaborative Innovation Center for Grain and Oil Crops in China and College of Agronomy, Hunan Agricultural University, Changsha, Hunan 410128, China
Full list of author information is available at the end of the article

(Fujino et al. 2008), and *qRC10-2* for root cold tolerance (Xiao et al. 2014). Two QTLs for rice cold tolerance, *Ctb1* and *COLD1*, have been cloned and functionally characterized (Saito et al. 2010; Ma et al. 2015). *Ctb1* is the first cloned QTL for rice cold tolerance and confers enhanced cold tolerance at the booting stage. *Ctb1* encodes a F-box protein and physically associates with Skp1, a subunit of the E3 ubiquitin ligase, suggesting the potential involvement of the ubiquitin–proteasome pathway in rice cold resistance (Saito et al. 2010). The newly identified *COLD1* gene confers cold tolerance in japonica rice at the seedling stage. Molecular characterization revealed that *COLD1* functions as a GTPase-accelerating factor and regulates G-protein signaling by sensing cold in order to trigger Ca^{2+} signaling for cold tolerance (Ma et al. 2015).

Genome-wide association analysis (GWAS) was applied for QTL mapping using large germplasm collections (Huang et al. 2010; Zhao et al. 2011). Many QTLs for multiple traits were identified, such as traits associated with agronomic characteristics (Huang et al. 2010; Zhao et al. 2011; Yang et al. 2014), and with responses to abiotic stresses (Famoso et al. 2011; Pan et al. 2015; Lv et al. 2016), and to biotic stresses (Jia et al. 2012; Wang et al. 2014; Kang et al. 2016; Wang et al. 2015). Using GWAS, Pan et al. (2015) recently mapped 51 QTLs for cold tolerance at the germination and booting stages with 174 Chinese rice accessions that were genotyped with 273 SSR markers. Fujino et al. (2015) also identified 17 QTLs responsible for rice low temperature germinability in 63 Japanese varieties genotyped with 115 SSR markers and two other markers. In addition, Lv et al. (2016) used 527 rice cultivars to identify 132 QTLs for both rice natural chilling and cold shock stresses.

In this study, we used GWAS to map QTLs associated with rice cold tolerance at the seedling stage (CTS). The GWAS involved 295 rice cultivars in the publically available rice diversity panel 1 (RDP1), these cultivars were collected from 82 countries and genotyped with a 44 K SNP chip (Zhao et al. 2011). The cold tolerance evaluations showed that both temperate and tropical japonica rice cultivars are more tolerant of cold stress than indica and AUS rice cultivars. A total of 67 QTLs associated with CTS were mapped on 11 chromosomes in the rice genome. These QTLs explained from 3.8 to 8.2% of the CTS. The mapped QTLs with corresponding linked SNP markers will be useful for the improvement of rice cold tolerance.

Results

Phenotypic Variation among RDP1 Seedlings in Response to Cold Treatment

To assess the phenotypic variation in the cold tolerance of RDP1 cultivars, we evaluated 295 cultivars at the 3-leaf seedling stage. The cold tolerance scores of these cultivars are listed in Additional file 1: Table S1. About 60% of the cultivars were tolerant (scores 1–4) and about 40% were sensitive (scores 5–9) (Fig. 1a; Additional file 2: Table S2). The RDP1 collection consists of 6 subpopulations including 64 tropical japonica (TRJ), 58 temperate japonica (TEJ), 45 ADMIX, 11 AROMATIC, 52 AUS and 65 indica (IND) cultivars (Fig. 1b). The TEJ and TRJ groups had the highest CTS with average scores of 2.5 and 2.5, respectively. The AUS and IND groups had the lowest CTS with average scores of 6.4 and 6.1, respectively. The CTS was intermediate for the ADMIX and AROMATIC groups (Fig. 1b). High cold tolerance (score ≤ 3) was exhibited by 57 cultivars in the TRJ group and by 51 in the TEJ group, whereas low cold tolerance (score > 4.0) was exhibited by 51 cultivars in the IND group and by 45 in the AUS group (Fig. 1c). These results suggested that both temperate and tropical japonica rice cultivars are tolerant of cold stress and that the indica and AUS cultivars are sensitive to cold stress.

To further investigate the relationship between the cold tolerance level of world rice accessions and their geographical distribution, we performed a correlation coefficient analysis between the latitude of the cultivars and their cold tolerance scores. The analysis showed that a negative correlation existed between cultivars' latitude and the level of their cold tolerance with a correlation coefficient at −0.188. In addition, the linear regression analysis confirmed the negative correlation (Fig. 1d).

Mapping of QTLs for CTS by GWAS

We performed GWAS using the CTS data and the 44 K SNP dataset published by Zhao et al. (2011) following the criteria of one associated locus between any two significant SNPs within 200 kb interval by Lv et al. (2016). The analysis revealed that 67 QTLs within 181 SNPs on 11 chromosomes were significantly associated with CTS with a well-fitted Quantile-Quantile (Q-Q) plots (Fig. 2a and b; Table 1 and Additional file 3: Table S3). The phenotypic contribution of each QTLs varied from 3.82 to 8.20% (Table 1). A comparison of previous results of cold tolerance gene mapping or functional characterization indicated that 11 associated loci were co-localized with the known, mapped QTLs or characterized genes (Additional file 3: Table S3). Among the co-localized loci, six known functional genes were *OsDREB1F* on chromosome 1 (Wang et al. 2008), *qLTG3-1* on chromosome 3 (Fujino et al. 2008), *OsRAN2* (Chen et al. 2011), *OsSPX1* on chromosome 6 (Zhao et al. 2009), *OsFAD8* on chromosome 7 (Nair et al. 2009), and *OsCYL4a* on chromosome 9 (Qin et al. 2015) (Fig. 2a; Additional file 3: Table S3). The remaining 56 QTLs were considered to be novel.

Next, we evaluated the allele frequency of the 181 significant SNPs in the 40 highly tolerant rice cultivars

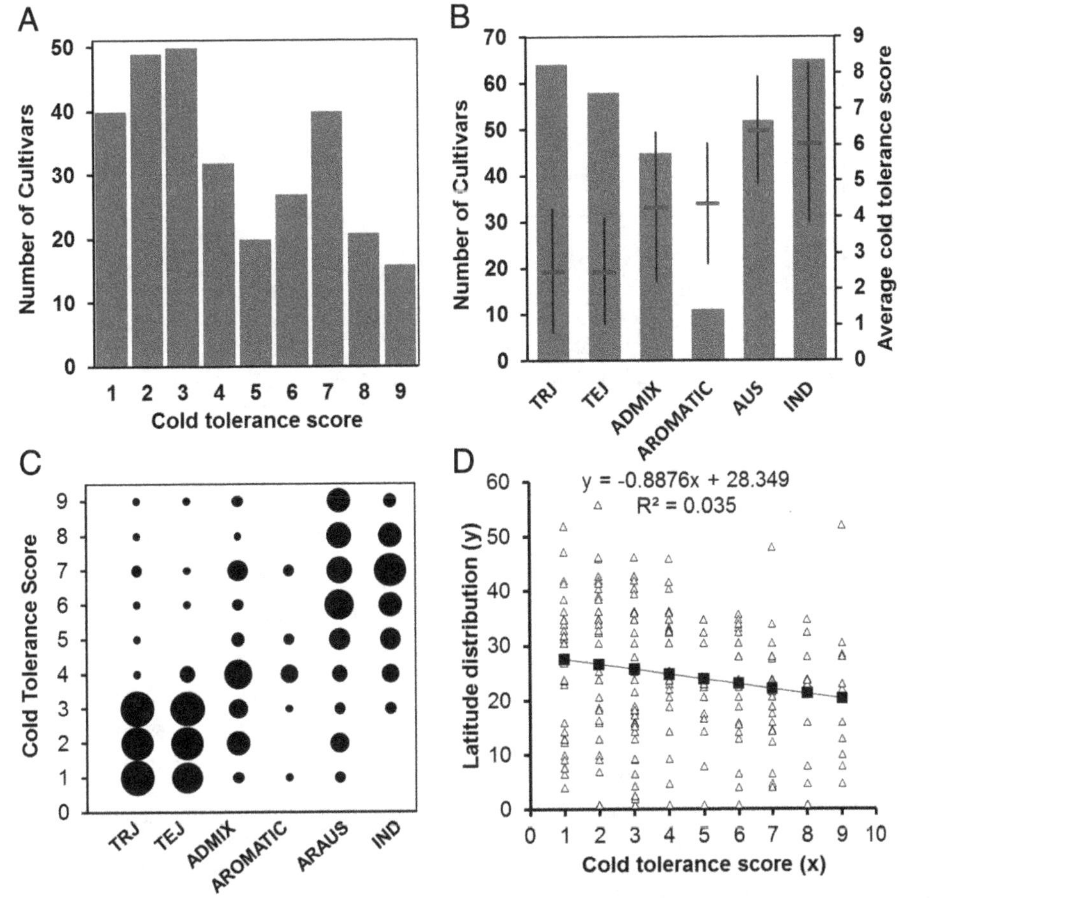

Fig. 1 Phenotypic structure of the rice diversity panel1 (RDP1) in response to cold treatment. **a** The phenotype distribution of 295 rice accessions. The X-axis is cold tolerance scores (1–9); Y-axis represents the numbers of cultivars of each score. **b** The rice accession numbers (*left Y-axis*) of each subspecies (X-axis) used in the study and the average cold tolerance scores (*right Y-axis*) with error bars (STDE) for each subspecies. **c** The cold tolerance score structure (Y-axis) of rice accessions in each subspecies (X-axis). The circle size represents the number of rice cultivars for the corresponding cold tolerance scores (Y-axis) in different subspecies (X-axis). **d** Correlation analysis between rice cold tolerance scores and its latitude distribution by linear regression analysis. X-axis is the cold tolerance score of tested rice cultivars. Y-axis is latitude of the cultivars. Light triangle is the latitude value of a cultivar in Y-axis with the corresponding cold tolerance score in X-axis. Black square linked with the line is the predicted latitude value. The predicted formula and R^2 are shown on the top

(score = 1) and the 19 extremely sensitive cultivars (score = 9) (Additional file 4: Table S4). The analysis showed that the highly tolerant rice accessions contained an average of 64.3% of the tolerant alleles and 33.1% of the sensitive types (Fig. 2c). In contrast, the highly sensitive cultivars only contained 26.9% of the tolerant alleles and 66.3% of the sensitive types (Fig. 2c). These results demonstrate that the identified 181 significant SNPs have a large contribution to the cold tolerance of the tested RPD1 cultivars.

Identification of Candidate Genes Responsible for CTS

To identify candidate genes that are localized in the CTS QTL regions, we analyzed a 500 kb genomic region by comparing the QTL regions with the Nipponbare reference genome. Interestingly, we found that the candidate gene LOC_Os03g09140 on chromosome 3, designate *qCTS3-9* (Fig. 2a and Additional file 3: Table S3), is *Osryh1*, which is linked to the NP:rs19079958 ($p = 9.61 \times 10^{-5}$) (Fig. 3a). *Osryh1* encodes a GTP-binding protein and is a homolog of the yeast *ryh1* gene, which is responsible for temperature sensing in yeast (Bednarek et al. 1994). However, the function of *Osryh1* in cold tolerance is not clear.

To confirm the association between *Osryh1* and the CTS phenotype, we sequenced *Osryh1* in 10 cold-tolerant and 10 cold-sensitive cultivars. The sequence analysis showed that the tolerant cultivars contained three deletions in the intron 3 with a total length of 27 bp comparing with the corresponding sequence of sensitive cultivars (Fig. 3b, c). Furthermore, we designed an indel marker using the 27-bp deletion sequence to genotype 153 RDP1 accessions (78 tolerant and 75 sensitive cultivars). The results showed that 89.74% (70 out of 78) tolerant

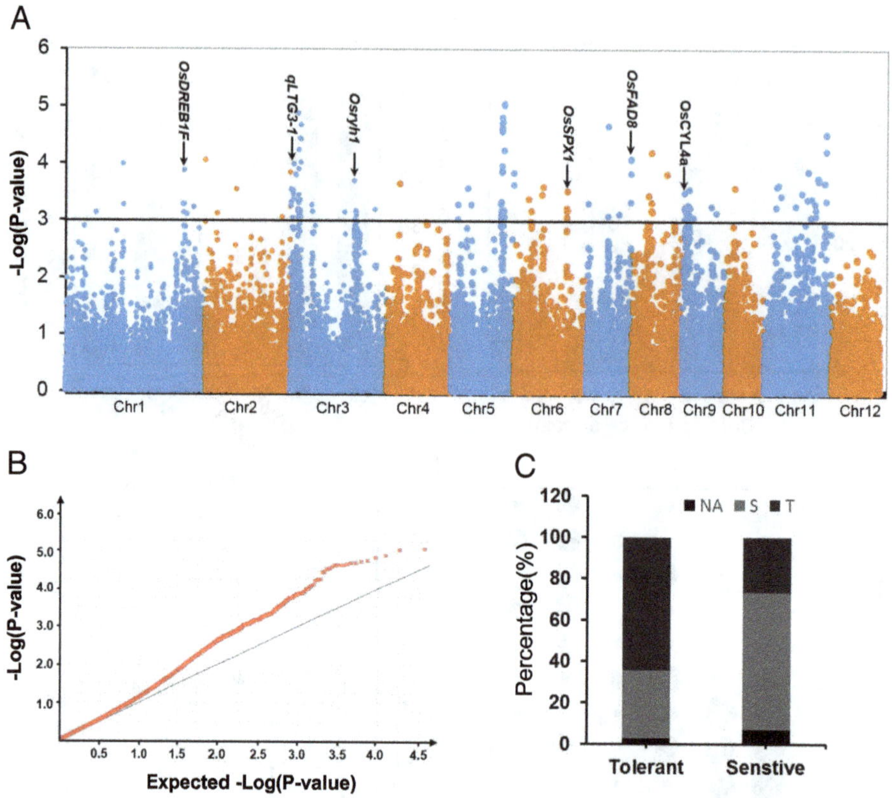

Fig. 2 GWAS mapping of rice cold tolerance at seedling stage (CTS). **a** The Manhattan plots of CTS on rice chromosomes analyzed with the data of the RPD1 rice accessions. X-axis is the genomic coordinates. Y-axis is the LOD score for each SNP. **b** The corresponding Q-Q plots for GWAS mapping. **c** The average allele frequency of the 181 significant SNPs in the 40 highly tolerant and the 19 extremely sensitive rice cultivars. X-axis is the tolerant or sensitive rice group and Y-axis means the average SNP frequency of the corresponding allele genotype. T is tolerant associated SNP genotype, S represents sensitive SNP genotype, NA means unknown genotype

cultivars (score less than or equal to 3) had tolerant alleles, and 85.3% (64 out of 75) sensitive cultivars (score more than or equal to 4) had sensitive alleles (Fig. 4a, b). In addition, 83.3% (25 out of 30) tolerant temperate japonica cultivars and 81.6% (31 out of 38) tolerant tropical had the deletions, and 78.4% (29 out of 37) sensitive IND cultivars and 78.9% (15 out of 19) sensitive AUS cultivars did not contain the deletions (Fig. 4c).

Discussion

In this study, we evaluated the genetic structure of cold tolerance in 295 RDP1 rice cultivars at the seedling stage. Consistent with a previous study (Pan et al. 2015), we found that both temperate and tropical japonica cultivars are more tolerant to cold stress than the IND and AUS cultivars (Fig. 1 and Additional file 1: Table S1). Interestingly, we found that the tropical rice cultivars, also known as javanica, show a high level of tolerance to cold stress. Although a study indicated that tropical japonica cultivars are tolerant to cold stress (Suh et al. 2010), a comprehensive evaluation of tropical japonica rice cultivars have not been reported. Our results revealed that 89% of the 64 TRJ rice cultivars from 40 countries are highly tolerant to cold stress (Additional file 2: Table S2), suggesting that tropical rice is also an important germplasm for rice cold tolerance improvement.

By evaluating the natural chilling and cold shock stresses with a panel of rice accessions from China, Lv et al. (2016) recently found that the latitudinal distribution of rice is correlated with natural cold adaptability. In this study, we also observed a negative correlation between latitudinal distribution and rice cold tolerance level among the RDP1 accessions (Fig. 1d). This result suggests that high latitude could be one of the selected environmental factors for domestication of cold tolerant rice cultivars. Additonaly, 40 RDP1 rice cultivars showed extremely high tolerance to cold stress (Additional file 2: Table S2), and these will be useful for breeding cold tolerance in rice.

Unlike traditional QTL mapping, GWAS enables the identification of rare alleles in natural populations and of high-resolution SNP markers (Gupta et al. 2014). Although Pan et al. (2015) and Fujino et al. (2015) recently reported the association mapping of cold tolerance QTLs

Table 1 The mapped QTLs for rice cold tolerance at seedling stage

QTLs	Chromosome	linked SNP loci[a]	linked tolerant SNP marker	LOD	Phenotype contribution	Candidate genes
qCTS1-1	1	7392085	CC	3.13	5.15	NA
qCTS1-2	1	19180885	TT	3.99	5.14	NA
qCTS1-3	1	20261058	AA	3.27	4.16	NA
qCTS1-4	1	41385263	TT	3.28	4.43	NA
qCTS1-5	1	41859177	AA	3.88	4.92	NA
qCTS1-6	1	42438945	CC	3.23	4.11	OsDREB1F, LOC_Os01g73770
qCTS2-1	2	74241	GG	4.05	5.45	NA
qCTS2-2	2	4797319	AA	3.12	3.85	NA
qCTS2-3	2	15096404	AA	3.54	4.44	NA
qCTS2-4	2	35929494	TT	3.84	5.07	NA
qCTS3-1	3	604512	CC	3.54	4.47	qLTG3-1, LOC_Os03g01320 Chr03:219977-221070
qCTS3-2	3	995211	TT	3.52	4.62	NA
qCTS3-3	3	1243161	AA	3.99	5.12	NA
qCTS3-4	3	2110407	TT	3.81	4.83	NA
qCTS3-5	3	2680925	CC	4.88	8.20	NA
qCTS3-6	3	3032938	TT	3.44	4.43	NA
qCTS3-7	3	3211019	CC	3.91	5.60	NA
qCTS3-8	3	3668304	CC	4.68	6.14	NA
qCTS3-9	3	3763778	GG	4.43	5.74	*Osryh1*, LOC_ Os03g019140
qCTS3-10	3	8192574	AA	3.26	4.04	NA
qCTS3-11	3	22540833	TT	3.14	3.90	NA
qCTS3-12	3	27608937	AA	3.69	5.96	NA
qCTS3-13	3	31924530	AA	3.19	3.97	NA
qCTS4-1	4	5850082	TT	3.65	4.59	NA
qCTS5-1	5	1970193	GG	3.03	4.44	NA
qCTS5-2	5	6171822	CC	3.58	4.51	NA
qCTS5-3	5	8040996	AA	3.29	4.51	NA
qCTS5-4	5	24616513	GG	3.33	4.18	NA
qCTS5-5	5	25098060	TT	4.56	6.25	NA
qCTS5-6	5	25154581	AA	4.82	6.39	NA
qCTS5-7	5	25720662	AA	3.1	4.00	NA
qCTS5-8	5	26092015	TT	5.05	6.77	NA
qCTS5-9	5	29432186	TT	3.84	4.97	NA
qCTS6-1	6	4733861	TT	3.39	4.25	NA
qCTS6-2	6	6228690	AA	3.18	4.02	NA
qCTS6-3	6	10452388	CC	3.46	5.81	NA
qCTS6-4	6	10964887	GG	3.6	4.55	NA
qCTS6-5	6	23876111	TT	3.52	4.42	OsSPX1, LOC_Os06g40120
qCTS7-1	7	914685	CC	3.3	4.13	NA
qCTS7-2	7	17875297	GG	3.09	3.82	NA
qCTS7-3	7	18202939	CC	4.67	7.06	NA
qCTS7-4	7	23783174	CC	3.12	3.92	NA
qCTS7-5	7	29220680	AA	4.08	5.34	OsFAD8 LOC_Os07g49310

Table 1 The mapped QTLs for rice cold tolerance at seedling stage *(Continued)*

qCTS8-1	8	8883364	AA	3.44	4.34	NA
qCTS8-2	8	10142772	CC	4.21	5.48	NA
qCTS8-3	8	10506862	GG	3.16	3.91	NA
qCTS8-4	8	20385371	AA	3.82	4.97	NA
qCTS9-1	9	1141448	CC	3.49	4.40	OsCYL4a, LOC_Os09g02270
qCTS9-2	9	1706118	CC	3.25	4.33	NA
qCTS9-3	9	3375476	CC	3.29	4.97	NA
qCTS9-4	9	3869544	GG	3.11	4.57	NA
qCTS9-5	9	4289803	GG	3.56	4.48	NA
qCTS9-6	9	4767843	TT	3.04	4.20	NA
qCTS9-7	9	6403821	AA	3.06	4.65	NA
qCTS9-8	9	16181745	CC	3.26	4.04	NA
qCTS9-10	9	17697057	GG	3.15	3.89	NA
qCTS10-1	10	5690202	CC	3.59	4.63	NA
qCTS11-1	11	5902436	AA	3.25	4.15	NA
qCTS11-2	11	6478793	GG	3.6	5.02	NA
qCTS11-3	11	7249274	GG	3.64	4.61	NA
qCTS11-4	11	9163345	GG	3.45	4.33	NA
qCTS11-5	11	17519565	AA	3.27	4.68	NA
qCTS11-6	11	19897968	AA	3.49	4.95	NA
qCTS11-7	11	21779484	AA	3.87	6.75	NA
qCTS11-8	11	22115297	GG	3.21	5.11	NA
qCTS11-9	11	24958836	CC	3.67	4.68	NA
qCTS11-10	11	25432012	CC	4.52	6.48	NA

Note: [a]only the linked SNP loci with highest LOD scores are shown. NA means not available

in rice, the low resolution of SSR marker genotyping data affects the interval distance of the mapped QTLs. In this study, we mapped 67 QTLs associated with CTS in 295 rice accessions. By comparing the current and previously published results (Yang et al. 2015; Lv et al. 2016), we found that 56 of the 67 QTLs are newly identified and that only 11 are co-localized with genes of known function or with mapped QTLs (Fig. 2 and Additional file 3: Table S3). Within these mapped QTLs, we sequenced and genotyped the predicted candidate gene *Osryh1* on QTL *qCTS3-8*, and found a high correlation between the indel marker genotype and CTS phenotype, demonstrating that our mapping results are suitable for further characterization of CTS candidate genes.

Furthermore, we compared the QTL loci for CTS in both japonica and indica cultivars. The results showed that 34 loci in the indica group and 91 loci in the japonica group were mapped (data not shown), suggesting that more QTLs for CTS are accumulated in the japonica cultivars than that in the indica ones. This could be one of reasons why japonica cultivars are more tolerant to cold stress than other type of rice (Additional file 1: Table S1).

In the collection of RPD1 rice accessions, we found that 40 rice cultivars are highly tolerant to cold stress at the seedling stage (Additional file 1: Table S1 and Additional file 4: S4). In 40 selected highly tolerant cultivars, 87.5% (35/40) of cultivars are japonica rice including 20 TRJ and 15 TEJ types (Additional file 4: Table S4). Further allele frequency analysis with the 181 associated SNPs revealed that over 60% loci in the 40 cultivars are tolerant SNP types, and this is a 2-fold more than that in the 19 extremely sensitive cultivars (Fig. 2c and Additional file 4: Table S4), suggesting that cold tolerant alleles are highly enriched in the 40 cultivars.

Cold stress is one of the major limitations for rice production, especially in the double-season rice cropping regions. For example, in the early season in South China, farmers would like to grow rice earlier to gain more yield. However, cold weather in early spring usually forces the delay of first rice planting. Consequently this causes the delay of the second crop planting and thus affect flowering when cold weather starts in middle or late September. In our study, the identified cultivars with high cold tolerance will be useful germplasm for rice cold tolerance breeding, and some of the associated SNPs can

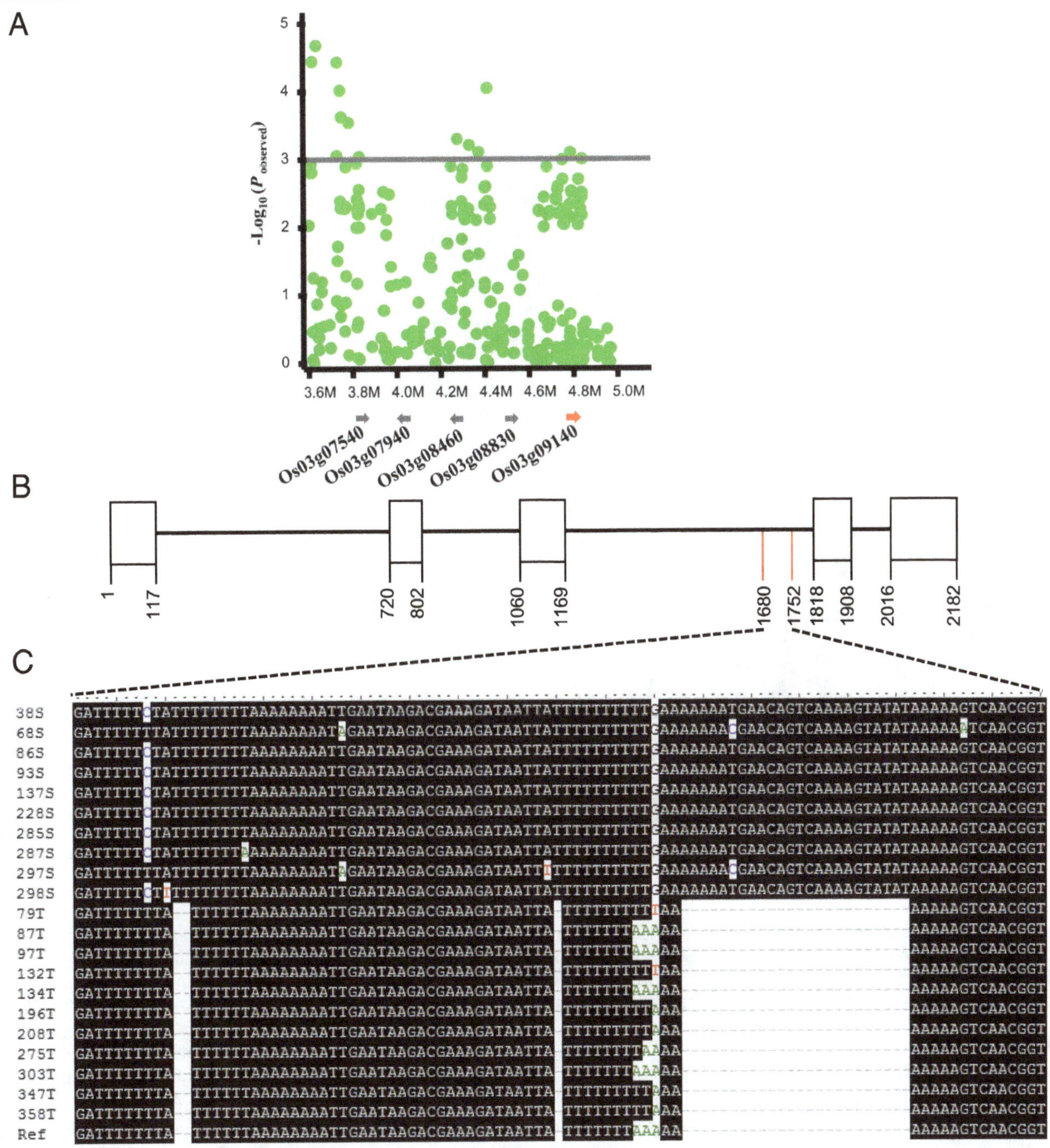

Fig. 3 Validation of the rice cold tolerance QTL *qCTS3-9*. **a** Candidate gene prediction of *qCTS3-9*. **b** The gene encoding structure of candidate gene LOC_Os03g019140. **c** Sequence alignment of LOC_Os03g019140 gene in the indel region between cold tolerant and sensitive accessions. T means tolerant, S is sensitive, Ref is the reference sequence of Nipponbare genome with an tolerant phenotype

be designed as molecular markers for cold tolerance breeding or used for fine mapping and cloning of cold tolerance QTLs.

Conclusions

We evaluated the cold tolerance of 295 rice accessions in RDP1 collected from 82 countries at the seedling stage, and performed GWAS to map QTLs associated with CTS. The evaluations indicated that both temperate and tropical japonica rice cultivars are more tolerant to cold stress than IND and AUS rice. GWAS revealed that 67 QTLs are associated with CTS. Among them, 56 are novel loci. Our study has provided new information on the genetic structure of cold tolerance in rice, identified highly cold tolerant cultivars and CTS-associated SNP markers for rice improvement.

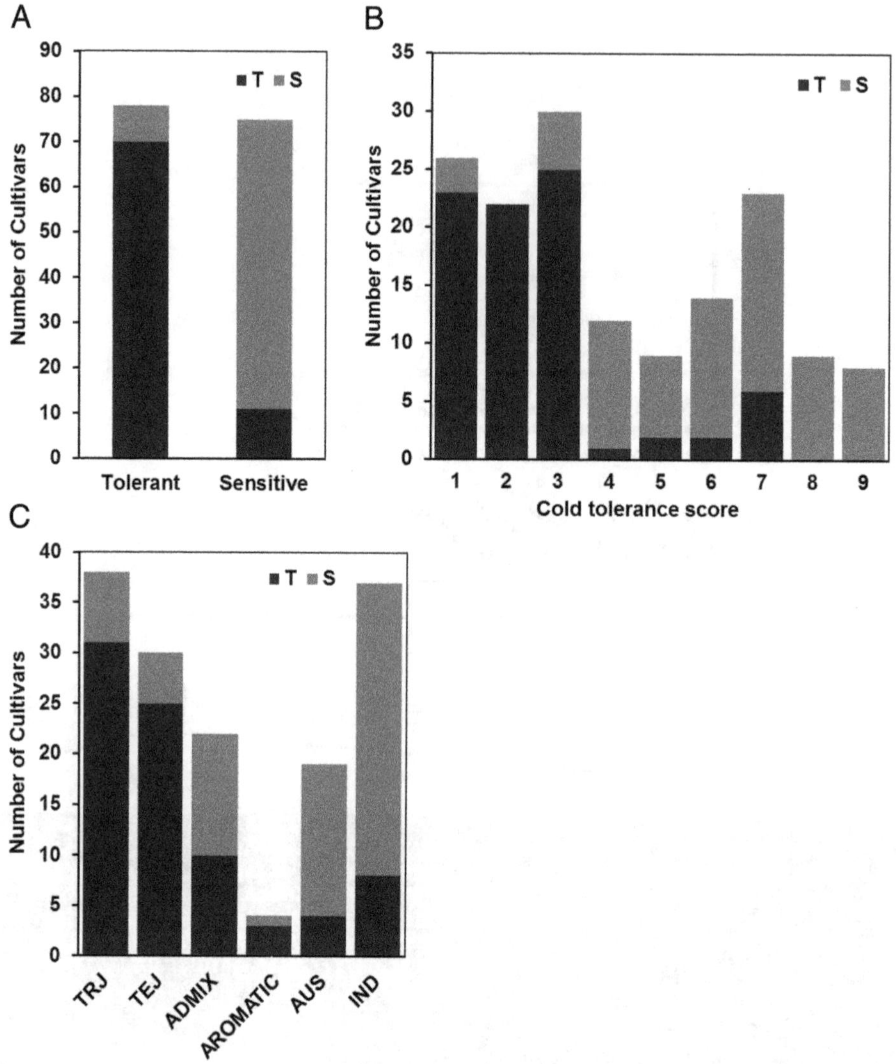

Fig. 4 Linkage confirmation of *qCTS3-9* in the rice collection using a newly developed indel marker. **a** The distribution of tolerant (T) genotype (*black box*) without the 27 bp sequence and sensitive (S) genotype (*gray box*) with the 27 bp sequence in 153 rice accessions. **b** The distribution of T and S types in each cold tolerance score. C. The distribution of T and S types in each subspecies

Methods

Plant Material and Plant Growth

About 30 seeds of each cultivar were germinated in Petri dishes in an incubator. The germinated seedlings were sown in pots containing soil and kept in a growth chamber. The standard growth conditions were 12 h of light at 25 °C with 75% humidity, and 12 h of dark at 20 °C with 70% humidity.

Cold-stress Treatment and Evaluation of Cold Tolerance

Rice seedlings at 3rd leaf stage were subjected to cold-stress treatment at 8 °C for 3 days with 75% humidity and 12 h light/12 h dark. After seven days under standard growth conditions, the seedlings were scored for cold tolerance. The experiment was repeated three times under the same conditions.

The seedlings were scored for cold tolerance on a 1–9 scale as previously reported (Han and Zhang 2004; Li et al. 2006). For scores of 1 2, 3, 4, 5, 6, 7, 8, and 9, the percentage of dry and red leaves is ≤20, between 21 and 30, between 31 and 40, between 41 and 50, between 51 and 60, between 61 and 70, between 71 and 80, between 81 and 90, and > 90, respectively. The average of three replications was used for GWAS.

Correlation Coefficient and Linear Regression Analysis

The cold tolerance scores of all the tested cultivars and their corresponding latitude values were used for the correlation analysis. The correlaton coefficient and linear regression programs in the Microsoft Excel software were used for the calculation.

GWAS Mapping of Rice CTS QTLs

Tassel 3.0 software was used for GWAS of rice CTS QTLs. The public 44 K SNP genotype and cold-tolerance phenotype dataset were input into Tassel 3.0 software and assessed with the MLM (mixed linear model). The analysis procedure was same as that reported by Kang et al. (2016).

Candidate Gene Prediction, Sequencing, and Sequence Alignment

The 500-kb reference sequence of the mapped CTS QTL regions was downloaded for gene annotation. Based on the annotation, the genes related to stresses were selected for selection of cold-stress candidate genes. Then, the corresponding candidate genes were cloned by PCR and sequenced in 20 cold-tolerant and 20 cold-sensitive rice cultivars. Sequence alignment was performed with the Clustal W program with the genes in the Nipponbare genome as a reference.

Validation of Candidate Genes with Molecular Markers

The 27-bp indel difference between the cold-tolerant and cold-sensitive rice cultivars for the candidate gene LOC_Os03g0191400 was designed to be an indel maker using the conserved flanking sequences as primers (forward primer: AGAATGGTCCTGACAT CG; reversed primer: TTTGGTGGCTCCTCTTACGGG). The marker was further used to genotype 153 rice cultivars by PCR with the following PCR protocol: 40 cycles at 95 °C for 15 s, 50 °C for 30 s, and 72 °C for 20 s. PCR products were run on a 7% SDS-PAGE gel and stained with $AgNO_3$.

Additional files

Additional file 1: Table S1. The cold tolerance scores of the evaluated rice accessions.

Additional file 2: Table S2. Cold tolerance distribution of the subpopulations in the rice diversity panel 1 (RDP1).

Additional file 3: Table S3. QTL lists for rice cold tolerance at the seedling stage in the RDP1 accessions.

Additional file 4: Table S4. Allele frequency of the 181 significant SNPs in the 40 highly tolerant and the 19 extremely sensitive rice accessions.

Acknowledgements

This work was supported by the National Natural Science Foundation of China to WD (grant number 31501628) and to LCG (grant number 31471755), the Aid Program for Science and Technology Innovative Research Team in Higher Educational Institutions of Hunan Province to WGL, and the Science Research Program of the Education Department of Hunan Province to JL (grant number 14B094). We thank the Genetic Stocks Oryza (GSOR) collection laboratory, the USDA Dale Bumpers National Rice Research Center, for providing the seeds of the Rice Diversity Panel 1 cultivars.

Authors' Contributions

WD: Designed and conducted experiments, original data organization and manuscript preparation; JL: Designed the experiments, analyzed the GWAS data and manuscript writing; LCG: Predicated candidate genes and wrote manuscript; KHX: Analyzed whole-genome sequence data; WY, TXQ, LMH, and DYF: Prepared the rice seeds, grew rice plants and performed cold treatment; WZL, LY and ZDY: Wrote the manuscript; XYH and WGL: Supervised the experiments and edited the manuscript. All authors read and approved the final manuscript.

Competing Interests

The authors declare that they have no competing interests.

Author details

[1]Southern Regional Collaborative Innovation Center for Grain and Oil Crops in China and College of Agronomy, Hunan Agricultural University, Changsha, Hunan 410128, China. [2]Institute of Plant Protection, Hunan Academy of Agricultural Sciences, Changsha, Hunan 410128, China. [3]State Key Laboratory for Biology of Plant Diseases and Insect Pests, Institute of Plant Protection, Chinese Academy of Agricultural Sciences, Beijing 100193, China. [4]Department of Plant Pathology, Ohio State University, Columbus, OH 43210, USA.

References

Andaya VC, Tai TH (2006) Fine mapping of the qCTS12 locus, a major QTL for seedling cold tolerance in rice. Theor Appl Genet 113:467–475

Andaya VC, Tai TH (2007) Fine mapping of the qCTS4 locus associated with seedling cold tolerance in rice (Oryza sativa L.). Mol Breed 20:349–358

Bednarek SY, Reynolds TL, Schroeder M, Grabowski R, Hengst L, Gallwitz D, Raikhel NV (1994) A small GTP-binding protein from Arabidopsis thaliana functionally complements the yeast YPT6 null mutant. Plant Physiol 104(2):591–596

Chen N, Xu Y, Wang X, Du C, Du J, Yuan M, Xu Z, Chong K (2011) OsRAN2, essential for mitosis, enhances cold tolerance in rice by promoting export of intranuclear tubulin and maintaining cell division under cold stress. Plant Cell Environ 34:52–64

Famoso AN, Zhao K, Clark RT, Tung CW, Wright MH, Bustamante C, Kochian LV, McCouch SR (2011) Genetic architecture of aluminum tolerance in rice (Oryza sativa) determined through genome-wide association analysis and QTL mapping. PLoS Genet 7:e1002221

Fujino K, Sekiguchi H, Matsuda Y, Sugimoto K, Ono K, Yano M (2008) Molecular identification of a major quantitative trait locus, qLTG3-1, controlling low-temperature germinability in rice. Proc Natl Acad Sci U S A 105:12623–12628

Fujino K, Obara M, Shimizu T, Koyanagi KO, Ikegaya T (2015) Genome-wide association mapping focusing on a rice population derived from rice breeding programs in a region. Breed Sci 65:403–410

Gupta PK, Kulwal PL, Jaiswal V (2014) Association mapping in crop plants: opportunities and challenges. Adv Gen et 85:109–147

Han L, Zhang S (2004) Methods of characterization and evaluation of cold tolerance in Rice. J Plant Genet Resour 5:75–80

Huang X, Wei X, Sang T, Zhao Q, Feng Q, Zhao Y, Li C, Zhu C, Lu T, Zhang Z, Li M, Fan D, Guo Y, Wang A, Wang L, Deng L, Li W, Lu Y, Weng Q, Liu K, Huang T, Zhou T, Jing Y, Lin Z, Buckler ES, Qian Q, Zhang QF, Li J, Han B (2010) Genome-wide association studies of 14 agronomic traits in rice landraces. Nat Genet 42:961–967

Jia L, Yan W, Zhu C, Agrama HA, Jackson A, Yeater K, Li X, Huang B, Hu B, McClung A, Wu D (2012) Allelic analysis of sheath blight resistance with association mapping in rice. PLoS One 7:e32703

Kang H, Wang Y, Peng S, Zhang Y, Xiao Y, Wang D, Qu S, Li Z, Yan S, Wang Z, Liu W, Ning Y, Korniliev P, Leung H, Mezey J, McCouch SR, Wang GL (2016) Dissection of the genetic architecture of rice resistance to the blast fungus Magnaporthe oryzae. Mol Plant Pathol 17:959–972

Kim SM, Suh JP, Lee CK, Lee JH, Kim YG, Jena KK (2014) QTL mapping and development of candidate gene-derived DNA markers associated with seedling cold tolerance in rice (Oryza sativa L.). Mol Genet Genomics 289:333–343

Koseki M, Kitazawa N, Yonebayashi S, Maehara Y, Wang ZX, Minobe Y (2010) Identification and fine mapping of a major quantitative trait locus originating from wild rice, controlling cold tolerance at the seedling stage. Mol Genet Genomics 284:45–54

Kuroki M, Saito K, Matsuba S, Yokogami N, Shimizu H, Ando I, Sato Y (2007) A quantitative trait locus for cold tolerance at the booting stage on rice chromosome 8. Theor Appl Genet 115:593–600

Li X, Dai CC, Cheng R (2006) Identification for cold tolerance at different growth stages in rice (Oryza sativa L.) and physiological mechanism of differential cold tolerance. Acta Agron Sin 32:76–83

Lv Y, Guo Z, Li X, Ye H, Xiong L (2016) New insights into the genetic basis of natural chilling and cold shock tolerance in rice by genome-wide association analysis. Plant Cell Environ 39:556–570

Ma Y, Dai X, Xu Y, Luo W, Zheng X, Zeng D, Pan Y, Lin X, Liu H, Zhang D, Xiao J, Guo X, Xu S, Niu Y, Jin J, Zhang H, Xu X, Li L, Wang W, Qian Q, Ge S, Chong K (2015) COLD1 confers chilling tolerance in rice. Cell 160:1209–1221

Mao D, Yu L, Chen D, Li L, Zhu Y, Xiao Y, Zhang D, Chen C (2015) Multiple cold resistance loci confer the high cold tolerance adaptation of Dongxiang wild rice (Oryza rufipogon) to its high-latitude habitat. Theor Appl Genet 128:1359–1371

Pan Y, Zhang H, Zhang D, Li J, Xiong H, Yu J, Li J, Rashid MAR, Li G, Ma X, Cao G, Han L, Li Z (2015) Genetic analysis of cold tolerance at the germination and booting stages in rice by association mapping. PLoS ONE 10:e0120590

Nair PMG, Kang IS, Moon BY, Lee CH (2009) Effects of low temperature stress on rice (Oryza sativa L.) plastid ω-3 desaturase gene, OsFAD8 and its functional analysis using T-DNA mutants. Plant Cell, Tissue and Organ Culture 98:87–96

Qin Y, Shen X, Wang N, Ding X (2015) Characterization of a novel cyclase-like gene family involved in controlling stress tolerance in rice. J Plant Physiol 181:30–41

Saito K, Hayano-Saito Y, Maruyama-Funatsuki W, Sato Y, Kato A (2004) Physical mapping and putative candidate gene identification of a quantitative trait locus Ctb1 for cold tolerance at the booting stage of rice. Theor Appl Genet 109:515–522

Saito K, Hayano-Saito Y, Kuroki M, Sato Y (2010) Map-based cloning of the rice cold tolerance gene Ctb1. Plant Sci 79:97–102

Shirasawa S, Endo T, Nakagomi K, Yamaguchi M, Nishio T (2012) Delimitation of a QTL region controlling cold tolerance at booting stage of a cultivar, 'Lijiangxintuanheigu', in rice, Oryza sativa L. Theor Appl Genet 124:937–946

Suh JP, Jeung JU, Lee JI, Choi YH, Yea JD, Virk PS, Mackill DJ, Jena KK (2010) Identification and analysis of QTLs controlling cold tolerance at the reproductive stage and validation of effective QTLs in cold-tolerant genotypes of rice (Oryza sativa L.). Theor Appl Genet 120:985–995

Wang Q, Guan Y, Wu Y, Chen H, Chen F, Chu C (2008) Overexpression of a rice OsDREB1F gene increases salt, drought, and low temperature tolerance in both Arabidopsis and rice. Plant Mol Biol 67:589–602

Wang X, Jia MH, Ghai P, Lee FN, Jia Y (2015) Genome-wide association of rice blast disease resistance and yieldrelated components of rice. Molecular Plant-Microbe Interactions 28 (12):1383–1392.

Wang C, Yang Y, Yuan X, Xu Q, Feng Y, Yu H, Wang Y, Wei X (2014) Genome-wide association study of blast resistance in indica rice. BMC Plant Biol 14:311

Xiao N, Huang WN, Zhang XX, Gao Y, Li AH, Dai Y, Yu L, Liu GQ, Pan CH, Li YH, Dai ZY, Chen JM (2014) Fine mapping of qRC10-2, a quantitative trait locus for cold tolerance of rice roots at seedling and mature stages. PLoS One 9: e96046

Xiao N, Huang WN, Li AH, Gao Y, Li YH, Pan CH, Ji H, Zhang XX, Dai Y, Dai ZY, Chen JM (2015) Fine mapping of the qLOP2 and qPSR2-1 loci associated with chilling stress tolerance of wild rice seedlings. Theor Appl Genet 128: 173–185

Yang W, Guo Z, Huang C, Duan L, Chen G, Jiang N, Fang W, Feng H, Xie W, Lian X, Wang G, Luo Q, Zhang Q, Liu Q, Xiong L (2014) Combining high-throughput phenotyping and genome-wide association studies to reveal natural genetic variation in rice. Nat Commun 5:5087

Yang T, Zhang S, Zhao J, Huang Z, Zhang G, Liu B (2015) Meta-analysis of QTLs Underlying Cold Tolerance in Rice (Oryza sativa L.). Mol Plant Breed 13:1–15

Zhao L, Liu F, Xu W, Di C, Zhou S, Xue Y, Yu J, Su Z (2009) Increased expression of OsSPX1 enhances cold/subfreezing tolerance in tobacco and Arabidopsis thaliana. Plant Biotechnol J 7:550–561

Zhao K, Tung CW, Eizenga GC, Wright MH, Ali ML, Price AH, Norton GJ, Islam MR, Reynolds A, Mezey J, McClung AM, Bustamante CD, McCouch SR (2011) Genome-wide association mapping reveals a rich genetic architecture of complex traits in Oryza sativa. Nat Commun 2:467

Zhao XQ, Wang WS, Zhang F, Zhang T, Zhao W, Fu BY, Li ZK (2013) Temporal profiling of primary metabolites under chilling stress and its association with seedling chilling tolerance of rice (Oryza sativa L.). Rice (N Y) 6:23

Zhou L, Zeng Y, Zheng W, Tang B, Yang S, Zhang H, Li J, Li Z (2010) Fine mapping a QTL qCTB7 for cold tolerance at the booting stage on rice chromosome 7 using a near-isogenic line. Theor Appl Genet 121:895–905

Zhu Y, Chen K, Mi X, Chen T, Ali J, Ye G, Xu J, Li Z (2015) Identification and fine mapping of a stably expressed QTL for cold tolerance at the booting stage using an interconnected breeding population in rice. PLoS One 10:e0145704

6

Distribution, Genetic Diversity and Potential Spatiotemporal Scale of Alien Gene Flow in Crop Wild Relatives of Rice (*Oryza* spp.) in Colombia

Evert Thomas[1*], Eduardo Tovar[2], Carolina Villafañe[3], José Leonardo Bocanegra[4] and Rodrigo Moreno[4]

Abstract

Background: Crop wild relatives (CWRs) of rice hold important traits that can contribute to enhancing the ability of cultivated rice (*Oryza sativa* and *O. glaberrima*) to produce higher yields, cope with the effects of climate change, and resist attacks of pests and diseases, among others. However, the genetic resources of these species remain dramatically understudied, putting at risk their future availability from *in situ* and *ex situ* sources. Here we assess the distribution of genetic diversity of the four rice CWRs known to occur in Colombia (*O. glumaepatula*, *O. alta*, *O. grandiglumis*, and *O. latifolia*). Furthermore, we estimated the degree of overlap between areas with suitable habitat for cultivated and wild rice, both under current and predicted future climate conditions to assess the potential spatiotemporal scale of potential gene flow from GM rice to its CWRs.

Results: Our findings suggest that part of the observed genetic diversity and structure, at least of the most exhaustively sampled species, may be explained by their glacial and post-glacial range dynamics. Furthermore, in assessing the expected impact of climate change and the potential spatiotemporal scale of gene flow between populations of CWRs and GM rice we find significant overlap between present and future suitable areas for cultivated rice and its four CWRs. Climate change is expected to have relatively limited negative effects on the rice CWRs, with three species showing opportunities to expand their distribution ranges in the future.

Conclusions: Given (i) the sparse presence of CWR populations in protected areas (ii) the strong suitability overlap between cultivated rice and its four CWRs; and (iii) the complexity of managing and regulating areas to prevent alien gene flow, the first priority should be to establish representative *ex situ* collections for all CWR species, which currently do not exist. In the absence of studies under field conditions on the scale and extent of gene flow between cultivated rice and its Colombian CWRs, effective *in situ* conservation might best be achieved through tailor-made management plans and exclusion of GM rice cultivation in areas holding the most genetically diverse CWR populations. This may be combined with assisted migration of populations to suitable areas where rice is unlikely to be cultivated under current and future climate conditions.

Keywords: Alien gene flow, Crop wild relatives, Genetic diversity, Habitat suitability, Rice cultivation, Colombia, Climate change

* Correspondence: evert.thomas@gmail.com
[1]Bioversity International, Lima, Peru
Full list of author information is available at the end of the article

Background

There is increasing global interest in crop wild relatives (CWRs) for the important contribution they can make to current endeavors to sustain and increase global food production (Vincent et al. 2013; Khoury et al. 2015; Castañeda-Álvarez et al. 2016). CWRs are wild plant taxons related to crops and their original progenitors, which have potential to contribute adaptive, nutritional, agronomic and other useful traits for crop improvement. In light of climate change, the need to improve the adaptive capacity of crop taxa for resisting acute and chronic biotic and abiotic stress, has never been more pressing. CWRs offer a treasure trove that can serve such capacity improvement (Atwell et al. 2014; Dempewolf et al. 2014). Many CWRs are adapted to conditions of environmental hardship, making them less susceptible to the effects of climate change, or even allowing them to prosper when climate gets hotter and dryer (Thomas et al. 2016). CWRs have permitted breeders to build in pest and disease resistance, abiotic stress tolerance, and quality traits in a growing number of food crops (Hajjar and Hodgkin 2007; Brar and Singh 2011). Over the past decades there has been a stable increase in the rate of release of cultivars containing genes from CWRs (Hajjar and Hodgkin 2007).

The *Oryza* genus is one of the most extensively studied genepools regarding the actual and potential use of CWR genes for crop improvement (Jena 2010; Brar and Singh 2011; Sanchez et al. 2013). The genus is widely distributed in tropical and sub-tropical regions around the globe (Atwell et al. 2014) and consists of two cultivated (*O. sativa* L. and *O. glaberrima* Steud.) and 22 wild species (Vaughan et al. 2003; Jena 2010). Four species complexes have been identified: the *O. sativa, O. officinalis, O. ridleyi,* and *O. meyeriana* complexes (Vaughan 1989; Vaughan et al. 2003). Both cultivated species pertain to the AA genome *(O. sativa* complex) making gene transfer from wild relatives in the same complex relatively easy. However, many beneficial genes have also been successfully transferred to cultivated rice from the more distantly related species (Jena 2010; Brar and Singh 2011; Sanchez et al. 2013).

Central and South America are home to four rice CWR species: the diploid *O. glumaepatula* Steud (2n = 24, AA), included in the *O. sativa* complex, and the three allotetraploid species *O. alta* Swallen, *O. grandiglumis* (Doell) Prod., and *O. latifolia* Desv. included in the *O. officinalis* complex (2n = 48, CCDD) (Veasey et al. 2008a). All these species have either been identified as having high potential for use in cultivated rice improvement programs, or have already been the subject of actual gene transfer (Vaughan et al. 2003; Jena 2010; Brar and Singh 2011; Niroula et al. 2012; Sanchez et al. 2013; Pak et al. 2013). However the genetic diversity and structure, as well as the *in situ* conservation status of these species remains severely understudied, jeopardizing the conservation and future use of their genetic resources. To date most studies have focused on *O. glumaepatula* (Akimoto et al. 1998; Buso et al. 1998; Brondani et al. 2005; Karasawa et al. 2007a; Karasawa et al. 2007b; Veasey et al. 2008b; Karasawa et al. 2012), and much less on the other three species (Quesada et al. 2002; Arrieta-Espinoza et al. 2005; Veasey et al. 2008b; Veasey et al. 2011).

The distribution pattern of genetic diversity across a species′ range can be influenced by climate change, notably through alteration of geographical connectivity and associated modification of gene flow patterns within and across populations (Huang and Schaal 2012). Rapid changes in population size during past glacial cycles have been related to processes of strong genetic drift (Knowles and Richards 2005). When populations get isolated in areas with variable environmental conditions, for example as a consequence of range contraction, divergent selection pressures for different populations can lead to local adaptation (Chen et al. 2012). The last glaciation (22,000–13,000 years before present [BP]) is the most important period of climate change in recent history that had a profound impact on the genetic distribution of many species across the world, including *O. rufipogon*, the progenitor of cultivated Asian rice (Huang and Schaal 2012). Owing to lower air temperatures, lower precipitation levels and increased water stress associated with lower atmospheric CO_2 concentrations, many species have experienced range contractions, leading to patterns of genetic differentiation that remain evident in their genetic profiles (Thomas et al. 2012; Galluzzi et al. 2015; Thomas et al. 2015). Here we assess the potential impact of past climate change on the genetic profiles of the Colombian rice CWRs.

One of the potential future threats to the *in situ* conservation of rice CWRs in Colombia may relate to introgression of alien genes. GM rice has been legally permitted in the country for use as food since 2008 (CERA 2012). Although its cultivation is currently not allowed, the potential future introduction of GM rice in areas where wild species occur may lead to a risk of introgression of GMO genes into the gene pools of wild populations with potential effects on their adaptive potential, population dynamics and ecological roles, among others (Chen et al. 2004; Garcia and Altieri 2005). Snow et al. (2005) argued that the environmental effects of the introgression of GMO genes in wild species are generally neutral or insignificant. In rice CWRs, high rates of self-pollination can be expected to result in low risk of alien gene transfer (Stewart et al. 2003). However, low risk is not zero risk and even at very low cross-pollination rates, the likelihood of a transgene being quickly dispersed through wild receptor populations increases when it grants a significant adaptive advantage in resistance to pests and diseases, higher tolerance to abiotic stresses (drought, salinity), or higher yields (Ellstrand 2003).

The purpose of this study was threefold. First, we investigated the genetic diversity and structure of the four rice CWRs in part of their natural distribution range in Colombia. Following other studies (Waltari et al. 2007; Thomas et al. 2012; Thomas et al. 2015; Thomas et al. 2017), we carried out suitability modeling at the last glacial maximum (LGM; ~21,000 years BP) and spatially overlaid the results with the outcomes of genetic sampling, to evaluate the potential impact of the last glacial period on the current distribution of genetic diversity in all four species. Second, we estimated the degree of overlap between areas with suitable habitat for cultivated and wild rice, both under current and predicted future climate conditions to assess the potential spatiotemporal scale of potential gene flow from GM rice to its CWRs. This also allowed us to evaluate the potential effect of climate change on the *in situ* persistence of the rice CWRs. Third and last, we discuss the implications of our findings for the design of tailor-made *in situ* and *ex situ* conservation strategies aimed at maintaining and enhancing the natural genetic variability of Colombian rice CWRs, and ensuring the availability of their genetic resources for future use.

Results

Genetic diversity and spatial structure

For the tetraploid species, highest allelic richness was observed in *O. latifolia* with 117 alleles and an average of 10.6 per locus, followed by *O. alta* and *O. grandiglumis* with 46 and 43 alleles at averages of 4.2 and 3.9 alleles per locus, respectively (Additional file 1: Table S1). The diploid species *O. glumaepatula*, yielded a total of 30 alleles with an average of 2.7 alleles per locus (Additional file 2: Table S2). The latter value is in agreement with a study conducted by Karasawa et al. (2007a) with SSR markers in *O. glumaepatula* populations from Brazil which reported values of 3.09. High values of expected heterozygosity (H_E) were observed in the tetraploid species. Average H_E across all loci was 0.57 for *O. latifolia*, 0.43 for *O. grandiglumis*, and 0.40 for *O. alta*. The lowest H_E value was found for *O. glumaepatula* (0.21).

A scatter plot of the first two axes of a PCoA undertaken for all species samples (Fig. 1), shows a clear separation of the diploid and the tetraploid species groups, with some degree of overlap between species within groups. The partial overlap between species samples in Fig. 1 may be partly due to the limitations of the Bruvo genetic distance metric to clearly separate all the different species in three-dimensional ordination space. However, Fig. 1 does distinctively group the Colombian CWR samples according to their taxonomic affiliation and separates them from reference samples pertaining to the same species, confirming their divergent genetic make-up.

Species-specific PCoAs allowed visualization of the differentiation between the spatially coherent genetic groups identified by hierarchical cluster analysis in all species (Fig. 2 and Additional file 3: Figure S1). Bayesian cluster analysis yielded very similar results for *O. grandiglumis* (two clusters) and provided more detail on the small-scale genetic structure of *O. alta*, *O. glumaepatula* and *O. latifolia*. ΔK values are shown in Additional file 4: Figure S2. For *O. alta* (K = 3) two subclusters were identified in cluster 2, corresponding with the two isolated groups of points in the upper and lower parts in the PCoA diagram, respectively. However, these two subgroups were not geographically coherent. Results of ΔK computation for *O. glumaepatula* showed support for K = 2 and 4. For K = 2 (results not shown) part of the individuals from Casanare department (cluster 1 in Fig. 2b and Additional file 4: Figure S2) were grouped together with cluster 2 in spite of their spatial isolation. For K = 4 the individuals of cluster 2 were assigned to a separate group, while three subgroups were identified among the individuals from cluster 1, even though these were separated by less than 3 km as the crow flies. Highest cluster richness was found for *O. latifolia* that was also the most extensively sampled species. Hierarchical cluster analysis identified four main clusters largely according to the geographical locations of sampled individuals (Fig. 3). However, Fig. 2d suggests a high degree of affinity between clusters 1 and 2. Results of ΔK computation for *O. latifolia* showed support for K = 5 and 7 (Additional file 4: Figure S2). The most parsimonious scenario (K = 5), grouped the individuals in the same clusters as the hierarchical cluster analysis, but identified two subgroups in cluster 3. For K = 7, cluster 4 was further separated in three subclusters. Overall, the large majority of individuals were consistently assigned to discrete genetic groups with little evidence of admixture within clusters. Furthermore, the subgroups identified in clusters 3 and 4 were arranged sequentially along the more or less linear gradients shown in Fig. 3.

All main clusters identified in the PCoA plots were separated by large geographical distances in *O. latifolia*, *O. glumaepatula* and *O. grandiglumis*, and by the Vita River, one of the tributaries of the Orinoco River, in *O. alta* (Fig. 3, Additional file 5: Figure S3, Additional file 6: Figure S4, and Additional file 7: Figure S5). These barriers can be expected to have constrained gene flow between the different groups, and hence are likely to have contributed to their genetic differentiation. This is corroborated by the significant isolation by distance correlations we found for

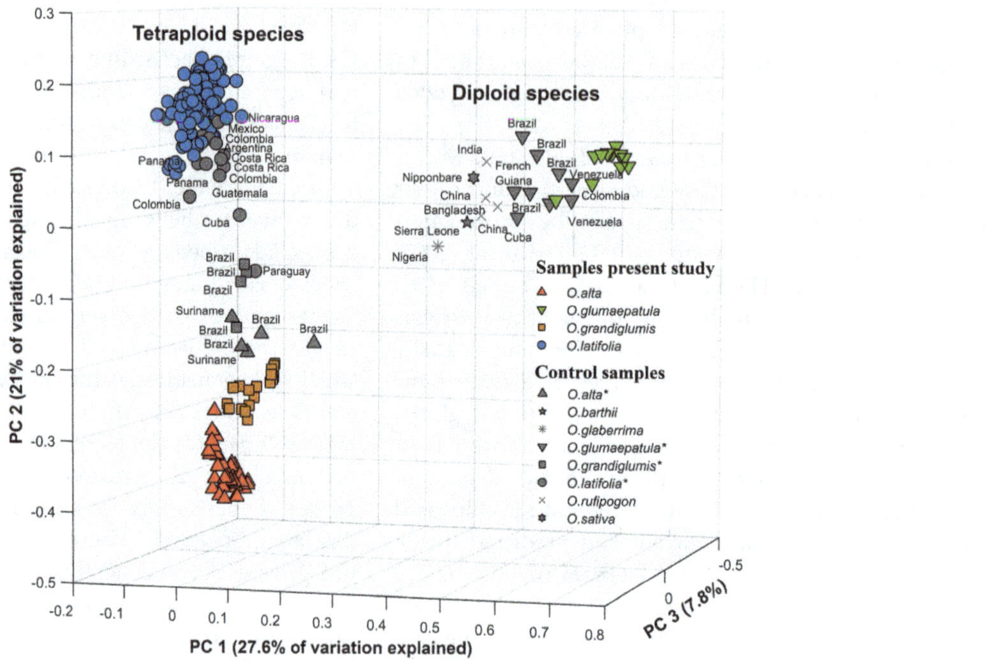

Fig. 1 Principal coordinates analysis of all *Oryza* individuals sampled for the present study and reference samples, based on the Bruvo genetic distance

all four species when considering all individuals sampled ($p < 0.0001$ in all cases; Additional file 8: Table S3). Significant correlations were also found at genetic cluster level, except for cluster 2 in *O. latifolia*, which was expected as plant individuals from this clusters where collected at very short distances (Additional file 8: Table S3).

Genetic distance between genetic groups as measured by G_{ST} (Nei 1973) was 0.129, 0.303 and 0.655 for the two clusters identified in *O. alta*, *O. grandiglumis* and *O. glumaepatula*, respectively, and varied between 0.068 and 0.184 for the four clusters identified in *O. latifolia*.

LGM habitat suitability and relation with genetic diversity distribution

To explore how the last glacial period may have contributed to structuring the distribution of genetic diversity of the wild relatives, we overlaid maps of their genetic diversity and LGM habitat suitability. For all species suitable habitat conditions seemed to have prevailed in the vicinities of the identified genetic clusters (Fig. 3, Additional file 5: Figure S3, Additional file 6: Figure S4, and Additional file 7: Figure S5), which could imply that the origin of these clusters may partly be due to genetic differentiation in isolated refugia during the last glacial period. This pattern is most apparent for *O. latifolia*. Combined interpretation of the distribution of genetic groups (Fig. 3a; beta diversity) and diversity scores at individual sampling sites of this species (Fig. 3b; alpha diversity) can help unraveling potential glacial and post-glacial range dynamics.

Figure 3b shows that the highest levels of genetic diversity observed in clusters 1, 2 and 3 are geographically closest to areas which may have been suitable during the LGM, with a trend of decreasing levels of genetic diversity in populations when moving away from these areas, in line with observed isolation-by-distance patterns (Additional file 8: Table S3). This could suggest that genetic diversity may have been concentrated in these suitable areas during the last glacial period, leading to genetic differentiation of the different groups, and that range expansion during the warming Holocene may have occurred along the gradient of decreasing diversity, as indicated by the arrows in Fig. 3. The limited genetic differentiation between the cluster from the Pacific coast and the central Colombian valleys (clusters 2 and 1, respectively), evident from the PCoA scatter plot (Fig. 2d), suggests that the former may represent a recent introduction, possibly even of anthropogenic nature. If correct, it is possible that although habitat conditions may have been suitable in the Pacific coast area, the species was not (yet) present there; otherwise one would expect higher diversity scores.

Current and future habitat suitability of cultivated rice and its wild relatives

A comparison of habitat suitability under current and future climate conditions suggests that all CWRs except *O. latifolia* might be able to expand their distribution ranges in the mid-term future (Fig. 4), implying that their genetic resources would not be seriously threatened by the effects of climate changes on their range sizes. For *O. latifolia* a

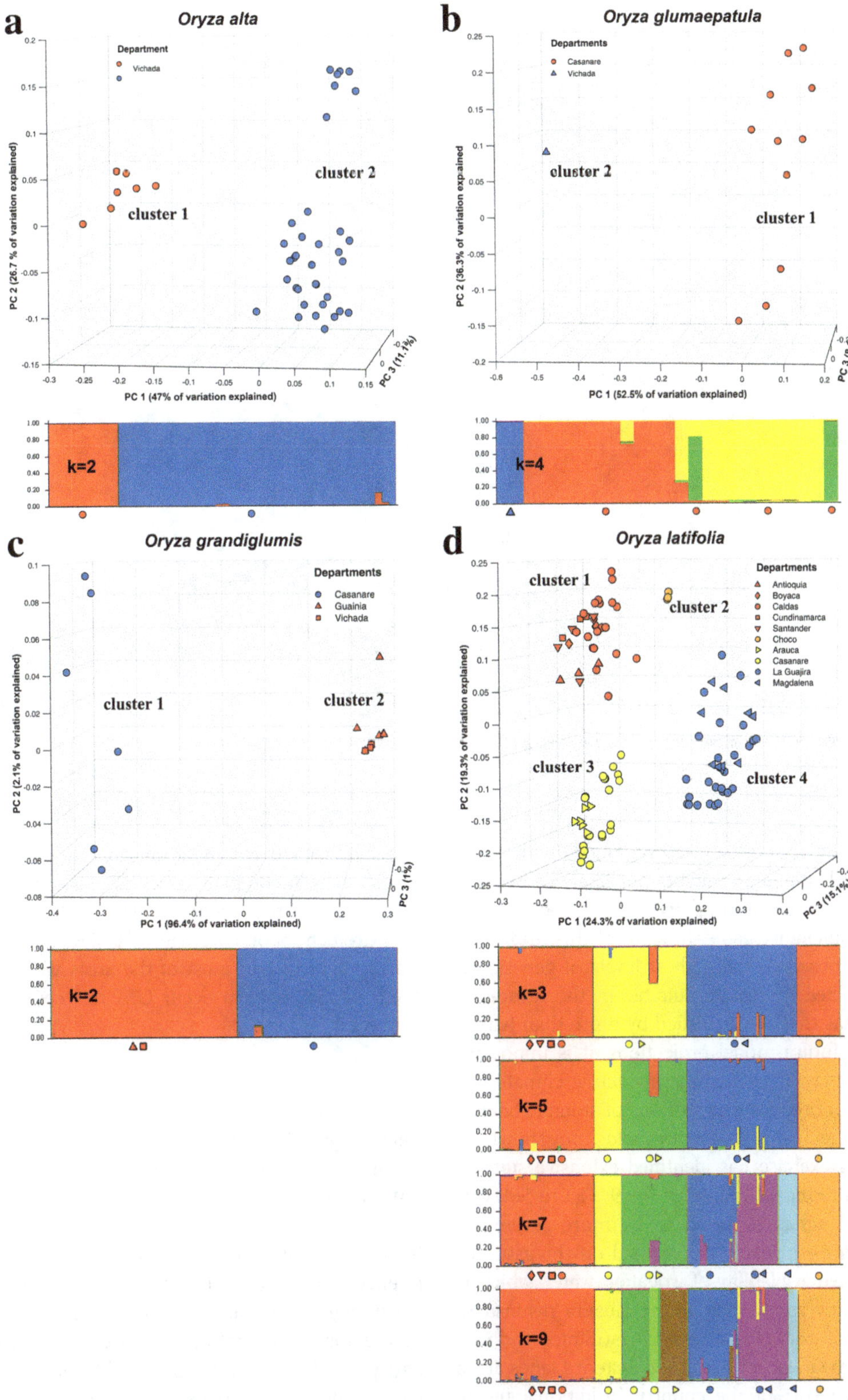

Fig. 2 Results of species-specific principal coordinates analysis based on the Bruvo genetic distance and Bayesian admixture proportions identified by STRUCTURE (Pritchard et al. 2000) of individual plants of **a** *O. alta*, **b** *O. glumaepatula*, **c** *O. grandiglumis*, **d** *O. latifolia*. The genetic groups in the PCoA plots were identified through hierarchical cluster analysis (>95% branch support; see Additional file 3: Figure S1). *Symbols* and *colors* used in the PCoA plots are associated with STRUCTURE clusters for different values of K

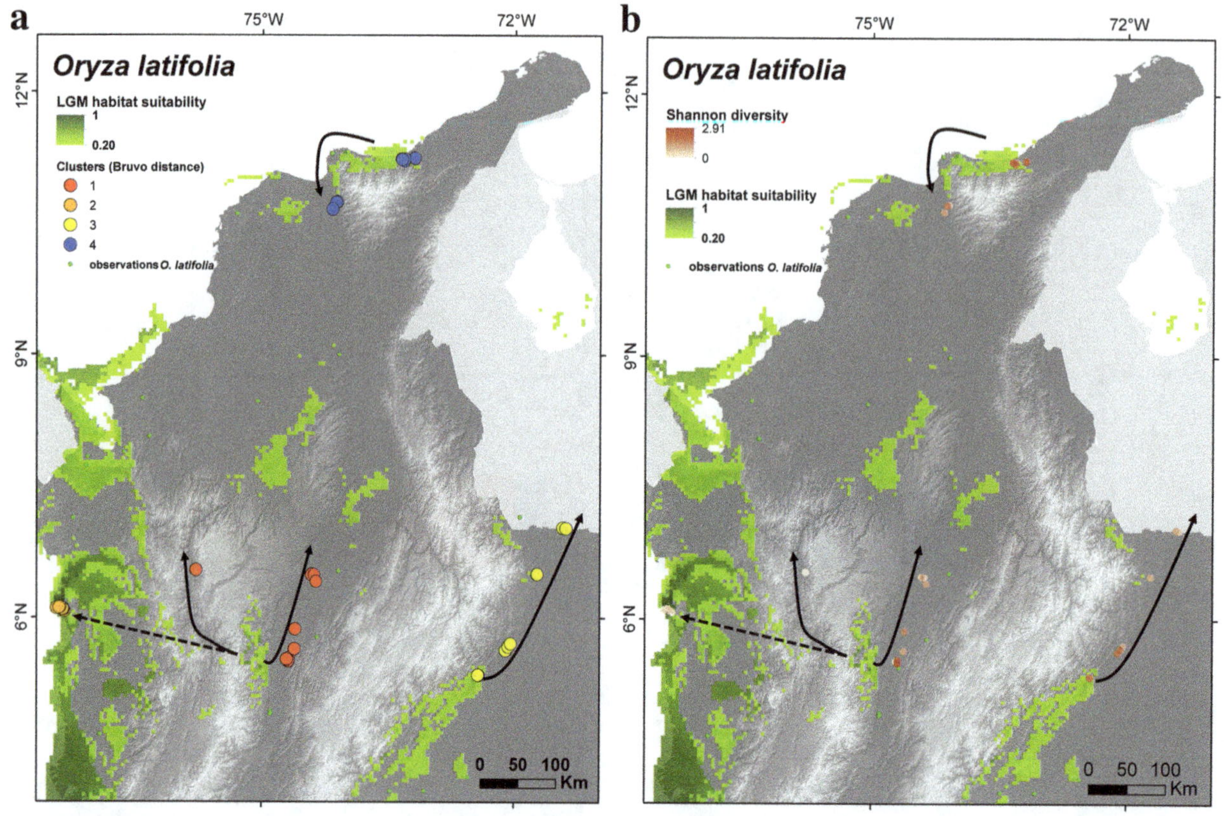

Fig. 3 Combination of LGM suitability map of *Oryza latifolia* and (**a**) the geographical distribution of genetic groups (**b**) Shannon diversity index measured at sampling sites. The *arrows* indicate the potential direction of range expansion from putative glacial refugia. The *dotted arrow* suggests that range expansion is likely to be relatively recent, and possibly of anthropogenic origin

slightly less optimistic result was obtained, with most areas currently identified as suitable but likely to become unsuitable in the future, and relatively limited opportunities for future suitability gains. In spite of this, nearly all areas where we observed the highest levels of genetic diversity are expected to remain suitable in the future. In only judging areas that are identified by more than half of the 30 'suitable' future model projections, it is important to note that we may be overestimating actual climate impacts, since much broader distributions of suitable habitat were obtained when using lower thresholds (not shown).

Our modelling exercise has identified extensive suitable areas for rice cultivation (Additional file 9: Figure S6). Climate change impact is expected to be more severe for rainfed rice than for irrigated rice, (based on the assumption of unrestricted availability of irrigation water, which is unrealistic in many areas, such as the Guajira peninsula), while the overall impact on cultivated rice (rainfed and irrigation rice) may be most serious in the eastern Llanos areas and the northern part of the country (Additional file 9: Figure S6). Most of the areas where the different CWR species are currently known to occur, or might occur, based on suitable habitat conditions, overlap with areas where rice cultivation is already ongoing, or likely to be possible either under either current or future climate conditions (Fig. 5). Additionally, a combination of the distribution of field observations and the results of our modeling exercise suggest an extremely poor coverage of CWRs in protected areas. This implies that in most of the areas where these species occur or are likely to occur GM rice could potentially be cultivated.

Discussion

Genetic diversity and spatial structure

Results suggest that the Colombian populations of rice CWRs we studied hold relatively high levels of unique diversity compared to samples (Fig. 1) and studies (see below) from other parts in their distribution ranges. Differences in allelic richness found for the four species partly reflect sampling intensity, but higher values observed for the allotetraploid species are also very likely partly due to the higher probability of encountering different alleles per locus compared to diploid species (maximum of four versus two per locus, respectively). The chance of observing a (partially) heterozygous locus is also greater in allotetraploids than in diploids for the same reason.

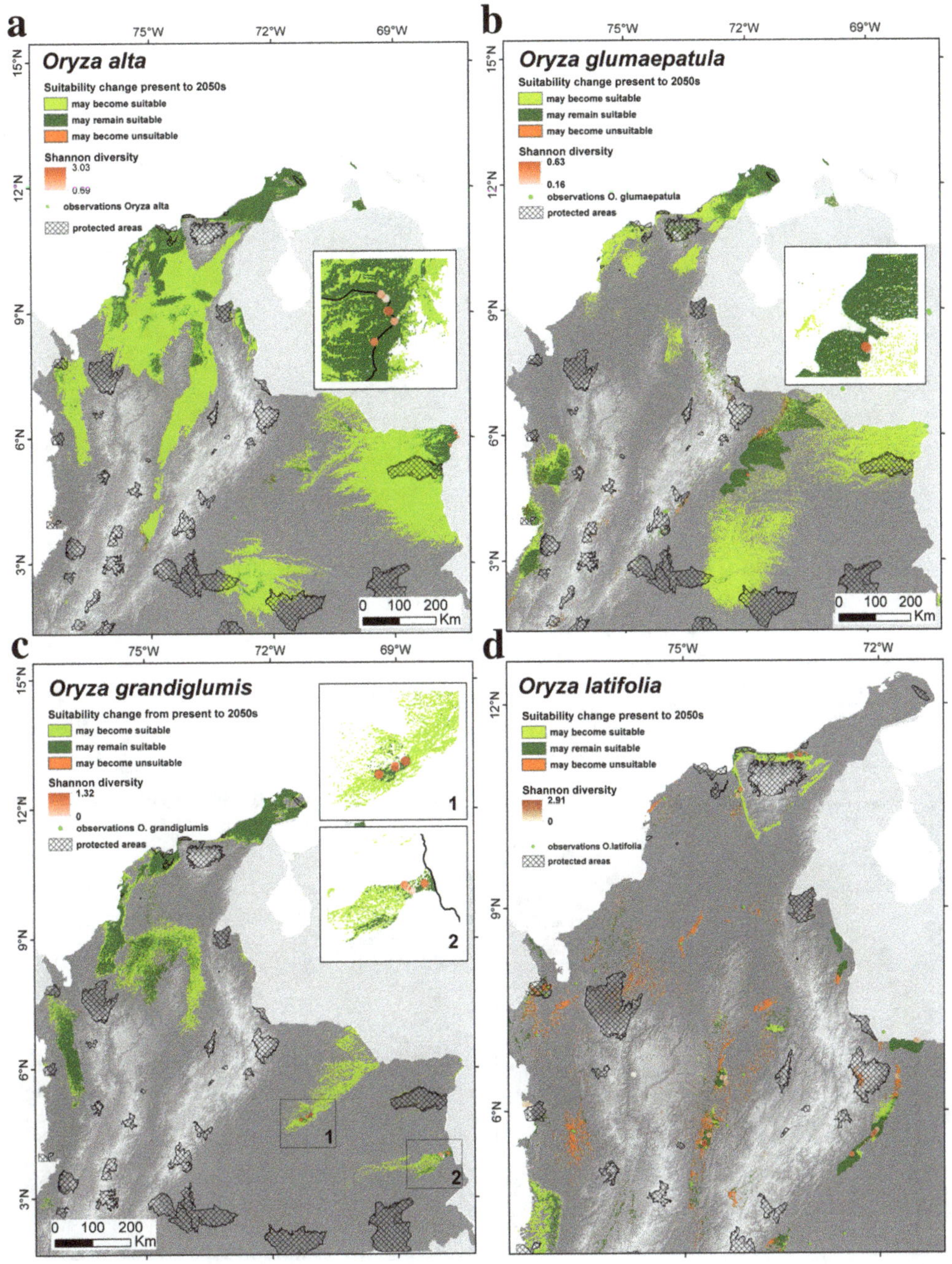

Fig. 4 Predicted changes in habitat suitability from present to the 2050s of **a** *O. alta*, **b** *O. glumaepatula*, **c** *O. grandiglumis*, and **d** *O. latifolia*, compared with Shannon diversity index at sampling sites

Compared to other studies (Additional file 10: Table S4), the H_O and H_E values (0.14 and 0.21, respectively) we found for *O. glumaepatula* could be considered relatively high and the inbreeding coefficient relatively low (0.37), when taking into account the small number of samples studied ($n = 25$). Karasawa et al. (2007b) found H_O and H_E to range from 0.10 to 0.23 and from 0.37 to 0.57, respectively, in populations of 100 individuals each along a latitudinal gradient in Brazil, genotyped with SSR markers. The fixation index varied between 0.37 and 0.82. In another SSR study of 414 individuals from Brazil, Brondani et al. (2005) found values for Ho and H_E at the low end of the ranges found by Karasawa et al. (2007b) (0.03 and 0.12, respectively). They accordingly found a high inbreeding coefficient ($F_{IS} = 0.79$), indicating a predominantly autogamous reproduction in

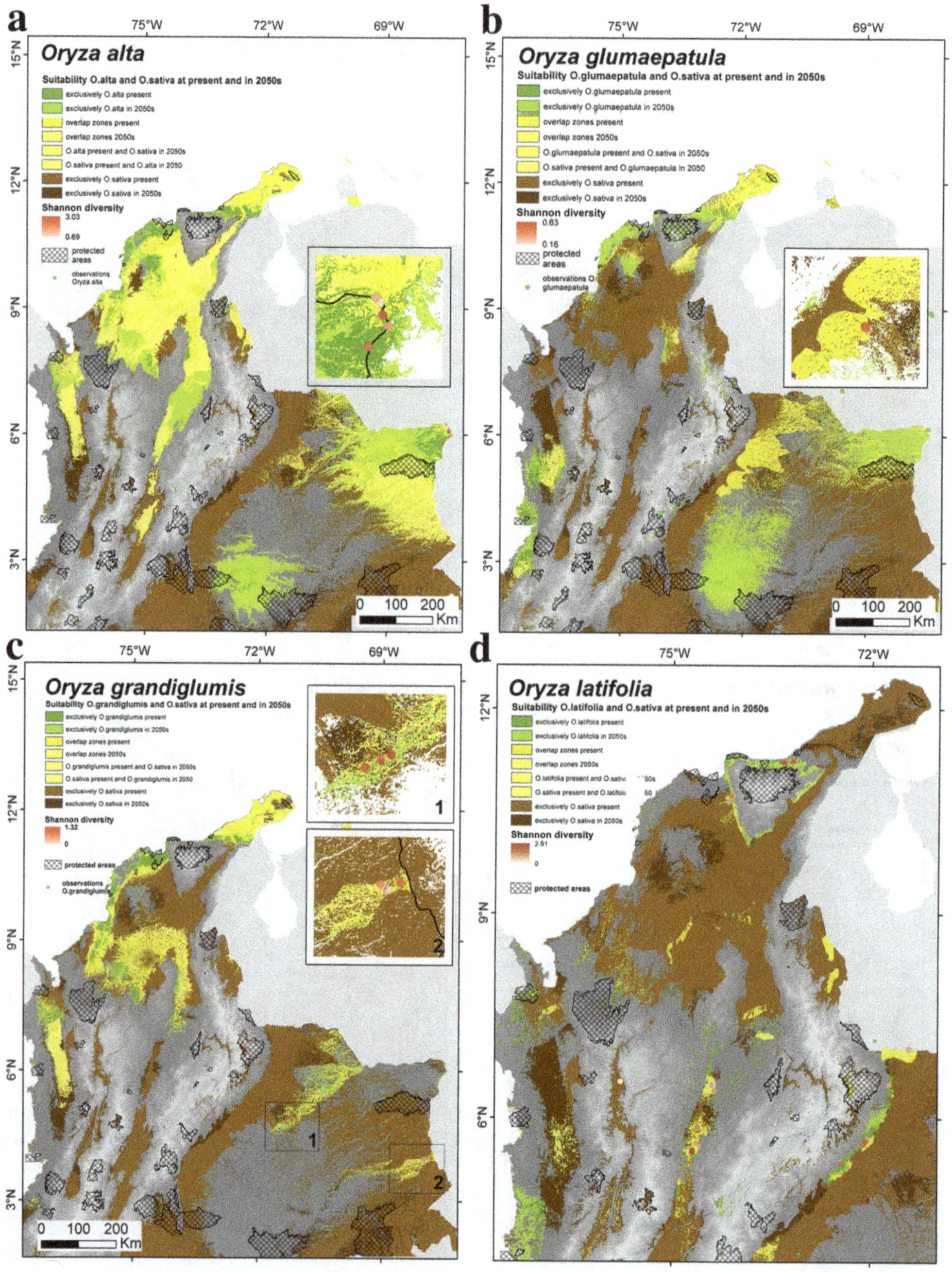

Fig. 5 Distribution and overlap of suitable areas of cultivated rice *O. sativa* (irrigated and rainfed) and its wild relatives **a** *O. alta*, **b** *O. glumaepatula*, **c** *O. grandiglumis* and **d** *O. latifolia* under current and expected future (2050s) climate conditions. Please note that overlap zones of suitable areas of cultivated rice and its wild relatives whereby habitat suitability of at least one of both is expected to occur under predicted future climate conditions (three different possible combinations) are all highlighted in *pale yellow*

their study populations. Similarly, isoenzyme marker studies –known to result in lower H_E scores than with SSRs (Gao et al. 2002)– from Brazil revealed low H_E values (0.04 and 0.06) and high F_{IS} values (>0.9) in a large number of individuals (333 and 1090) (Akimoto et al. 1998; Veasey et al. 2008a), respectively), pointing to a predominantly autogamous reproduction.

Karasawa et al. (2007b) argued that the reproductive system of *O. glumaepatula* may be variable both among and within populations, ranging from complete selfing to a mixed system with predominance of selfing, which could explain the variable observations of the inbreeding coefficient across populations. If correct, the Colombian populations we sampled may have a mixed mating system, characterized by a low but significant degree of outcrossing, evidenced by the relatively low inbreeding-coefficient compared to literature (Additional file 10: Table S4). One possibility that merits further research is whether the mating system of *O. glumaepatula* may be influenced by the ecological conditions of its habitat. Karasawa et al. (2007b) found the lowest values of inbreeding in a population from the Brazilian savanna, the same ecosystem where also the Colombian samples were collected. Possibly more open vegetation (savanna) may stimulate mixed mating systems while selfing may be an ecological response to more closed vegetation types (rainforest).

To our knowledge, diversity studies based on SSR markers are non-existent for the *Oryza* tetraploid species, complicating comparisons with other studies which are mainly based on isoenzymes (Karasawa et al. 2012). However, high values of H_E observed in the three species (0.40–0.57) suggest the Colombian populations hold high genetic diversity. For example, Sun et al. (1998) found a H_E value of *only* 0.25 across ten populations of *Elymus fibrosus*, a wild Poaceae species. Previous studies, based on different molecular techniques have repeatedly observed a genetic relatedness between *O. alta* and *O. grandiglumis* (Wang et al. 1992; Federici et al. 2002; Bao and Ge 2003 using Restriction Fragment Length Polymorphism; Aggarwal et al. 1997; Aggarwal et al. 1999 using total DNA hybridization and Amplified Fragment Length Polymorphism; and Joshi et al. 2000 using Inter Simple Sequence Repeats). The Colombian populations we studied did show some affiliation, but could nonetheless clearly be separated from each other in ordination space. Overall, our findings confirm the clear genetic differences in species pertaining to the CCDD genome observed by Aggarwal et al. (Aggarwal et al. 1996) and supported by others (Ge et al. 1999; Zamora et al. 2003; Veasey et al. 2008a).

LGM habitat suitability and relation with genetic diversity distribution

As suggested by the relatively elevated G_{ST} values we found (Hedrick 2005; Jost 2008; Meirmans and Hedrick 2011), genetic groups within all four species showed high degrees of genetic differentiation. All the genetic groups could be linked to areas that are likely to have remained suitable for the respective species in the late Pleistocene. This could suggest that they genetically differentiated as a consequence of prolonged isolation of their source populations owing to range contractions that took place in response to the drying and cooling effects of the last glacial period. Similar evidence exists for the genetic differentiation of *O. rufipogon* in Southeast Asia (Fuller et al. 2010; Huang and Schaal 2012). Here this trend was most apparent for three of the four main genetic groups distinguished in *O. latifolia*. For each of these groups the probable directions of range expansion from the putative glacial refugia in which they are likely to have differentiated could be inferred from decreasing diversity trends when moving away from the putative refugia. Similar findings have been documented for other neotropical species (Thomas et al. 2012; Thomas et al. 2015).

Within clusters, gene flow seemed to occur at relatively short distances in all species, based on our finding of high G_{ST} values and strong isolation by distance patterns (cf. Schaal et al. 1998) in all but one of the clusters, even those with a relatively narrow geographic distribution (notably *O. alta* and *O. grandiglumis*). These observations are in line with the findings of (Veasey et al. 2011) and may be due to the reduced probability of long-distance pollination events resulting from the combination of the species′ predominantly autogamous reproduction systems and short gene flow distances expected from findings for other species in the *Oryza* genus (see below). Also the subgroups detected by Bayesian cluster analysis for clusters 3 and 4 of *O. latifolia* (Fig. 2d) might be explained by the same gene flow dynamics whereby sequential populations along the almost linear gradients of range expansion have been engaging in processes of genetic differentiation ever since their establishment. Further research is necessary to either confirm or refute this hypothesis.

Potential current and future niche overlap and risk of alien gene transfer

Suitability modeling under current and future climate conditions suggests that there is a strong spatiotemporal overlap between suitable areas of cultivated rice (whether rainfed or irrigated), and its four CWRs. Although niche suitability is not necessarily followed by cultivation practice, this outcome is indicative of the potential scale of gene flow between GM rice and its CWRs. Measures of gene flow frequencies are lower than 1% between cultivated rice and below 3% between cultivated and wild rice (Chen et al. 2004). The chance of actual transfer of GM genes to wild populations is highest for *O. glumaepatula* which has an AA genome type and hence is genetically compatible with *O. sativa* (Naredo et al. 1997; Naredo et al. 1998). Hybridization between *O. sativa* and other *Oryza* AA genome types is feasible and has been evidenced in several studies (Song et al. 2003; Chen et al. 2004; Wang et al. 2006), including *O. glumaepatula* (Jena 2010). However, it does not seem to occur very frequently under natural conditions. Usually, this type of interspecific hybridization depends on an extensive embryo rescue and retro-crossing efforts in order to obtain fertile hybrids (Vaughan et al. 2003). Gene flow between cultivated rice and the tetraploid CWRs is even less probable. Crosses between AA and CCDD genomes are feasible (Brar and Singh 2011), but are very difficult to obtain, even under artificial conditions (Mathias Lorieux, pers. comm.).

In addition to genetic compatibility issues, some additional barriers to gene flow prevail. Importantly, pollination in cultivated rice is mainly autogamous, due to flower morphology and the short-term viability of pollen (<30 min) (Oka 1988; Song et al. 2001), thereby reducing the probability of gene flow and hybridization events. Furthermore, the bulk of cross-pollination between adjacent plants or fields of cultivated rice usually does not seem to exceed a few meters (at the 1.0% threshold (Messeguer et al. 2001; Rong et al. 2004; Rong et al. 2005)), although under certain circumstances pollen-mediated gene flow might potentially occur at distances of tens to hundreds of meters (Song et al. 2002; Song et al. 2003; Yao et al. 2008; Kanya et al. 2009). Consequently, spatial buffer zones ranging from a few meters to more than 250 m wide have been proposed to avoid alien gene flow from genetically modified to conventional rice or its wild relatives (Rong et al. 2007; Kanya et al. 2009). While theoretically feasible for cultivated rice, this is much more difficult for rice CWRs, for which dispersal is conditioned by natural processes. Wild species can appear spontaneously along the margins, or even inside GM rice fields, and hence be exposed to alien gene flow, largely beyond human control.

Effective gene flow between GM rice and wild species is scarcely documented, possibly owing to the fact that GM rice is not yet commercially cultivated worldwide. The introgression of alien alleles in the sink population genome will ultimately depend on the manifestation, or not, of adaptive advantages. Alleles that grant some evolutionary advantage (reproduction and survival) may overcome almost any reproductive barrier, and may eventually become fixed in all exposed gene pools (Piálek and Barton 1997; Morjan and Rieseberg 2004). On the other hand, if an allele or gene does not improve the reproduction or survival of the plant, the probability that they get fixed is very low, and if they do succeed they will most likely remain as rare alleles. While the introgression of GM genes in rice CWRs might at best result in a competitive advantage for the respective sink populations, the impact on the ecological system they are part of may be less favorable and in the worst case lead to cascading effects. To date, few studies have been conducted on the scale, dynamics and potential detrimental environmental effects of gene flow between GM rice and its wild relatives under actual field conditions. Further research is clearly needed to properly assess or anticipate possible risks of alien gene flow in areas where GM rice and its CWRs overlap as is highlighted by the recent findings of Pu et al. (2014). In a large scale study across China, they found that several hundreds of insect species engage in pollen-mediated gene flow, some of which, like the European honey bee *Apis mellifera*, carry pollen at least 500 m from the pollen source. Furthermore, they showed that *A. mellifera*-mediated transgene flow from genetically modified to conventional varieties of cultivated rice was much higher than when the bees did not participate in gene flow.

Future research on gene transfer between GM rice and rice CWR in Colombia and beyond should therefore focus on identifying potential insect pollinators, and understanding CWRs' phenology, gene flow distances, seeds and pollen longevity, and the adaptive advantage of any transgenic traits introduced. For example, our hypothesis that the *O. glumaepatula* populations we sampled may have a mixed mating system characterized by a significant degree of outcrossing (implying that these populations may be more susceptible to introgression of alien genes), requires exhaustive field testing to allow developing adequate *in situ* conservation strategies. In the absence of detailed knowledge on the actual probabilities of gene flow between GM rice and its wild relatives, including the tetraploid ones, a precautionary approach may be appropriate.

In situ and *ex situ* conservation

The potential uses of CWR genes for improvement of cultivated rice is well documented and include: elongation ability, source of cytoplasmic male sterility, tolerance to heat (*O. glumaepatula*); resistance to brown plant hopper and bacterial blight (*O. latifolia*); resistance to striped stem-borer (*O. alta*); and high biomass production (*O. alta, O. latifolia, O. grandiglumis*) (Jena 2010; Brar and Singh 2011; Sanchez et al. 2013). Our finding that all species, except *O. latifolia*, are expected to predominantly experience range expansion and almost no contraction under climate change (Fig. 4), may indicate they contain useful genes permitting this adaptedness and plasticity (e.g. heat tolerance in *O. glumaepatula*), which may in turn have potential for introgression in cultivated rice, for which less optimistic results were obtained (Additional file 6*: Figure S6). Similar results were obtained in a study of ten Mesoamerican crop genepools and their CWRs (Thomas et al. 2016).

The use of CWR genes for rice improvement is contingent on the effective conservation and availability of their genetic resources, either through the conservation of viable populations under *in situ* conditions, or through the establishment and maintenance of *ex situ* collections. The fact that nearly all of the areas where the rice CWRs are known to occur in Colombia are located outside of protected areas and are also suitable for (GM) rice cultivation, complicates the possibility of *in situ* conservation in areas exempt from risks of alien gene flow. The current *ex situ* conservation status of Colombian rice CWR is similarly problematic. In spite of their high breeding potential and distinctive genetic makeup, Colombian rice CWRs are very poorly conserved in international and national gene banks. At the time of this research, the International Rice Research Institute (IRRI), held only two *O. latifolia* and one *O. glumaepatula* accessions from Colombia, while the International Center for Tropical Agriculture (CIAT) whose international gene bank is located in Cali, Colombia, conserved none. To our knowledge, national gene bank collections for any of the four CWRs do not exist in Colombia.

In light of the strong overlap between present and future suitable areas of cultivated rice and its four CWRs, the poor coverage of populations in protected areas, and the complexity of managing and regulating areas which are exempted from the possibility of alien gene flow (e.g. through a ban on cultivation of GM rice), a first priority should be to establish representative *ex situ* collections for all wild species. Priority populations of rice CWRs for the collection of accessions for *ex situ* conservation and the development of *in situ* conservation strategies are a combination of populations that hold the highest levels of genetic diversity (alpha diversity) and are complementary in terms of adaptive traits and their degree of genetic differentiation (beta diversity). In practical terms this would mean *in situ* conservation strategies should focus on those areas holding the highest genetic diversity for each of the genetic clusters. Additionally, the feasibility of establishing valuable populations, through assisted migration or assisted colonization, in areas holding favorable habitat conditions under current and/or future habitat for CWR, but not cultivated rice could be explored. According to our models such areas are likely to exist for all species.

However, it is clear that our genetic characterization data are not complete and other highly diverse and genetically differentiated populations may exist that were not included in our sample but nonetheless deserve attention for conservation. Based on the results obtained here, identification of areas where CWRs are either known or likely to occur (through botanical collections and suitability mapping, respectively) and that overlap with, or are spatially close to, areas that may have held suitable habitat conditions during the LGM, could, in the absence of genetic characterization data, be a useful first step to prioritize areas deserving further attention. Areas that have acted as glacial refugia are likely to contain higher levels of unique genetic diversity compared to expansion areas within species' ranges (Thomas et al. 2012; Thomas et al. 2015; Thomas et al. 2016). While this hypothesis is corroborated most notably by our observations for *O. latifolia*, for all species there are multiple areas that correspond to these characteristics and hence should be targeted for future collection and characterization missions. Logically, the development and implementation of in situ conservation plans should be preceded by genetic (and phenotypic) characterization and the collection of representative germplasm samples for ex situ conservation. This will allow establishing a baseline of the genetic diversity and structure of the CWR populations under consideration to which observations from future monitoring efforts can be compared.

Methods

Plant materials

Between 2009 and 2012, leaf tissue of a total of 291 individuals was collected across the four Colombian rice CWRs: 175 belonged to *O. latifolia*, 41 to *O. grandiglumis*, 50 to *O. alta*, and 25 to *O. glumaepatula*. Field sampling was guided by suitability maps based on occurrence data obtained from botanical collection records. No collection permits were required since the Alexander von Humboldt Biological Resources Research Institute, ascribed to the Ministry of Environment and Sustainable Development, is exempted from this requirement (Presidential Decree 302 of 2003). Taxonomic identification of the different species was carried out based on a combination of morphological and molecular methods. The ploidy of the species was identified through RFLP (Restriction Fragment Length Polymorphism) analysis and comparison with reference samples of known taxonomic classification, grown from seeds obtained from the International Rice Research Institute (IRRI). Identification based on morphological traits was based on Vaughan (Vaughan 1989; Vaughan 2003) and discussed in Villafañe et al. (2007). Photographs of morphological traits distinguishing the species with CCDD genome are presented in Additional file 11: Figure S7.

To allow evaluating the genetic differentiation of Colombian samples compared with accessions from the same and other species collected elsewhere in the world, we included 40 reference samples: 23 from the *O. officinalis* complex comprising the allotetraploid species *O. alta*, *O. grandiglumis* and *O. latifolia*, and 17 from the *O. sativa* complex with representatives from the diploid species *O. rufipogon*, *O. sativa*, *O. bartthii*, *O. glaberrima*, and *O. glumaepatula* (Fig. 1; Additional file 12: Table S5).

DNA extraction and microsatellite markers

DNA extraction was carried out from foliar material using the DNeasy Plant Mini Kit (QIAGEN). Total DNA quality was verified through visualization by horizontal electrophoresis in 0.8% w/v agarose gels stained with SYBR Safe (Invitrogen). DNA concentration was determined through quantification using the BioPhotometer spectrophotometer (Eppendorf).

Molecular characterization of the samples was undertaken with 11 previously evaluated Simple Sequence Repeats (SSR) markers (McCouch et al. 2002) selected from the rice database (http://www.gramene.org/markers/microsat) (Additional file 13: Table S6). A recent meta-analysis of 127 species (Mittell et al. 2015) has shown that this marker number is sufficient for characterizing a population's molecular genetic variation at comparable markers. Polymerase chain reactions (PCR) were carried out in a PTC-100TM thermal cycler (MJ Research) and consisted of a first cycle of denaturation at 94 °C for 3 min; followed by 30 cycles of 94 °C for 30 s, 55 °C for 30 s, and 72 °C for 1 min; and a final extension cycle at 72 °C for 5 min. The reaction mixture with a final volume of 50 μl had 100 ng of DNA template, 1× buffer (Tris-HCl 100 mM, KCl 500 mM), 3.0 mM of $MgCl_2$, 0.2 mM of each dNTP, 0.3 μM of each primer, and 1 U of *Taq* polymerase. PCR products were visualized in 6% w/v polyacrylamide gels, 7.5 M of urea, stained with silver nitrate ($AgNO_3$) (Bassam et al. 1991). Allele sizes were determined with the Quantity One software version 4.6.3 (Bio-Rad) through comparison with the 10 bp standard weight marker (Invitrogen™).

Suitability modeling

We characterized the spatial distribution of favorable habitat conditions of cultivated rice and its four CWRs in Colombia, under different climatic conditions, by means of suitability mapping based on ensembles of modeling algorithms, implemented in the R package *BiodiversityR* (Kindt and Coe 2005). Habitat suitability of the CWRs during the LGM was modeled to obtain insights of the potential impact of the last glacial period on the current distribution of their genetic diversity. Habitat suitability under present and predicted future climate conditions was modeled for all species, to identify potential overlap zones between cultivated rice and its wild relatives, as well as to assess the expected impact of climate change on the *in situ* conservation status of the CWRs.

Both rainfed and irrigated rice cultivation is practiced in Colombia. Distribution data of both cultivation types were obtained from FEDEARROZ, the Colombian national rice federation (http://www.fedearroz.com.co). Suitability mapping was carried out separately for both rainfed and irrigated rice (236 and 349 presence cells, respectively), and posteriorly merged to obtain overall suitability maps of cultivated rice in Colombia. Presence data of the four CWRs collected during field sampling were complemented with records extracted from numerous sources (www.gbif.org; www.cwrdiversity.org; www.tropicos.org). Given the low number of presence points available for Colombia, we carried out model calibrations based on all species records available for Latin America, from Mexico to Argentina (Additional file 14: Figure S8). Record numbers per species are given in Additional file 15: Table S7.

We applied two different strategies for suitability modeling under past and future climate conditions. Model calibrations for projections to LGM climate conditions (only for CWR) were carried out at 2.5 arc minute resolution using only WorldClim climate layers (Hijmans et al. 2005) as explanatory variables. Model calibrations intended for projections to future climate scenarios (period 2040–2069; referred to as 2050s), were carried out at 30 arc sec resolution, using aside from climate layers also altitude, slope, aspect, terrain roughness, direction of water flow and soil type (FAO/IIASA/ISRIC/ISS-CAS/JRC 2012). Collinear explanatory variables were removed based on iterative calculations of variance inflation factors (VIF), retaining only variables with VIFs smaller than 5. The resulting sets of explanatory variables, as well as presence, background and absence points used for model calibrations are given in Additional file 15: Table S7.

Background points (an overall maximum of 10,000 and maximum one per grid cell) for LGM model calibrations were randomly selected from the area enclosed by a convex hull polygon constructed around all presence points and extended with a buffer corresponding to 10% of the polygon's largest axis. For future model calibrations we additionally limited the selection of background points to areas of the extended convex hull intersected by the vegetation units from Olson et al. (Olson et al. 2001) with at least one species presence.

Modeling algorithms considered in the ensembles were maximum entropy (MAXENT), boosted regression trees (BRT, including a stepwise implementation), random forests (RF), generalized linear models (GLM; including stepwise selection of explanatory variables), generalized additive models (GAM; including stepwise selection of explanatory variables), multivariate adaptive regression splines (MARS), regression trees (RT), artificial neural networks (ANN), flexible discriminant analysis (FDA), support vector machines (SVM), and the BIOCLIM algorithm. As spatial autocorrelation among species presence points is known to bias model evaluations based on cross-validation, presence data permitting, we evaluated the ability of all individual modeling algorithms to cope with spatial autocorrelation by calculating calibrated Area Under Curve (cAUC) values and comparing these with a geographical null model (Hijmans 2012). We compared the cAUCs of each of the distribution models with the cAUCs of the geographical null model resulting from ten iterations, by means of Mann-Whitney tests. Only models that gave cAUC values that were significantly higher than the null model were retained in the ensemble model used for projections. Each ensemble combination was constructed as the weighted average of its individual composing models, using the cAUC values as weights. In cases where it was not possible to calculate cAUC values (i.e. when less than 10 presences available for model testing), we used AUC in a similar manner as described above, but included all models in the ensemble that yielded AUC values equal or higher than the null model.

To assess habitat suitability under LGM climate conditions we carried out projections to two climate models (MIROC and CCSM; (Braconnot et al. 2007)). For characterizing future climate conditions, we used 30 downscaled climate models for the period 2040–2069 based on the Representative concentration pathway (RCP) 4.5 scenario of greenhouse gas emissions, prepared for the Fifth Assessment IPCC report (CMIP5) and obtained from the WorldClim website (http://www.worldclim.org/CMIP5v1). The RCP4.5 scenario assumes that greenhouse gas emissions will peak around 2040 and will then decline. It is expected that under scenario global surface temperature change for the end of the 21st century is more likely than not to exceed 2 °C (Thomson et al. 2011).

We limited model projections to areas where suitability scores were at least as high as the score of the least suitable presence cell. This is considered a liberal approach in distribution modeling, as the likeliness of false positives and overestimation of suitable areas increases at lower thresholds. Since our purpose here was to identify areas where each of the species has a realistic chance of surviving and reproducing, either naturally (rice CWRs) or through human intervention (cultivated rice), potential overestimations are justified. In fact, the use of the highest possible threshold that includes all presence locations in the area identified as suitable might still be too conservative as there is no certainty that our presence data include representative samples at the extremes of each species' fundamental niche. To obtain summarizing maps for the two LGM climate models we averaged the threshold-limited suitability maps constructed for both individual climate scenarios.

Future suitability maps (period 2040–2069) were limited to areas that were identified as suitable by at least half of all 30 possible threshold-limited climate projections. To identify areas where suitable habitat of cultivated rice (irrigated and rainfed) can realistically spatially overlap with that of its four wild relatives in Colombia, we overlaid maps of the combined current and future habitat suitability for cultivated rice and each of the wild relatives.

Statistical analysis and genetic diversity mapping

For the allotetraploid species, diversity parameters H_E (expected heterozygosity) and G_{ST} (proportion of genetic diversity attributed to the differentiation among the populations) were estimated with ATetra v1.2 (Van Puyvelde et al. 2010). Due to the large number of partial heterozygotes present in the species analyzed (which significantly increases the number of iterative substitutions to generate the tetraploid genotypes), we carried out 10,000 Monte Carlo simulations for obtaining approximate diversity indices. The concept of partial heterozygote refers to the situation where for a given locus three alleles are observed and one of these (undetermined) is homozygous. For example, an ABC microsatellite pattern for one locus could represent a genotype of AABC, ABBC, or ABCC. For the diploid species, *O. glumaepatula*, the number of alleles per locus (A) and the estimators of H_E, observed heterozygosity (H_O), overall fixation index (F_{IT}), fixation index or genetic differentiation of subpopulations (F_{ST}), and inbreeding coefficient (F_{IS}), were calculated with FSTAT v2.9.3 (Goudet 2001). Correlations between genetic and geographical distance matrices were assessed by means of Mantel tests with 10,000 permutations, using the R (R Development Core Team 2011) *ade4* package (Dray and Dufour 2007).

We carried out Principal Coordinates Analyses (PCoA) for visualizing genetic differentiation between and within species samples, using the R package *Polysat* (Clark and Jasieniuk 2011). Genetic distance matrices for both diploid and tetraploid species were based on the Bruvo distance (Bruvo et al. 2004) which is applicable to organisms of any ploidy level. The Bruvo distance is similar to band-sharing indices used with dominant data, but takes into account mutational distances between alleles, to address the fact that allele copy numbers are frequently unknown in polyploid microsatellite data (Clark and Jasieniuk 2011). PCoA of the diploid *O. glumaepatula* showed the same structure using either the Nei (not shown) or the Bruvo distance. We identified different genetic groups for each of the species by submitting the Bruvo distance matrices to hierarchical cluster analysis with bootstrap resampling (10,000 replicas) to provide indications of branch support, implemented in the R package *pvclust* (Suzuki and Shimodaira 2015). We tested different clustering methods and selected the method generating the highest cophenetic correlation coefficient, being the Unweighted Pair Group Method with Arithmetic Mean in all cases (cophenetic coefficients >0.92). Only clusters with branch support higher than 95% were considered. We additionally used STRUCTURE version 2.3.4 (Pritchard et al. 2000) to probabilistically assign individuals to genetic clusters (K) in each of the rice CWR, under admixture model assumptions and without consideration of sampling localities. For *O. alta*, *O. grandiglumis* and *O. glumaepatula* the number of groups (K) tested varied from 1 to 6, and for *O. latifolia* from 1 to 13. We used burnin periods of one million steps followed by ten million additional MCMC repetitions. For each value of K, we carried out 10 independent iterations. We used the method of Evanno et al. (2005) for detection of the most probable number of genetically homogeneous clusters (K), through use of the STRUCTURE HARVESTER software v0.6.94 (Dent and VonHoldt 2011) as implemented on the Web server http://taylor0.biology.ucla.edu/structureHarvester.

To visualize geographical patterns in SRR diversity of all species, we carried out grid-based spatial analyses at a resolution of 30 arc sec (~1 km at the equator). To obtain sufficient and more evenly distributed data points for constructing high resolution maps, we constructed circular neighborhoods of 5 arc minutes diameter (~10 km at the equator), following (Thomas et al. 2012). For each species we calculated values of allelic richness and Shannon diversity of individual raster cells based on all individuals included in the circular area of 5 arc minutes diameter. Calculation of parameters for tetraploid and diploid species was carried out with functions from the *Polysat* package and custom-made code for R (Thomas et al. 2012), respectively. The circular neighborhood diameter was chosen to allow comparison of the genetic profile of adjacent populations and hence detect spatial patterns at landscape level across the species distributions. All maps were edited in ArcMap v10 (ESRI).

Additional files

Additional file 1: Table S1. Genetic diversity estimators for the tetraploid species *O. alta*, *O. grandiglumis* and *O. latifolia*.

Additional file 2: Table S2. Genetic diversity estimators for the diploid species *O. glumaepatlula*.

Additional file 3: Figure S1. Hierarchical cluster dendrograms based on the Bruvo and Nei (the latter for *O. glumaepatula* only) genetic distances and the Unweighted Pair Group Method with Arithmetic Mean. To types of p-values are provided to quantify branch support: the AU (Approximately Unbiased) p-value and BP (Bootstrap Probability) value. According to Suzuki and Shimodaira (2015) *"AU p-value, which is computed by multiscale bootstrap resampling, is a better approximation to unbiased p-value than BP value computed by normal bootstrap resampling"*. Clusters with branch support >95% are highlighted in boxes with different colors.

Additional file 4: Figure S2. ΔK values of all rice wild relatives for different values of K tested, following Evanno et al. (2005), implemented in the STRUCTURE HARVESTER software (Dent and VonHoldt 2011). (TIF 1302 kb)

Additional file 5: Figure S3. Habitat suitability of *O. alta* during the Last Glacial Maximum (~21,000BP), compared with the distribution of genetic clusters and Shannon diversity index of sampled populations.

Additional file 6: Figure S4. Habitat suitability of *O. glumaepatula* during the Last Glacial Maximum (~21,000BP), compared with the distribution of genetic clusters and Shannon diversity index of sampled populations.

Additional file 7: Figure S5. Habitat suitability of *O. grandiglumis* during the Last Glacial Maximum (~21,000BP), compared with the distribution of genetic clusters and Shannon diversity index of sampled populations.

Additional file 8: Table S3. Results of Mantel tests for all rice CWR between genetic and geographical distances for all individuals sampled and individuals per genetic cluster.

Additional file 9: Figure S6. Predicted changes in habitat suitability from present to the 2050s of cultivated rice *Oryza sativa*, considering both rainfed and irrigated rice.

Additional file 10: Table S4. Metrics obtained from genetic diversity studies of *O. glumaepatula* in Latin America.

Additional file 11: Figure S7. Details of panicles (top), ligules and auricles (bottom) of (A) *Oryza latifolia* with flower size smaller than 7 mm; (B) *Oryza grandiglumis* with approximately equal-sized sterile and fertile lemmas; and (C) *Oryza alta* with flower size larger than 7 mm (photo credits: Oscar Andrés Estrada).

Additional file 12: Table S5. List of analyzed *Oryza* spp. samples with genetic characterization data.

Additional file 13: Table S6. PCR primer sequences, allele size ranges and chromosome position on the Nipponbare genome of the SSR loci used in this study.

Additional file 14: Figure S8. Distribution of species records used in suitability model calibrations.

Additional file 15: Table S7. Suitability model statistics and metrics.

Acknowledgments

The authors wish to thank the Ministry of the Environment and Sustainable Development and the Alexander von Humboldt Biological Resources Research Institute for their support and cooperation to the Biosafety Line in Colombia. Thanks are due to agronomist-botanist Francisco Castro for his great cooperation during field expeditions and to Diana Carolina López of the Laboratory of Molecular Biology and Plant Tissue Bank of the Humboldt Institute for her assistance in the technical conduct of the study and conservation of the samples. We are grateful to the biodiversity group at CIAT for their valuable collaboration and to Dr. Zaida Lentini for literature suggestions and to the communities of the municipality of Puerto Carreño, and in general to all those who, in one way or another, contributed to this study. We are thankful to the editor and two anonymous reviewers whose comments and suggestions considerably helped improving the manuscript, and to Vincent Johnson for proofreading a final version.

Authors' contributions

CV, JLB and RM designed the study. CV and JLB carried out field work and lab analyses Evert Thomas and Eduardo Tovar analyzed the data and led the writing. All authors revised and approved the paper.

Competing interests

The authors declare that they have no competing interests.

Author details

[1]Bioversity International, Lima, Peru. [2]The Alexander von Humboldt Biological Resources Research Institute, Laboratory of Conservation Genetics, Bogota, Colombia. [3]Ministry of Environment and Sustainable Development, Genetic Resources Group, Bogota, Colombia. [4]The Alexander von Humboldt Biological Resources Research Institute, International Affairs, Policy and Cooperation Office, Bogota, Colombia.

References

Aggarwal RK, Brar DS, Huang N, Khush GS (1996) Differentiation within CCDD genome species in the genus Oryza as revealed by total genomic hybridization and RFLP analysis. Rice Genet Newsl 13:54–57

Aggarwal RK, Brar DS, Khush GS (1997) Two new genomes in the Oryza complex identified on the basis of molecular divergence analysis using total genomic DNA hybridization. Mol Gen Genet 254:1–12. doi:10.1007/s004380050384

Aggarwal RK, Brar DS, Nandi S et al (1999) Phylogenetic relationships among Oryza species revealed by AFLP markers. Theor Appl Genet 98:1320–1328. doi:10.1007/s001220051198

Akimoto M, Shimamoto Y, Morishima H (1998) Population genetic structure of wild Oryza glumaepatula distributed in the Amazon flood area influenced by its life history traits. Mol Ecol 7:1371–1381

Arrieta-Espinoza G, Sánchez E, Vargas S et al (2005) The weedy rice complex in Costa Rica. I. Morphological Study of relationships between commercial rice varieties, wild Oryza relatives and weedy types. Genet Resour Crop Evol 52: 575–587. doi:10.1007/s10722-004-6109-x

Atwell BJ, Wang H, Scafaro AP (2014) Could abiotic stress tolerance in wild relatives of rice be used to improve Oryza sativa? Plant Sci 215–216:48–58. doi:10.1016/j.plantsci.2013.10.007

Bao Y, Ge S (2003) Identification of Oryza species with the CD genome based on the RFLP analysis of nuclear ribosomal ITS sequences. Acta Bot Sin 45:762–765

Bassam B, Caetano-Anollés G, Gresshoff P (1991) Fast and sensitive silver staining of DNA in polyacrylamide gels. Anal Biochem 196:80–83

Braconnot P, Otto-bliesner B, Harrison S et al (2007) Results of PMIP2 coupled simulations of the mid-holocene and last glacial maximum – part 2: feedbacks with emphasis on the location of the ITCZ and mid- and high latitudes heat budget. Clim Past 3:279–296

Brar D, Singh K (2011) Oryza. In: Kole C (ed) Wild crop relatives: Genomic and breeding resources: cereals. Springer, Heidelberg, Dordrecht, London, New York, pp 321–336

Brondani RPV, Zucchi MI, Brondani C et al (2005) Genetic structure of wild rice Oryza glumaepatula populations in three Brazilian biomes using microsatellite markers. Genetica 125:115–123

Bruvo R, Michiels NK, D'souza TG, Schulen-Burg H (2004) A simple method for the calculation of microsatellite genotype distances irrespective of ploidy level. Mol Ecol 13:2101–2106

Buso GSC, Rangel PH, Ferreira ME (1998) Analysis of genetic variability of South American wild rice populations (Oryza glumaepatula) with isozymes and RAPD markers. Mol Ecol 7:107–117

Castañeda-Álvarez NP, Khoury CK, Achicanoy HA, et al. (2016) Global conservation priorities for crop wild relatives. Nat Plants 16022. doi: 10.1038/nplants.2016.22

CERA (Center for Environmental Risk Assessment) (2012) GM Crop Database. ILSI Research Foundation, Washington D.C. http://cera-gmc.org/GMCropDatabase . Accessed 20 June 2015

Chen LJ, Lee DS, Song ZP et al (2004) Gene flow from cultivated rice (Oryza sativa) to its weedy and wild relatives. Ann Bot 93:67–73. doi:10.1093/aob/mch006

Chen J, Kallman T, Ma X et al (2012) Disentangling the roles of history and local selection in shaping clinal variation of allele frequencies and gene expression in Norway spruce (Picea abies). Genetics 191:865–881

Clark LV, Jasieniuk M (2011) Polysat: an R package for polyploid microsatellite analysis. Mol Ecol Resour 11:562–566

Dempewolf H, Eastwood RJ, Guarino L et al (2014) Adapting agriculture to climate change: a global initiative to collect, conserve, and use crop wild relatives. Agroecol Sustain Food Syst 38:369–377, http://dx.doi.org/10.1080/21683565.2013.870629

Dent EA, VonHoldt BM (2011) STRUCTURE HARVESTER: a website and program for visualizing STRUCTURE output and implementing the Evanno method. Conserv Genet Resour 4:359–361. doi:10.1007/s12686-011-9548-7

Dray S, Dufour AB (2007) The ade4 package: implementing the duality diagram for ecologists. J Stat Softw 22:1–20

Ellstrand NC (2003) Dangerous liaisons? When cultivated plants mate with their wild relatives. John Hopkins University Press, Baltimore

Evanno G, Regnaut S, Goudet J (2005) Detecting the number of clusters of individuals using the software STRUCTURE: a simulation study. Mol Ecol 14: 2611–2620. doi:10.1111/j.1365-294X.2005.02553.x

FAO/IIASA/ISRIC/ISS-CAS/JRC (2012) Harmonized world soil database (version 1.2). FAO/IIASA, Rome, Luxemburg

Federici MT, Shcherban AB, Capdevielle F, et al (2002) Analysis of genetic diversity in the Oryza officinalis complex. Electron J Biotechnol 5:0–0. doi: 10.2225/vol5-issue2-fulltext-9

Fuller DQ, Sato YI, Castillo C et al (2010) Consilience of genetics and archaeobotany in the entangled history of rice. Archaeol Anthropol Sci 2: 115–131. doi:10.1007/s12520-010-0035-y

Galluzzi G, Dufour D, Thomas E et al (2015) An Integrated Hypothesis on the Domestication of Bactris gasipaes. PLoS One 10, e0144644. doi:10.1371/journal.pone.0144644

Gao L, Schaal BA, Zhang C et al (2002) Assessment of population genetic structure in common wild rice Oryza rufipogon Griff. using microsatellite and allozyme markers. Theor Appl Genet 106:173–180. doi:10.1007/s00122-002-1027-9

Garcia MA, Altieri MA (2005) Transgenic Crops: Implications for Biodiversity and Sustainable Agriculture. Bull Sci Technol Soc 25:335–353. doi:10.1177/0270467605277293

Ge S, Sang T, Lu B-R, Hong D-Y (1999) Phylogeny of rice genomes with emphasis on origins of allotetraploid species. Proc Natl Acad Sci 96:14400–14405. doi:10.1073/pnas.96.25.14400

Goudet J (2001) FSTAT, a program to estimate and test gene diversities and fixation indices (version 2.9.3). Available at: https://www.unil.ch/dee/en/home/menuinst/softwares–dataset/softwares/fstat.html

Hajjar R, Hodgkin T (2007) The use of wild relatives in crop improvement: a survey of developments over the last 20 years. Euphytica 156:1–13. doi:10.1007/s10681-007-9363-0

Hedrick PW (2005) A standardized genetic differentiation measure. Evolution 59: 1633–1638. doi:10.1111/j.0014-3820.2005.tb01814.x

Hijmans RJ (2012) Cross-validation of species distribution models: removing spatial sorting bias and calibration with a null model. Ecology 93:679–688

Hijmans RJ, Cameron SE, Parra JL et al (2005) Very high resolution interpolated climate surfaces for global land areas. Int J Climatol 25:1965–1978. doi:10.1002/joc.1276

Huang P, Schaal B a (2012) Association between the geographic distribution during the last glacial maximum of Asian wild rice, Oryza rufipogon (Poaceae), and its current genetic variation. Am J Bot 99:1866–1874. doi:10.3732/ajb.1200348

Jena KK (2010) The species of the genus Oryza and transfer of useful genes from wild species into cultivated rice, O. sativa. Breed Sci 60:518–523. doi:10.1270/jsbbs.60.518

Joshi SP, Gupta VS, Aggarwal RK et al (2000) Genetic diversity and phylogenetic relationship as revealed by inter simple sequence repeat (ISSR) polymorphism in the genus Oryza. Theor Appl Genet 100:1311–1320. doi:10.1007/s001220051440

Jost L (2008) GST and its relatives do not measure differentiation. Mol Ecol 17: 4015–4026. doi:10.1111/j.1365-294X.2008.03887.x

Kanya JI, Kinyamario JI, Amugune NO, Hauser TP (2009) Dispersal distance of rice (Oryza Sativa L.) pollen at the Tana River delta in the coast province, Kenya. African Journal Biotechnol 8:2265–2270

Karasawa MMG, Vencovsky R, Silva CM et al (2007a) Genetic structure of Brazilian wild rice (Oryza glumaepatula Steud., Poaceae) populations analyzed using microsatellite markers. Genet Mol Biol 30:400–410. doi:10.1590/S1415-47572007000300017

Karasawa MMG, Vencovsky R, Silva CM et al (2007b) Mating system of Brazilian Oryza glumaepatula populations studied with microsatellite markers. Ann Bot 99:245–253. doi:10.1093/aob/mcl248

Karasawa MMG, Vencovsky R, Silva CM et al (2012) Comparison of microsatellites and isozymes in genetic diversity studies of Oryza glumaepatula (Poaceae) populations. Rev Biol Trop 60:1463–1478

Khoury CK, Castañeda-Alvarez NP, Achicanoy H a et al (2015) Crop wild relatives of pigeonpea [Cajanus cajan (L.) Millsp.]: distributions, ex situ conservation status, and potential genetic resources for abiotic stress tolerance. Biol Conserv 184:259–270. doi:10.1016/j.biocon.2015.01.032

Kindt R, Coe R (2005) A manual and software for common statistical methods for ecological biodiversity studies. World Agroforestry Centre (ICRAF), Nairobi

Knowles LL, Richards CL (2005) Importance of genetic drift during Pleistocene divergence as revealed by analyses of genomic variation. Mol Ecol 14:4023–4032

McCouch S, Teytelman L, Xu Y et al (2002) Development and mapping of 2240 new SSR markers for rice (Oryza sativa L.). DNA Res 9(6):199–207, CrossRef. DNA Res 9:199–207

Meirmans PG, Hedrick PW (2011) Assessing population structure: FST and related measures. Mol Ecol Resour 11:5–18. doi:10.1111/j.1755-0998.2010.02927.x

Messeguer J, Fogher C, Guiderdoni E et al (2001) Field assessment of gene flow from transgenic to cultivated rices (Oryza sativa L.) using a herbicide resistance genes as tracer marker. Theor Appl Genet 103:1151–1159

Mittell EA, Nakagawa S, Hadfield JD (2015) Are molecular markers useful predictors of adaptive potential? Ecol Lett 18:n/a-n/a. doi: 10.1111/ele.12454

Morjan CL, Rieseberg LH (2004) How species evolve collectively: implications of

gene flow and selection for the spread of advantageous alleles. Mol Ecol 13: 1341–1356. doi:10.1111/j.1365-294X.2004.02164.x
Naredo M, Juliano A, Lu B, Jackson M (1997) Hybridization of AA genome rice species from Asia and Australia, I: crosses and development of hybrids. Genet Resour Crop Evol 44:17–23. doi:10.1023/A:1008670923628
Naredo M, Juliano A, Lu B, Jackson M (1998) Taxonomic status of Oryza glumaepatula Steud, II: hybridization between new world diploids and AA genome species from Asia and Australia. Genet Resour Crop Evol 45:205–214
Nei M (1973) Analysis of gene diversity in subdivided populations. Proc Natl Acad Sci 70:3321–3323
Niroula RK, Pucciariello C, Ho VT et al (2012) SUB1A-dependent and -independent mechanisms are involved in the flooding tolerance of wild rice species. Plant J 72:282–293. doi:10.1111/j.1365-313X.2012.05078.x
Oka HI (1988) Origin of cultivated rice. Elsevier, Amsterdam
Olson DM, Dinerstein E, Wikramanaya ED et al (2001) Terrestrial ecoregions of the world: a new map of life on earth. Bioscience 51:933–938
Pak JH, Oh JS, Kim HJ et al (2013) Development of near-isogenic transgenic rice lines harboring wild rice (Oryza grandiglumis)-derived fungal resistance gene (OgPR1). Plant Breed Biotechnol 1:122–130. doi:10.9787/PBB.2013.1.2.122
Piálek J, Barton NH (1997) The spread of an advantageous allele across a barrier: the effects of random drift and selection against heterozygotes. Genetics 145:493–504
Pritchard JK, Stephens M, Donnelly P (2000) Inference of population structure using multilocus genotype data. Genetics 155:945–959. doi:10.1111/j.1471-8286.2007.01758.x
Pu D, Shi M, Wu Q et al (2014) Flower-visiting insects and their potential impact on transgene flow in rice. J Appl Ecol 51:1357–1365. doi:10.1111/1365-2664.12299
Quesada T, Lobo J, Espinoza AM (2002) Isozyme diversity and analysis of the mating system of the wild rice Oryza latifolia Desv. in Costa Rica. Genet Resour Crop Evol 49:633–643. doi:10.1023/A:1021240924383
R Development Core Team (2011) R: A language and environment for statistical computing. version 2.14. R Foundation for Statistical Computing, Vienna
Rong J, Xia H, Zhu Y et al (2004) Asymmetric gene flow between traditional and hybrid rice varieties (Oryza sativa) indicated by nuclear simple sequence repeats and implications for germplasm conservation. New Phytol 163:439–445. doi:10.1111/j.1469-8137.2004.01100.x
Rong J, Song Z, Su J et al (2005) Low frequency of transgene flow from Bt/CpTI rice to its non-transgenic counterparts planted at close spacing. New Phytol 168:559–566. doi:10.1111/j.1469-8137.2005.01539.x
Rong J, Lu BR, Song Z et al (2007) Dramatic reduction of crop-to-crop gene flow within a short distance from transgenic rice fields. New Phytol 173:346–353. doi:10.1111/j.1469-8137.2006.01906.x
Sanchez PL, Wing RA, Brar DS (2013) The wild relative of rice: genomes and genomics. In: Zhang Q, Wing RA (eds) Genetics and genomics of rice, plant gene. Springer, New York, pp 616–626
Schaal BA, Hayworth DA, Olsen KM et al (1998) Phylogeographic studies in plants: problems and prospects. Mol Ecol 7:465–474. doi:10.1046/j.1365-294x.1998.00318.x
Snow A, Andow D, Gepts P et al (2005) Genetically engineered organisms and the environment: Current status and recommendations. Ecol Appl 15:377–404. doi:10.1890/04-0539
Song Z, Lu B, Chen J (2001) A study of pollen viability and longevity in Oryza rufipogon, O. sativa and their hybrid. Int Rice Res Notes 26:31–32
Song Z, Lu B, Zhu Y, Chen J (2002) Pollen competition between cultivated and wild rice species (Oryza sativa and O. rufipogon). New Phytol 153:289–296. doi:10.1046/j.0028-646X.2001.00319.x
Song ZP, Lu B-R, Zhu YG, Chen JK (2003) Gene flow from cultivated rice to the wild species Oryza rufipogon under experimental field conditions. New Phytol 157:657–665. doi:10.1046/j.1469-8137.2003.00699.x
Stewart CN, Halfhill MD, Warwick SI (2003) Transgene introgression from genetically modified crops to their wild relatives. Nat Rev Genet 4:806–817. doi:10.1038/nrg1179
Sun GL, Diaz O, Salomon B, Von Bothmer R (1998) Microsatellite variation and its comparison with allozyme and RAPD variation in Elymus fibrosus (Schrenk) Tzvel. (Poaceae). Hereditas 129:275–282. doi:10.1111/j.1601-5223.1998.00275.x
Suzuki R, Shimodaira H (2015) pvclust: hierarchical clustering with p-values via multiscale bootstrap resampling. R package version 2.0-0. http://CRAN.R-project.org/package=pvclust
Thomas E, van Zonneveld M, Loo J et al (2012) Present spatial diversity patterns of theobroma cacao L. in the neotropics reflect genetic differentiation in pleistocene refugia followed by human-influenced dispersal. PLoS One 7: e47676. doi:10.1371/journal.pone.0047676
Thomas E, Alcazar Caicedo C, McMichael CH et al (2015) Uncovering spatial patterns in the natural and human history of Brazil nut (Bertholletia excelsa) across the Amazon basin. J Biogeogr 42:1367–1382
Thomas E, Ramirez M, Van Zonneveld M et al (2016) An assessment of the conservation status of Mesoamerican crop species and their wild relatives in light of climate change. In: Maxted N, Dulloo E, Ford-LLoyd BV (eds) Enhancing crop genepool utilization: capturing wild relative and landrace diversity for crop improvement. CAB International, UK, pp 248–270
Thomas E, Gil Tobón C, Gutierrez JP, Alcazar Caicedo C, Moscoso Higuita LG, Becerra LA, Loo J, Gonzales MA (2017) Genetic diversity of Enterolobium cyclocarpum in Colombian seasonally dry tropical forest: implications for conservation and restoration. Biodivers Conserv 26:825–842
Thomson AM, Calvin KV, Smith SJ et al (2011) RCP4.5: a pathway for stabilization of radiative forcing by 2100. Clim Change 109:77–94. doi:10.1007/s10584-011-0151-4
Van Puyvelde K, Van Geert A, Triest L (2010) ATETRA, a new software program to analyse tetraploid microsatellite data: comparison with TETRA and TETRASAT. Mol Ecol Resour 10:331–334
Vaughan DA (1989) The genus Oryza L. Current status of taxonomy. IRRI Research Paper Series No. 138, Manila
Vaughan DA (2003) Revised key to species in genus Oryza - appendix 1. In: Nanda JS, Sharma SD (eds) Monograph on genus Oryza. Science Publishers Inc., Enfield New Hampshire, pp 347–351
Vaughan D, Morishima H, Kadowaki K (2003) Diversity in the Oryza genus. Curr Opin Plant Biol 6:139–146
Veasey EA, Cardin DC, Silva RM et al (2008a) Assessing the genetic structure of Oryza glumaepatula populations with isozyme markers. Brazilian Arch Biol Technol 51:873–882
Veasey EA, Da Silva EF, Schammass EA et al (2008b) Morphoagronomic genetic diversity in American wild rice species. Brazilian Arch Biol Technol 51:95–104. doi:10.1590/S1516-89132008000100012
Veasey EA, Bressan EDA, Zucchi MI et al (2011) Genetic diversity of American wild rice species. Sci Agric 68:440–446. doi:10.1590/S0103-90162011000400008
Villafañe C, Estrada OA, Lentini Z et al (2007) Diagnóstico y fortalecimiento de la línea base del conocimiento del género Oryza (arroz) en Colombia, un aporte para la toma de decisiones en el ámbito de bioseguridad. In: Hodson de Jaramillo E, Carrizosa P (eds) Desarrollo de capacidades para evaluación y gestión de riesgos y monitoreo de organismos genéticamente modificados (OGM). Tomo I. Resultados de proyectos específicos. Instituto de Investigación de Recursos Biológicos Alexander von Humboldt, Bogotá, D.C, pp 33–49
Vincent H, Wiersema J, Kell S et al (2013) A prioritized crop wild relative inventory to help underpin global food security. Biol Conserv 167:265–275. doi:10.1016/j.biocon.2013.08.011
Waltari E, Hijmans RJ, Peterson T et al (2007) Locating pleistocene refugia: comparing phylogeographic and ecological niche model predictions. PLoS One 2, e563. doi:10.1371/journal.pone.0000563
Wang ZY, Second G, Tanksley SD (1992) Polymorphism and phylogenetic relationships among species in the genus Oryza as determined by analysis of nuclear RFLPs. Theor Appl Genet 83:565–581. doi:10.1007/BF00226900
Wang F, Yuan Q-H, Shi L et al (2006) A large-scale field study of transgene flow from cultivated rice (Oryza sativa) to common wild rice (O. rufipogon) and barnyard grass (Echinochloa crusgalli). Plant Biotechnol J 4:667–676. doi:10.1111/j.1467-7652.2006.00210.x
Yao K, Hu N, Chen W et al (2008) Establishment of a rice transgene flow model for predicting maximum distances of gene flow in southern China. New Phytol 180:217–228. doi:10.1111/j.1469-8137.2008.02555.x
Zamora A, Barboza C, Lobo J, Espinoza AM (2003) Diversity of native rice (Oryza Poaceae:) species of Costa Rica. Genet Resour Crop Evol 50:855–870. doi:10.1023/A:1025998801637

7

Identification of QTLs for Yield and Agronomic Traits in Rice under Stagnant Flooding Conditions

Anshuman Singh[1,4], Jerome Carandang[4], Zennia Jean C. Gonzaga[4], Bertrand C. Y. Collard[2,4], Abdelbagi M. Ismail[4] and Endang M. Septiningsih[3,4*]

Abstract

Background: Stagnant flooding, where water of 25–50 cm remains until harvest time, is a major problem in rainfed lowland areas. Most of the Sub1 varieties, which can withstand around 2 weeks of complete submergence, perform poorly in these conditions. Hence, varieties tolerant of stagnant flooding are essential.

Results: This paper presents the first study to map QTLs associated with tolerance to stagnant flooding, along with a parallel study under normal irrigation, using an F_7 mapping population consisting of 148 RILs derived from a cross of Ciherang-Sub1 and the stagnant-flooding tolerant line IR10F365. Phenotypic data was collected for 15 key traits under both environments. Additionally, survival rate was measured under stress conditions. Genotyping was performed using the Illumina Infinium genotyping platform with a 6 K SNP chip, resulting in 469 polymorphic SNPs. Under stress and irrigated conditions, 38 and 46 QTLs were identified, respectively. Clusters of QTLs were detected in both stress and normal conditions, especially on chromosomes 3 and 5.

Conclusions: Unique and common QTLs were identified and their physiological consequences are discussed. These beneficial QTLs can be used as targets for molecular breeding and can be further investigated to understand the underlying molecular mechanisms involved in stagnant flooding tolerance in rice.

Keywords: Irrigated rice, *Oryza sativa*, QTL, Rainfed rice, Stagnant flooding, Sub1 varieties

Background

Most of rainfed lowlands and some irrigated areas in South and Southeast Asia are prone to flooding which causes huge crop losses worldwide. This situation may worsen under future climate change scenarios (Redfern et al. 2012). Floods are largely caused by incessant rainfall during the monsoon season from mid-June to early October, often causing complete submergence followed by stagnant flooding in areas with poor drainage, resulting in huge financial losses (Singh et al. 2011). Rice has developed adaptive mechanisms to survive under a wide range of water regimes; however, extreme weather leading to too much or too little water may cause serious yield reductions or total crop loss.

Different types of flooding can reduce rice production. Flash flooding is the most common, which occurs when rice plants are completely submerged for up to 2 weeks during the vegetative stage. If submergence prevails for more than 5 days, susceptible rice plants start to die and recovery is hampered (Mackill et al. 2010). Fortunately, a number of Sub1 varieties have been developed and released as remedies for this type of flooding, as the *Sub1* gene enables the submerged plants to become dormant and conserve resources for a rapid recovery after the floods recede. The first generation of Sub1 lines were the first six varieties developed by the International Rice Research Institute (IRRI), including Swarna-Sub1, IR64-Sub1, Samba Mahsuri-Sub1, BR11-Sub1, TDK1-Sub1, and CR1009-Sub1. While the second generation of Sub1 lines developed by IRRI were Ciherang-Sub1 and PSB Rc18-Sub1 in which IR64-Sub1 was used as the donor for the *SUB1* gene (Iftekharuddaula et al. 2011;

* Correspondence: eseptiningsih@tamu.edu
[3]Present address: Department of Soil and Crop Sciences, Texas A&M University, College Station, TX 77843, USA
[4]International Rice Research Institute, DAPO 7777, Metro Manila, Philippines

2016; Ismail et al. 2013; Neeraja et al. 2007; Septiningsih et al. 2009, 2015). Another type of flooding stress is flooding during germination, also known as anaerobic germination (AG). This can be due to unlevelled fields in direct-seeded rainfed or irrigated environments, especially in flood affected areas, when rain falls just after seeding causing poor seedling establishment (Ismail et al. 2009). Several major QTLs for tolerance to AG have been identified (Angaji et al. 2010; Baltazar et al. 2014; Septiningsih et al. 2013b) and a few lines with tolerance to AG alone or combined with *SUB1* have been developed through molecular and conventional breeding (Kretzschmar et al. 2015; Septiningsih et al. 2013a; Toledo et al. 2015). Some rainfed areas are also affected by deep water flooding where water stagnation is more than 1 m. In this situation, deepwater adapted varieties with fast internode elongation are required to escape from water while their shoot tips and tillers remain above the water (Catling 1992; Hattori et al. 2009). QTLs and genes have been identified and some tolerant lines have been developed for this type of flooding (Hattori et al. 2008, 2009).

The last type of flooding stress, which is the focus of this study, is stagnant flooding (SF), where water of 25 to 50 cm depth stagnates in the field for several weeks to few months (Mackill et al. 2010; Singh et al. 2011). In this stress environment, varieties with facultative elongation are needed; however, deepwater rice performs poorly under stagnant flooding conditions mainly due to lodging and the consequent reductions in yield and grain quality (Kato et al. 2014; Vergara et al. 2014). At the same time, most irrigated rice varieties also perform poorly under these conditions, as water depth greater than 25 cm greatly inhibits the growth of rice. The detrimental effect of SF can be seen as reduced number of tillers, poor fertility, smaller panicles and excessive lodging; together leading to poor grain yield and quality (Kato et al. 2014; Singh et al. 2014). The situation worsens when floods result in complete submergence followed by stagnant flooding until harvest time (Singh et al. 2011). Therefore, it is essential to develop varieties having tolerance of complete submergence during vegetative stage and stagnant flooding in the same genetic background. Farmers mostly use traditional low-yielding landraces in areas that are prone to both stresses within the same season. These landraces have partial tolerance of complete submergence and can resist stagnant flooding stress by stem and leaf elongation, but they have low yield of around 0.5–1.5 t ha^{-1} and low quality (Singh et al. 2011). The introgression of *SUB1* QTL in several mega varieties successfully improved submergence tolerance with a yield advantage of 2–3.5 t ha^{-1} in farmers' fields (Singh et al. 2009). However, these varieties perform poorly under stagnant flooding conditions, especially since most of the Sub1 varieties are relatively short, making the plants more prone to damage if water depth stays up to canopy level for longer than 2 weeks (Singh et al. 2011).

Conventional breeding to develop improved varieties with stagnant flooding has long been conducted at IRRI, and some tolerant lines have been developed (Collard et al. 2013; Mackill et al. 2010). Thus far, however, studies on genetic and molecular basis of stagnant flooding had not been initiated. In this study we report for the first time, the identification of QTLs under stagnant flooding conditions from an F_7 RIL population derived from a cross of the improved version of the popular high-yielding Indonesian variety, Ciherang-Sub1 (IR09F436) (Septiningsih et al. 2015) and an IRRI submergence and stagnant flooding tolerant breeding line, IR10F365 (Collard et al. 2013) using an Illumina 6 K SNP chip platform (Thomson 2014). A subset of this RIL population was recently used in our study to identify non-Sub1 QTLs for tolerance to submergence during vegetative stage (Gonzaga et al. 2017). In the current study, 15 traits related to yield and important agronomic traits were investigated in both stagnant flooding and normal irrigated field conditions. The parallel study under control irrigated conditions is also valuable since the RIL population was developed using a popular cultivar (Ciherang-Sub1) and an elite line (IR10F365) which are both high yielding lines. Additionally, it allows detecting QTLs associated with responses induced by stagnant flooding versus inherent responses, with the former being most important for breeding varieties adapted to stagnant flooding conditions.

Methods

Plant material and development of the mapping population

A RIL mapping population was developed comprising of 156 F_7 lines derived from a cross of Ciherang-Sub1 with IR10F365. Both parents carry the tolerant allele of the *SUB1* gene (i.e. Sub1 gene was fixed in this population), which contributes to tolerance to submergence during vegetative stage. A subset of 115 RILs from this population was used to map non-Sub1 QTLs for submergence tolerance during vegetative stage (Gonzaga et al. 2017); however, in this current study the entire population of 156 RILs was used. During the final analysis, however, 8 lines were removed due to high levels of missing genotype or phenotype data, resulting in a final set of 148 RILs. Other than the two parents, Swarna-Sub1 and IRRI154 were also included as susceptible and tolerant checks, respectively. IRRI154 (also known as NSIC Rc222) is a popular high-yielding irrigated rice variety released in the Philippines. This variety is highly tolerant to stagnant flooding but susceptible to complete submergence during vegetative stage. Swarna-Sub1 is the first submergence tolerant variety released in South Asia (Neeraja et al. 2007).

Experimental field conditions

The mapping population was evaluated in 2014 wet season (WS) in the field plots at the International Rice Research Institute (IRRI), Los Baños, Laguna, Philippines (14°10'11.81"N, 121°15'39.22"E) under stagnant flooding stress and normal irrigated conditions. A row-column design was used with two replications, with planting distance of 20 × 20 cm in individual plots of 5 × 0.8 m under stress condition. The same population was planted under controlled irrigation using the same experimental design and planting conditions following standard agronomic practices. Besides the two parents, IRRI154 and Swarna-Sub1 were also included as checks. Fertilizer was applied at three stages as recommended at basal, maximum tillering and panicle initiation. Seeds were sown in raised seed bed nursery and 16 d-old seedlings were transplanted in the submergence plot with one seedling per hill. The water in the SF plot was raised to 10 cm at 2 weeks after transplanting (WAT) and was gradually increased by 5 cm each week until the 6th WAT. Thereafter, water was raised gradually by 10 cm until the 8th WAT, for a final depth of 50 cm. During this period plants were at maximum tillering stage, and this condition was maintained up to harvesting time. Water depth in the control field was kept at about 5 cm through maturity.

Phenotyping and data analysis

Sixteen traits selected based on previous physiological studies of stagnant flooding tolerance or related studies were investigated (Hattori et al. 2007; Hattori et al. 2008; Kato et al. 2014; Nagai et al. 2014; Singh et al. 2011; Vergara et al. 2014). Those traits were: days to flowering (DTF), plant height (PH), shoot elongation rate (SER), number of tillers (TN), number of panicles (PN), 100 grain weight (GW), shoot biomass (BM), flag leaf length (FLL), flag leaf width (FLW), panicle length (PL), harvest index (HI), leaf sheath length for the first, second and third internodes (LSL_1, LSL_2, and LSL_3, respectively), grain yield per plot (GY), and survival rate (SR). The same traits were measured under irrigated condition for direct comparisons, except SR. Details of phenotyping are provided in Additional file 1: Table S1. Data were analyzed using PB Tools 1.4 and STAR (http://bbi.irri.org/products). Analysis of variance (ANOVA) and correlation analysis were performed. Means were estimated using linear mixed models and broad-sense heritability (H^2) was calculated using PBTools.

Genotyping

Total genomic DNA was extracted using a modified CTAB technique (Murray and Thompson 1980), and the final concentration was normalized to 50 ng/μL. Genotyping was performed using the Illumina Infinium genotyping platform with a 6 K SNP chip designed by Susan McCouch at Cornell University and run at the Genotyping Services Laboratory, IRRI (Thomson 2014), as described previously (Gonzaga et al. 2017).

QTL analysis and linkage map construction

Map Manager QTX, vQTXb20 (Manly et al. 2001) was used for linkage map construction with Kosambi map function. Due to high rates of missing data in 8 lines, only 148 RILs were used in the final analysis. QTL analysis was performed using QTL Cartographer v2.5 (Wang et al. 2010) with 1,000 permutations in both interval mapping (IM) and composite interval mapping (CIM) methods to determine the threshold at $P \leq 0.05$ and $P \leq 0.01$ to declare the significance of QTLs. Forward-backward stepwise regression of F-in = 0.01 and F-out = 0.01 was used in CIM. For comparison, QGene (Nelson 1997) was also used based on IM and CIM with permutation of 10,000 iterations. Standard rice QTL nomenclature was used (McCouch 2008).

QTL comparison

QTLs identified in this study were compared with similar QTLs identified in previous studies using the Gramene QTL database (http://archive.gramene.org/qtl/) and QTL Annotation Rice Online (Q-TARO) database (http://qtaro.abr.affrc.go.jp/) (Yonemaru et al. 2010). QTLs for the same traits sharing similar regions between the irrigated and stagnant flooding environments and clusters of QTLs governing various traits located in similar regions are also reported.

Results and Discussion

Performance of parents and checks

SUB1A suppresses shoot elongation during complete submergence to limit carbohydrate consumption and increase chances of survival after water recedes; but this can lead to a disadvantage if submergence is followed by a longer duration of stagnant flooding (Sarkar et al. 2009; Singh et al. 2009). It was also reported that the first six Sub1 lines developed by IRRI (Septiningsih et al. 2009) were susceptible to stagnant flooding (Vergara et al. 2014). The susceptibility of Sub1 lines was worse with varieties having short stature, as in the case of Swarna-Sub1—which was used as a susceptible check in our stagnant flooding stress phenotyping, showing very low survival by the end of the experiment. On the other hand, the tolerant check IRRI154 performed well, confirming our previous results (Kato et al. 2014). There is no significant difference of survival rates among Ciherang-Sub1 (69%), IRRI154 (62%), and IR10F365 (56%) (Table 1). The grain yield under stagnant flooding was also higher for IRRI154 (3,394 kg ha^{-1}) and Ciherang-Sub1 (2,933 kg ha^{-1}), but

Table 1 Descriptive statistics of each trait in the mapping population, parents, and checks under stagnant flooding

Traits[a)]	Ciherang-Sub1	IR10F365	IRRI154	Swarna-Sub1	Population Mean	Highest line	Lowest line	LSD .05	$H^{2b)}$	$P^{c)}$
DTF (d)	96	104	104	117	96	107	88	4.20	0.88	***
PH (cm)	122.8	118.2	123.9	57.0	129.3	154.7	94.6	19.00	0.62	***
TN	5	7	10	4	7	11	4	3.40	0.21	ns
PN	5	6	10	4	6	11	4	3.20	0.33	**
FLL (cm)	27.5	30.8	40.4	12.7	29.6	38.6	22.0	7.40	0.5	***
FLW (cm)	1.9	1.6	1.8	0.9	1.8	2.1	1.5	0.28	0.4	***
PL (cm)	25.2	27.6	25.8	12.2	25.5	29.7	20.0	4.26	0.45	***
BM (g/m^2)	768.5	853.7	1230.8	520.5	985.0	1544.7	513.2	527.80	0.18	ns
SER (cm/d)	1.2	1.2	1.4	0.6	1.3	1.8	0.8	0.28	0.73	***
HI	0.36	0.14	0.29	0.01	0.28	0.56	0.10	0.20	0.39	**
GW (g)	2.8	2.5	2.4	0.7	2.7	3.2	1.6	0.28	0.85	***
LSL_1 (cm)	12.6	13.8	11.0	3.2	14.2	24.4	9.2	4.40	0.45	***
LSL_2 (cm)	15.2	17.8	13.1	6.4	17.4	24.0	12.0	5.60	0.45	**
LSL_3 (cm)	29.2	21.7	29.7	13.4	27.9	35.4	17.8	7.40	0.43	***
GY (kg/Ha)	2933	1246	3394	154	2685	5184	49	1543.80	0.64	***
SR (%)	69	56	62	1	68	89	40	27.40	0.22	*

[a)]*DTF* days to flowering, *PH* plant height, *TN* number of tillers, *PN* number of panicles, *FLL* flag leaf length, *FLW* flag leaf width, *PL* panicle length, *BM* shoot biomass, *SER* shoot elongation rate, *HI* harvest index, *GW* 100 grain weight, *LSL_1, LSL_2, and LSL_3, respectively* leaf sheath length for the first, second and third internodes, *GY* grain yield per plot, and *SR* survival rate

[b)]Heritability

[c)]ns: non significant; *significant at $P \le 0.05$; **significant at $P \le 0.01$; ***significant at $P \le 0.001$

significantly lower for IR10F365 (1,246 kg ha^{-1}) and much lower for Swarna-Sub1 (154 kg ha^{-1}). The yield reduction of Ciherang-Sub1 under SF compared to the control was comparable to that of IRRI154 (49.5 vs. 51.3%). In addition, Ciherang-Sub1 matured 5 days earlier under non-stress and 8 days earlier under stress compared with IRRI154 and IR10F365 (Tables 1 and 2). Under non-stress conditions, Ciherang-Sub1 had similar numbers of tillers and panicles compared to IRRI154 and IR10F365; while Swarna-Sub1 had a higher number of tillers and panicles. However, under stress tiller number and panicles of both Ciherang-Sub1 and IR10F365 was significantly reduced compared to IRRI154. Likewise, Ciherang-Sub1 and IR10F365 biomass was significantly affected under the stress (reduced by 40.9 and 40.7%, respectively), while IRRI154 biomass was hardly affected and Swarna-Sub1 was severely affected (reduced by 71.1%). It was also observed that IRRI154 had the highest shoot elongation rate under stress among the four varieties used in this study. The trend was also demonstrated in our earlier finding in which IRRI154 had the least biomass reduction and the fastest elongation compared with the other four varieties used in the study, including another tolerant check, IRRI119 or PSB Rc68 (Kato et al. 2014).

The current study showed that even though Ciherang-Sub1 elongation was less than that of IRRI154 (4.0 vs. 9.8%) and its biomass reduction was more affected compared to IRRI154, Ciherang-Sub1 had better tolerance to stagnant flooding compared to the first generation of Sub1 lines, which were the first six Sub1 lines developed by IRRI (Vergara et al. 2014). The good performance of Ciherang-Sub1 could partly be the result of its intermediate height (118.1 cm), which was similar to IR10F365 (122.8 cm), slightly taller than IRRI154 (112.8 cm) and significantly taller than Swarna-Sub1 (106 cm) under control condition. While under stagnant flooding its height (122.8 cm) was similar to IRRI154 (123.9 cm) and IR10F365 (118.2 cm), but significantly taller than that of Swarna-Sub1 (57 cm). Shoot elongation rate of Ciherang-Sub1 (1.2 cm/d) was similar to IR10F365 (1.2 cm/d) and IRRI154 (1.4 cm/d), but significantly faster than that of Swarna-Sub1 (0.6 cm/d). As shown in our previous study, IRRI154 tolerance was mostly due to its medium elongation rate, which enabled its canopy to keep up with the water surface (Kato et al. 2014). In summary, the advantage of Ciherang-Sub1 under stagnant flooding is its partial elongation and inherent intermediate height.

IR10F365, a breeding line from IRRI's submergence breeding team, has consistently shown strong submergence and stagnant flooding tolerance in field trials. This line has also performed well in trials from the National Coordinated Trials of the Philippines in flood-prone areas. In our current study, however, IR10F365 slightly underperformed compared to Ciherang-Sub1 in both control and stagnant flooding. This might be due to the

Table 2 Descriptive statistics of each trait in the mapping population, parents, and checks under irrigated control condition

Traits[a)]	Ciherang-Sub1	IR10F365	IRRI154	Swarna-Sub1	Population Mean	Highest line	Lowest line	LSD .05	$H^{2b)}$	$P^{c)}$
DTF (d)	87	92	92	106	89	96	80	2.81	0.95	***
PH (cm)	118.1	122.8	112.8	109.0	125.3	139.3	99.92	8.32	0.88	***
TN	10	11	12	16	10	13	7	2.68	0.54	***
PN	10	11	11	16	10	18	7	3.41	0.32	*
FLL (cm)	29.5	38.0	36.4	26.3	32.0	43.09	23.91	6.57	0.61	***
FLW (cm)	2.1	1.8	1.7	1.9	1.9	2.35	1.54	0.31	0.53	***
PL (cm)	25.01	25.99	23.46	23.49	26.09	30.5	21.54	2.22	0.78	***
BM (g/m^2)	1301.33	1438.67	1262.24	1803.67	1329.34	1762.8	757.2	399	0.25	*
SER (cm/d)	1.4	1.1	1.3	1.0	1.3	1.8	0.9	0.31	0.55	***
HI	0.45	0.31	0.56	0.18	0.43	0.6	0.22	0.16	0.45	**
GW (g)	2.7	2.4	2.4	2.0	2.7	3.12	2.39	0.17	0.89	***
LSL_1 (cm)	9.8	14.0	11.9	11.0	14.1	23.96	5.65	4.73	0.55	***
LSL_2 (cm)	20.3	20.0	18.4	17.0	22.1	28.1	12.6	4.82	0.53	***
LSL_3 (cm)	35.1	30.9	31.1	27.3	32.7	41.16	25	5.08	0.74	***
GY (kg/Ha)	5809	4680	6974	3391	5605	7398	3600	1322.80	0.59	***

[a)]*DTF* days to flowering, *PH* plant height, *TN* number of tillers, *PN* number of panicles, *FLL* flag leaf length, *FLW* flag leaf width, *PL* panicle length, *BM* shoot biomass, *SER* shoot elongation rate, *HI* harvest index, *GW* 100 grain weight, LSL_1, LSL_2, *and* LSL_3, *respectively* leaf sheath length for the first, second and third internodes, and *GY* grain yield per plot
[b)]Heritability
[c)]ns: non significant; *significant at $P \leq 0.05$; **significant at $P \leq 0.01$; ***significant at $P \leq 0.001$

use of older seed stock of this particular line (due to unavailability of fresh seeds from the same batch) which may have caused a slower rate of germination and growth, which subsequently affected yield. For our mapping population, however, other than the parental lines we also included the susceptible check Swarna-Sub1 and the tolerant check IRRI154 which can also be used as negative and positive checks, respectively, when we investigate the performance of the selected individuals in the mapping population.

Mapping population performance under stagnant flooding vs. control conditions

The mean grain yield (GY) of the population decreased significantly, by 52.1% under SF (Tables 1 and 2). Kato et al. (2014) reported similar results where yield under stress was reduced by 47% across genotypes. The best performing line under SF (RIL-214) yielded 5,184 kg ha^{-1} with a yield advantage of 43.5% over Ciherang-Sub1 and 34.5% over the best tolerant check, IRRI154. The large yield increase under stress, however, was not reflected under irrigated conditions since the yield of the same was 5,676 kg ha^{-1} under control conditions compared to the most productive line (RIL-186) at 7,398 kg ha^{-1} (Additional file 2: Table S2). However, RIL-186 yield was only 2,584 kg ha^{-1} under SF, suggesting that this line is less tolerant. Interestingly, the sixth best line under SF was also at the same rank under control conditions, with grain yield of 4,067 and 6,681 kg ha^{-1}, respectively. Therefore, it is possible to develop high-yielding lines that thrive in both environments out of this population, even though they might not be the best performers in either environment. Likewise, the tolerant check IRRI154 performs well in both control and SF conditions, but the downside of this variety is its low grain quality and sensitivity to complete submergence. The worst performer under SF only yielded 921 kg ha^{-1}. Compared to this line, the best performer under stress was 24.0% taller, had more than double its biomass, with 44% faster rate of shoot elongation and had more than triple harvest index. It was worth noting that under SF, 26 RILs yielded higher than IRRI154 and 60 RILs yielded higher than Ciherang-Sub1. Under control conditions, only one line has higher yield than IRRI154, while 66 lines were better than Ciherang-Sub1 (data not shown). For 100 grain weight (GW), the mean of the population in both conditions was unchanged (Tables 1 and 2). The averages biomass (BM) and harvest index (HI) of the population were reduced by 25.9 and 34.9% under stress, respectively. Among these traits, heritability (H^2) was highest for grain weight in both stress and control conditions (0.85 vs. 0.89), while heritability for grain yield was moderate (0.64 vs. 0.59), and low for HI and BM.

The mean of plant height (PH) under SF similar is similar compared to the control (only 3.4% increased); while the mean shoot elongation rate (SER) stayed the same under both environments (Tables 1 and 2). Under SF, mean days to flowering (DTF) was 96 days, which was equal to that of Ciherang-Sub1 but earlier than

IR10F365 and IRRI154 by 8 days and Swarna-Sub1 by 21 days. The mean of DTF under control condition was 89 days, which was not significantly difference compared to those of Ciherang-Sub1, IR10F365 and IRRI154, but 17 days earlier than Swarna-Sub1. The overall mean delay caused by SF across the mapping population was 7 days. It was previously reported that stagnant flooding delays flowering.

Under SF, tillers and panicle number were reduced by 30 and 40%, respectively, compared to the control, which was similar to that of IR10F365 and Ciherang-Sub1 (Tables 1 and 2). Interestingly, the average panicle length of the mapping population, along with those of Ciherang-Sub1, IR10F365, and IRRI154 were hardly affected. Likewise, the population averages of flag leaf length (FLL) and flag leaf width (FLW) were only slightly reduced. Average leaf sheath length for the 1st internode (LSL_1) of the mapping population was less affected; however, average leaf sheath length for the 2nd and 3rd internodes (LSL_2 and LSL_3) of the population were reduced by 21.3 and 15.3%, respectively. This trend was similar to those of the parents and checks, with Swarna-Sub1 being the most affected. Overall, the population has high mean of survival rate under SF, ranging from 40 to 89%, with an average of 68%, which was similar to that of Ciherang-Sub1 (69%).

To investigate the effect of individual traits on yield stability across stagnant flooding and irrigated conditions, the percentage difference for each trait was calculated for the 25 highest and 25 lowest yielding lines under SF and the control (Additional file 3: Figure S1 and Additional file 4: Figure S2). Under SF, the two traits with biggest differences were GY and HI; this was followed by moderate differences in PN, BM, SER, SR, all LSL, TN, PH, and FLL. These traits largely contribute to higher HI and yield. Under control condition, the two largest contrasting traits were also GY and HI, even though the differences were smaller compared to those under stress. It was also shown that the lines with higher shoot elongation rate had higher survival rate but not all performed well. The lines having elongation rate between 1.2 and 1.6% were high yielding and had good survival rate of 60–80% (Additional file 5: Figure S3). This was likely due to high carbohydrate consumption in fast-elongating lines to avoid complete submergence, but at the expense of yield. However, moderate SER was strongly and positively correlated with grain yield under stress, as previously reported (Kato et al. 2014; Vergara et al. 2014).

Correlation among traits

Under stagnant flooding, grain yield of the mapping population correlated positively with all the traits, except DTF. Traits having very highly significant correlation (***; p-value ≤ 0.001) with GY were HI (0.70), PH (0.58), SER (0.44), BM (0.43), LSL_3 (0.43), FLW (0.40) and LSL_1 (0.40), LSL_2 (0.40), PN (0.36), PL (0.31), FLL (0.29), and GW (0.29); while TN (0.25) had highly significant correlation (**; p-value ≤ 0.01; Additional file 6: Table S3). Survival rate was correlated with all the traits, except HI. Traits having very highly significant correlation with SR were PH (0.61), BM (0.47), GW (0.47), LSL_2 (0.43), PL (0.42), FLW (0.41), LSL_1 (0.40), LSL_3 (0.39), SER (0.39), PN (0.38), and FLL (0.30); while traits having highly significant correlation were TN (0.25) and DTF (−0.24). All LSS traits were positively very highly significant correlation with PH and SER, but negatively very highly significant correlation with DTF (Additional file 6: Table S3). Under SF, all traits were correlated with plant height. Traits having very highly significant correlation with PH were LSL_2 (0.72), BM (0.67), LSL_1(0.63), SER (0.62), SR (0.61), GW (0.59), PL (0.59), GY (0.58), LSL_3 (0.55), FLW (0.54), FLL (0.37), and PN (0.28); while traits having highly significant correlation were TN (0.24) and DTF (−0.24), and trait having significant correlation (*; p-value ≤ 0.05) was HI (0.18). Additionally, DTF were negatively very highly significant correlation with SER (−0.56), LSL_{1-3} (−0.41, −0.41, −0.42, respectively), and FLL (−0.33); while it was negatively highly significant correlated with SR (−0.24) and PH (−0.24), and negatively significant correlated with HI (−0.21), FLW (−0.21) and TN (−0.16).

In general, the correlations among traits were weaker under control conditions than under SF. Under control conditions, the only trait that was very highly significant correlated (***) with GY was HI (0.71); it was also highly significantly correlated (**) with LSL_2 (0.25) and DTF (−0.25), and significantly correlated (*) with SER (0.17) (Additional file 7: Table S4). All LSL were very highly significantly correlated with PH. Under control conditions, DTF had either positive or negative correlation with some of the traits (Additional file 7: Table S4).

Identification of QTLs

Out of 4606 high quality SNP markers on the 6 K SNP chip, 10% (469) were polymorphic between Ciherang-Sub1 and IR10F365. The rice physical map of Nipponbare (MSU v.7) was used to order the markers. The marker distances were calculated from the genotype data using MapManager QTX vQTXb20 (Manly et al. 2001) and had a total length of 1,250.4 cM with an average of 2.74 cM between markers. QTL analysis using both QTL Cartographer and QGene identified a total of 38 and 46 QTLs under stagnant flooding and the control, respectively (Tables 3 and 4; Figs. 1 and 2). There were 16 QTLs detected in both environments; in addition several QTL clusters were observed. Under SF, QTLs were detected from 13 out of 16 traits (except PN, TN, and BM), while under control, QTLs were detected from all 15

Table 3 QTLs for yield in agronomic traits from a RIL population of Ciherang-Sub1/IR10F365 identified under stagnant flooding condition

Trait[a)]	QTL	Chr	Flanking markers	Source[b)]	QTL Cart. CIM			QTL Cart. IM[c)]			QGENE CIM			QGENE IM		
					LOD	R^2	Add	LOD	R^2	Add	LOD	R^2	Add	LOD	R^2	Add
GW	*q GW1.1*	1	214137-id1007975	C	**4.36**	**8.7**	**−0.06**	**4.21**	**12.3**	**−0.07**	**3.35**	**10.0**	**0.05**	**3.24**	**10.0**	**0.07**
	q GW2.1	2	id2004418-1845605	C				**2.70**	**8.0**	**−0.05**						
	q GW3.1	3	2499734-id3002805	C				**2.80**	**10.2**	**−0.06**						
	qGW5.1	5	5428382-ud5000983	C	**4.07**	**8.3**	**−0.06**	**3.80**	**11.0**	**−0.06**	**3.73**	**11.0**	**0.08**	**4.17**	**12.0**	**0.07**
	qGW10.1	10	10603169-10703329	I	**6.30**	**13.1**	**0.07**	**3.15**	**10.0**	**0.06**						
	qGW10.2	10	9958372-id10003608	C	**4.14**	**9.0**	**−0.06**				**3.41**	**10.0**	**−0.07**	*2.50*	*8.0*	*−0.06*
DTF	*qDTF1.1*	1	801364-id1016674	I	**3.08**	**6.0**	**0.88**	*2.78*	*8.2*	*1.03*	*3.40*	*10.0*	*−0.89*			
	qDTF3.1	3	2499734-2560888	C	**12.50**	**26.0**	**−1.80**	**6.81**	**20.0**	**−1.57**	**5.43**	**16.0**	**1.20**	**6.00**	**17.0**	**1.49**
	qDTF5.1	5	5515384-id5013231	C	**4.11**	**7.0**	**−1.01**	**3.17**	**9.3**	**−1.13**	**3.65**	**11.0**	**1.27**	**3.63**	**11.0**	**1.23**
	qDTF6.1	6	id6015097-6906738	I				**3.95**	**12.0**	**1.54**				**4.37**	**13.0**	**−1.73**
	qDTF10.1	10	id10002842-10586997	I	**5.22**	**9.0**	**1.07**	**3.01**	**9.0**	**1.04**	*2.68*	*8.0*	*−0.82*			
FLW	*qFLW3.1*	3	2499734-2560888	C	**4.00**	**11.3**	**−0.04**	**3.47**	**11.0**	**−0.04**				**3.22**	**10.0**	**0.04**
	qFLW7.1	7	7869914-7949610	C				**3.11**	**14.0**	**−0.04**						
FLL	*qFLL4.1*	4	*id4000574-4295290*	I	**4.11**	**9.3**	**1.10**	**2.78**	**8.2**	**1.03**	**3.79**	**11.1**	**−1.12**	**2.87**	**8.5**	**−1.08**
	qFLL5.1	5	ud5000983-5747652	I	**6.18**	**14.0**	**1.36**	**3.69**	**11.0**	**1.17**				**3.53**	**10.4**	**−1.18**
	qFLL6.1	6	6887046-6906738	C				*2.71*	*8.5*	*−1.28*				**2.98**	**8.9**	**1.42**
	qFLL9.1	9	9592671-id9007287	I	**2.94**	**13.0**	**1.23**				**2.99**	**8.9**	**−1.65**	*2.60*	*7.8*	*−1.85*
GY	*qGY3.1*	3	2499734-2560888	C	**6.20**	**14.3**	**−321.60**	*2.83*	*9.0*	*−246.10*	*2.64*	*8.0*	*224.3*	**2.84**	**8.4**	**250.46**
	qGY5.1	5	id5003312-ud5000983	I	**3.66**	**12.2**	**301.7**				*2.15*	*7.0*	*−190.7*			
	qGY6.1	6	id6000402-5903052	C	**3.46**	**7.4**	**−282.1**				**3.04**	**9.0**	**263.8**			
HI	*qHI2.1*	2	1489783-1845605	I	**3.41**	**72.4**	**0.03**				*2.18*	*6.6*	*−0.02*			
	qHI5.1	5	id5003312-5515384	I	**3.34**	**72**	**0.03**									
	qHI7.1	7	7768382-7949610	C	**8.67**	**20.4**	**−0.05**	**3.64**	**10.6**	**−0.03**	**7.07**	**19.7**	**0.05**	**3.80**	**11.1**	**0.03**
	qHI7.2	7	7102234-id7002749	I	**4.65**	**10**	**0.03**									
LSL_1	*$qLSL_15.1$*	5	ud5000983-5747652	I	**4.40**	**11.0**	**0.74**	**4.00**	**12.0**	**0.76**				**4.04**	**12.0**	**−0.79**
LSL_2	*$qLSL_25.1$*	5	ud5000983-5747652	I	**4.73**	**10.2**	**0.82**	**4.86**	**15.0**	**0.96**	*2.57*	*7.7*	*−0.82*	**4.96**	**14.3**	**−0.98**
	$qLSL_26.1$	6	6228054-6585321	I	**4.10**	**9.0**	**0.91**							**3.66**	**11.0**	**−1.38**
LSL_3	*$qLSL_32.1$*	2	1529216-1845605	C	**4.16**	**10.0**	**−1.06**	**2.99**	**9.0**	**−1.00**	*2.37*	*7.1*	*0.85*	**2.78**	**8.3**	**0.97**
	$q\ LSL_35.1$	5	5515384-5747652	I	**3.49**	**8.0**	**1.00**	*2.72*	*9.0*	*1.04*				**3.17**	**9.4**	**1.11**
SER	*qSER5.1*	5	ud5000983-id5013231	I	**12.5**	**32.0**	**0.11**	**8.63**	**24.0**	**0.09**	**4.12**	**12.0**	**−0.06**	**9.20**	**24.9**	**−0.10**
	qSER6.1	6	6619487-6816224	C	**3.05**	**6.0**	**−0.05**	*2.82*	*8.3*	*−0.06*	**3.79**	**11.1**	**0.05**	**2.87**	**8.5**	**0.06**
PH	*qPH3.1*	3	2499734-2560888	C	**4.04**	**8.3**	**−2.77**	**3.10**	**9.4**	**−2.88**	*3.34*	*10.0*	*2.67*	*2.57*	*8.00*	*2.69*
	qPH5.1	5	ud5000983-5747652	I	**5.59**	**12.1**	**3.45**	**4.73**	**14.0**	**3.65**				**4.97**	**14.3**	**−3.81**
	qPH6.1	6	6228054-6585321	I							**3.40**	**10.0**	**−3.74**			
PL	*qPL7.1*	7	id7004922-7949610	I	**3.21**	**11.2**	**0.61**									
	qPL9.1	9	9641863-9869869	I	**5.47**	**22.0**	**0.82**	**4.42**	**21.2**	**0.81**	*3.29*	*10.0*	*−0.53*	**3.64**	**11.0**	**−0.61**
SR	*qSR1.1*	1	20215-id1001821	I	**2.89**	**10.0**	**3.18**				*2.63*	*8.0*	*−2.77*			
	qSR6.1	6	id6008688-6766627	C	**2.97**	**7.3**	**−2.99**				*2.50*	*7.4*	*4.01*			

[a)] *DTF* days to flowering, *FLW* flag leaf width, *FLL* flag leaf length, *GY* grain yield per plot, *HI* harvest index, *LSL_1, LSL_2, and LSL_3, respectively* leaf sheath length for the first, second and third internodes, *SER* shoot elongation rate, *PH* plant height, *PL* panicle length, and *SR* survival rate

[b)] C: Ciherang-Sub1; I: IR10F365

c)underlined and bold numbers: significant at $P \leq 0.01$; bold numbers: significant at $P \leq 0.05$; italic: not significant but with LOD ≥ 2.0

Table 4 QTLs for yield in agronomic traits from a RIL population of Ciherang-Sub1/IR10F365 identified under irrigated control condition

Trait	QTL	Chr	Flanking markers	Source	QTL Cart. CIM			QTL Cart. IM			QGENE CIM			QGENE IM		
					LOD	R^2	Add	LOD	R^2	Add	LOD	R^2	Add	LOD	R^2	Add
GW	*qGW1.1*	1	214137-id1007975	C	**4.74**	**11.4**	**−0.05**	**5.34**	**15.2**	**−0.05**	**4.40**	**12.8**	**0.06**	**4.35**	**12.7**	**0.05**
	qGW2.1	2	1661393-1981019	C	**3.29**	**6.5**	**−0.04**	**4.21**	**12.2**	**−0.05**	**3.13**	**9.3**	**0.04**	**3.14**	**9.3**	**0.04**
	qGW2.1	2	2153125-2422788	C							**2.74**	**8.2**	**−0.04**			
	qGW5.1	5	5428382-ud5000983	C	*2.39*	*5.0*	*−0.03*	**3.60**	**10.5**	**−0.04**	*2.51*	*7.5*	*0.04*	**3.13**	**9.3**	**0.04**
DTF	*qDTF3.1*	3	2499734-2519460	C	**21.81**	**42.2**	**−2.44**	**13.53**	**34.9**	**−2.20**	**17.36**	**41.7**	**2.35**	**13.14**	**33.6**	**2.22**
	qDTF6.1	6	id6015421-6906738	I				**4.09**	**11.9**	**1.70**				**3.70**	**10.9**	**−1.70**
	qDTF10.1	10	id10003608-10579252	I	**5.70**	**9.0**	**1.13**	**2.99**	**8.8**	**1.11**	**5.43**	**15.5**	**−1.18**			
FLW	*qFLW3.1*	3	2499734-2519460	C	**5.18**	**11.0**	**−0.05**				**3.49**	**10.3**	**0.10**			
	qFLW4.1	4	4683923-4728960	C	**8.88**	**20.1**	**−0.09**	**3.73**	**10.9**	**−0.06**	**3.98**	**11.6**	**0.06**	**3.29**	**9.7**	**0.06**
	qFLW5.1	5	id5001564-id5003312	C	*2.67*	*5.4*	*−0.04*							**2.72**	**8.1**	**−0.05**
	qFLW12.1	12	c12p5738988-id12005832	I	**4.80**	**10.3**	**0.06**									
	qFLW12.2	12	12740505-12939027	C	**3.44**	**7.3**	**−0.05**									
FLL	*qFLL1.1*	1	id1016971-1305247	C	**3.81**	**25.2**	**−1.79**									
	qFLL2.1	2	1543608-1902547	I	**3.00**	**6.8**	**0.96**	*2.97*	*8.8*	*1.06*	**3.85**	**11.3**	**−1.19**	**3.54**	**10.4**	**−1.17**
	qFLL4.1	4	id4002348-4253547	I	**3.05**	**6.8**	**0.99**				**3.15**	**9.3**	**−1.15**			
	qFLL5.1	5	5719825-5747652	I	**3.90**	**9.1**	**1.12**									
GY	*qGY12.1*	12	12648838-12931835	I	**3.39**	**9.2**	**247.20**				*2.27*	*6.8*	*−206.30*			
HI	*qHI3.1*	3	2499734-2560888	I	**3.50**	**8.0**	**0.02**				**4.91**	**14.2**	**−0.03**	*2.25*	*6.7*	*−0.02*
	qHI7.1	7	id7002749-7730100	C							**3.50**	**10.3**	**0.05**			
LSL_1	$qLSL_{1}1.1$	1	275443-id1007975	C	**4.37**	**9.8**	**−0.78**	**3.18**	**9.4**	**−0.77**						
	$qLSL_{1}3.1$	3	fd9-2519460	C	**5.33**	**13.1**	**0.93**	**3.78**	**11.0**	**−0.82**	**3.95**	**11.6**	**0.83**	**3.23**	**9.6**	**0.76**
	$qLSL_{1}5.1$	5	ud5000983-5747652	I				**3.93**	**11.8**	**0.88**	**2.79**	**8.3**	**−0.86**	**3.80**	**11.1**	**0.88**
LSL_2	$qLSL_{2}3.1$	3	id3003846-3236543	C	**3.70**	**8.1**	**−0.67**	*2.59*	*7.8*	*−0.65*	**3.23**	**9.6**	**0.94**			
	$qLSL_{2}5.1$	5	5515384-id5013231	I	**4.95**	**11.2**	**0.84**	*2.77*	*9.4*	*0.76*				**3.03**	**9.0**	**−0.83**
	$qLSL_{2}9.1$	9	9592671-id9007287	I	**4.33**	**16.8**	**0.97**	*2.84*	*17.6*	*0.99*	**4.45**	**12.9**	**−1.59**	**3.29**	**9.7**	**−1.45**
LSL_3	$qLSL_{3}2.1$	2	1562092-id2007797	C				**3.92**	**11.4**	**−1.06**	**4.41**	**12.8**	**1.10**	**3.89**	**11.4**	**1.08**
	$qLSL_{3}3.1$	3	2573346-2610698	C	*2.63*	*4.1*	*−0.69*	**3.60**	**10.5**	**−1.03**	**4.40**	**12.8**	**1.24**	**3.18**	**9.4**	**1.03**
	$qLSL_{3}5.1$	5	ud5000983-5747652	I	**6.73**	**12.8**	**1.17**	**2.96**	**10.2**	**1.03**				**3.12**	**9.3**	**−1.02**
	$qLSL_{3}6.1$	6	6619487-6816224	C	**5.83**	**10.4**	**−1.10**									
	$qLSL_{3}8.1$	8	8803052-9049928	I	**7.41**	**14.7**	**1.29**	**4.18**	**22.1**	**1.49**	**6.24**	**17.6**	**−2.47**	**5.59**	**16.0**	**−2.38**
	$qLSL_{3}11.1$	11	id11007523-11608239	C	**5.22**	**9.3**	**−0.95**	**3.40**	**10.0**	**−0.97**	**4.03**	**11.8**	**1.19**	**3.69**	**10.9**	**1.05**
SER	*qSER5.1*	5	5515384-id5013231	I	**8.10**	**20.0**	**0.08**	**6.53**	**20.0**	**0.08**	**3.02**	**9.0**	**−0.06**	**7.08**	**19.8**	**0.09**
	qSER6.1	6	6585321-6816224	C	**3.70**	**8.0**	**−0.05**									
PH	*qPH1.1*	1	197232-id1007975	C				**3.13**	**10.2**	**−2.38**				*2.55*	*7.6*	*2.09*
	qPH3.1	3	2519460-id3002805	C	**3.15**	**7.0**	**−1.96**				**3.12**	**9.3**	**2.31**			
	qPH5.1	5	ud5000983-5747652	I	**9.90**	**23.9**	**3.71**	**8.22**	**22.4**	**3.57**	**3.98**	**11.7**	**−2.99**	**7.62**	**21.1**	**−3.58**
TN	*qTN3.1*	3	2519460-id3003846	I	**3.31**	**9.3**	**0.41**	**3.85**	**11.7**	**0.45**	**3.64**	**10.7**	**−0.49**	**4.29**	**12.5**	**−0.49**
PL	*qPL1.1*	1	id1015390-id1016971	C	**3.51**	**6.7**	**−0.44**									
	qPL2.1	2	1529216-1845605	I	**7.11**	**14.3**	**0.66**	**4.66**	**13.4**	**0.60**	**4.80**	**13.9**	**−0.62**	**4.59**	**13.3**	**−0.61**
	qPL6.1	6	6619487-6816224	C	**4.91**	**10.3**	**−0.55**	**5.92**	**18.4**	**−0.72**	**5.01**	**14.4**	**0.66**	**4.81**	**13.9**	**0.64**
	qPL7.1	7	id7003853-id7004922	I				*2.84*	*8.4*	*0.49*	**3.29**	**9.7**	**−0.56**	**3.62**	**10.7**	**−0.58**
	qPL9.1	9	9592671-9869869	I				**3.36**	**20.4**	**0.73**	**3.27**	**9.7**	**−0.83**	**2.92**	**8.7**	**−0.73**

Table 4 QTLs for yield in agronomic traits from a RIL population of Ciherang-Sub1/IR10F365 identified under irrigated control condition *(Continued)*

PN	*qPN1.1*	1	147907-395936	I				*2.55*	*7.6*	*0.39*	**3.49**	**10.3**	**−0.55**	**3.16**	**9.4**	**−0.55**
	qPN3.1	3	2573346-2610698	I				**2.88**	**8.9**	**0.43**	**2.74**	**8.2**	**−0.47**	**2.83**	**8.4**	**−0.44**
BM	*qBM1.1*	1	147907-275443	I	**3.92**	**10.2**	**51.73**	*2.44*	*72.0*	*43.16*	**3.68**	**10.8**	**−50.83**	*2.30*	*6.9*	*−42.4*
	qBM3.1	3	2499734-2560888	C	**2.89**	**7.0**	**−45.20**				**3.33**	**9.9**	**49.64**			

[a)]*GW* 100 grain weight, *DTF* days to flowering, *FLW* flag leaf width, *FLL* flag leaf length, *GY* grain yield per plot, *HI* harvest index, *LSL*$_1$, *LSL*$_2$, and *LSL*$_3$, respectively leaf sheath length for the first, second and third internodes, *SER* shoot elongation rate, *PH* plant height, *TN* number of tillers, *PL* panicle length, *PN* number of panicles, and *BM* shoot biomass

[b)]C: Ciherang-Sub1; I: IR10F365

c)underlined and bold numbers: significant at $P \leq 0.01$; bold numbers: significant at $P \leq 0.05$; italic: not significant but with LOD ≥ 2.0

traits. Most of the QTLs were in similar positions as previously reported QTLs (Tables 5 and 6).

QTLs for GY, GW, HI, and SR

Three QTLs were detected for grain yield under stagnant flooding, *qGY3.1*, *qGY5.1* and *qGY6.1*. (Table 3, Fig. 1). It is noteworthy that the largest QTL, *qGY3.1* with an LOD score of 6.2 and phenotypic variation (R^2) of 14.3% was detected on the top of chromosome 3S as a cluster with four other QTLs: *qDTF3.1*, which was detected in both environments, *qGW3.1*, *qFLW3.1*, and *qPH3.1*. Interestingly, the alleles that increase the effects

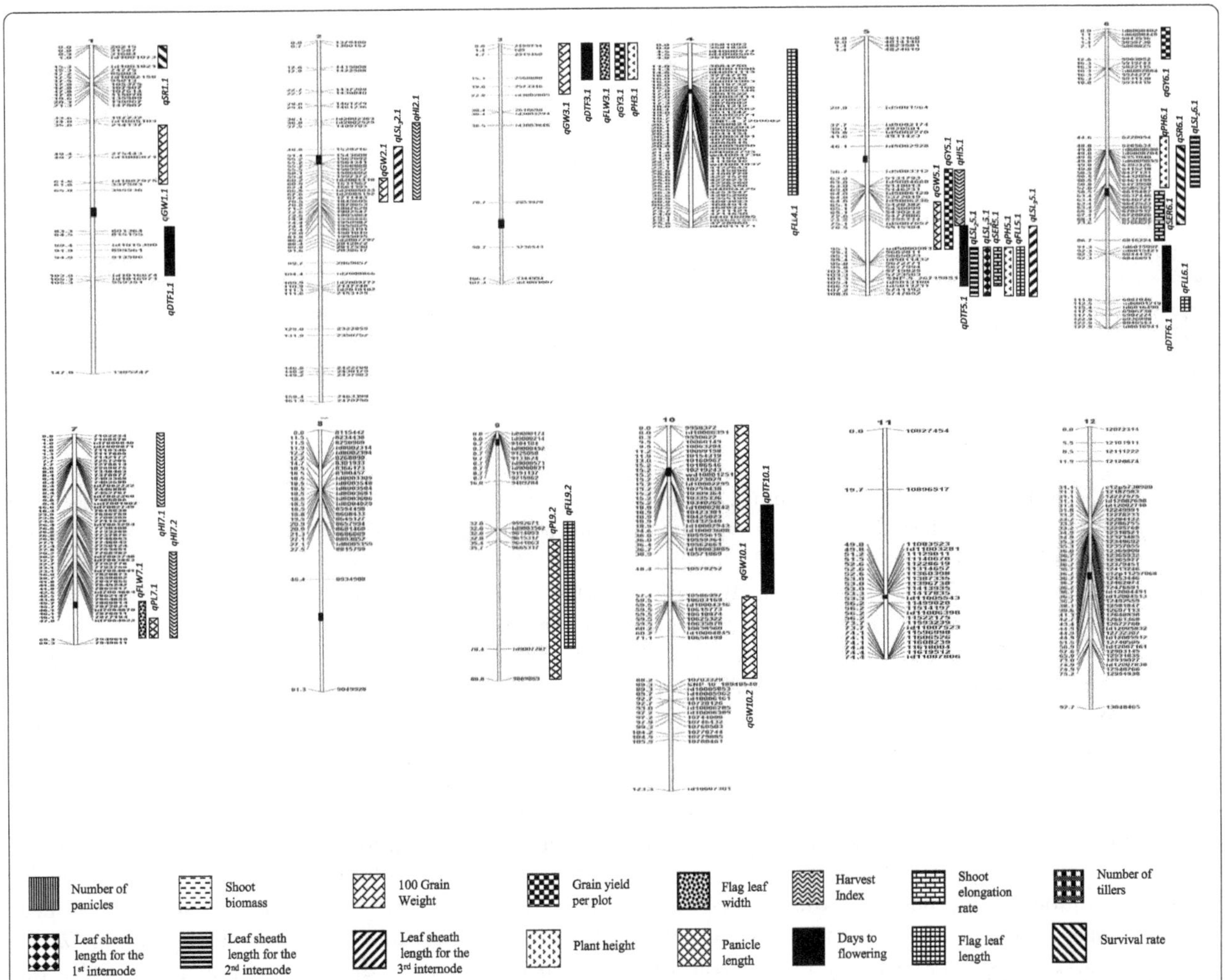

Fig. 1 Linkage map of a RIL population (F_7) derived from a cross between Ciherang-Sub1 and IR10F365 under stagnant flooding along with the positions of QTLs for 16 traits. QTLs identified are indicated above each bar, traits are indicated by the patterns of the bars. Centromeres are shown as *black* boxes in each chromosome

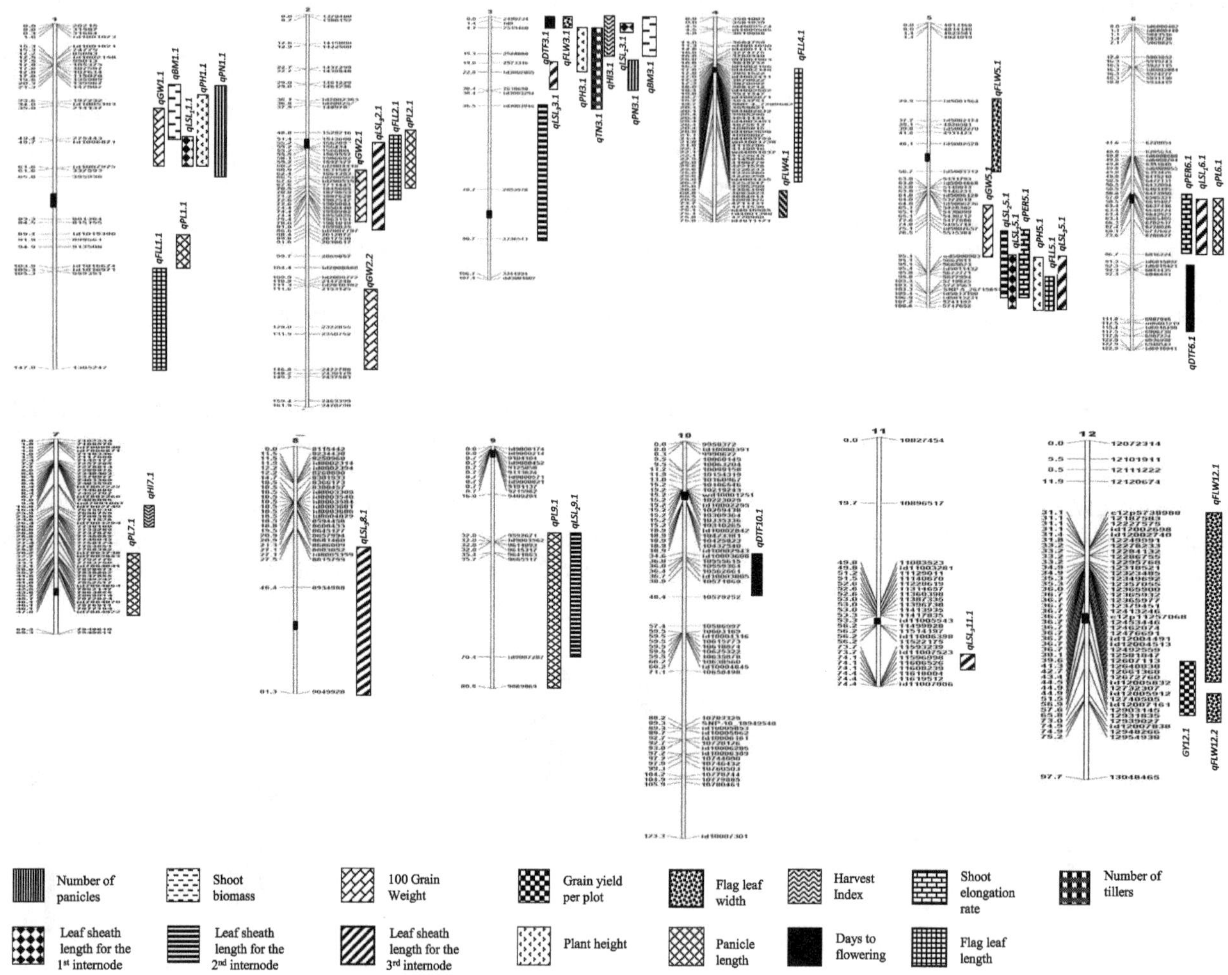

Fig. 2 Linkage map of a RIL population (F_7) population derived from a cross between Ciherang-Sub1 and IR10F365 under control irrigated condition along with the positions of QTLs for 15 traits. The QTL boundaries are indicated by the closest flanking markers, traits are indicated by the patterns of the bars. Centromeres are shown as *black* boxes in each chromosome

of all the QTLs in this cluster came from Ciherang-Sub1. *qGY3.1* was also detected in similar regions as previous reported by Xiao et al. (1996)) under irrigated condition, (Moncada et al. 2001) under upland cultivation, and (Bernier et al. 2007) under drought condition (Table 5). Another QTL, *qGY5.1* (LOD = 3.66, R^2 = 12.2%) was in a cluster with two other QTLs of different traits on the upper part of chromosome 5 L, *qGW5.1*, which was detected in both environments, and *qHI5.1*. However, the increased-alleles of *qGY5.1* and *qHI5.1* were from IR10F365, while that of *qGW5.1* was from Ciherang-Sub1 (Table 3, Fig. 1). *qGY5.1* was also detected in the similar region as *qYl-5* (Cho et al. 2003). Under control conditions, only one QTL for grain yield, *qGY12.1* was detected (LOD = 3.39 and R^2 = 9.3) (Table 4, Fig. 2); this QTL partially overlapped with *qFLW12.1* and in the same region as a major QTL for yield under drought (Bernier et al. 2007) (Table 6). It is interesting to note that some of the GY QTLs detected in this study were in similar positions as previously identified GY QTLs, under different environments. However, further research is needed to determine whether they are governed by the same gene(s) or they are simply closely linked.

Six QTLs were detected for 100-grain weight under SF on chromosomes 1, 2, 3, 5, and 10 (2 QTLs). All alleles with increasing effects were from Ciherang-Sub1 except for the largest QTL, *qGW10.2* (LOD = 6.30 and R^2 = 13.1%) on chromosome 10S (Table 3, Fig. 1). *qGW1.1* was also detected as the largest QTL for GW under control conditions in this study (LOD = 5.34 and R^2 = 15.2%) and co-localized with *GW1-1* (Yu et al. 2008) (Table 5). *qGW2.1* was in the similar region as *tgw2* (Yoon et al. 2006) and in a cluster with *qLSI3-2.1* and *qHI2*; however, the increased-allele for *qHI2* was from IR10F365. *qGW3.1* was co-localized with *QKw3a* (Li et al. 1997) and in a cluster with other four QTLs detected on chromosome 3S. *qGW5.1* was in the

Table 5 Genomic positions of QTLs identified in stagnant flooding with its QTL cluster and comparison to the QTLs identified in control irrigated condition in this study and previously published QTLs

QTL	Studies shared common region
qGY3.1	(Xiao et al. 1996); *dth3.1*: (Moncada et al. 2001); *qtl3.1*: (Bernier et al. 2007); *qPH3.1, qFLW3.1, qGW3.1*: this study under SF; *qDTF3.1*: this study under SF and control
qGY5.1	*qYI-5*: (Cho et al. 2003); *qGW5.1*: this study under SF and control; *qHI5.1*: this study under SF
qGY6.1	*qDTY6.1*: (Venuprasad et al. 2012)
qGW1.1	*GW1-2*: (Yu et al. 2008); *qGW1.1*: this study under control
qGW2.1	*tgw2*: (Yoon et al. 2006); *qHI2.1*, $qLSL_32.1$: this study under SF
qGW3.1	*QKw3a*: (Li et al. 1997); *qPH3.1, qFLW3.1, qGY3.1*: this study under SF; *qDTF3.1*: this study under SF and control
qGW5.1	*gw-5*: (Lu et al. 1997); *qGW5.1*: this study under control *qHI5.1* and *qGY5.1*: this study under SF
qGW10.1	*QKw10*: (Li et al. 1997)
qGW10.2	(Ishimaru 2003); *gw10.1*: (Thomson et al. 2003)
qHI2.1	*qGW2.1* and $qLSL_32.1$: this study under SF
qHI5.1	*qGW5.1*: this study SF and control; *qGY5.1*: this study under SF
qHI7.1	none
qHI7.2	*qPL7.1* and *qFLW7.1*: this study under SF
qSR1.1	none
qSR6.1	*qSER6.1*: this study under SF and control
qPL7.1	(Mei et al. 2003); *qHI7.2, qFLW7.1*: this study under SF
qPL9.1	*qPL-9*: (Cho et al. 2003); *pl9.1*: (Septiningsih et al. 2003); *pl9.1*: (Thomson et al. 2003); (Marri et al. 2005); (Li et al. 2006); *qPL9.1*: this study under control
qPH3.1	(Li et al. 1997); *qPh3a*: (Li et al. 2003); *qGY3.1, qFLW3.1, qGW3.1*: this study under SF; *qDTF3.1*: this study under SF and control
qPH5.1	(Mei et al. 2003); *ph5.1*: (Moncada et al. 2001); *ph5.1*: (Thomson et al. 2003); *qPH5.1*: this study under control; $qLSL_15.1$, *qLSI2-5.1, qLSI3-5.1, qSER5.1, qFLL5.1* and *qDTF5.1*: this study under SF and control
qPH6.1	*ph6*: (Xiao et al. 1996); *qPH2-6-1*: (Cui et al. 2004); $qLSL_26.1$: this study under SF
qSER5.1	*qSER5.1*: this study under control; *qPH5.1*, $qLSL_15.1$, $qLSl_25.1$, $qLSl_35.1$, *qSER5.1, qFLL5.1*, and *DTF5.1*: this study under SF and control
qSER6.1	*qSR6.1*: this study under SF and control
$qLSL_15.1$	$qLSL_15.1$: this study under control; *qPH5.1*, $qLSL_25.1$; $qLSL_35.1$, *qSER5.1, qFLL5.1*, and *DTF5.1*: this study under SF and control
$qLSL_25.1$	$qLSL_25.1$: this study undercontrol; *qPH5.1*, $qLSL_15.1$ and $qLSL_35.1$, *qSER5.1, qFLL5*: this study under SF and control; *DTF5.1*: this study under SF
$qLSL_26.1$	*qPH6.1*: this study under SF
$qLSL_32.1$	*qHI2.1* and *qGW2.1*: this study under SF
$qLSL_35.1$	$qLSL_35.1$: this study under control; *qPH5.1*, $qLSL_15.1$, $qLSL_25.1$, *qSER5.1*, and *qFLL5.1*: this study under SF and control; *DTF5.1*: under SF
qFLL4.1	(Mei et al. 2005)
qFLL5.1	*QFll7*: (Bing et al. 2006); (Yan et al. 2003); *qFLL5.1*: this study under irrigated; *qPH5.1, qLSI1-5.1*, $qLSL_25.1$, $qLSL_35.1$, *qSER5.1*: this study under SF and control; *DTF5.1*: this study under SF
qFLL6.1	(Mei et al. 2005); *qDTF6.1*: this study under SF and control
qFLL9.1	*fll9*: (Yan et al. 1999); *qPL9.1*: this study under SF and control
qFLW3.1	*qPH3.1, qGW3.1, qGY3.1*: this study under SF; *qDTF3.1*: this study SF and control
qFLW7.1	*qHI7.2, qPL7.1*: this study under SF
qDTF1.1	*qDTH-1*: (Cho et al. 2003); *QHd1b*: (Li et al. 2003); (Mei et al. 2003); *qFLL6.1* : this study under SF
qDTF3.1	(Albar et al. 1998); *dth3.1*: (Moncada et al. 2001); *qHD-3-1*: (Takeuchi et al. 2001); *qHDD3-1*: (Hittalmani et al. 2003); *QHd3a*: (Li et al. 2003); (Mei et al. 2003); *qEMF3*: (Hirabayashi et al. 2014); *qDTF3.1*: this study under control; *qGW3.1, qFLW3.1, qPH3.1, qGY3.1*: this study under SF
qDTF5.1	(Li et al. 1998); *QHd5b*: (Li et al. 2003); *qPH5.1*, $qLSL_15.1$, $qLSL_25.1$, $qLSL_35.1$, *qSER5.1*, and *qFLL5.1*: this study under SF and control
qDTF6.1	*QHd6b*: (Li et al. 2003); (Mei et al. 2003); *qDTF6.1*: this study under control; *qFLL6.1*: this study under SF
qDTF10.1	(Price et al. 1997); (Li et al. 1998); *qDTF10.1*:this study under control

Table 6 Genomic positions of QTLs identified in control irrigated condition with its QTL cluster and comparison to the QTLs identified in stagnant flooding in this study and previously published QTLs

QTL	Studies shared common region
qGY12.1	*qtl12.1*: (Bernier et al. 2007)
qGW1.1	*GW1-2*: (Yu et al. 2008); *qGW1.1*: this study under SF
qGW2.1	*gw-2*: (Lu et al. 1997); *QKw2b*: (Li et al. 1997); *qGW-2*: (Cho et al. 2003); *Kw2-2*: (Gao et al. 2004); *qLSL*$_3$*2.1*: this study under control
qGW2.2	*QKw2a*: (Li et al. 1997); *qTGWT-2-2*: (Zhuang et al. 2002); *tgw2*: (Ishimaru 2003); *gw2.1*: (Thomson et al. 2003); *qTGW-2a*: (Zou et al. 2005)
qGW5.1	*gw-5*: (Lu et al. 1997); *QKw5*: (Li et al. 1997); *qGW5.1*: this study under SF
qHI3.1	*qhi3.3/qhi3.4/qhi3.5*: (Lanceras et al. 2004); *qFLW3.1*, *qDTH3.1*: this study under control and SF; *qLSL*$_1$*3.1*, *qBM3.1*: this study under control
qHI7.1	none
qTN3.1	*TN(R3)*: (Kobayashi et al. 2003); *qPN3.1*, *qLSL*$_3$*3.1*: this study under control
qPN1.1	*PnN(R3-1)*: (Kobayashi et al. 2003); *npp(1.1)*: Nagata et al. 2002; *qBM1.1*: this study under control
qPN3.1	*PN(3)*: (Jiang et al. 2004); *qFLW3.1*, *qDTH3.1*: this study under control and SF; *qTN3.1*, *qLSL*$_3$*3.1*: this study under control
qBM1.1	*PW(1)*: (Lian et al. 2005); *qPN1.1*: this study under control
qBM3.1	*qHI3.1*; *qLSL*$_1$*3.1*: this study under control
qPL1.1	(Mei et al. 2005)
qPL2.1	*pl2*: (Xing et al. 2001); *qPL2-1*: (Cui et al. 2002); (Mei et al. 2003) 3; *qFLL2.1*: this study under control
qPL6.1	(Mei et al. 2003); *PL(6)*: (Jiang et al. 2004) J; *qSER6.1*, *qLSL*$_3$*6.1*: this study under control
qPL7.1	(Mei et al. 2005)
qPL9.1	*pl9*: (Xing et al. 2001); *qPL-9*: (Cho et al. 2003); *pl9.1*: (Septiningsih et al. 2003); *pl9.1*: (Thomson et al. 2003); (Marri et al. 2005); (Li et al. 2006); *qPL9.1*: this study under SF
qPH1.1	(Li et al. 1998); (Mei et al. 2003); *qLSL*$_1$*1.1*: this study under control
qPH3.1	(Li et al. 1997); *QPh3a*: (Li et al. 2003)
qPH5.1	(Mei et al. 2003); *ph5.1*: (Moncada et al. 2001); (Thomson et al. 2003); *qPH5.1*: this study under SF; *qLSL*$_1$*_5.1*, *qLSL*$_2$*5.1*, *qLSL*$_3$*5.1*, *qSER5.1*, *qFLL5.1*: this study under control and SF
qSER5.1	*qSER5.1*: this study under SF; *qPH5.1*, *qLSL*$_1$*_5.1*, *qLSI2-5.1*, *qLSL*$_2$*5.1*, *qFLL5.1*: this study under control and SF
qSER6.1	*qSER6.1*: this study under SF; *qPL6.1*, *qLSL*$_3$*6.1*: this study under control
qLSL$_1$*1.1*	*qPH1.1*: this study under control
qLSL$_1$*3.1*	*qBM3.1*, *qHI3.1*: this study under control
qLSL$_1$*5.1*	*qLSL*$_1$*5.1*: this study under SF; *qPH5.1*, *qLSL*$_2$*5.1*, *qLSL*$_3$*5.1*, *qSER5.1*, *qFLL5.1*: this study under control and SF
qLSL$_2$*3.1*	none
qLSL$_2$*5.1*	*qLSL*$_2$*5.1*: this study under SF; *qPH5.1*, *qLSL*$_1$*5.1*, *qLSL*$_3$*5.1*, *qSER5.1*, *qFLL5.1*: this study under control and SF
qLSI$_2$*9.1*	none
qLSL$_3$*2.1*	*qGW2.1*: this study under control
qLSL$_3$*3.1*	*qTN3.1*, *qPN3.1*: this study under control
qLSL$_3$*5.1*	*qLSL*$_3$*5.1*: this study under SF; *qPH5.1*, *qSER5.1*, *qLSL*$_1$*5.1*, *qLSL*$_2$*5.1*, *qFLL5.1*: this study under control and SF
qLSL$_3$*6.1*	*qSER6.1*: this study under control and SF; *qPL6.1*: this study under control
qLSI$_3$*8.1*	none
qLSL$_3$*11.1*	none
qFLL1.1	(Mei et al. 2005)
qFLL2.1	*qPL2.1*: this study under irrigated
qFLL4.1	(Mei et al. 2005)
qFLL5.1	*QFll7*: Yue et al. 2006; Yan et al. 2003; *qFLL5.1*: this study under SF; *qPH5.1*, *qLSL*$_1$*5.1*, *qLSL*$_2$*5.1*, *qLSL*$_3$*5.1*, *qSER5.1*: this study under irrigated and SF
qFLW3.1	(Mei et al. 2005); *qFLW3.1*: this study under SF; *qHI3.1*, *qBM3.1*, and *qLSL*$_1$*3.1*: this study under irrigated); *qDTF3.1*: this study under irrigated and SF
qFLW4.1	*QFlw4*: (Bing et al. 2006); *flw4*: (Yan et al. 1999)
qFLW5.1	none

Table 6 Genomic positions of QTLs identified in control irrigated condition with its QTL cluster and comparison to the QTLs identified in stagnant flooding in this study and previously published QTLs *(Continued)*

qFLW12.1	none
qFLW12.2	(Mei et al. 2005)
qDTF3.1	(Albar et al. 1998); *dth3.1*: (Moncada et al. 2001); *qHD-3-1*: (Takeuchi et al. 2001); *Hd9*: (Lin et al. 2002) 2; *qHDD3-1*: (Hittalmani et al. 2003); *QHd3a*: (Li et al. 2003); (Mei et al. 2003); *qEMF3*: (Hirabayashi et al. 2014); *qDTF3.1*: this study under SF; *qFLW3.1*: this study under control and SF; *qHI3.1*, *qBM3.1*, and $qLSL_13.1$: this study under control
qDTF6.1	*QHd6b*: (Li et al. 2003); (Mei et al. 2003); *qDTF6.1*: this study under SF
qDTF10.1	(Price et al. 1997); Li et al. (1998)); *qDTF10.1*: this study under SF

similar region as *gw-5* (Lu et al., 1997); the same QTL was also detected in the same position in the control and in a cluster with other two QTLs on the upper part of chromosome 5 L. *qGW10.1* was co-localized with *QKw10* (Li et al. 1997); while *qGW10.2* was in the similar regions of the QTLs detected by Ishimaru (2003) and Thomson et al. (2003). Under control conditions, four QTLs were detected on chromosomes 1, 2 (2 QTLs) and 5. In all cases the source of the increased alleles were from Ciherang-Sub1 (Table 4, Fig. 2). The two QTLs on chromosome 2 were co-localized with many previous reported QTLs; in addition, *qGW2.1* was in a similar region of $qLSI_32.1$ (Table 6).

Four harvest index QTLs were detected under stagnant flooding located on chromosomes 2, 5 and 7 (2 QTLs). The largest QTL, *qHI7.2* (LOD = 8.67 and R^2 = 20.4%) was the only one that had the increased allele from IR10F365 (Table 3, Fig. 1). This QTL was co-localized with two other QTLs, *qPL7.1* and *qFLW7.1*; however, the increased-allele of *qFLW7.1* was from Ciherang-Sub1. Under control conditions, two QTLs were identified on chromosomes 3 and 7, with the largest one, *qHI3.1*, detected with a LOD of 4.91 and R^2 of 14.2% (Table 4, Fig. 2). On the other hand, only two minor QTLs for survival rate under stagnant flooding were identified on chromosomes 1 and 6, *qSR1.1* (LOD = 2.89 and R^2 = 10.0%) with the beneficial allele from IR10F365 and *qSR6.1* (LOD = 2.97 and R^2 = 7.3%) with that of Ciherang-Sub1, respectively (Table 3, Fig. 1). *qSR6.1* was in a cluster with three other QTLs, *qSER6.1*, which was detected in both environments, *qPH6.1*, and $qLSL_26.1$; however, the increased-allele in *qPH6.1* was from IR10F365.

QTLs for DTF, PH, SER and PL

Five QTLs were detected for days to flowering under SF on chromosomes 1, 3, 5, 6, and 10 (Table 3, Fig. 1). Interestingly, the three QTLs on chromosomes 3, 6 and 10 were also mapped in similar regions under control conditions (Table 4, Fig. 2). The largest QTL was *qDTF3.1* with an increased allele or lateness from Ciherang-Sub1 (LOD = 12.51 and R^2 = 26.0% in stress; LOD = 21.81 and R^2 = 42.19% in the control); while the other two QTLs, *qDTF6.1* and *qDTF10.1*, with lateness from IR10F365. *qDTF5.1* was strongly detected by both software but only under SF, on the lower part of chromosomes 5L (LOD = 4.11 and R^2 = 7.0%) in a cluster with six other QTLs, $qLSL_15.1$, $qLSL_25.1$, $qLSL_35.1$, *qSER5.1*, *qPH5.1*, and *qFLL5.1* (Table 5). All of the QTLs for DTF detected in this study were previously reported in similar positions by other researchers (Tables 5 and 6).

Three QTLs were detected for plant height on chromosomes 3, 5 and 6 under SF, with the largest, *qPH5.1* (LOD = 5.59 and R^2 = 12.1%) on the lower arm of chromosome 5 L with an increased-allele from IR10F365 (Table 3, Fig. 1). Under control conditions, the same QTL was detected with larger effect (LOD = 9.90 and R^2 = 23.9%) (Table 4, Fig. 2). The QTL on chromosome 3, *qPH3.1*, was also detected under control condition with partly overlapping position; with an increased effect from Ciherang-Sub1. A minor QTL, *qPH1.1* was also detected on chromosome 1 under control conditions, with an increased allele effect from Ciherang-Sub1. All of the PH QTLs detected in this study were previously reported in similar positions by other researchers (Tables 5 and 6). On a related trait, two QTLs were detected for shoot elongation rate under SF on chromosomes 5 and 6. The largest QTL, *qSER5.1* (LOD = 12.46 and R^2 = 32%) positioned in a cluster of QTLs on the lower part of chromosome 5L (Table 3, Fig. 1), with an increased allele from IR10F365. The second QTL, *qSER6.1* was a minor one with the increased-allele effect from Ciherang-Sub1. It was worth noting that under control, both QTLs were mapped at the same positions on chromosomes 5 and 6 (Table 4, Fig. 2).

Under SF, two QTLs for panicle length *qPL7.1* and *qPL9.1* were identified with increase alleles from IR10F365 (Table 3, Fig. 1). *qPL7.1* was the largest with an LOD value of 5.47 and R^2 of 22%. Under control condition, five QTLs were detected on chromosomes 1, 2, 6, 7 and 9 (Table 4, Fig. 2). Three of them had increased alleles from IR10F365, including the largest QTL *qPL2.1* (LOD = 7.11 and R^2 = 14.31%). *qPL9.1* was detected under both SF and control, sharing the same region on the lower arm of chromosome 9 L. The panicle length QTLs detected in this study have also been previously mapped in similar positions (Tables 5 and 6).

QTLs for FLL and FLW

Four QTLs were identified for flag leaf length under SF on chromosomes 4, 5, 6, and 9 (Table 3, Fig. 1). Three of the QTLs had increased alleles from IR10F365, including the largest QTL *qFLL5.1* (LOD = 6.18; R^2 = 14.0%) positioned in a cluster with other QTLs on the lower part of chromosome 5L; the same QTL was also detected under control conditions. Under the control, four QTLs were identified on chromosomes 1, 2, 4, and 5, with three of them had increased alleles from IR10F365 (Table 4, Fig. 2). For flag leaf width, two QTLs, *qFLW3.1* (LOD = 4.00 and R^2 = 11.3%) and *qFLW7.1* (LOD = 3.11 and R^2 = 14.0%) were detected under SF with their increased allele from Ciherang-Sub1 (Table 3, Fig. 1). Under the control, 5 QTLs were detected on chromosomes 3, 4, 5 and 12 (2 QTLs), *qFLW12.1* being the only QTL having increased allele from IR10F365 (Table 4, Fig. 2). *qFLW4.1* was the largest under the control, with an LOD score of 8.88 and R^2 of 20.11%. The positions of *qFLW3.1* mostly overlap under SF and control conditions. Most of the FLL and FLW QTLs were previously mapped in similar regions (Tables 5 and 6).

QTLs for LSL

Under SF, only one QTL for leaf sheath length for the first internode, $qLSL_15.1$ with increased allele from IR10F365, was detected (LOD = 4.44 and R^2 = 11.0%) in a cluster of QTLs, including $qLSL_25.1$ *and* $qLSL_35.1$, on the lower arm of chromosome 5 L (Table 3, Fig. 1). Three QTLs were identified under control on chromosomes 1, 3, and 5 (Table 4, Fig. 2). $qLSL_15.1$ was detected in the same position as in SF condition as the largest QTL in irrigated (LOD = 5.33 and R^2 = 13.1%). The other two QTLs had increased alleles from Ciherang-Sub1. Meanwhile for LSL_2, two QTLs were identified on chromosomes 5 and 6 under SF, with the increased alleles from IR10F365 (Table 3, Fig. 1). $qLSL_25.1$ was the largest QTL (LOD = 4.96 and R^2 = 14.3%). Under control, three QTLs were detected on chromosomes 3, 5, and 9 (Table 4, Fig. 2). Two of the QTLs had increased alleles effects from IR10F365, including $qLSL_25.1$, which was also identified at the same position as under SF. This QTL was also the largest QTL under control (LOD = 4.95 and R^2 = 11.2%). Two QTLs for LSL_3 were detected under SF, located on chromosomes 2 and 5 with the increased allele from Ciherang-Sub1 and IR10F365, respectively (Table 3, Fig. 1). The largest QTL was $qLSL_32.1$ on chromosome 2 (LOD = 4.16 and R^2 = 10.0%). $qLSL_35.1$ was mapped on a cluster of QTLs on the lower part of chromosome 5L. Meanwhile, under control, six QTLs were identified and four of the QTLs had increased alleles from Ciherang-Sub1 (Table 4, Fig. 2). The LOD values of the QTLs ranged from 4.40 – 7.41 and R^2 of 9.26 – 14.72%. The largest QTL was $qLSL_38.1$, with the increased allele from IR10F365. $qLSL_32.1$ and $qLSL_35.1$ were also mapped in similar positions on chromosomes 2 and 5, respectively, as in SF environment. Some of these LSL traits were mapped in a cluster with other traits (Tables 5 and 6).

QTLs for TN, PN, and BM

There were no QTLs detected for tiller number, panicle number and shoot biomass under stagnant flooding. While under control conditions, only one QTL detected for tiller number on top of chromosome 3S, *qTN3.1* (LOD = 4.29, R^2 = 12.5%), with increased allele from IR10F365, in a cluster with some other QTLs, *qPH3.1, qHI3.1, qPN3.1, qBM3.1,* and $qLSL_33.1$ (Table 4, Fig. 2). Two QTLs were identified for panicle number under control on chromosomes 1 and 3 with the increased alleles from IR10F365. *qPN1.1* (LOD = 3.49 and R^2 = 10.3%) was detected in a cluster with some other QTLs *qPH1.1, qGW1.1, qBM1.1,* and $qLSL_11.1$, on the lower part of chromosome 1S. Two QTLs were detected under control for shoot biomass, *qBM1.1* (LOD = 3.92 and R^2 = 10.2%) and *qBM3.1* (LOD = 3.33 and R^2 = 9.9%), with the increased-alleles from IR10F365 and Ciherang-Sub1, respectively. Similar QTLs for the three traits had been previously detected in the similar regions (Tables 5 and 6).

Our results demonstrated that combinations of three or two yield QTLs gave better performances under stagnant flooding conditions (Table 7). Under control conditions, however, *qGY12* only had a small contribution for increasing yield (7.2%; Table 8). A group of RILs having combinations of three QTLs gave the best performance under stagnant flooding with an increased yield of 89.2, 56.1, 44.3, 51.5, 30.6, 17.2, and 37.6% compared to the RIL that did not possess any of the yield QTLs, only *qGY6* from Ciherang-Sub1, only *qGY5* from IR10F365, only *qGY3* from Ciherang-Sub1 (detected in a cluster with four different QTLs referred as C3S QTLs in Table 7), *qGY5* and *qGY6*, *qGY3* and *qGY6*, and *qGY3* and *qGY5*, respectively. In general the groups having higher yield under SF also are taller, have more tillers and panicles, longer flag leaf, more biomass, higher shoot elongation rate, longer leaf sheaths, and higher survival rate. A similar cluster of QTLs on the top of chromosome 3 (C3S QTLs; Table 8), however, hardy had any effect in yield under control; either alleles contributed similarly to yield. This might be due to the fact that the beneficial alleles of those QTLs detected in the control were mix from both parents (Tables 7 and 8; Fig. 2). On the other hand, a cluster of seven different QTLs on the lower part of chromosome 5L just below *qGY5* (C5L QTLs; Table 7) revealed that these QTLs indirectly contribute to a yield increase of 19.5%

Table 7 Phenotypic effect of grain yield QTLs and QTL clusters identified in stagnant flooding condition

qGY3 and C3S QTLs[a]	*qGY5*	*qGY6*	C5L QTLs[c]	C6C QTLs[d]	#RIL	DTF (d)	PH (cm)	TN	PN	FLL (cm)	FLW (cm)	PL (cm)	BM (g/m^2)	SER (cm/d)	HI	GW (g)	LSL_1 (cm)	LSL_2 (cm)	LSL_3 (cm)	GY (kg/ha)	SR (%)
+	+	+	NA	NA	28	96	134.2	7	7	29.8	1.83	25.8	1037.1	1.35	0.35	2.67	14.9	18.0	28.6	3408	71.2
+	+	-	NA	NA	12	97	129.8	7	6	28.7	1.77	26.1	907.4	1.32	0.30	2.66	14.6	17.8	28.6	2476	67.0
+	-	+	NA	NA	25	97	134.9	7	6	29.3	1.83	26.3	1070.3	1.39	0.28	2.76	14.6	18.0	28.2	2908	71.7
-	+	+	NA	NA	35	93	127.7	7	6	30.5	1.76	25.1	984.8	1.42	0.28	2.60	14.2	17.5	28.4	2610	68.6
+	-	-	NA	NA	8	97	128.2	6	5	27.5	1.86	25.6	914.2	1.31	0.25	2.79	13.1	16.7	29.5	2249	71.3
-	+	-	NA	NA	8	97	122.6	6	6	28.4	1.74	25.8	888.8	1.23	0.29	2.56	14.3	16.3	25.2	2361	66.9
-	-	+	NA	NA	19	95	126.6	7	6	29.8	1.79	25.0	967.7	1.30	0.26	2.68	14.0	17.8	27.3	2183	64.0
-	-	-	NA	NA	1	97	119.2	4	4	27.4	1.84	27.3	544.7	1.03	0.28	3.14	12.5	14.3	21.0	1801	58.3
NA[b]	NA	NA	+	NA	83	95	132.7	7	6	30.2	1.80	25.7	1013.6	1.43	0.30	2.65	14.8	18.2	28.7	2856	69.3
NA	NA	NA	-	NA	32	97	125.2	7	6	28.1	1.78	25.5	951.6	1.21	0.26	2.72	13.1	16.4	26.0	2390	67.3
NA	NA	NA	NA	+	74	95	130.2	7	7	30.2	1.81	26.0	1001.6	1.39	0.30	2.68	14.4	17.6	28.3	2879	69.7
NA	NA	NA	NA	-	32	97	131.1	7	6	28.7	1.77	25.6	997.1	1.30	0.28	2.64	14.7	17.6	28.0	2572	67.7

[a] a cluster of QTLs on the top of chromosome 3S
[b] not applied
[c] a cluster of QTLs on the lower part of chromosome 5L just below *qGY5*
[d] a cluster of QTLs close to the centromeric region of chromosome 6

Table 8 Phenotypic effect of grain yield QTL and QTL clusters identified in irrigated control conditions

qGY12	C3S QTLs[b)]	C5L QTLs[c)]	C6C QTLs[d)]	#RIL	DTF	PH	TN	PN	FLL	FLW	PL	BM	PER	HI	GW	LSL_1	LSL_2	LSL_3	GY
+	NA	NA	NA	32	90	124.0	11	11	31.5	1.88	25.8	1372.2	1.33	0.44	2.71	13.9	21.6	31.5	5910
-	NA	NA	NA	80	89	126.3	10	10	32.2	1.93	26.2	1323.3	1.35	0.43	2.74	14.1	22.6	33.2	5511
NA[a)]	+	NA	NA	77	91	126.7	10	10	32.5	1.89	25.9	1358.8	1.31	0.42	2.73	14.9	21.9	33.2	5588
NA	-	NA	NA	67	87	124.2	10	10	31.4	1.97	26.4	1290.1	1.37	0.45	2.73	13.4	22.6	32.2	5671
NA	NA	+	NA	83	88	128.3	10	10	32.6	1.96	26.4	1321.8	1.40	0.44	2.72	14.9	22.6	33.4	5729
NA	NA	-	NA	32	90	120.4	10	10	30.8	1.85	25.5	1359.0	1.22	0.42	2.77	13.2	20.8	31.3	5399
NA	NA	NA	+	74	88	125.9	10	10	32.8	1.95	26.6	1337.2	1.37	0.44	2.75	14.2	22.3	32.8	5749
NA	NA	NA	-	32	91	126.2	10	10	31.1	1.85	25.5	1350.7	1.26	0.42	2.70	14.7	22.1	32.0	5470

[a)]not applied
[b)]a cluster of QTLs on the top of chromosome 3S
[c)]a cluster of QTLs on the lower part of chromosome 5L just below *qGY5*
[d)]a cluster of QTLs close to the centromeric region of chromosome 6

under stagnant flooding; which is partly due to increases in plant height, elongation rate, leaf sheath length, flag leaf length, and earliness from IR10F365 alleles. A similar cluster of six QTLs was also detected in the same region under control; in which this genomic region of IR10F365 indirectly increased yield by 6.1% (C5L QTLs; Table 8). Additionally, a smaller cluster of QTLs close to the centromeric region of chromosome 6 with increased alleles from Ciherang-Sub1 (C6C QTLs; Tables 7 and 8) contributed to a yield increase of 11.9 and 5.1% under SF and control, respectively.

Conclusions

This study reports the first effort for mapping beneficial QTLs under stagnant flooding conditions. In contrast to most Sub1-lines developed to date, Ciherang-Sub1 was found to be tolerant to stagnant flooding and many beneficial QTLs were derived from this variety. Future research in stagnant flooding tolerance should target these valuable QTLs after validation of trait data by conducting additional field trials across years and locations, especially on the distal part of chromosome 3 (*qGY3* and 4 other QTLs), *qGY5*, *qGY6*, clusters of QTLs on the lower part of chromosome 5L and around the centromeric region of chromosome 6. These QTLs can be further investigated for molecular studies and for use in molecular breeding. In addition, since the two parents are elite breeding material, the best yielding lines that performed well in each environment can be further evaluated through multi-environmental trials. It should be noted that all of these RILs are already fixed for the *SUB1* gene, which provides additional protection from complete submergence in both stagnant flooding and irrigated environments.

Additional files

Additional file 1: Table S1. List of traits measured in both stagnant flooding and irrigated control conditions.

Additional file 2: Table S2. Performances of selected RILs in stagnant flooding and irrigated control IRRI fields in 2014 WS.

Additional file 3: Figure S1. Percentage difference between means of trait values of 25 highest and lowest yielding lines relative to the mean of the population under stagnant flooding condition.

Additional file 4: Figure S2. Percentage difference between means of trait values of 25 highest and lowest yielding lines relative to the mean of the population under irrigated control condition.

Additional file 5: Figure S3. Relationship between shoot elongation rate (SER) under stagnant flooding condition, relative grain yield (ratio of yield under SF to that in the control) and survival rate (SR).

Additional file 6: Table S3. Correlations among traits under stagnant flooding condition.

Additional file 7: Table S4. Correlations among traits under irrigated control condition.

Acknowledgements
We thank J. Borgonia and R. Formaran for field work assistance and the IRRI GSL team for genotyping assistance. The work reported here was supported in part by a grant from the Bill and Melinda Gates Foundation (BMGF) through the project "Stress-tolerant rice for Africa and South Asia (STRASA)" and also by the Global Rice Science Partnership (GRiSP), the National Institute of Food and Agriculture, U. S. Department of Agriculture, Hatch project 1009300.

Authors' contributions
EMS, AS, AMI designed the experiments; BCYC developed the RIL population using rapid generation advance (RGA); JC performed seed multiplication of the mapping population, ZJCG and AS performed DNA extraction, and AS and JC performed phenotyping under supervision of EMS; AS, EMS, and ZJCG analyzed the data; EMS and AS wrote the manuscript; all authors read and approved the manuscript.

Competing interests
The authors declare that they have no competing interests.

Author details
[1]Present address: Rani Lakshmi Bai Central Agricultural University, Jhansi 284003, India. [2]Present address: Sugar Research Australia, 71378 Bruce Highway, Meringa, PO Box122, Gordonvale, QLD 4865, Australia. [3]Present address: Department of Soil and Crop Sciences, Texas A&M University, College Station, TX 77843, USA. [4]International Rice Research Institute, DAPO 7777, Metro Manila, Philippines.

References

Albar L, Lorieux M, Ahmadi N, Rimbault I, Pinel A, Sy A, Ghesquière A (1998) Genetic basis and mapping of the resistance to rice yellow mottle virus. I. QTLs identification and relationship between resistance and plant morphology. Theor Appl Genet 97(7):1145–1154

Angaji SA, Septiningsih EM, Mackill DJ, Ismail AM (2010) QTLs associated with tolerance of flooding during germination in rice (Oryza sativa L.). Euphytica 172(2):159–168. doi:10.1007/s10681-009-0014-5

Baltazar MD, Ignacio JCI, Thomson MJ, Ismail AM, Mendioro MS, Septiningsih EM (2014) QTL mapping for tolerance of anaerobic germination from IR64 and the aus landrace Nanhi using SNP genotyping. Euphytica 197(2):251–260

Bernier J, Kumar A, Ramaiah V, Spaner D, Atlin G (2007) A large-effect QTL for grain yield under reproductive-stage drought stress in upland rice. Crop Sci 47(2):507–516

Bing Y, Wei-Ya X, Li-Jun L, Yong-Zhong X (2006) QTL analysis for flag leaf characteristics and their relationships with yield and yield traits in rice. Acta Genet Sin 33(9):824–832

Catling D (1992). Gas Exchange and Nutrition *Rice in Deep Water*. Springer, London, p 153–157

Cho Y, Suh J, Choi I, Hong H, Baek M, Kang K, Hwang H (2003). QTLs analysis of yield and its related traits in wild rice relative Oryza rufipogon. Treat Crop Res 4(1): 19–29.

Collard B, Septiningsih E, Das S, Carandang J, Sanchez D, Kato Y, AULA K (2013). Developing new flood-tolerant varieties at the International Rice Research Institute (IRRI). SABRAO J Breed Genet 45(1):42–56

Cui K, Peng S, Xing Y, Yu S, Xu C (2002) Genetic analysis of the panicle traits related to yield sink size of rice. Yi Chuan Xue Bao 29(2):144–152

Cui K, Peng S, Ying Y, Yu S, Xu C (2004) Molecular dissection of the relationships among tiller number, plant height and heading date in rice. Plant Prod Sci 7(3):309–318

Gao Y-M, Zhu J, Song Y-S, He C-X, Shi C-H, Xing Y-Z (2004) Analysis of digenic epistatic effects andQE interaction effects QTL controlling grain weight in rice. J Zhejiang Univ (Sci) 5(4):371–377

Gonzaga ZJC, Carandang J, Singh A, Collard BCY, Thomson MJ, & Septiningshih EM (2017). Mapping QTLs for submergence tolerance in rice using a population fixed for SUB1A tolerant allele. Molecular Breeding. doi:10.1007/s11032-017-0637-5.

Hattori Y, Miura K, Asano K, Yamamoto E, Mori H, Kitano H, Ashikari M (2007) A major QTL confers rapid internode elongation in response to water rise in deepwater rice. Breed Sci 57(4):305–314

Hattori Y, Nagai K, Mori H, Kitano H, Matsuoka M, Ashikari M (2008) Mapping of three QTLs that regulate internode elongation in deepwater rice. Breed Sci 58(1):39–46

Hattori Y, Nagai K, Furukawa S, Song X-J, Kawano R, Sakakibara H, Kitano H (2009) The ethylene response factors SNORKEL1 and SNORKEL2 allow rice to adapt to deep water. Nature 460(7258):1026–1030

Hirabayashi H, Sasaki K, Kambe T, Gannaban RB, Miras MA, Mendioro MS, Takemoto-Kuno Y (2014). qEMF3, a novel QTL for the early-morning flowering trait from wild rice, Oryza officinalis, to mitigate heat stress damage at flowering in rice, O. sativa. J Exp Bot 66(5):1227–1236

Hittalmani S, Huang N, Courtois B, Venuprasad R, Shashidhar H, Zhuang J, Sidhu J (2003) Identification of QTL for growth-and grain yield-related traits in rice across nine locations of Asia. Theor Appl Genet 107(4):679–690

Iftekharuddaula K, Newaz M, Salam M, Ahmed H, Mahbub M, Septiningsih E, Mackill D (2011) Rapid and high-precision marker assisted backcrossing to introgress the SUB1 QTL into BR11, the rainfed lowland rice mega variety of Bangladesh. Euphytica 178(1):83–97

Iftekharuddaula KM, Ahmed HU, Ghosal S, Amin A, Moni ZR, Ray BP, Septiningsih EM (2016) Development of early maturing submergence-tolerant rice varieties for Bangladesh. Field Crop Res 190:44–53. doi:10.1016/j.fcr.2015.12.001

Ishimaru K (2003) Identification of a locus increasing rice yield and physiological analysis of its function. Plant Physiol 133(3):1083–1090

Ismail AM, Ella ES, Vergara GV, Mackill DJ (2009) Mechanisms associated with tolerance to flooding during germination and early seedling growth in rice (Oryza sativa). Ann Bot 103(2):197–209. doi:10.1093/aob/mcn211

Ismail AM, Singh US, Singh S, Dar MH, Mackill DJ (2013) The contribution of submergence-tolerant (Sub1) rice varieties to food security in flood-prone rainfed lowland areas in Asia. Field Crop Res 152:83–93

Jiang G, Xu C, Li X, He Y (2004) Characterization of the genetic basis for yield and its component traits of rice revealed by doubled haploid population. Yi Chuan Xue Bao 31(1):63–72

Kato Y, Collard BC, Septiningsih EM, Ismail AM (2014) Physiological analyses of traits associated with tolerance of long-term partial submergence in rice. AoB Plants 6:plu058

Kobayashi S, Fukuta Y, Sato T, Osaki M, Khush G (2003) Molecular marker dissection of rice (Oryza sativa L.) plant architecture under temperate and tropical climates. Theor Appl Genet 107(8):1350–1356

Kretzschmar T, Pelayo MAF, Trijatmiko KR, Gabunada LFM, Alam R, Jimenez R, Septiningsih EM (2015) A trehalose-6-phosphate phosphatase enhances anaerobic germination tolerance in rice. Nat Plants 1(9):15124. doi:10.1038/nplants.2015.124

Lanceras JC, Pantuwan G, Jongdee B, Toojinda T (2004) Quantitative trait loci associated with drought tolerance at reproductive stage in rice. Plant Physiol 135(1):384–399

Li Z, Pinson S, Park W, Paterson A, Stansel J (1997) Quantitative trait Loci associated with drought tolerance at reproductive stage in rice. Plant Physiol 135:384–399

Li J, He P, Zheng X, Lu R, Zhu L (1998) Identification and interaction analysis of six agronomic trait loci of rice based on a recombinant inbred population. Acta Bot Sin 41(11):1199–1203

Li Z-K, Yu S-B, Lafitte H, Huang N, Courtois B, Hittalmani S, Shashidhar H (2003) QTL× environment interactions in rice. I. Heading date and plant height. Theor Appl Genet 108(1):141–153

Li C, Zhou A, Sang T (2006) Genetic analysis of rice domestication syndrome with the wild annual species, Oryza nivara. New Phytol 170(1):185–194

Lian X, Xing Y, Yan H, Xu C, Li X, Zhang Q (2005) QTLs for low nitrogen tolerance at seedling stage identified using a recombinant inbred line population derived from an elite rice hybrid. Theor Appl Genet 112(1):85–96

Lin H, Ashikari M, Yamanouchi U, Sasaki T, Yano M (2002) Identification and characterization of a quantitative trait locus, Hd9, controlling heading date in rice. Breed Sci 52(1):35–41

Lu C, Shen L, He P, Chen Y, Zhu L, Tan Z, Xu Y (1997) Comparative mapping of QTLs for agronomic traits of rice across environments by using a doubled-haploid population. Theor Appl Genet 94(1):145–150

Mackill DJ, Ismail AM, Pamplona AM, Sanchez DL, Carandang JJ, Septiningsih EM (2010) Stress tolerant rice varieties for adaptation to a changing climate. Crop Environ Bioinformatics 7:250–259

Manly KF, Cudmore RH Jr, Meer JM (2001) Map Manager QTX, cross-platform software for genetic mapping. Mamm Genome 12(12):930–932

Marri PR, Sarla N, Reddy LV, Siddiq E (2005) Identification and mapping of yield and yield related QTLs from an Indian accession of Oryza rufipogon. BMC Genet 6(1):33

McCouch SR (2008) Gene nomenclature system for rice. Rice 1(1):72–84

Mei H, Luo L, Ying C, Wang Y, Yu X, Guo L, Li Z (2003) Gene actions of QTLs affecting several agronomic traits resolved in a recombinant inbred rice population and two testcross populations. Theor Appl Genet 107(1):89–101

Mei H, Li Z, Shu Q, Guo L, Wang Y, Yu X, Luo L (2005) Gene actions of QTLs affecting several agronomic traits resolved in a recombinant inbred rice population and two backcross populations. Theor Appl Genet 110(4):649–659

Moncada P, Martinez C, Borrero J, Chatel M, Gauch H Jr, Guimaraes E, McCouch S (2001) Quantitative trait loci for yield and yield components in an Oryza sativa × Oryza rufipogon BC2F2 population evaluated in an upland environment. Theor Appl Genet 102(1):41–52

Murray M, Thompson WF (1980) Rapid isolation of high molecular weight plant DNA. Nucleic Acids Res 8(19):4321–4326

Nagai K, Kondo Y, Kitaoka T, Noda T, Kuroha T, Angeles-Shim RB, Ashikari M (2014) QTL analysis of internode elongation in response to gibberellin in deepwater rice. AoB Plants 6:plu028

Nagata K, Fukuta Y, Shimizu H, Yagi T, Terao T (2002) Quantitative trait loci for sink size and ripening traits in rice (Oryza sativa L.). Breed Sci 52(4):259–273.

Neeraja CN, Maghirang-Rodriguez R, Pamplona A, Heuer S, Collard BC, Septiningsih EM, Mackill DJ (2007) A marker-assisted backcross approach for developing submergence-tolerant rice cultivars. Theor Appl Genet 115(6): 767–776. doi:10.1007/s00122-007-0607-0

Nelson JC (1997) QGENE: software for marker-based genomic analysis and breeding. Mol Breed 3(3):239–245

Price A, Young E, Tomos A (1997) Quantitative trait loci associated with stomatal conductance, leaf rolling and heading date mapped in upland rice (Oryza sativa). New Phytol 137(1):83–91

Redfern SK, Azzu N, Binamira JS (2012) Rice in Southeast Asia: facing risks and vulnerabilities to respond to climate change. Build Resilience Adaptation Climate Change Agric Sect 23:295

Sarkar R, Panda D, Reddy J, Patnaik S, Mackill DJ, Ismail AM (2009) Performance of submergence tolerant rice (Oryza sativa) genotypes carrying the Sub1 quantitative trait locus under stressed and non-stressed natural field conditions. Indian J Agric Sci 79(11):876–883

Septiningsih E, Prasetiyono J, Lubis E, Tai T, Tjubaryat T, Moeljopawiro S, McCouch S (2003) Identification of quantitative trait loci for yield and yield components in an advanced backcross population derived from the Oryza sativa variety IR64 and the wild relative O. rufipogon. Theor Appl Genet 107(8):1419–1432

Septiningsih EM, Pamplona AM, Sanchez DL, Neeraja CN, Vergara GV, Heuer S, Mackill DJ (2009) Development of submergence-tolerant rice cultivars: the Sub1 locus and beyond. Ann Bot 103(2):151–160. doi:10.1093/aob/mcn206

Septiningsih EM, Collard BC, Heuer S, Bailey-Serres J, Ismail AM, Mackill DJ (2013a). Applying genomics tools for breeding submergence tolerance in rice. Transl Genomics Crop Breed 2:9–30.

Septiningsih EM, Ignacio JCI, Sendon PM, Sanchez DL, Ismail AM, Mackill DJ (2013b). QTL mapping and confirmation for tolerance of anaerobic conditions during germination derived from the rice landrace Ma-Zhan Red. Theor Appl Genet 126(5):1357–1366.

Septiningsih EM, Hidayatun N, Sanchez DL, Nugraha Y, Carandang J, Pamplona AM, Mackill DJ (2015) Accelerating the development of new submergence tolerant rice varieties: the case of Ciherang-Sub1 and PSB Rc18-Sub1. Euphytica 202(2):259–268

Singh S, Mackill DJ, Ismail AM (2009) Responses of SUB1 rice introgression lines to submergence in the field: yield and grain quality. Field Crop Res 113(1):12–23

Singh S, Mackill DJ, Ismail AM (2011) Tolerance of longer-term partial stagnant flooding is independent of the SUB1 locus in rice. Field Crop Res 121(3):311–323

Singh S, Mackill DJ, Ismail AM (2014). Physiological basis of tolerance to complete submergence in rice involves genetic factors in addition to the SUB1 gene. AoB Plants 6. doi:10.1093/aobpla/plu060

Takeuchi Y, Hayasaka H, Chiba B, Tanaka I, Shimano T, Yamagishi M, Yano M (2001) Mapping quantitative trait loci controlling cool-temperature tolerance at booting stage in temperate japonica rice. Breed Sci 51(3):191–197

Thomson MJ (2014) High-throughput SNP genotyping to accelerate crop improvement. Plant Breed Biotechnol 2(3):195–212

Thomson M, Tai T, McClung A, Lai X, Hinga M, Lobos K, McCouch S (2003) Mapping quantitative trait loci for yield, yield components and morphological traits in an advanced backcross population between Oryza rufipogon and the Oryza sativa cultivar Jefferson. Theor Appl Genet 107(3):479–493

Toledo AMU, Ignacio JCI, Casal C, Gonzaga ZJ, Mendioro MS, Septiningsih EM (2015) Development of improved Ciherang-Sub1 having tolerance to anaerobic germination conditions. Plant Breed Biotechnol 3(2):77–87

Venuprasad R, Bool M, Quiatchon L, Cruz MS, Amante M, Atlin G (2012) A large-effect QTL for rice grain yield under upland drought stress on chromosome 1. Mol Breed 30(1):535–547

Vergara GV, Nugraha Y, Esguerra MQ, Mackill DJ, Ismail AM (2014) Variation in tolerance of rice to long-term stagnant flooding that submerges most of the shoot will aid in breeding tolerant cultivars. AoB Plants 6:plu055

Wang S, Basten C, Zeng Z (2010) Windows QTL Cartographer 2.5. 2007. Department of Statistics, North Carolina State University, Raleigh

Xiao J, Li J, Yuan L, Tanksley S (1996) Identification of QTLs affecting traits of agronomic importance in a recombinant inbred population derived from a subspecific rice cross. Theor Appl Genet 92(2):230–244

Xing Y-Z, Xu C-G, Hua J-P, Tan Y-F (2001) Analysis of QTL x environment interaction for rice panicle characteristics. Yi Chuan Xue Bao 28(5):439–446

Yan J, Zhu J, He C, Benmoussa M, Wu P (1999) Molecular marker-assisted dissection of genotype × environment interaction for plant type traits in rice (Oryza sativa L.). Crop Sci 39(2):538–544

Yan C-J, Liang G-H, Chen F, Li X, Tang S, Yi C, Gu M (2003) Mapping quantitative trait loci associated with rice grain shape based on an indica/japonica backcross population. Yi Chuan Xue Bao 30(8):711–716

Yonemaru J-i, Yamamoto T, Fukuoka S, Uga Y, Hori K, Yano M (2010) Q-TARO: QTL annotation rice online database. Rice 3(2–3):194–203

Yoon D-B, Kang K-H, Kim H-J, Ju H-G, Kwon S-J, Suh J-P, Ahn S-N (2006) Mapping quantitative trait loci for yield components and morphological traits in an advanced backcross population between Oryza grandiglumis and the O. sativa japonica cultivar Hwaseongbyeo. Theor Appl Genet 112(6):1052–1062

Yu S, Yang C, Fan Y, Zhuang J, Li X (2008) Genetic dissection of a thousand-grain weight quantitative trait locus on rice chromosome 1. Chin Sci Bull 53(15): 2326–2332

Yue B, Xue WY, Luo LJ, Xing YZ (2006) QTL analysis for flag leaf characteristics and their relationships with yield and yield traits in rice. Acta Genet Sin 33(9): 824–832.

Zhuang J-Y, Fan Y-Y, Rao Z-M, Wu J-L, Xia Y-W, Zheng K-L (2002) Analysis on additive effects and additive-by-additive epistatic effects of QTLs for yield traits in a recombinant inbred line population of rice. Theor Appl Genet 105(8):1137–1145

Zou G, Mei H, Liu H, Liu G, Hu S, Yu X, Luo L (2005) Grain yield responses to moisture regimes in a rice population: association among traits and genetic markers. Theor Appl Genet 112(1):106–113

8

Silencing of the Rice Gene *LRR1* Compromises Rice *Xa21* Transcript Accumulation and XA21-Mediated Immunity

Daniel F. Caddell[1], Chang-Jin Park[1,2], Nicholas C. Thomas[1,3], Patrick E. Canlas[1] and Pamela C. Ronald[1*]

Abstract

Background: The rice immune receptor XA21 confers resistance to *Xanthomonas oryzae pv. oryzae* (*Xoo*), the causal agent of bacterial leaf blight. We previously demonstrated that an auxilin-like protein, XA21 BINDING PROTEIN 21 (XB21), positively regulates resistance to *Xoo*.

Results: To further investigate the function of XB21, we performed a yeast two-hybrid screen. We identified 22 unique XB21 interacting proteins, including LEUCINE-RICH REPEAT PROTEIN 1 (LRR1), which we selected for further analysis. Silencing of *LRR1* in the XA21 genetic background (XA21-LRR1Ri) compromises resistance to *Xoo* compared with control XA21 plants. XA21-LRR1Ri plants have reduced *Xa21* transcript levels and reduced expression of genes that serve as markers of XA21-mediated activation. Overexpression of *LRR1* is insufficient to alter resistance to *Xoo* in rice lines lacking XA21.

Conclusions: Taken together, our results indicate that LRR1 is required for wild-type *Xa21* transcript expression and XA21-mediated immunity.

Keywords: Rice, Immunity, *Xanthomonas oryzae* pv. *oryzae*, XA21, XB21, LRR1, Leucine rich repeat

Background

Cell surface immune receptors, which often perceive conserved microbial signatures present outside host cells, represent a first line of defense against potential pathogens. Well characterized plant cell surface immune receptors include *Arabidopsis* FLAGELLIN SENSING 2 (FLS2), which recognizes a 22 amino acid epitope derived from bacterial flagellin (flg22) (Boller and Felix 2009), EF-TU RECEPTOR (EFR) which recognizes an 18 amino acid epitope derived from bacterial elongation factor-Tu (elf18) (Zipfel et al. 2006), and rice XA21, which recognizes a 21 amino acid tyrosine-sulfated epitope derived from the bacterial protein REQUIRED FOR ACTIVATION OF XA21-MEDIATED IMMUNITY X (RaxX; RaxX21-sY) (Pruitt et al. 2015). These immune receptors fall into the leucine-rich repeat receptor-like kinase class (LRR-RLK) consisting of an N-terminal signal peptide, predicted extracellular LRR domain, transmembrane domain and an intracellular kinase domain. Upon microbial recognition, the kinase domain is hypothesized to initiate a signaling cascade that leads to resistance against specific pathogens. These immune receptors require the presence of LRR-RLK co-receptors that are part of the SOMATIC EMBRYOGENESIS RECEPTOR KINASE (SERK) family. For example, FLS2-mediated immune response requires the BRI1-ASSOCIATED RECEPTOR KINASE (BAK1) (Chinchilla et al. 2007) and XA21-mediated immune response requires OsSERK2 (Chen et al. 2014).

Despite the importance of immune receptors to plant survival, few components that mediate the downstream signaling cascade have yet been characterized. One method that has been successful at identifying such proteins is the yeast two-hybrid system. We used this

* Correspondence: pcronald@ucdavis.edu
[1]Department of Plant Pathology and the Genome Center, University of California, Davis, Davis, CA 95616, USA

approach to generate an XA21 interactome (Seo et al. 2011). One of the isolated proteins was named XA21 BINDING PROTEIN 21 (XB21). XB21 encodes an auxilin-like protein that regulates resistance to *Xoo* and is predicted to function in clatharin-mediated endocytosis (Park et al. manuscript in preparation). To investigate the function of XB21, we performed a yeast two-hybrid screen utilizing XB21 as bait with a rice cDNA library. From this analysis, we identified LEUCINE RICH REPEAT PROTEIN 1 (LRR1). LRR1 shares 55% identity and 68% similarity with the extracellular domain of OsSERK2. To determine if LRR1 is required for XA21-mediated immunity, we assessed its function in rice plants in the presence and absence of XA21.

Transgenic rice plants expressing *Xa21* that were silenced for *LRR1* (XA21-LRR1Ri) displayed enhanced susceptibility to *Xoo*. XA21-LRR1Ri plants also have decreased expression of *Xa21* transcripts. These results demonstrate that *LRR1* is required for XA21 –mediated immune responses and for *Xa21* transcription. These results also highlight the diverse roles played by LRR containing proteins in regulating plant responses against pathogens.

Results

LRR1 was Identified Through a Yeast two-Hybrid Screen to Identify Components of XA21-Mediated Immunity

Because full-length XB21 displayed auto-activity in the yeast two-hybrid system, an N-terminal deleted XB21 (XB21ΔNT) was expressed in yeast and used as bait to screen a rice cDNA library pool derived from leaf mRNA of 7-week old rice plants containing *Xa21* (Seo et al. 2011). This screen identified 22 unique rice gene IDs that we named XB21-INTERACTING PROTEINS (XB21IPs) (Table 1). We isolated two independent clones (clone 3 and 16) that encoded LRR1 (Os01g59440). To validate the interaction between XB21 and LRR1, His-XB21 was co-incubated with GST or GST-LRR1 that were purified from *E. coli*. His-XB21 was pulled down with GST-LRR1, but not GST, indicating that LRR1 and XB21 interact in vitro (Additional file 1: Figure S1). We then selected LRR1 for further characterization due to its similarity with OsSERK2 (Fig. 1), which is a required component of XA21-mediated immunity (Chen et al. 2014). LRR1 and OsSERK2 share 55% identity and 68% similarity over the region spanning the length of LRR1 (Kearse et al. 2012). LRR1 is predicted to encode a 213 amino acid protein containing an N-terminal signal peptide (Amino acids 1–21), N-terminal LRR capping domain (AA 24–63) (Kolade et al. 2006), and five consecutive leucine-rich repeats (AA 65–184) following the plant-specific LRR consensus sequence LxxLxxLxLxxNxLSGx IPxxLGx (McAndrew et al. 2014) (Fig. 1). LRR1 is predicted to have two N-glycosylation sites compared to six in the OsSERK2 ectodomain (Zhang et al. 2004). Unlike OsSERK2, LRR1 does not contain a transmembrane domain or intracellular domain.

Table 1 List of identified XB21 Interacting Proteins (XB21IPs)

Clone ID	MSU Annotation[a]	Locus ID
1	NAD dependent epimerase/dehydratase family protein	LOC_Os03g23980
2	thioredoxin	LOC_Os12g08730
3	BRI1-associated kinase 1 precursor	LOC_Os01g59440
4	NAD dependent epimerase/dehydratase family protein	LOC_Os03g23980
5	Glutaminyl-tRNA synthetase	LOC_Os01g09000
6	Beta-D-glucan exohydrolase-like protein	LOC_Os05g37700
7	FHA domain containing protein	LOC_Os11g03390
8	OsCML24 - Calmodulin-related calcium sensor protein	LOC_Os07g48340
9	metallothionein-like protein	LOC_Os01g10400
10	peptidyl-tRNA hydrolase	LOC_Os01g49900
11	photosystem I reaction center subunit	LOC_Os07g05480
12	phosphoesterase family protein	LOC_Os03g61130
13	peroxiredoxin	LOC_Os02g33450
14	dehydrin	LOC_Os03g45280
15	universal stress protein domain containing protein	LOC_Os07g36600
16	BRI1-associated kinase 1 precursor	LOC_Os01g59440
17	glutamine synthetase	LOC_Os02g50240
18	ZIM domain containing protein	LOC_Os10g25290
19	transmembrane protein 56	LOC_Os03g46410
20	AAA-type ATPase family protein	LOC_Os11g47970
21	WRKY53	LOC_Os05g27730
22	ABC1 family domain containing protein	LOC_Os07g12530
23	bZIP transcription factor domain containing protein	LOC_Os01g58760
24	metallothionein-like protein	LOC_Os01g10400
25	Ser/Thr receptor-like kinase	LOC_Os01g02560

[a]MSU Isa1 Release 7 (http://rice.plantbiology.msu.edu/)

Silencing of *LRR1* in the *Xa21* Genetic Background Enhances Susceptibility to *Xoo*

To investigate the role of LRR1 in XA21-mediated immunity, RNAi was used to study the effects of *LRR1* silencing in vivo. Kitaake plants carrying *Xa21* with its native promoter, XA21 23A-1-14 (referred to as XA21 rice plants in this paper) (Park et al. 2010), were transformed with the *LRR1-RNAi* (*LRR1Ri*) construct by *Agrobacterium*-mediated plant transformation. Five independent XA21-LRR1Ri T_0 lines were inoculated with *Xoo* strain PXO99Az (PXO99). Independently transformed lines 1, 3, and 4 showed an enhanced susceptibility phenotype and were advanced to the next generation

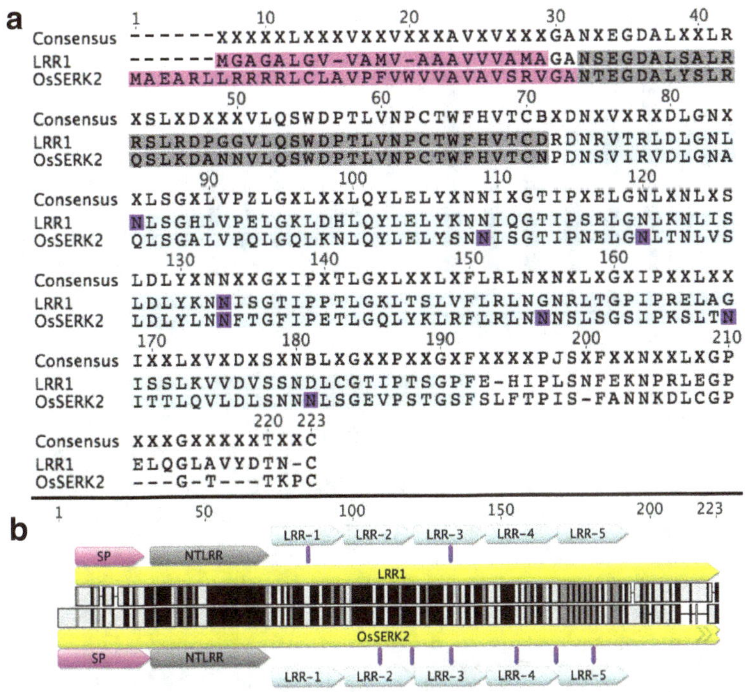

Fig. 1 LRR1 has sequence identity similar to the extracellular region OsSERK2. **a** Amino acid alignment of *LRR1* (Os01g59440) and the consensus region of *OsSERK2* (Os04g38480). Alignments were performed using Geneious R6.1.8 (Kearse et al. 2012). 'X' in the consensus sequence signifies a non-conserved amino acid. *Dashes* indicate gaps. **b** Alignment of LRR1 and OsSERK2 showing predicted structural motifs. SP: Secretion peptide (*pink*), NTLRR: N-terminal LRR capping domain (grey), LRR: Leucine rich repeat (*blue*), PG: Predicted glycosylation site (*purple*). Colors of the amino acid sequences indicate conservation. *Black*: conserved, *grey*: synonymous, *white*: nonsynonymous, *dashes* indicate gaps

(T_1) (Additional file 2: Figure S2). A fourth line, 13, which was not inoculated in the T_0 generation, was also advanced to the T_1 generation. *Xoo* inoculations were repeated in the T_1 generation and enhanced susceptibility was observed in lines 1 and 4. Enhanced susceptibility was not statistically significant in transgenic plants of line 3, although the mean lesion length was longer (combined average of 5.6 ± 2.4 cm) compared to line 13 (4.0 ± 1.7 cm) and XA21 (4.1 ± 1.3 cm) (Additional file 3: Figure S3). In the T_2 generation, 6-weeks old rice from lines 1–1, 1–5, 3–1, 3–8, 4–1, 4–3, 13–1, and 13–7 were inoculated with PXO99. Fourteen days after inoculation, XA21-LRR1Ri rice lines 1–1, 1–5, 3–1, 3–8, 4–1, 4–3 displayed enhanced susceptibility to PXO99 compared to *Xa21*, line 13–1, and 13–7. Kitaake plants were fully susceptible to PXO99 (Fig. 2b and Additional file 4: Figure S4). Silencing of LRR1 was confirmed in three independently transformed lines, 1–5-1, 3–8-6, and 4-1-2 (Fig. 2a), and three additional plants, 1-6-22, 3-2-22, and 4-3-25 (Additional file 5: Figure S5). However, line 13-6-1, which does not show the enhanced susceptibility phenotype did not have reduced expression of *LRR1* (Fig. 2a). This result indicates that silencing of *LRR1* impairs XA21-mediated immunity.

XA21-LRR1Ri Lines Have Reduced Levels of *Xa21* Transcripts

The expression level of *Xa21* transcripts was assayed in XA21-LRR1Ri lines 1-5-1, 3-8-6, 4-1-2, and 13-6-1 (Fig. 3). XA21-LRR1Ri lines 1-5-1, 3-8-6, and 4-1-2, had significant reductions in *Xa21* transcript accumulation. However, XA21-LRR1Ri line 13-6-1 does not reduce expression of *LRR1* and did not have reduction in *Xa21* transcripts (Fig. 3). A pairwise alignment of LRR1Ri and XA21 was performed. The shared nucleotide identity across the consensus region was 56.8% with no consecutive stretch longer than 8 nucleotides, suggesting that XA21 silencing was not caused by off-target silencing of the LRR1Ri construct (Additional file 6: Figure S6). In contrast, the expression of Os11g36180, the Kitaake XA21 homolog, did not have statistically significant reduction in expression in any tested XA21-LRR1Ri line compared to XA21 (Additional file 7: Figure S7). Additionally, we tested if *LRR1* silencing affected other immune receptors or components of XA21-mediated immunity. Transcript abundance was not reduced in two other immune receptors, rice CHITIN ELICITOR RECEPTOR KINASE 1 (*OsCERK1*) and *OsFLS2* (Additional file 7: Figure S7). Similarly, reduced expression was not observed in the XA21 signaling components OsSERK2,

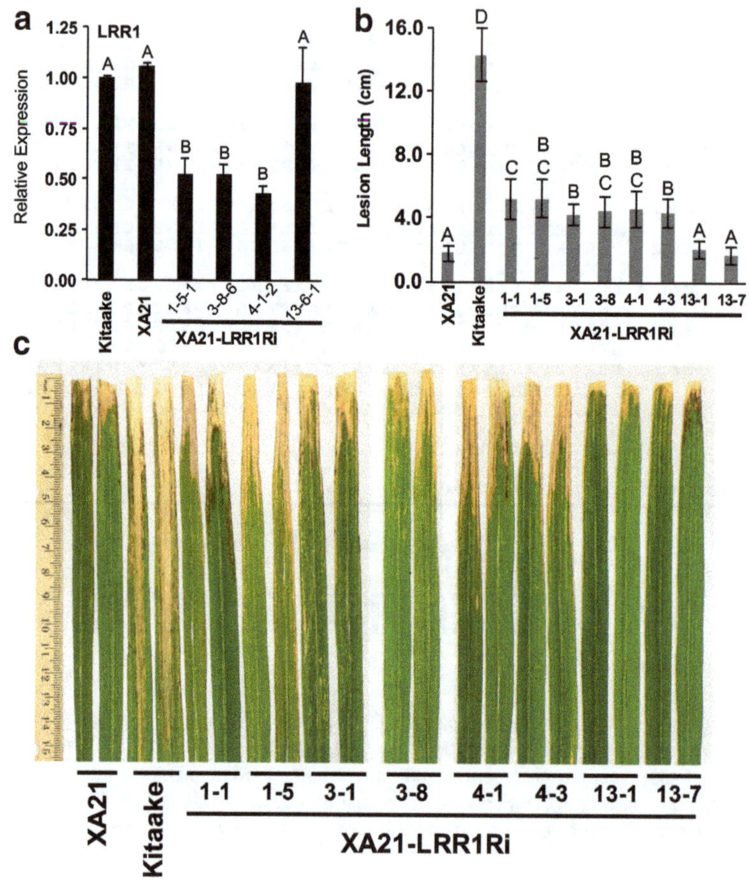

Fig. 2 Silencing of *LRR1* in the XA21 genetic background enhances susceptibility to *Xoo*. **a** Relative expression level of *LRR1* in four independent transgenic rice lines. Bars depict the average and standard deviation of *LRR1* expression normalized to Kitaake of two technical replicates. Different letters indicate a significant difference in gene expression ($P < 0.05$, ANOVA, Tukey-HSD). This experiment was repeated three times with similar results. **b** Lesion length of T_2 generation XA21-LRR1Ri plants 14 days after inoculation with PXO99. Bars indicate the average lesion length and standard deviation of ten to 15 combined plants that had one to six inoculated leaves each. Different letters indicate a significant difference in lesion length ($P < 0.05$, Kruskal-Wallis test, Dunn's *post-hoc* test with Benjamini–Hochberg correction). Lesion length of individual T_2 plants from this experiment are shown in Additional file 4: Figure S4. Experimental results were repeated three times, with similar results. **c** Two representative inoculated leaves from each line were photographed at the time of lesion scoring

XB3, XB15, XB21, or XB24, indicating that *LRR1Ri* specifically reduced *Xa21* transcript levels. To determine if silencing of OsSERK2 also reduces *Xa21* transcript levels, we compared *Xa21* expression between XA21 rice and a homozygous line expressing an OsSERK2 silencing construct (XA21-OsSERKRi) (Chen et al. 2014). No statistical difference in *Xa21* expression was detected, indicating that silencing of *OsSERK2* does not significantly reduce *Xa21* transcript levels (Additional file 8: Figure S8).

To assess if *Xa21* transcript reduction was specific to the *LRR1Ri* construct, two sibling LRR1 silenced lines from line 3 that had lesion lengths measured in the T_0 and T_1 generation were analyzed for *Xa21* expression in the T_2 generation. XA21-LRR1Ri line 3–2 is homozygous for the transgene. Sibling line, XA21-LRR1Ri 3–3, is a null-segregant lacking the *LRR1Ri* transgene. Reduction in *Xa21* transcripts was specific to line 3–2, which contains the *LRR1Ri* construct, while line 3–3 did not display similarly reduced levels of *Xa21* transcripts (Additional file 9: Figure S9). Taken together, these results indicate that *Xa21* transcript reduction correlates with the *LRR1* silencing and is not due to aberrant expression of *Xa21* in the genetic background used for transformation.

XA21-LRR1Ri Plants Fail to Express XA21 Genetic Markers After RaxX21-sY Peptide Treatment

We previously demonstrated that treatment of XA21 leaves with sulfated RaxX21-sY peptides induces expression of specific genetic markers. In contrast, treatment of XA21 plants with non-sulfated RaxX21 peptides do not induce these marker genes (Pruitt et al. 2015). To determine if the XA21-LRR1Ri plants fail to respond to RaxX, rice leaves from *LRR1* silencing lines were treated

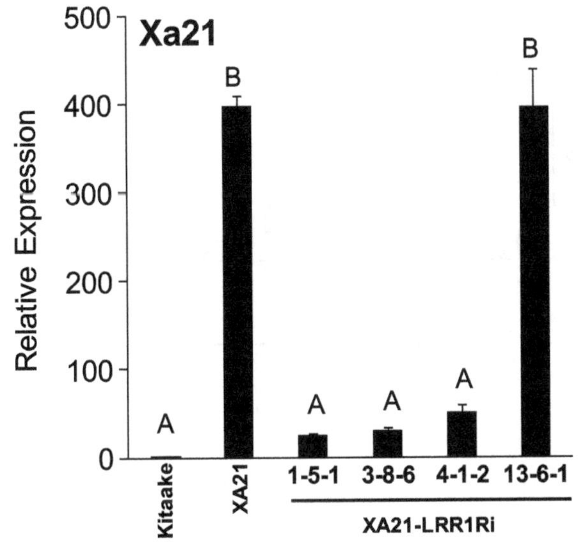

Fig. 3 *Xa21* expression is reduced in XA21-LRR1Ri rice lines. Relative expression of *Xa21* in four independent transgenic rice lines. Bars depict the average and standard deviation of *Xa21* expression normalized to Kitaake of two technical replicates. Different letters indicate a significant difference in gene expression ($P < 0.05$, ANOVA, Tukey-HSD). This experiment was repeated three times with similar results

with RaxX21-sY peptides. Six hours after RaxX21-sY peptide application, the expression level of *PR10b* (Fig. 4a), *Os10g28299* (Fig. 4b), or *Os04g10010* (Fig. 4c) were assayed in Kitaake, XA21, and XA21-LRR1Ri lines. We observed that RaxX21-sY treatment of XA21 plants induced expression of these marker genes as previously reported (Pruitt et al. 2015). In contrast, RaxX21-sY peptide treatment failed to induce the expression of the three stress-related genes in the XA21-LRR1Ri lines. These results suggest that a reduced ability to initiate an XA21-mediated immune response contributes to the enhanced susceptibility observed in XA21-LRR1Ri plants.

Overexpression of *LRR1* in the Kitaake Background Does not Confer Resistance to *Xoo*

Previously it was reported that overexpression of *LRR1* did not enhance rice resistance to *Xoo* strain LN44 in the Japonica rice variety SN1033 (Zhou et al. 2009). Because we observed that *LRR1* expression is required for XA21-mediated immunity, we examined if overexpression of *LRR1* is sufficient to enhance resistance to PXO99 in the rice variety Kitaake, which lacks *Xa21*. The overexpression of LRR1 in the XA21 background was not tested because XA21 plants are already resistant to *Xoo*. We generated sixteen transgenic rice plants expressing *LRR1* under the control of the maize ubiquitin promoter (LRR1ox) by *Agrobacterium*-mediated plant transformation. Fourteen of 16 transgenic lines contained the transgene as determined by PCR analysis. Kitaake, XA21, and ten T_0 lines were inoculated with PXO99. Four of ten lines displayed significantly shorter lesion lengths than Kitaake (Additional file 10: Figure S10). Two additional T_0 lines, LRR1ox-15 and LRR1ox-16, were inoculated at a later date and also displayed shorter lesions than Kitaake (Additional file 10: Figure S10). Four T_0 Lines, LRR1ox-7, –10, –12, and –16 were advanced to the next generation. Inoculations were repeated in the T_1 generation. LRR1ox lines 7, 10, 12, and 16 were all fully susceptible to *Xoo* (Additional file 11: Figure S11).

Overexpression of *LRR1* in LRR1ox lines 7–10, 7–12, 10–2, 10–3, 12–1, 12–3, 16–1 and 16–2 was confirmed in the T_2 generation by quantitative RT-PCR (Fig. 5a) and the plants were inoculated with PXO99. Lesion development was measured 14 days after inoculation and no significant difference in lesion development was found compared with the nontransgenic Kitaake control plants (Fig. 5b). The partial resistance phenotype of LRR1ox lines in the T_0 generation was not observed in the T_1 or T_2 generation inoculations. Because T_0 plants are regenerated from rice calli, we hypothesize that the shorter lesions are due to differences in the leaf developmental stage at the time of inoculation compared to Kitaake, which was germinated from seed. Together, these results indicate that overexpression of *LRR1* in Kitaake is not sufficient to enhance resistance to *Xoo*, thus confirming the previous report by Zhou et al. 2009, but in a different genetic background.

Discussion

To elucidate the function of XA21 and the XA21 binding protein XB21, we characterized LRR1, which interacts with XB21. LRR1 is a leucine rich repeat containing protein with similarity to the LRR domain of OsSERK2 (Chen et al. 2014) (Fig. 1), which is a co-receptor for XA21. Silencing of *LRR1* in the resistant XA21 genetic background conferred enhanced susceptibility to *Xoo* and reduced the expression of *Xa21* (Fig. 2b). These results suggest that LRR1 is required for *Xa21* transcription.

LRR1 Functions in Plant Immunity

Our results support previous studies demonstrating a role for LRR1 in plant immunity. For example, overexpression of rice *LRR1* in *Arabidopsis thaliana* and Chinese cabbage is sufficient to confer enhanced resistance against the bacterial pathogens *Pseudomonas syringae* pv. *tomato* DC3000 (*Pst* DC3000) (Zhou et al. 2009; Zhou et al. 2010) and *Pectobacterium carotovorum* subsp. *carotovorum* (Park et al. 2012) respectively. In contrast to these results, we found overexpression of *LRR1* in the Kitaake rice variety did not enhance resistance to *Xoo* (Fig. 5b). A similar result was reported by Zhou et al. 2009, who found that overexpression of *LRR1* did not enhance resistance to *Xoo*

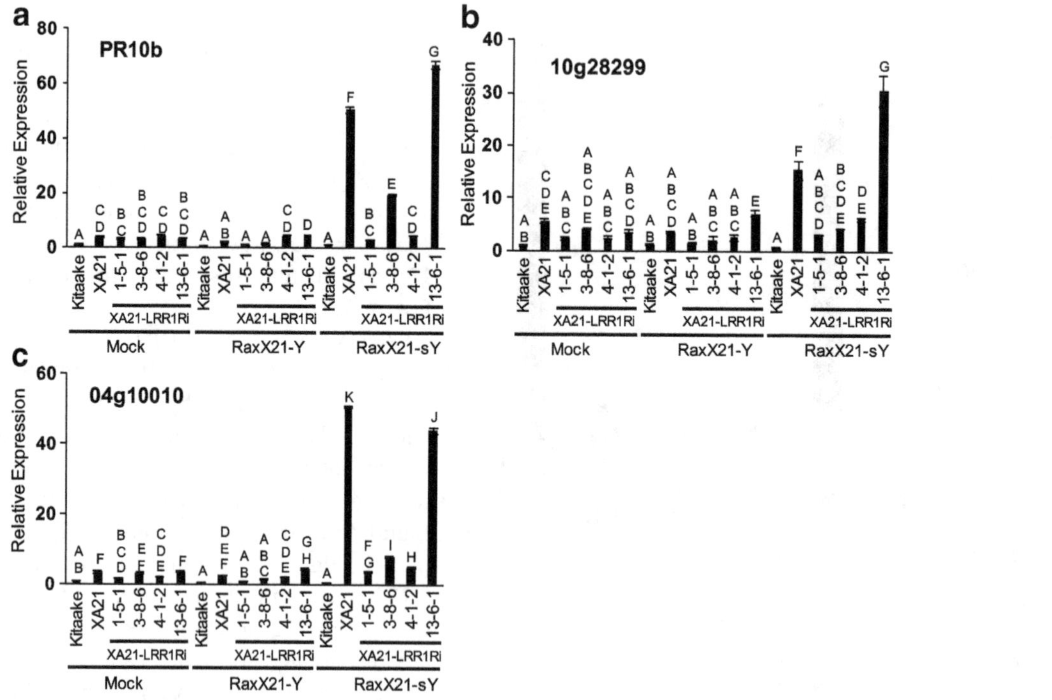

Fig. 4 XA21-LRR1Ri rice plants are unresponsive to RaxX21-sY peptide treatment. Gene expression of *PR10b* (**a**), *10 g28299* (**b**), and *04 g10010* (**c**) 6 h after 500nM application of RaxX21, RaxX21-sY, or mock treatment in mature leaves of the four independent XA21-LRR1Ri rice lines. Bars depict the average and standard deviation of expression level normalized to Kitaake mock treatment of two technical replicates. Different *letters* indicate a significant difference in gene expression ($P < 0.05$, ANOVA, Tukey-HSD). Experiment was performed two times with similar results

strain LN44 in the rice variety SN1033 (Zhou et al. 2009). The *LRR1* ortholog of pepper, *CaLRR1*, which is induced upon infection with *Xanthomonas campestris* pv. *vesicatoria* (*Xcv*) (Jung et al. 2004), did not enhance resistance to *Pst* DC3000 when overexpressed in *Arabidopsis* (Choi et al. 2012). However, overexpression enhanced callose deposition and reduced conidiospore production during *Hyaloperonospora arabidopsidis* infection (Choi et al. 2012). *CaLRR1* silenced pepper plants have enhanced susceptibility to *Xcv* and reduced H_2O_2 production (Hwang et al. 2014). They also have reduced production of defense gene expression and total SA accumulation (Choi et al. 2012). Together, our study and previous reports demonstrate that the LRR1 family of proteins have diverse effects on plant immune responses.

The Function of LRR1 is Distinct from OsSERK2

OsSERK2 is required for OsBRI1-mediated development and regulating XA21, XA3 and OsFLS2-mediated immune responses (Chen et al. 2014). XA21 rice plants with reduced levels of OsSERK2 (XA21-OsSERK2Ri) have compromised resistance to *Xoo* (Chen et al. 2014). Despite the structural similarity between the LRR domains of LRR1 and the OsSERK2, it is likely that their functions are distinct. The intracellular domain of OsSERK2 is essential for the interaction between OsSERK2 and the intracellular domains of XA21, XA3, OsFLS2, and OsBRI1 (Chen et al. 2014). Direct interaction with these cell-surface receptors is thought to be critical for OsSERK2 functionality. Unlike OsSERK2, LRR1 lacks an intracellular domain. When XA21-LRR1Ri rice lines expressed *LRR1Ri*, we consistently detected a significant reduction in *Xa21* transcript abundance (Fig. 3). Because *LRR1Ri* causes reduced *Xa21* expression, we determined whether silencing *OsSERK2* had a similar effect. However, XA21-OsSERK2Ri rice plants do not have reduced *Xa21* transcript levels (Additional file 8: Figure S8). This suggests that the enhanced susceptibility conferred by *LRR1Ri* is mechanistically different than *OsSERK2* silencing in rice plants.

LRR1 Interacts With OsHIR1

The results of our study suggest a new role of LRR1 in regulating the expression of a rice immune receptor. Previous studies demonstrated that LRR1 interacts with HYPERSENSITIVE INDUCED REACTION PROTEIN 1 (OsHIR1) in rice (Zhou et al. 2010; Zhou et al. 2009). While the role of OsHIR1 in rice immunity has not been extensively studied, proteomic and gene expression analysis of XA21 rice plants indicate that *OsHIR1* expression increased after inoculation with *Xoo*. OsHIR1 is also thought to be post-translationally modified in XA21

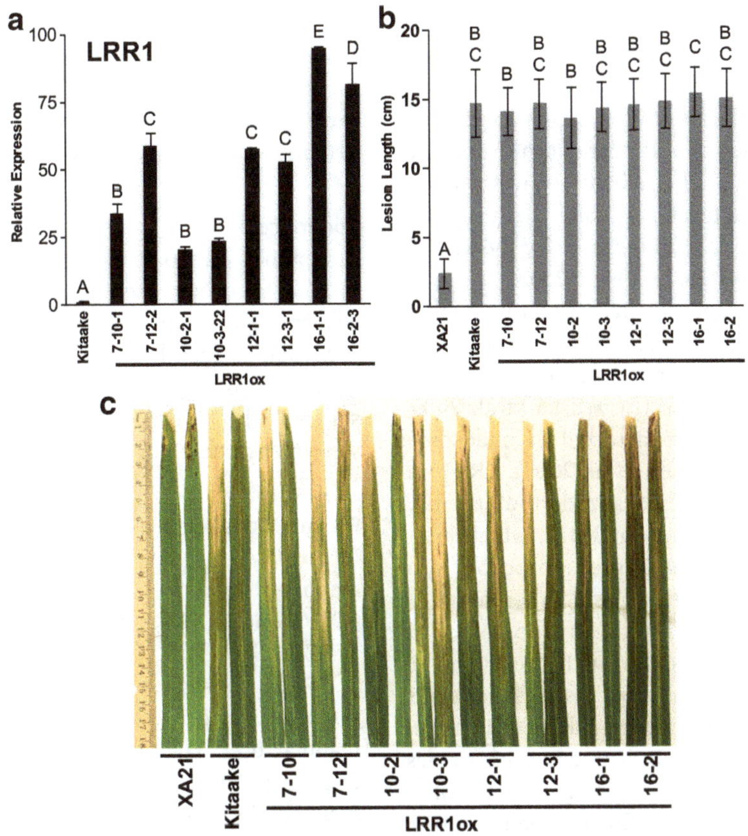

Fig. 5 Kitaake rice overexpressing *LRR1* are susceptible to *Xoo*. **a** Bars depict the average and standard deviation of *LRR1* expression normalized to Kitaake of two technical replicates. Different letters indicate a significant difference in gene expression ($P < 0.05$, ANOVA, Tukey-HSD). This experiment was repeated three times with similar outcomes. **b** Lesion length of LRR1ox 14 days after inoculation with PXO99. Bars indicate the mean lesion length and standard deviation on four to nine plants that had two to nine inoculated leaves. Different letters indicate a significant difference in lesion length ($P < 0.05$, Kruskal-Wallis test, Dunn's *post-hoc* test with Benjamini–Hochberg correction). This experiment was repeated 4 times with similar results. **c** Two representative inoculated leaves from each line were photographed at the time of lesion scoring

plants challenged with *Xoo* (Chen et al. 2007), suggesting that OsHIR1 might function in XA21-mediated immunity.

The orthologous proteins CaLRR1 and CaHIR1 also interact in pepper (Jung and Hwang 2007). Overexpression of *OsHIR1* (Zhou et al. 2010) or *CaHIR1* (Jung and Hwang 2007) in *Arabidopsis* both promoted the formation of cell death lesions (Jung and Hwang 2007). The cell death symptom of *CaHIR1* overexpressing plants was alleviated by simultaneously overexpressing *CaLRR1* (Jung and Hwang 2007). CaHIR1 positively regulates RESISTANCE TO PSEUDOMONAS SYRINGAE PROTEIN 2 (RPS2)-mediated immunity to *Xcv* (Choi et al. 2013; Jung and Hwang 2007; Jung et al. 2004; Choi et al. 2011). During infection by *Xcv*, the virulence factor FILAMENTOUS HAEMAGGLUTININ ADHESIN 1 (Fha1) targets CaHIR1 where it modulates cell death and suppresses pepper *PR* gene expression (Choi et al. 2013). Under normal conditions, CaLRR1 and CaHIR1 both co-localize at the plasma membrane (Choi et al. 2011), while CaHIR1-Fha1 heterodimers were mostly found in endosomes during *Xcv* infection (Choi et al. 2013). These results suggest that the plasma membrane localization of CaHIR1 is important for the CaHIR1-mediated resistance.

LRR1 is Required for XA21 Expression

The transcriptional regulation of plant immune receptors is not well studied, and we do not know how LRR1 expression is linked to the transcription of *Xa21*. A recent report identified four receptor-like cytoplasmic kinases that also modulate *proA-Xa21* expression in the Kitaake genetic background (Zhou et al. 2016). Future experiments are needed to understand whether LRR1 is regulating *Xa21* through the same pathway or parallel pathways.

While we did not observe high complementarity between Xa21 and LRR1 (Additional file 6: Figure S6), we cannot rule out the possibility that the primary siRNA derived from processing of *LRR1Ri* or a transitive secondary siRNA produced upon cleavage of *LRR1* have sufficient complementarily with *Xa21* to trigger mRNA cleavage. However, it is known that the expression of plant immune receptors must be tightly regulated for optimal growth and overexpression of plant immune receptors can decrease plant fitness (Bergelson and Purrington 1996; Huot et al. 2014). For example, a 9% reduction in fitness was observed in *Arabidopsis* ecotype Bla-2 after the resistance gene RESISTANCE TO P. SYRINGAE PV MACULICOLA 1 (*RPM1*) was inserted (Tian et al. 2003).

Previous studies have reported that proteins involved in membrane trafficking including clatharin adapters, accessory proteins, and endosomal proteins facilitate the modification of gene expression (Pyrzynska et al. 2009). In rice, OsHIR1 and LRR1 are detected in endosomes and the plasma membrane (Zhou et al. 2009). LRR1 enhances the plasma membrane localization of OsHIR1 (Zhou et al. 2010). It is possible that XB21, which interacts with LRR1 in yeast, facilitates the transport of proteins such as OsHIR1 from the plasma membrane to endosomes. XB21 contains a C-terminal J domain. This class of protein has been associated with the uncoating of clatharin-coated vesicles, which occurs after endocytic vesicles bud off from the plasma membrane (Lemmon 2001; Kampinga and Craig 2010). These studies suggest that endosomal accumulation of OsHIR1 and other plasma membrane proteins may influence the expression of genes encoding receptors such as XA21.

Conclusions

Rice LRR1 and its pepper ortholog were previously reported to enhance resistance to bacterial pathogens in *Arabidopsis* and Chinese cabbage. In this study, we identified a new role of LRR1: transcriptional regulation of the rice immune receptor *Xa21*.

Methods

Plant Material and Growing Conditions

Rice plants (*Oryza sativa* L.) were germinated in water filled petri dishes at 28 °C and transferred to the greenhouse (~28–30 °C, 75–85% humidity) after 7 days. Rice plants were grown in 80/20 (sand/peat) soil mixture. Greenhouse grown rice plants were supplemented with artificial light from November to April to obtain 14 h days. Six-week old greenhouse grown plants were transferred to growth chambers for inoculation. Growth chambers for inoculation were set at 28 °C Day/26 °C Night, 85% humidity, 16 h days. Inoculations were performed a minimum of 3 days after transferring plants to reduce plant stress from transfer. For gene expression assays, rice plants were transferred to growth chambers (22 °C, 80% humidity, 16 h days) after 7 days and grown in hydroponics conditions using Hoagland solution changed twice weekly. Fully expanded leaves from 4-week old hydroponics grown plants were cut into 1.5 cm strips and floated overnight in Milli-Q H_20 to reduce wounding. Mock, RaxX21-Y or RaxX21-sY peptide treatments were applied at a final concentration of 500nM for 6 h. Leaf fragments were flash frozen in liquid nitrogen and stored at −80 °C for further analysis.

Xoo Inoculations

Philippines race 6 strain of *Xoo* (PXO99Az) was plated on peptone sucrose agar plates for 3 days then suspended in Milli-Q H_2O to a concentration of OD_{600} = 0.5. Fully expanded and healthy rice leaves were inoculated approximately 2 cm from the leaf tip using the scissors- clip method (Chern et al. 2005). Lesion lengths were measured 12 or 14 days after inoculation.

Plasmid Construction

Rice cDNA encoding full-length *LRR1* (Os01g59440) was amplified from rice cDNA using the primers 5′- CACCATGGGGGCGGGGGCGCTGGG -3′ (LRP-FL-S) and 5′- CTAGCAGTTGGTGTCATATACAGC-3′ (LRP-FL-A S). PCR products were cloned into pENTR/D-TOPO vector (Invitrogen) using the instructions provided by the manufacturer. The proper sequence was confirmed by sequencing. The *LRR1* overexpression construct was constructed by recombining the LRR1/pENTR/D-TOPO construct into the gateway-compatible Ubi-CAMBIA-1300 (Chern et al. 2005). For *E. coli* expression, XB21/pENTR/D-TOPO, which encodes full-length XB21 protein (Os12g36180), was amplified from rice cDNA using primers 5′-CACCATGGACGACTTCCAGGGCCTCCT GGCC-3′ and 5′-TTAGAAGAGTTCCTCTGAGTTGA ATTTG-3′ (Park et al. manuscript in preparation). LRR1/pENTR/D-TOPO and XB21/pENTR/D-TOPO were recombined into pDEST15 and pDEST17 expression vectors respectively. The pGEX-1 vector (Chen et al. 2010) was used for expression of GST. For silencing of *LRR1*, the 357 bp fragment shared by three predicted isoforms of *LRR1* was amplified from rice cDNA using the primers 5′-CACCTTGGGAATTTGAACTTATCT GGTC-3′ (LRP-RNAi-S) and 5′-AGTTGCTTAGGGG AATGTGCTC-3′ (LRP-RNAi-AS), and cloned into pENTR/D-TOPO vector to produce LRR1Ri/pENTR/D-TOPO. The LRR1Ri/pENTR/D-TOPO was recombined into the pANDA silencing vector (Miki and Shimamoto 2004) using gateway LR clonase (Invitrogen).

Rice Transformation

Kitaake rice plants and homozygous Kitaake plants carrying *Xa21* with its native promoter, XA21 23A-1-14 (referred to as XA21 in this paper) (Park et al. 2010), were transformed as described previously (Chern et al. 2005) using *Agrobacterium* strain EHA105 to infect rice calli. Transformants carrying the transgenes were selected using hygromycin and later confirmed by PCR.

Yeast two-Hybrid

The N-terminal deleted XB21 (XB21△NT) (Os12g36180) covering amino acids 433–926 was amplified from rice cDNA using the primers 5′-CACCATGTCAATAGATG AACTGGAAGATTTT-3′ (XB21-A-S) and 5′-TTAGA AGAGTTCCTCTGAGTTGAATTTG-3′ (XB21-FL-AS), cloned into the yeast two-hybrid bait vector pMC86, and transformed into yeast strain HF7c MATa. Transformation of target yeast (Y187) used cDNA obtained from a Hybrizap (Stratagene) yeast two-hybrid library derived from leaf MRNA of 7-week-old Indica rice cultivar IRBB21. Screening was performed as described previously (Seo et al. 2011).

Gene Expression Assays and Quantitative RT-PCR

RNA was extracted from rice plants using Spectrum Plant Total RNA Kit (Sigma) followed by DNase treatment performed using Turbo DNA-free DNase (Ambion). RNA integrity was confirmed by agarose gel electrophoresis (with 0.1% SDS). 2 μg of RNA was used as template for cDNA synthesis, which was performed using the High-Capacity cDNA Reverse Transcription Kit (Applied Biosystems). Quantitative RT-PCR using SsoFast EvaGreen Supermix (Bio-Rad) was performed on a CFX96 Real-Time system coupled to a C1000 Thermal Cycler (Bio-Rad) with the conditions: 95 °C for 10 s, 60 °C for 10 s, 40 cycles. Amplicon specificity was confirmed by checking the melting curve at the completion of 40 cycles. Expression of actin was used to normalize samples for analysis. Technical replication was performed by amplifying the same cDNA sample 2 or 3 times as described in the individual figure legends.

Statistical Analyses

The normal distribution of data was tested using the Shapiro-Wilk test. Normally distributed data was tested using a one-way analysis of variance (ANOVA) test. Statistically significant results were followed up with a Tukey's honest significance (Tukey-HSD) *post-hoc* test using JMP v.13.1 software (SAS Institute Inc.). A Kruskal-Wallis non-parametric test was applied to data that violated normality assumptions. Statistically significant results were followed up with a Dunn *post-hoc* test with Benjamini–Hochberg correction (Ogle 2017) using R v.3.3.2 (Team 2016). Significance for all statistical tests was predetermined at $p < 0.05$.

Primers

A complete list of primers used in this study can be found in Additional file 12: Table S1.

Additional files

Additional file 1: Figure S1. His-XB21 and GST-LRR1 interact in vitro. His-XB21, GST, and GST-LRR1 proteins were immunoprecipitated from *E. coli* using Ni-NTA resin and Glutathione Sepharose 4B respectively (lanes 1,2,3). His-XB21 was added to bound GST and GST-LRR1. After co-incubation and centrifugation, unbound supernatant (lanes 4,5) and co-immunoprecipitated (Co-IP) (lanes 5,6) fractions were obtained. The samples were then subjected to SDS-PAGE for western analysis using anti-HIS and anti-GST antibodies. This experiment was performed twice with similar results.

Additional file 2: Figure S2. T_0 generation *Xoo* inoculation of XA21-LRR1Ri lines. Lesion length of XA21-LRR1Ri plants 12 days after inoculation with PXO99. Bars indicate the average lesion length and standard deviation on individual rice plants that had three to 19 inoculated leaves. Different letters indicate a significant difference in lesion length ($P < 0.05$, Kruskal-Wallis test, Dunn's post-hoc test with Benjamini–Hochberg correction).

Additional file 3: Figure S3. T_1 generation *Xoo* inoculation of XA21-LRR1Ri lines. Lesion length of XA21-LRR1Ri plants 12 days after inoculation with PXO99. Bars indicate the average lesion length and standard deviation on individual rice plants that had one to four inoculated leaves. Different letters indicate a significant difference in lesion length ($P < 0.05$, Kruskal-Wallis test, Dunn's *post-hoc* test with Benjamini–Hochberg correction). Gray bars indicate the presence of the LRR1Ri construct, white bars indicate null-segregants.

Additional file 4: Figure S4. T_2 generation *Xoo* inoculation of XA21-LRR1Ri lines. Lesion length of XA21-LRR1Ri plants 14 days after inoculation with PXO99. Bars indicate the average lesion length and standard deviation on individual rice plants that had one to six inoculated leaves. Gray bars indicate the presence of the LRR1Ri construct, white bars indicate null-segregants. Different letters indicate a significant difference in gene expression ($P < 0.05$, ANOVA, Tukey-HSD). (PDF 28 kb)

Additional file 5: Figure S5. *LRR1* expression is reduced in XA21-LRR1Ri rice lines. Relative expression of *LRR1* in three independent transgenic rice lines. Bars depict the average and standard deviation of *LRR1* expression normalized to XA21 of two technical replicates.

Additional file 6: Figure S6. Pairwise alignment of LRR1Ri and XA21. The region of LRR1 used to develop the LRR1Ri vector was aligned to XA21 using Geneious R6.1.8 (Kearse et al. 2012). Nucleotide numbers refer to the original sequences. Highlighted in black are the conserved nucleotides between the two sequences. The shared identity within the consensus region is 56.8%.

Additional file 7: Figure S7. *LRR1* silencing does not reduce the expression of three rice receptor kinases or components of XA21-mediated immunity. Relative expression of *OsSERK2, Os11g36180, OsFLS2, OsCERK1, XB3, XB15, XB21*, and *XB24* in four independent XA21-LRR1Ri transgenic rice lines. Bars depict the average and standard deviation of expression level normalized to XA21 of three technical replicates. Different letters indicate a significant difference in gene expression ($P < 0.05$, ANOVA, Tukey-HSD). This experiment was repeated at least two times with similar results.

Additional file 8: Figure S8. XA21-OsSERK2Ri plants do not reduce *Xa21* expression. Relative expression of *OsSERK2* (A) and *Xa21* (B) in Kitaake, XA21, and XA21-OsSERK2Ri (homozygous line A814) transgenic rice. Bars depict the average and standard deviation of expression level normalized to XA21 of two technical replicates. Different letters indicate a significant difference in gene expression ($P < 0.05$, ANOVA, Tukey-HSD). This experiment was repeated three times with similar results.

Additional file 9: Figure S9. XA21-LRR1Ri null segregants do not have reduced levels of *Xa21* expression. *Xa21* expression of six individual plants per line. Line 3–2 is homozygous for the *LRR1Ri* transgene. Line 3–3 is a null segregant for *LRR1Ri*. Bars depict the average and standard deviation of *Xa21* expression normalized to Kitaake of two technical replicates. Different letters indicate a significant difference in gene expression ($P < 0.05$, ANOVA, Tukey-HSD).

Additional file 10: Figure S10. T_0 generation *Xoo* inoculation of LRR1ox lines. Lesion length of LRR1ox plants 14 days after inoculation with PXO99. Bars indicate the average lesion length and standard deviation on individual rice plants that had four to 13 inoculated leaves. Different letters indicate a significant difference in lesion length ($P < 0.05$, Kruskal-Wallis test, Dunn's *post-hoc* test with Benjamini–Hochberg correction).

Additional file 11: Figure S11. T_1 generation *Xoo* inoculation of LRR1ox lines. Lesion length of LRR1ox plants 14 days after inoculation with PXO99. Bars indicate the average lesion length and standard deviation on individual rice plants that had one to eight inoculated leaves. Different letters indicate a significant difference in lesion length ($P < 0.05$, Kruskal-Wallis test, Dunn's *post-hoc* test with Benjamini–Hochberg correction). Gray bars indicate the presence of the LRR1ox construct, white bars indicate null-segregants.

Additional file 12: Table S1. Full list of primers used in this study.

Abbreviations

ANOVA: Analysis of variance test; BAK1: BRI1-ASSOCIATED RECEPTOR KINASE 1; BRI1: BRASSINOSTEROID-INSENSITIVE 1; CERK1: CHITIN ELICITOR RECEPTOR KINASE 1; EFR: EF-TU RECEPTOR; elf18: 18 amino acid epitope of elongation factor-Tu; Fha1: FILAMENTOUS HAEMAGGLUTININ ADHESIN 1; flg22: 22 amino acid epitope of bacterial flagellin; FLS2: FLAGELLIN SENSING 2; HIR1: HYPERSENSITIVE INDUCED REACTION PROTEIN 1; LRR1: LEUCINE-RICH REPEAT PROTEIN 1; LRR1ox: LRR1 overexpression construct driven by the maize ubiquitin promoter; LRR1Ri: LRR1 RNAi silencing construct; LRR-RLK: leucine-rich repeat receptor-like kinase; *Pst* DC3000: *Pseudomonas syringae* pv. *tomato* DC3000; PXO99: Philippine *Xoo* strain PXO99Az; RaxX: REQUIRED FOR ACTIVATION OF XA21-MEDIATED IMMUNITY X; RaxX21-sY: 21 amino acid tyrosine sulfated epitope of RaxX; RPM1: RESISTANCE TO P. SYRINGAE PV MACULICOLA 1; RPS2: RESISTANCE TO PSEUDOMONAS SYRINGAE PROTEIN 2; SERK: SOMATIC EMBRYOGENESIS RECEPTOR KINASE; Tukey-HSD: Tukey's honest significance test; XA21: *XOO* RESISTANCE 21; XA21-LRR1Ri: XA21 rice plants containing the LRR1Ri construct; XB21: XA21 BINDING PROTEIN 21; XB21IPs: XB21 interacting proteins; XB21ΔNT: N-terminal deleted XB21; *Xcv*: *Xanthomonas campestris* pv. *vesicatoria*; *Xoo*: *Xanthomonas oryzae* pv. *oryzae*

Acknowledgements

We thank Xia Xu for her contribution to the yeast two-hybrid screen.

Funding

This work was supported by the National Institute of Health (NIH, GM55962) and the National Science Foundation (NSF, IOS-0817738). DC was supported by the Monsanto Beachell-Borlaug International Scholars Program.

Authors' contributions

PC generated the transgenic rice plants. DC, CP, NT performed the experiments. DC, CP, PR wrote the manuscript. All authors discussed the research and approved the manuscript.

Competing interests

The authors declare that they have no competing interests.

Author details

[1]Department of Plant Pathology and the Genome Center, University of California, Davis, Davis, CA 95616, USA. [2]Present Address: Department of Bioresources Engineering and PERI, Sejong University, Seoul 05006, Republic of Korea. [3]Present Address: Department of Plant Pathology and Microbiology, University of California, Riverside, Riverside, CA 92521, USA.

References

Bergelson J, Purrington CB (1996) Surveying Patterns in the Cost of Resistance in Plants. Am Nat 148(3):536–558

Boller T, Felix G (2009) A renaissance of elicitors: perception of microbe-associated molecular patterns and danger signals by pattern-recognition receptors. Annu Rev Plant Biol 60:379–406. doi:10.1146/annurev.arplant.57.032905.105346

Chen F, Yuan Y, Li Q, He Z (2007) Proteomic analysis of rice plasma membrane reveals proteins involved in early defense response to bacterial blight. Proteomics 7(9):1529–1539. doi:10.1002/pmic.200500765

Chen X, Chern M, Canlas PE, Jiang C, Ruan D, Cao P, Ronald PC (2010) A conserved threonine residue in the juxtamembrane domain of the XA21 pattern recognition receptor is critical for kinase autophosphorylation and XA21-mediated immunity. J Biol Chem 285(14):10454–10463. doi:10.1074/jbc.M109.093427

Chen X, Zuo S, Schwessinger B, Chern M, Canlas PE, Ruan D, Zhou X, Wang J, Daudi A, Petzold CJ, Heazlewood JL, Ronald PC (2014) An XA21-associated kinase (OsSERK2) regulates immunity mediated by the XA21 and XA3 immune receptors. Mol Plant 7(5):874–892. doi:10.1093/mp/ssu003

Chern M, Canlas PE, Fitzgerald HA, Ronald PC (2005) Rice NRR, a negative regulator of disease resistance, interacts with Arabidopsis NPR1 and rice NH1. Plant J 43(5):623–635. doi:10.1111/j.1365-313X.2005.02485.x

Chinchilla D, Zipfel C, Robatzek S, Kemmerling B, Nurnberger T, Jones JD, Felix G, Boller T (2007) A flagellin-induced complex of the receptor FLS2 and BAK1 initiates plant defence. Nature 448(7152):497–500. doi:10.1038/nature05999

Choi HW, Kim YJ, Hwang BK (2011) The hypersensitive induced reaction and leucine-rich repeat proteins regulate plant cell death associated with disease and plant immunity. Mol Plant Microbe Interact 24(1):68–78. doi:10.1094/MPMI-02-10-0030

Choi DS, Hwang IS, Hwang BK (2012) Requirement of the cytosolic interaction between PATHOGENESIS-RELATED PROTEIN10 and LEUCINE-RICH REPEAT PROTEIN1 for cell death and defense signaling in pepper. Plant Cell 24(4):1675–1690. doi:10.1105/tpc.112.095869

Choi HW, Kim DS, Kim NH, Jung HW, Ham JH, Hwang BK (2013) Xanthomonas filamentous hemagglutinin-like protein Fha1 interacts with pepper hypersensitive-induced reaction protein CaHIR1 and functions as a virulence factor in host plants. Mol Plant Microbe Interact 26(12):1441–1454. doi:10.1094/MPMI-07-13-0204-R

Huot B, Yao J, Montgomery BL, He SY (2014) Growth-defense tradeoffs in plants: a balancing act to optimize fitness. Mol Plant 7(8):1267–1287. doi:10.1093/mp/ssu049

Hwang IS, du Choi S, Kim NH, Kim DS, Hwang BK (2014) Pathogenesis-related protein 4b interacts with leucine-rich repeat protein 1 to suppress PR4b-triggered cell death and defense response in pepper. Plant J 77(4):521–533. doi:10.1111/tpj.12400

Jung HW, Hwang BK (2007) The leucine-rich repeat (LRR) protein, CaLRR1, interacts with the hypersensitive induced reaction (HIR) protein, CaHIR1, and suppresses cell death induced by the CaHIR1 protein. Mol Plant Pathol 8(4):503–514. doi:10.1111/j.1364-3703.2007.00410.x

Jung EH, Jung HW, Lee SC, Han SW, Heu S, Hwang BK (2004) Identification of a novel pathogen-induced gene encoding a leucine-rich repeat protein expressed in phloem cells of Capsicum annuum. Biochim Biophys Acta 1676(3):211–222. doi:10.1016/S0167-4781(03)00120-9

Kampinga HH, Craig EA (2010) The HSP70 chaperone machinery: J proteins as drivers of functional specificity. Nat Rev Mol Cell Biol 11(8):579–592. doi:10.1038/nrm2941

Kearse M, Moir R, Wilson A, Stones-Havas S, Cheung M, Sturrock S, Buxton S, Cooper A, Markowitz S, Duran C, Thierer T, Ashton B, Mentjies P, Drummond A (2012) Geneious Basic: an integrated and extendable desktop software platform for the organization and analysis of sequence data. Bioinformatics 28(12):1647–1649

Kolade OO, Bamford VA, Ancillo Anton G, Jones JD, Vera P, Hemmings AM (2006) In vitro characterization of the cysteine-rich capping domains in a plant leucine rich repeat protein. Biochim Biophys Acta 1764(6):1043–1053. doi:10.1016/j.bbapap.2006.03.014

Lemmon SK (2001) Clathrin uncoating: Auxilin comes to life. Curr Biol 11(2):R49–R52

McAndrew R, Pruitt RN, Kamita SG, Pereira JH, Majumdar D, Hammock BD, Adams PD, Ronald PC (2014) Structure of the OsSERK2 leucine-rich repeat extracellular domain. Acta Crystallogr D Biol Crystallogr 70(Pt 11):3080–3086. doi:10.1107/S1399004714021178

Miki D, Shimamoto K (2004) Simple RNAi vectors for stable and transient suppression of gene function in rice. Plant Cell Physiol 45(4):490–495

Ogle DH (2017) FSA: Fisheries Stock Analysis. R package version 0.8.12, 0812th edn

Park CJ, Bart R, Chern M, Canlas PE, Bai W, Ronald PC (2010) Overexpression of the endoplasmic reticulum chaperone BiP3 regulates XA21-mediated innate immunity in rice. PLoS One 5(2):e9262. doi:10.1371/journal.pone.0009262

Park YH, Choi C, Park EM, Kim HS, Park HJ, Bae SC, Ahn I, Kim MG, Park SR, Hwang DJ (2012) Over-expression of rice leucine-rich repeat protein results in activation of defense response, thereby enhancing resistance to bacterial soft rot in Chinese cabbage. Plant Cell Rep 31(10):1845–1850. doi:10.1007/s00299-012-1298-9

Pruitt RN, Schwessinger B, Joe A, Thomas N, Liu F, Albert M, Robinson MR, Chan LJ, Luu DD, Chen H, Bahar O, Daudi A, De Vleesschauwer D, Caddell D, Zhang W, Zhao X, Li X, Heazlewood JL, Ruan D, Majumder D, Chern M, Kalbacher H, Midha S, Patil PB, Sonti RV, Petzold CJ, Liu CC, Brodbelt JS, Felix G, Ronald PC (2015) The rice immune receptor XA21 recognizes a tyrosine-sulfated protein from a Gram-negative bacterium. Sci Adv 1(6):e1500245. doi:10.1126/sciadv.1500245

Pyrzynska B, Pilecka I, Miaczynska M (2009) Endocytic proteins in the regulation of nuclear signaling, transcription and tumorigenesis. Mol Oncol 3(4):321–338. doi:10.1016/j.molonc.2009.06.001

Seo YS, Chern M, Bartley LE, Han M, Jung KH, Lee I, Walia H, Richter T, Xu X, Cao P, Bai W, Ramanan R, Amonpant F, Arul L, Canlas PE, Ruan R, Park CJ, Chen X, Hwang S, Jeon JS, Ronald PC (2011) Towards establishment of a rice stress response interactome. PLoS Genet 7(4):e1002020. doi:10.1371/journal.pgen.1002020

Team RC (2016) R: A language and environment for statistical computing, 332nd edn. R Foundation for Statistical Computing, Vienna

Tian D, Traw MB, Chen JQ, Kreitman M, Bergelson J (2003) Fitness costs of R-gene-mediated resistance in Arabidopsis thaliana. Nature 423(6935):74–77. doi:10.1038/nature01588

Zhang M, Gaschen B, Blay W, Foley B, Haigwood N, Kuiken C, Korber B (2004) Tracking global patterns of N-linked glycosylation site variation in highly variable viral glycoproteins: HIV, SIV, and HCV envelopes and influenza hemagglutinin. Glycobiology 14(12):1229–1246. doi:10.1093/glycob/cwh106

Zhou L, Cheung MY, Zhang Q, Lei CL, Zhang SH, Sun SS, Lam HM (2009) A novel simple extracellular leucine-rich repeat (eLRR) domain protein from rice (OsLRR1) enters the endosomal pathway and interacts with the hypersensitive-induced reaction protein 1 (OsHIR1). Plant Cell Environ 32(12): 1804–1820. doi:10.1111/j.1365-3040.2009.02039.x

Zhou L, Cheung MY, Li MW, Fu Y, Sun Z, Sun SM, Lam HM (2010) Rice hypersensitive induced reaction protein 1 (OsHIR1) associates with plasma membrane and triggers hypersensitive cell death. BMC Plant Biol 10:290. doi:10.1186/1471-2229-10-290

Zhou X, Wang J, Peng C, Zhu X, Yin J, Li W, He M, Wang J, Chern M, Yuan C, Wu W, Ma W, Qin P, Ma B, Wu X, Li S, Ronald P, Chen X (2016) Four receptor-like cytoplasmic kinases regulate development and immunity in rice. Plant Cell Environ 39(6):1381–1392. doi:10.1111/pce.12696

Zipfel C, Kunze G, Chinchilla D, Caniard A, Jones JD, Boller T, Felix G (2006) Perception of the bacterial PAMP EF-Tu by the receptor EFR restricts Agrobacterium-mediated transformation. Cell 125(4):749–760. doi:10.1016/j.cell.2006.03.037

9

Comparative Expression Analysis of Rice and *Arabidopsis* Peroxiredoxin Genes Suggests Conserved or Diversified Roles Between the Two Species and Leads to the Identification of Tandemly Duplicated Rice Peroxiredoxin Genes Differentially Expressed in Seeds

Yun-Shil Gho[1†], Sun-A Park[1†], Sung-Ruyl Kim[1,2†], Anil Kumar Nalini Chandran[1], Gynheung An[1] and Ki-Hong Jung[1*]

Abstract

Background: Peroxiredoxins (PRXs) have recently been identified as plant antioxidants. Completion of various genome sequencing projects has provided genome-wide information about PRX genes in major plant species. Two of these – *Oryza sativa* (rice) and *Arabidopsis* – each have 10 PRX members. Although significant progress has been made in understanding their biological roles in *Arabidopsis*, those functions in rice, a model crop plant, have not been well studied.

Results: We performed a comparative expression analysis of rice and *Arabidopsis* PRXs. Our phylogenetic analysis revealed that one subgroup contains three rice and three *Arabidopsis* Type-II PRXs that are expressed ubiquitously. This suggests that they are involved in housekeeping functions to process reactive oxygen species (ROS). Within the second subgroup, expression of *Os1-CysPrxA (LOC_Os7g44430)* and *AtOs1-CysPrx* is conserved in seeds while *Os1-CysPrxB (LOC_Os7g44440)* shows a root-preferential pattern of expression. We used transgenic plants expressing the GUS reporter gene under the control of the promoters of these two tandem duplicates to confirm their meta-expression patterns. Our GUS expression data from developing seeds and those that were germinating indicated that *Os1-CysPrxB* is involved in root development, as initiated from the embryo, while *Os1-CysPrxA* has roles in regulating endosperm development near the aleurone layer. For the third and fourth subgroups, the rice PRXs are more likely to show leaf/shoot-preferential expression, while those from *Arabidopsis* are significantly expressed in the flowers and seeds in addition to the leaf/shoot. To determine the biological meaning of those expression patterns that were dominantly identified in rice PRXs, we analyzed three rice genes showing leaf/shoot-preferential expression in a mutant of the light-responsive *1-deoxy-D-xylulose 5-phosphate reductoisomerase* (*dxr*) gene and found that two of them were significantly down-regulated in the mutant.

(Continued on next page)

* Correspondence: khjung2010@khu.ac.kr
†Equal contributors
[1]Graduate School of Biotechnology & Crop Biotech Institute, Kyung Hee University, Yongin 17104, Republic of Korea
Full list of author information is available at the end of the article

(Continued from previous page)

Conclusion: A global expression analysis of the PRX family in rice identified tandem duplicates, *Os1-CysPrxA* and *Os1-CysPrxB*, in the 1-CysPrx subgroup that are differentially expressed in developing seeds and germinating seeds. Analysis of the *cis*-acting regulatory elements (CREs) revealed unique CREs responsible for embryo and root or endosperm-preferential expression. In addition, the presence of leaf/shoot-preferential PRXs in rice suggests that they are required in that crop because those plants must tolerate a higher light intensity in their normal growth environment when compared with that of *Arabidopsis*. Downregulation of two *PRX*s in the *dxr* mutant causing an albino phenotype, implying that those genes have roles in processing ROS produced during photosynthesis. Network analysis of four PRXs allowed us to model regulatory pathways that explain the underlying protein interaction network. This will be a useful hypothetical model for further study.

Keywords: Arabidopsis, *cis*-acting regulatory elements, Rice, Peroxiredoxin family, Gus, Tandem duplication

Background

Peroxiredoxins (PRXs) catalyze the decomposition of peroxides and function in diverse compartments to protect cells against damage from reactive oxygen species (ROS) (Dietz 2003). The PRX family contains thiol-dependent peroxidases that are evolutionarily widespread in bacteria, fungi, animals, cyanobacteria, and plants (Umate 2010; Dietz et al. 2006; Bhatt and Tripathi 2011). Completion of several genome sequencing projects has provided genome-wide information about PRX genes in major plant species. According to their primary structure, they have been classified into four functional subgroups: PrxQ, 1-CysPrx, 2-CysPrx, and Type-II Prx (Umate 2010). In *Suaeda salsa*, PrxQ helps determine tolerance to abiotic stresses such as cold and salt (Jing et al. 2006), 1-CysPrx is involved in delaying seed germination under abiotic stress (Haslekas et al. 2003), PrxIIE represses protein nitration and is active in pathogen defenses (Romero-Puertas et al. 2007), while PrxIIF regulates root growth in the presence of cadmium and salicylhydroxamic acid (Haslekas et al. 2003). Two model systems, *Oryza sativa* (rice) and *Arabidopsis thaliana* (herein *Arabidopsis*), have 10 and 11 PRX members, respectively, in their genomes (Umate 2010). Based on their subcellular localization, the PRXs in *Arabidopsis* comprise four subgroups: four PRXs (2-CysPrxA, 2-CysPrxB, PrxQ, and PrxIIE), localized in the chloroplast; a PrxIIF, in the mitochondrion; a 1-CysPrx, in the nucleo-cytoplasm; and three (PrxIIB, PrxIIC, and PrxIID), in the cytosol (Dietz 2011). In *Arabidopsis*, 2-CysPrx is involved in growth and photosynthesis (Baier and Dietz 1999; Pulido et al. 2010). For rice, however, no genetic studies have revealed functional roles for any PRXs except rice 1 Cys-peroxiredoxin (R1C-Prx/Os1-CysPrxA), which acts as a dormancy regulator and an antioxidant, based on a study with transgenic plants of tobacco, an heterologous system, that constitutively express *R1C-Prx* (GenBank Accession Number C19186; Lee et al. 2000).

Expression profiles under diverse developmental stages and growth conditions are a simple and powerful tool for obtaining information about genes for which functions have not been characterized (Chandran et al. 2016a, 2016b). We recently suggested putative functions for most members of several gene families based on the integration of experimental and meta-expression data (Jung et al. 2013; Jin et al. 2012; Nguyen et al. 2013; Nguyen et al. 2014; Nguyen et al. 2015). Applying a similar approach with the rice PRX family could help elucidate the functions of those members.

The 2-C methyl-D-erythritol 4-phosphate (MEP) pathway is a unique and essential process for bacteria, algae, and plants (Proteau 2004). The final product, isopentenyl pyrophosphate, is used for the synthesis of diverse secondary metabolites such as isoprenoids, carotenoids, chlorophylls, and tocopherols, as well as for hormones such as gibberellins and abscisic acid (Jung et al. 2008). We recently identified a T-DNA insertional mutant in a gene encoding the second enzyme in this pathway, 1-deoxy-d-xylulose 5-phosphate reductoisomerase (DXR; EC 1.1.1.267). Under greenhouse conditions, this mutant exhibits an albino phenotype at the early seedling stage (Jung et al. 2008). Co-expression of genes in the MEP pathways with those in the pathways for chlorophyll and carotenoid biosynthesis implies a potential relationship among those pathways during photosynthesis (Jung et al. 2008). In addition, the photosystems are major sources for generating ROS during photosynthesis (Pospisil 2009). Because the four PRXs in rice are localized to the chloroplast (Dietz et al. 2006), we are very interested in learning how they are involved in light responses or photosynthesis.

Here, we performed a comparative expression analysis, within the context of a phylogenic tree, using PRX family genes from rice and *Arabidopsis*. Our results suggested probable functional orthologs between specific pairs. Tandemly duplicated *Os1-CysPrxA/LOC_Os07g44430* and *Os1-CysPrxB/LOC_Os07g44440* showed expression that was distinctly preferential in both developing seeds and germinating seeds. We then used transgenic plants expressing the GUS reporter gene under the control of their promoters to confirm the pattern of differential expression by these two tandem duplicates. Finally, we identified unique *cis*-acting regulatory elements (CREs) for root-, embryo-, or endosperm-preferential expression in rice. In the rice mutant *1-deoxy-D-xylulose 5-phosphate reductoisomerase* (*dxr*), which is defective in the light response, downregulation of two *PRX* genes indicated the involvement of those proteins in *DXR*-mediated light signaling.

Results

Identification of rice peroxiredoxin genes and comparative phylogenic analysis with *Arabidopsis* PRX genes

Umate (2010) recently performed a phylogenetic analysis of 11 PRX family members of rice and 10 in *Arabidopsis*. However, unlike for the latter, roles for the individual members in rice have not been well characterized. Therefore, our objective was to assign biological functions to rice PRX genes according to developmental stage or type of tissue/organ. Our phylogenic analysis with genes from each species indicated that one rice member (LOC_Os07g15670) was not clustered with the others (data not shown). After removing that one, we had 10 from each species to examine. In all, four subgroups were revealed: 1-CysPrx, 2-CysPrx, PrxQ, and Type-II Prx (Fig. 1). Five rice and six *Arabidopsis* PRXs were assigned to the Type-II Prx subgroup, two rice and one *Arabidopsis* to the 1-CysPrx subgroup, two rice and two *Arabidopsis* to the 2-CysPrx subgroup, and one each for rice and *Arabidopsis* to the PrxQ subgroup. We hypothesized that the PRXs of both species that clustered within the same subgroup could have similar biological functions. The tandem duplicates included three *Arabidopsis PRX*s (*AtPrxIIA*, *AtPrxIIB*, and *AtPrxIIC*) on Chromosome 1 and two rice *PRX*s (*Os1-CysPrxA* and *Os1-CysPrxB*). We believe that these duplicates are probably redundant. All of the remaining genes were independently located on different chromosomes (Additional file 1: Figure S1).

Functional assignment of rice and *Arabidopsis* PRX genes using meta-expression analysis of six tissues/organs

Although the results from our phylogenetic analysis allowed us to infer functional similarity among rice and *Arabidopsis* PRX genes clustered within the same subgroup, we performed additional experiments to determine whether those functions were conserved between

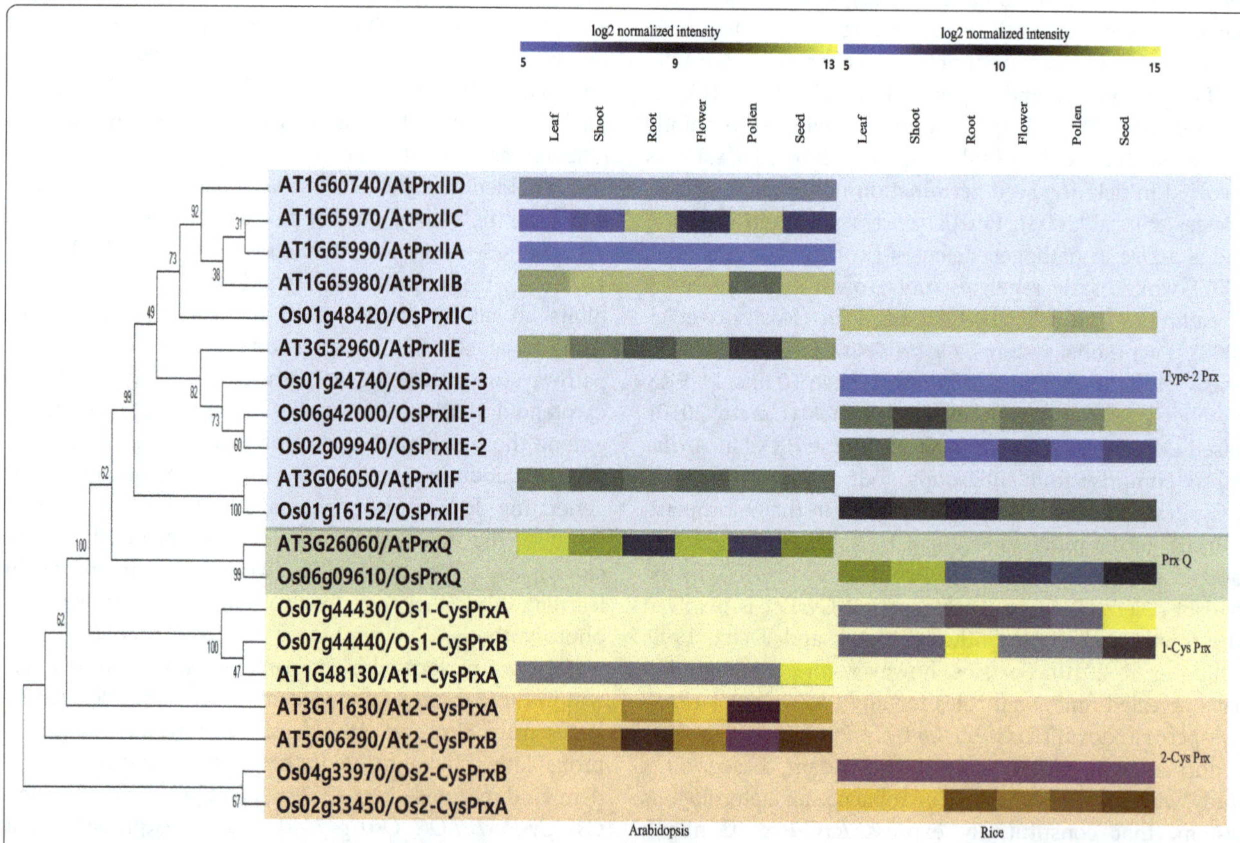

Fig. 1 Comparative meta-expression analysis of rice and *Arabidopsis* PRX family genes in six tissues/organs. Data from leaf, shoot, root, flower, pollen, and seed samples of both species were combined to produce phylogenetic tree. *Yellow, high level* of expression; *blue, low level.* Numeric values indicate average of normalized $\log_2$ intensity of microarray data. Names of subgroups are on *right side* of heatmap. Heatmaps in *middle* indicate meta-expression data for *Arabidopsis* PRX genes on *left* and rice PRX genes on *right*

species. Expression patterns for most of these *PRX*s were investigated based on a large collection of microarray data (Fig. 1 and Additional file 2: Figure S2). The exception was *AtPrxIID*, for which no probes were available on the Affymetrix array we used.

Within the Type-II Prx subgroup, OsPrxIIC clustered together with AtPrxIIA, AtPrxIIB, AtPrxIIC, and AtPrxIID. Because its expression pattern was the most similar to that of *AtPrxIIB*, we inferred that they are functional orthologs. We found it interesting that, although *AtPrxIIA*, *AtPrxIIB*, and *AtPrxIIC* are tandemly duplicated, their expression patterns were not similar. Unlike *AtPrxIIB*, which was ubiquitously expressed, *AtPrxIIC* showed root- and pollen-preferential expression, suggesting its possible role in the development of those organs in *Arabidopsis*. Expression of *AtPrxIIA* was suppressed overall, suggesting a loss-of-function during evolution or possible suppression through epigenetic regulation. Although OsPrxIIC, OsPrxIIE2, and OsPrxIIE3 clustered with AtPrxIIE, none of the first three showed a pattern of expression similar to that of *AtPrxIIE*. OsPrxIIF clustered with ATPRXIIF; both of these genes were ubiquitously expressed but differed in their patterns. In the Type-II Prx subgroup, we identified six *Arabidopsis* and five rice members. We concluded that those genes with ubiquitous expression patterns could have similar functions. However, the root- and pollen-preferential functioning of AtPrxIIC versus leaf- and shoot-preferential functioning of OsPrxIIE2 suggested that their expression is specific to one species only.

In the 1-Cys Prx subgroup, Os1-CysPrxA and Os2-CysPrxB clustered with At1-CysPrx. Genes in this cluster showed tissue- or organ-specific expression patterns. *Os1-CysPrxB* was root-preferential while *Os2-CysPrxA* and *At1-CysPrx* were preferentially expressed in the seed. This indicated that the seed-related functions of genes in this subgroup might be conserved between species. We found it interesting that expression of *Os1-CysPrxB* was root-preferential in rice, suggesting a functional role in root development in this crop. The promoters of *Os1-CysPrxA* or *Os1-CysPrxB* likely contain CREs that determine root- or seed-preferential expression. In the *PrxQ* subgroup, *OsPrxQ* and *AtPrxQ* presented similar patterns in various tissues/organs, but expression of the former was much higher in the flowers and seeds when compared with the latter. In the *2-Cys Prx* subgroup, expression in the pollen and seed was somewhat similar among *Os2-CysPrxA*, *At2-CysPrxA*, and *At2-CysPrxB*. Although *Os2-CysPrxB* was expressed at a very low level overall, it was most highly expressed in pollen, indicating that genes in that subgroup have unique functions in the development of rice pollen and *Arabidopsis* seed.

AtPrxIIC showed root- and pollen-preferential expression patterns. Although it is a possible functional ortholog of *Os1-CysPrxB* in root development, these two genes did not show a close evolutionary relationship. Leaf/shoot-preferential expression was only identified in the rice PRX family (three members), while five *Arabidopsis PRX*s and two rice *PRX*s were highly expressed in the seed.

Validation of meta-expression patterns of rice PRX genes using RT-PCR analysis

To confirm the meta-expression patterns we had observed for the rice PRX genes, we conducted RT-PCR analysis with seedling shoots, roots, flowers, and developing seeds at 10 or 15 days after pollination (DAP). For the five genes in the type-II Prx subgroup, three (*OsPrxIIC*, *OsPrxII E1*, and *OsPrxIIF*) were ubiquitously expressed in the tested tissues/organs (Fig. 2). *OsPrxIIE2* showed a shoot-preferential expression pattern but expression of *OsPrxIIE3* was very low overall. *OsPrxQ* was expressed at the highest level in the shoots (Fig. 2). The latter gene was also significantly expressed in flowers. *Os1-CysPrxA* showed the highest expression in developing seeds while *Os1-CysPrxB* in the *1-Cys Prx* subgroup was highly expressed in the roots. Although expression of *Os2-CysPrxA* was highest in the shoots, that of *Os2-CysPrxB* was very low overall. These data were consistent with our meta-expression data, indicating that the latter, when based upon a large collection of reference data, are reliable.

GUS expression analysis of two tandemly duplicated rice PRX genes

Expression patterns differed between the two tandemly duplicated rice PRX genes in the 1-Cys Prx subgroup. For closer examination, we constructed vectors expressing GUS under the control of the promoters of these two genes. Vector maps are presented in Fig. 3A and detailed information for the promoter analyses is provided in Additional file 3: Table S1. For the assays, we used whole seedlings, leaves, root cross sections, flowers, developing seeds, and cross sections of germinating seeds. As expected from the results of meta-expression analysis in those six tissues/organs, we confirmed root-preferential expression for *Os1-CysPrxB* and endosperm-preferential expression for *Os1-CysPrxA* (Fig. 3). We also found, unpredictably, that *Os1-CysPrxB* showed an embryo-preferential pattern in both developing and germinating seeds. This should have been expected of a root-preferential gene because developing rice seeds form precursors of embryonic roots at 10 or 12 DAP (Itoh et al. 2005). All of these data suggested that these closely linked genes have specialized roles in the development of two major seed organs. *Os1-CysPrxA* was

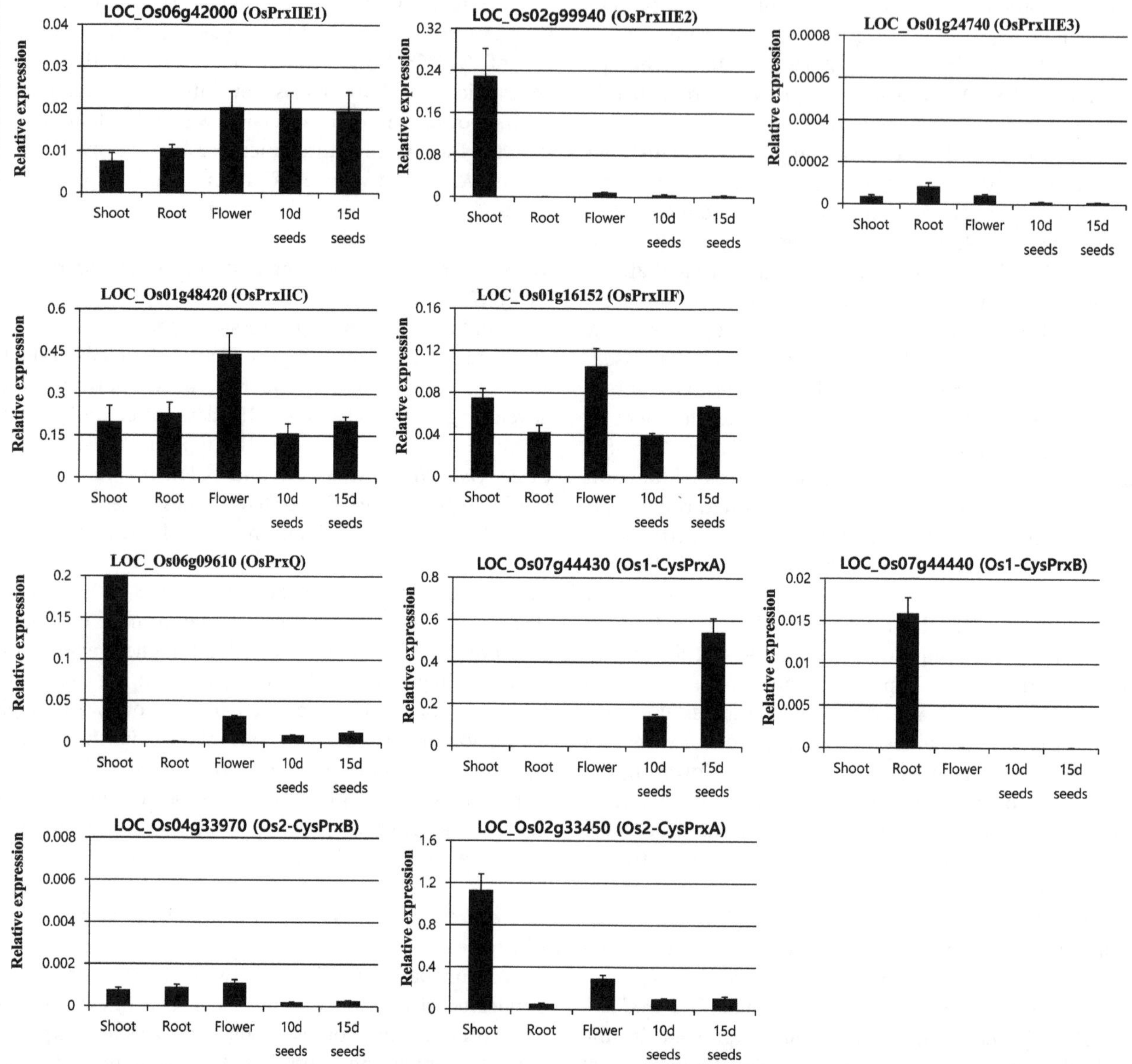

Fig. 2 Validation of meta-expression patterns in six tissues/organs for 10 PRX genes in rice based on qRT-PCR analysis of samples from seedling shoots, roots, flowers, seeds at 10 DAP, and seeds at 15 DAP. *Rice ubiquitin 5 (OsUbi5, LOC_Os01g22490)* was used as internal control. Y-axis, expression level relative to *OsUbi5*; X-axis, samples used for analyses

expressed in the aleurone layer of the endosperm in developing and germinating seeds. Expression of this gene expanded from the aleurone layer to the central region of the endosperm during development. Therefore, we hypothesized that these two genes contain CREs in their promoter regions that are essential for differential expression.

Analysis of CREs responsible for tissue- or organ-preferential expression patterns of tandemly duplicated *Os1-CysPrxA* and *Os1-CysPrxB*

To identify the unique CREs that determine root/embryo (*Os1-CysPrxB/LOC_Os07g44440*)- or endosperm (*Os1-CysPrxA/LOC_Os07g44430*)-preferential expression, we first found four more genes that showed expression patterns similar to *Os1-CysPrxA*, and five more with expression similar to *Os1-CysPrxB*. For this, we utilized top Pearson correlation coefficient (PCC) values obtained with the anatomy based-co-expression network tool installed in GENEVESTIGATOR (Hruz et al. 2008; Additional file 4: Figure S3). In silico analysis of CREs in promoter regions that included the first intron or exon (in the case of genes without an intron) of five genes in the *Os1-CysPrxA* group and six genes in the *Os1-CysPrxB* group revealed the presence of 40 CREs common to the promoter regions of both groups. This was accomplished

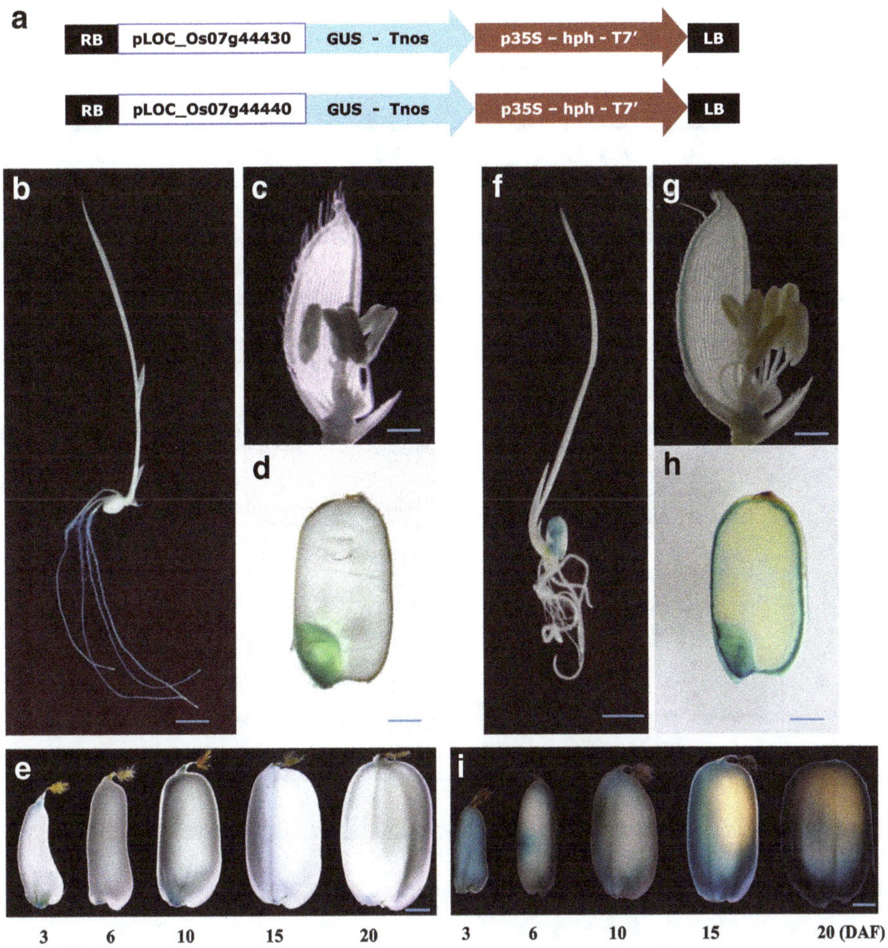

Fig. 3 Validation of differential expression patterns for two tandemly duplicated rice genes using *GUS* reporter system. **a** Maps of promoter-*GUS* vector for tandemly duplicated genes LOC_Os07g44430 (*upper*) and LOC_Os07g44440 (*lower*). **b-e** *GUS* activity in transgenic plant carrying pLOC_Os07g44440-*GUS* vector, using samples from whole seedling (**b**), flower (**c**), germinating seed (**d**), and developing seed at 3, 6, 10, 15, and 20 DAP. (**e**). **f-i** *GUS* activity in transgenic plant carrying pLOC_Os07g44430-*GUS* vector, using samples from whole seedling (**f**), flower (**g**), germinating seed (**h**), and developing seed at 3, 6, 10, 15, and 20 DAP (**i**)

using the PLANTPAN 2.0 database (Chow et al. 2016). The *Os1-CysPrxA* group contained 65 CREs in common, from which we selected 25 unique CREs in the promoters of that group as candidate CREs responsible for endosperm-preferential expression after 40 CREs were removed that are conserved in the promoters of the *Os1-CysPrxB* group (Additional files 5 and 6: Tables S2, S3). Of most interest to us were the four abscisic acid (ABA)-responsive CREs as well as those related to seed development, i.e., ABREDISTBBNNAPA/GCCACTTGTC, ABREAZMRAB28/GCCACGTGGG, ABREBZMRAB28/TCCACGTCTC, and ABREOSRAB21/ACGTSSSC. We also found three related to seed germination, i.e., GADOWNAT/ACGTGTC, TATCCAYMOTIFOSRAMY3D/TATCCAY, and TATCCAOSAMY/TATCCA, in the promoters of the Os1-CysPrxA group (Additional files 6 and 7: Tables S3, S4). Of the selected CREs for endosperm-preferential expression, TATCCAYMOTIFOSRAMY3D was more frequently identified than the others in those promoter regions (Fig. 4). The TATCCAYMOTIFOSRAMY3D motif was originally identified in the 5′ upstream regulatory region of rice α-*Amy3D* and is essential for the regulation of this gene by sugars (Lu et al. 1998; Lu et al. 2002). The TATCCAOSAMY motif (TATCCA) is part of TATCCAYMOTIFOSRAMY3D and is a binding site of OsMYBS2 and OsMYBS3, which function in sugar- and hormone-mediated regulation of alpha-amylase gene expression (Lu et al. 1998; Lu et al. 2002). The alpha-amylase in the aleurone layer plays an important role in hydrolyzing the endosperm starch into sugars that can be metabolized during seed germination, thereby providing energy for shoot growth (Murai et al. 1985; Mitsui et al. 1985). Whereas ABREDISTBBNNAPA is required for seed-specific expression and ABA-responsiveness (Ezcurra

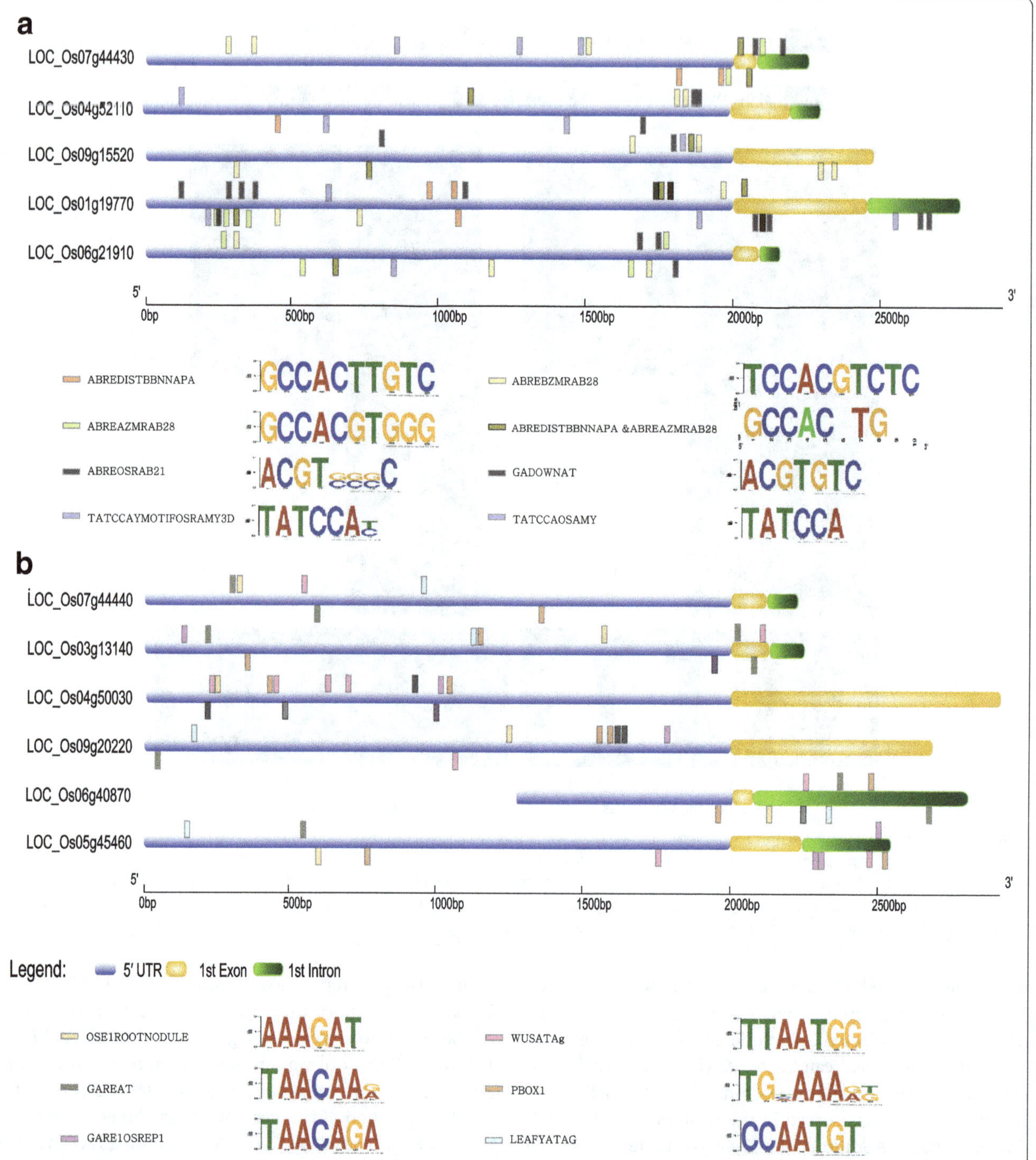

Fig. 4 Identification of CREs responsible for unique expression patterns of two tandemly duplicated rice PRX genes. Mapping, names, and logos **a** of seven CREs to promoters of *LOC_Os07g44430* and its co-expressed genes. Mapping, names, and logos **b** of six CREs to promoters of *LOC_Os07g44440* and its co-expressed genes

et al. 2000; Busk and Pages 1997, 1998), the other ABA-responsive CREs – ABREAZMRAB28 (GCCACGTGGG) and ABREBZMRAB28 (TCCACGTCTC) – are ABA-responsive element A (ABRE A) and ABA-responsive element B (ABRE B), which occur in the promoter of *Rab28* from *Zea mays* (maize) (Busk and Pages 1997, 1998). In addition, *Rab28* is highly expressed during the late stage of embryo development and is stimulated in response to ABA (Niogret et al. 1996; Busk et al. 1999). A role for ABRE A in the endosperm of developing or germinating seeds is easily estimated because ABA-mediating drought response is closely associated with seed ripening process. These ABA-responsive CREs might be involved in reducing the water content during the process of seed maturation.

Within the *Os1-CysPrxB* group, we identified 50 CREs common to their promoters (Fig. 4; Additional files 3 and 7: Tables S1, S4). Among these, 10, including PBOX1 (TGRAAG), GAREAT (TAACAAR), GARE1OSREP1 (TAACAGA), OSE1ROOTNODULE (AAAGAT), LEA FYATAG (CCAATGT), and WUSATAg (TTAATGG) were unique to that group and did not occur in the *Os1-CysPrxA* group (Additional files 8 and 9: Tables S5, S6). GAREAT and GARE1OSREP1 are involved in gibberellin (GA)-responsiveness, suggesting that they help regulate the GA signaling pathway in rice embryos (Sutoh and Yamauchi 2003; Ogawa et al. 2003). Furthermore, OSE1R OOTNODULE (AAAGAT) is an organ-specific element (OSE) for the regulation of an arbuscular mycorrhizal and nodule-inducing leghaemoglobin gene in root nodules (Stougaard et al. 1990; Vieweg et al. 2004; Fehlberg et al. 2005); and LEAFYATAG and WUSATAg are present in the promoter of a rice WUSCHEL-type homeobox gene that is expressed in the central cells of quiescent centers in the root apical meristem (Kamiya et al. 2003). Both GAREAT and GARE1OSREP1 might be expressed preferentially in the embryo while LEAFYATAG, WUSATAg, and OSE1ROOTNODULE show root-preferential expression as members of the *Os1-CysPrxB* group. The other CREs not mentioned here might have novel roles in driving root/embryo- or endosperm-preferential expression. Future experiments will be necessary to confirm our predictions (Additional files 5 and 8: Tables S2, S5).

Two PRXs showing leaf- or shoot-preferential expression patterns are involved in the *DXR*-mediated light-response pathway to remove ROS during photosynthesis

Analysis with various tissues/organs revealed three *PRX*s that are most highly expressed in the leaf and shoot. Recently, we used a T-DNA insertion to identify a rice *1-deoxy-D-xylulose 5-phosphate reductoisomerase* (*dxr*) mutant with an albino phenotype. That finding indicated that the defective gene is involved in the light-response pathway (Jung et al. 2008). The DXR enzyme is active in the second step of the MEP pathway for isoprenoid biosynthesis, which interconverts 1-deoxy-D-xylulose 5-phosphate and MEP. Of the three rice *PRX*s (*OsPrxIIE2, Os2-CysPrxA*, and *OsPrxQ*) that show leaf/shoot-preferential expression, both *OsPrxIIE2* and *OsPrxQ* were significantly down-regulated in the *dxr* mutant when compared with the wild type (WT). This was demonstrated in our previous analyses via quantitative RT-PCR and microarray data (Chandran et al. 2016c), and it further supports our conclusion that these rice PRX genes are involved in the *DXR*-mediated light-response pathway (Fig. 5, Additional file 10: Figure S4).

Regulatory network mediated by rice PRX genes

As part of our exploration of the regulatory mechanism mediated by rice PRX genes, we recently performed a protein–protein interaction network analysis in the Rice Interaction Viewer (RIV; http://bar.utoronto.ca/interactions/cgi-bin/rice_interactions_viewer.cgi) (Chandran et al. 2016a). Among the 10 rice PRXs, four (Os1-CysPrxA, Os2-CysPrxA, OsPrxQ, and OsPrxIIC) encode 82 predicted interacting proteins (Fig. 6): Os2-CysPrxA has 60; OsPrxIIC, 16; Os1-CysPrx A, 3; and OsPrxQ, 3 (Fig. 6).

To develop a more detailed protein–protein interaction network, we incorporated the MapMan classification data (indicated by node color) in Additional file 11: Table S7. This regulatory network provides insight into the molecular functions of the rice PRXs (Fig. 6). We also integrated PCC values based on meta-expression data for various tissues/organs in the network. In all, 12 genes had PCC values >0.5 for *Os2-CysPrxA*, the PRX gene most highly expressed in leaves and shoots; three genes had PCC values >0.5 for the ubiquitously expressed *OsPrxIIC*; and two genes had PCC values >0.5 for *OsPrxQ*, with a leaf-preferential expression pattern. We did not identify co-expressed genes for *Os1-CysPrxA*, which had a seed-preferential expression pattern (Fig. 6).

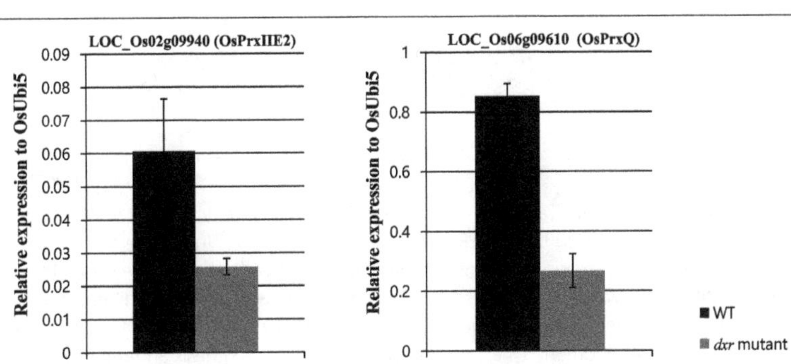

Fig. 5 RT-PCR analysis of *OsPrxQ/LOC_Os06g09610* and *OsPrxIIE2/LOC_Os02g09940* in *dxr* mutant. Expression level relative to *OsUbi5* was compared between WT segregant and *dxr* mutant

Functions were characterized for three genes in this network: *LOC_Os07g28280*, encoding a rice ubiquitin-related modifier 1 (RURM1) involved in mobilizing Active MITE mPing (Tsukiyama et al. 2013); *LOC_Os03g13970/RPN10*, encoding the 26S proteasome non-ATPase regulatory subunit 4, involved in canavanine resistance (Takase et al. 2004); and *LOC_Os06g04280*, encoding 5-enolpyruvoylshikimate-3-phosphate synthase (EPSPS), which is involved in the production of tillers, panicles, and seeds (Wang et al. 2014). These known genes linked to PRXs can now provide valuable clues about the functions of the unknown PRXs. Elements with multiple interactions in the network will also enable greater refinement of predicted protein–protein interaction networks when compared with other interactors. In fact, we identified four such elements in our network: *LOC_Os08g39140*, encoding heat shock protein 90 (HSP90) that interacts with OsPrxIIC and Os1-CysPrxA; *LOC_Os07g44710*, encoding a calcium-mediating protein kinase (CAMK); and *LOC_Os07g08840* and *LOC_Os03g58630*, which encode thioredoxins that interact with Os1-CysPrxA and Os2-CysPrxA. Among these interacting proteins in the primary network, we considered those showing co-expression or multiple interactions with PRX proteins to be more reliable candidates for constructing a regulatory network associated with PRXs in rice. Such a refined and integrated network will be a useful molecular frame for designing further studies.

Discussion

What is the functional similarity or diversity between rice and *Arabidopsis* PRXs?

Peroxiredoxins are universal among bacteria, fungi, animals, cyanobacteria, and plants (Dietz 2003). Among all of these organisms, PRXs are most abundant in plants, most likely because they must process large quantities of ROS that are produced during photosynthesis. We found that OsPrxIIC, a Type-II Prx, clustered with AtPrxIIA, AtPrxIIB, AtPrxIIC, and AtPrxIID, while OsPrxIIE1, OsPrxIIE2, and OsPrxIIE3 clustered with AtPrxII E (Fig. 1). Therefore, those three members of the Type-II subgroup mean that further expansion was initiated from AtPrxIID in *Arabidopsis* and OsPrxIIE3 in rice. Because those subgroup members show a ubiquitous expression pattern, we might infer that these genes primarily have housekeeping roles. For the newly expanded members, their tissue-preferred roles are to be expected. For example, AtPrxIIC in *Arabidopsis* might function in the processing of ROS produced from roots and pollen, while OsPrxIIE2 in rice might be active in processing ROS produced from leaves and shoots. Genes

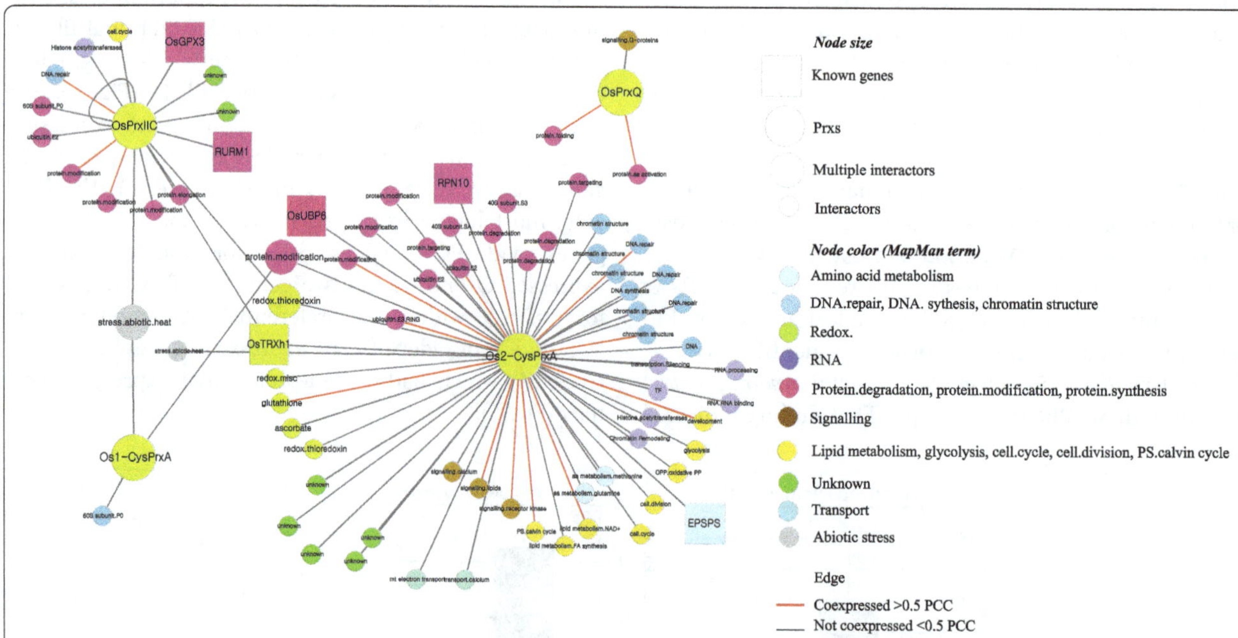

Fig. 6. Working model of rice PRXs for regulatory pathway using refined protein–protein interaction network. Predictions were made with RIV, which revealed interactions with 161 proteins mediated by four PRX proteins. *Red edges* (*lines*) indicate elements with PCC values >0.5; *large rectangles* in node, three interaction proteins with known functions identified from OGRO database; *large circles*, PRX family members; *medium circles*, interactors showing multiple interactions with PRX proteins. MapMan annotation is indicated by different node colors: *bright green*, redox class proteins, including peroxiredoxins and thioredoxins; *pale blue*, proteins involved in amino acid metabolism; *yellow*, proteins involved in cell cycle, lipid metabolism, photosynthesis, and development; *blue*, proteins involved in DNA synthesis and repair; *pink*, proteins involved in protein degradation and post-translational modification; *purple*, proteins involved with RNA, mostly transcriptional regulators; *brown*, signaling proteins; *gray*, abiotic stress proteins; *light blue*, transporter class proteins; *green*, proteins with unknown functions (see Additional file 11: Table S7).

in the *1-Cys Prx* subgroup in those two species were preferentially expressed in endosperm, based on their meta-expression patterns. However, the rice members of that subgroup might have unique roles, based on our finding that, unlike its fellow members, *Os1-CysPrxB* was preferentially expressed in the root and embryo. Although *AtPrxQ* and *OsPrxQ* have similar patterns, being highly expressed in leaves and shoots, the former showed greater expression in flowers and seeds than did the latter. All genes in the 2-CysPrx subgroup, except for *Os2-CysPrxB*, which had very low expression in most tested tissues/organs, were commonly expressed at high levels in leaves and shoots. However, expression in the flowers and seeds was higher for *Arabidopsis* members than for rice members in that subgroup. Rice genes in the *2-CysPrx* and *PrxQ* subgroups were more likely to function in leaves and shoots when compared with the *Arabidopsis* members, indicating that rice requires them for survival when growing under higher light intensities than are associated with *Arabidopsis*. All of these results suggest that functional diversity of some PRXs has developed evolutionarily between the two species.

A hypothetical model for the rice PRX-mediated pathway for ROS processing

Our predicted protein–protein interaction network analysis presented 82 interactors associated with four rice PRXs that showed ubiquitous or leaf/shoot-preferential expression patterns. Two of them already have known functions, including Os2-CysPrxA, which interacts with h-type thioredoxin 1 (OsTRXh1) and is involved in ABA sensitivity during seed germination and the early seedling stage (Zhang et al. 2011). This PRX also interacts with EPSPS, which participates in tolerance to drought and salinity (Tuteja et al. 2013); RPN10, which is related to canavanine resistance as a subunit of 26S proteasome (Takase et al. 2004), and the ubiquitin specific protease 6 (OsUBP6), which is related to growth rates during the seedling stage as well as the transport of iron into the mitochondria (Moon et al. 2009). These linkages suggest potential roles for Os2-CysPrxA in regulating ABA sensitivity, abiotic stress tolerance, normal plant development, and ubiquitin-dependent proteolysis at the cellular and molecular levels. Its higher expression in leaves than in other tissues suggests that its primary role is in leaf development, as mediated by the cellular and molecular functions mentioned above.

The second PRX with a known function is OsPrxIIC, which interacts with glutathione peroxidase 3 (OsGPX3) for hydrogen peroxide homeostasis; RURM1, involved in the mobilization of Active MITE mPing as a ubiquitin-related protein; and the previously discussed OsTRXh1. Based on its pattern of ubiquitous expression, OsPrxIIC may have a housekeeping function through its association with ROS-processing enzymes (including glutathione peroxidase and thioredoxin) or else in ubiquitin-related processes.

The signaling cascades of PRXs that occur in response to stresses have been analyzed for 1-CysPrx and 2-CysPrx in *Arabidopsis* (Dietz 2016). A redox-dependent retrograde signaling pathway from the chloroplast to the nucleus has been reported for 2-CysPrx (Baier and Dietz 2005). Because of its leaf/shoot-preferential expression pattern, Os2-CysPrxA might be a functional ortholog of *Arabidopsis* 2-CysPrx, thereby suggesting that rice has a potential ortholog for its redox-dependent retrograde signaling pathway. Here, we identified two candidate PRXs for the light-dependent signaling pathway to process ROS. Both *OsPrxIIE2* and *OsPrxQ* were down-regulated in the *dxr* mutant, which is defective in the light response. These PRXs might have separate roles in different retrograde signaling pathways or might combine with the signaling pathway of Os2-CysPrxA. This possibility remains to be elucidated through further studies.

Tandemly duplicated PRXs with differential expression patterns increase functional diversity in rice

In *Arabidopsis*, *AtPrxIIC*, *AtPrxIIB*, and *AtPrxIIA* are tandem duplicates that show different expression patterns. In particular, *AtPrxIIC* is expressed at a high level in pollen but at lower levels in flowers and seed, while the opposite is true for *AtPrxIIB*. This suggests mutually antagonistic regulation. In addition, expression of *AtPrxIIA* is very low overall, possibly because of suppression via epigenetic regulation. The tandemly duplicated PRX genes in rice, *Os1-CysPrxA* and *Os1-CysPrxB*, also have dissimilar expression patterns. A Blast2 alignment of these PRXs indicated 89% amino acid sequence similarity (Additional file 12: Figure S5), which might mean that they are functionally redundant. However, their meta-expression patterns differed among tissues/organs. Whereas *Os1-CysPrxB* showed a root/embryo-preferential expression pattern, *Os1-CysPrxA* was preferentially expressed in the endosperm. This indicated that their promoters contain unique *cis*-acting elements. In our study, we focused on six CREs unique to the promoters of genes present in the *Os1-CysPrxB* subgroup but not in the *Os1-CysPrxA* subgroup. Three of the six – LEAFYATAG, WUSATAg, and OSE1ROOTNODULE –appear necessary for root-preferential expression while two others – GAREAT and GARE1OSREP1 – are related to embryo-preferential expression. For those expressed preferentially in the endosperm and belonging to the *Os1-CysPrxA* subgroup, we focused on seven CREs. Of them, ABREDISTBBNNAPA, ABREAZMRAB28, ABREBZMRAB28, and ABREOSRAB21 might be required for the ABA response during seed maturation while GADOWNAT, TATCCAYMOTIFOSRAMY3D, and TATCCAO

SAMY might be needed for the processing of endosperm after germination. However, further investigations are necessary to evaluate the functionality of these unique CREs suggested here.

Studies of functional genomics that utilize gene-indexed mutants or genome-editing methods will provide more insight into the role of these rice PRXs. In addition, functional classifications retrieved from MapMan annotations of the interactors linked with four rice PRXs might help researchers elucidated the yet-unknown molecular mechanisms by which PRXs mediate stress responses.

Conclusions

We carried out comparative expression analysis of rice and Arabidopsis PRX family genes which suggests conserved or diversified roles between the two species, leading the identification of tandemly duplicated rice PRX genes in the 1-CysPrx subgroup, Os1-CysPrxA and Os1-CysPrxB, differentially expressed in seeds. Os1-CysPrxB showed embryo- or root-preferential expression, while Os1-CysPrxA showed endosperm-preferential expression. Analysis of the cis-acting regulatory elements (CREs) revealed unique CREs responsible for embryo and root or endospermpreferential expression. In addition, the presence of leaf/shoot-preferential PRXs in rice suggests their evolutional requirement to survive in the growth environment with a higher light intensity when compared with that of Arabidopsis. Downregulation of two PRXs in the dxr mutant causing an albino phenotype implies that those genes have roles in processing ROS produced during photosynthesis. Predicted protein-protein network associated with four PRXs suggests useful regulatory model for further study.

Methods

Multiple sequence alignment and phylogenetic analysis

To perform a phylogenomic analysis of PRXs in rice and *Arabidopsis*, we collected 11 family members from the Rice Genome Annotation Project using locus IDs (http://rice.plantbiology.msu.edu/), and 10 *Arabidopsis* family members previously reported (Umate 2010). Multiple alignment of amino acid sequences was conducted with the ClustalX program, version 2.0.11 (Higgins et al. 1996). The phylogenetic analysis included MEGA 5.2 and the following parameters: Neighbor-Joining tree method, complete deletion, and bootstrap with 500 replicates (Tamura et al. 2011). The resulting phylogenic tree comprised 11 rice and 10 *Arabidopsis* PRX proteins (Umate 2010). Those from rice were classified into four subgroups based on a comparative analysis between rice and *Arabidopsis* (Fig. 1). We also generated a phylogenic tree for rice PRXs to integrate meta-expression data for six tissues/organs (Additional file 2: Figure S2).

Meta-analysis of tissue-specific expression profiles

Integration of spatio-temporal expression profiles into a phylogenetic context can direct experimental strategies for further functional analysis (Jung et al. 2010). Therefore, we used meta-analysis of 17 tissue/organ-type expression profiles based on data from 983 Affymetrix arrays downloaded from the NCBI gene expression omnibus (http://www.ncbi.nlm.nih.gov/geo/) (Cao et al. 2012). We then uploaded the log_2-normalized intensity data in a tab-delimited text format into the Multi Experiment Viewer (http://mev.tm4.org/#/welcome) and illustrated these data using heat maps (Additional file 2: Figure S2). In addition, we analyzed the meta-expression patterns of *Arabidopsis* PRX genes in six tissues/organs using the *Arabidopsis* Affymetrix microarray data series GSE5630, GSE5633, GSE5631, GSE5632, GSE5634, GSM943445, and GSM943446. Similar to the rice data analysis, we generated meta-expression data. To compare gene expression, we examined the data from six tissue/organ types from both rice and *Arabidopsis* (Fig. 1). This resulted in meta-expression data for all PRX genes, except for AtPrxIID, which is not present in the *Arabidopsis* Affymetrix array platform (GPL2025). The log_2 intensities of microarray data differed between species, with the rice heat map ranging from 5 to 15 while the *Arabidopsis* heat map had a range from 5 to 13. Integrated meta-expression data were used to determine functional conservancy in terms of anatomy between rice and *Arabidopsis* PRX ortholog pairs.

Predicted protein–protein interactions and PCC analysis

To infer the regulatory network of PRXs in rice, we conducted a predicted protein–protein interaction analysis in the RIV (Chandran and Jung 2014). After copying and pasting the locus IDs of the 10 PRX rice proteins into the RIV query box, we were able to identify interactions with 161 proteins that were mediated by four of those PRX proteins (Fig. 6). The interaction information presented in Additional file 11: Table S7 was then uploaded to Cytoscape (Shannon et al. 2003). The large circles (nodes) indicated PRXs and elements with known functions acquired from the Overview of Functionally Characterized Genes in Rice Online database (OGRO) database (Yamamoto et al. 2012), while the small circles indicated proteins predicted to interact with PRXs in rice. We then performed a PCC analysis between targets and interactors in RIV, using affymetrix expression data (Cao et al. 2012). This enabled us to identify elements with PCC values >0.5. In addition, three interaction proteins with known functions from the OGRO database were identified. The MapMan annotation indicated redox class proteins (including peroxiredoxins and thioredoxins), as well as proteins involved in amino acid metabolism, the cell cycle, lipid metabolism, photosynthesis, normal plant development, DNA synthesis and repair, protein degradation and post-translational modification, RNA (mostly transcriptional regulators), signaling proteins, abiotic stress proteins, and transporter class proteins. The annotations also indicated proteins with yet-unknown functions (Additional file 13: Table S8). The refined network with integrating omics data was utilized to infer regulatory pathways (Fig. 6).

Plant materials

After sterilization with 50% (*w*/*v*) commercial bleach for 30 min with gentle shaking, rice seeds were germinated on a Murashige Skoog medium under controlled conditions (28 °C/25 °C day/night, 8-h photoperiod, and 78% relative humidity). In addition, we harvested flowers and pollen at the heading stage, and collected seeds at 10 and 15 DAP in greenhouse condition to extract total RNAs. All samples were directly frozen in liquid nitrogen and stored at −70 °C. For analyzing differential expression between the *dxr* mutant and the WT, we extracted total RNA from 7-day-old seedlings of both genotypes grown under greenhouse conditions.

Promoter analysis using the *GUS* reporter gene

The *Os1-CysPrxA* promoter region (−1 to −2076 bp from the initiation ATG codon) and the *Os1-CysPrxB* promoter region (−1 to −1758 bp) were amplified with the 5-GTACCAGTTCGCCTCTAGAATTGAG-3 (Xba I) / 5-TCTCGAGGCGACGAACGACTGTGCTGC-3 (Xho I) primer set and the 5-ATCTAGAGTGGGTGTTGTGG TTGG-3 (Xba I) / 5- ACTCGAGATGAGGAATCGAGG ATTAACC-3 (Xho I) primer set, respectively. The promoter DNA fragment was placed upstream of the *beta-glucuronidase* (*GUS*) reporter gene located in binary vector pGA3383 (Kim et al. 2009). Transgenic plants harboring the above construct were obtained in the *japonica* cultivar 'Dongjin' background through the *Agrobacterium tumefaciens* method of transformation (Lee et al. 1999).

To monitor *GUS* expression, we used 7-day-old whole seedlings, plus mature leaves, roots, flowers, developing seeds, and germinating seeds for two transgenic lines and performed GUS staining as described by Jefferson et al. (1987). As we have recently described (Hong et al., 2017) photographs of the GUS-assayed seedlings, mature leaves, flowers, developing seeds, and germinating seeds were produced using an EOS 560 digital camera (Canon, Tokyo, Japan) while those of hand-sectioned roots and germinating seeds were obtained using a BX61 microscope (Olympus, Tokyo, Japan).

Analysis of *cis*-regulatory elements

To compare the CREs responsible for differences in promoter-mediated spatial expression between *Os1CysPrxA* and *Os1-CysPrxB*, we first used the co-expression tool in GENEVESTIGATOR (Zimmermann et al. 2004) to identify four more genes showing similar expression patterns with *Os1-CysPrxA* and five for *Os1-CysPrxB*. We then extracted 2-kb sequences upstream of ATG for these 11 genes from PLANTPAN (http://plantpan2.itps.ncku.edu.tw) (Chow et al. 2016), and analyzed the CREs in promoters using PLACE (Higo et al. 1998). Those CREs were aligned with the Motif Alignment and Search Tool (Bailey et al. 2009; Fig. 4 and Additional file 3: Table S1). Known target motifs were selected based on *P*-values ≤0.05.

RNA extraction, semi-quantitative RT-PCR, and real-time PCR

Roots and shoots from rice seedlings were frozen in liquid nitrogen and ground with a Tissue Lyser II (Qiagen, Hilden, Germany). Their RNAs were extracted with the RNAiso Plus Kit according to the manufacturer's protocol (Takara Bio, Kyoto, Japan). To determine tissue-specific expression patterns by RT-PCR, we used a primer set for *rice ubiquitin 5* (*OsUbi5*, *LOC_Os01g22490*) (Jain et al. 2006). The PCR conditions included 22 to 38 cycles of 95 °C for 30 s, 57 °C for 30 s, and 72 °C for 1 min 30 s. For real-time PCR, the cycling conditions were 95 °C for 15 s, 57 °C for 30 s, and 72 °C for 60 s. This experiment was repeated three times using the same control primer sets. Relative transcript levels and fold-changes were calculated by the $2^{-\Delta Ct}$ and $2^{-\Delta\Delta Ct}$ methods (Schmittgen and Livak 2008), respectively.

To compare the expression of PRX genes between the *dxr* mutant and WT plants, we first balanced the transcripts between their leaves using *OsUbi5* as we have previously described (Chandran et al. 2016b). We then conducted an RT-PCR analysis with leaf samples to compare band intensities between the two genotypes. All primers used for RT-PCR analysis and vector construction are described in Additional file 14: Table S9.

Additional files

Additional file 1: Figure S1. Maps showing chromosomal localization of 10 rice (A) and 10 *Arabidopsis* (B) PRX genes.

Additional file 2: Figure S2. Detailed meta-expression analysis using 995 tissues/organs of rice PRX family genes.

Additional file 3: Table S1. Summary of selected promoter regions and PCC values for individual members in *Os1-CysPrxA* and *Os1-CysPrxB* groups.

Additional file 4: Figure S3. Meta-expression data for genes co-expressed with *Os1-CysPrxA* and *Os1-CysPrxB*, using anatomy tool in GENEVESTIGATOR.

Additional file 5: Table S2. Summary of 25 CREs uniquely identified in *Os1-CysPrxA* group compared with *Os1-CysPrxB* group.

Additional file 6: Table S3. Detailed promoter regions of 25 CREs uniquely identified in *Os1-CysPrxA* group compared with *Os1-CysPrxB* group.

Additional file 7: Table S4. Summary of CREs described in Fig. 4.

Additional file 8: Table S5. Summary of 10 CREs uniquely identified in *Os1-CysPrxB* group compared with *Os1-CysPrxA* group.

Additional file 9: Table S6. Detailed promoter regions of 10 CREs uniquely identified in *Os1-CysPrxB* group compared with *Os1-CysPrxA* group.

Additional file 10: Figure S4. Phenotype comparison between *dxr* mutant and WT, RT-PCR analysis of *OsPrxQ* and *OsPrxIIE2* in mutant, and microarray data analysis of *dxr* mutant/WT using Agilent 44 K array.

Additional file 11: Table S7. Detailed information about genes integrated in nodes of protein–protein interaction network.

Additional file 12: Figure S5. Blast2 alignment of amino acid sequences from *Os1-CysPrxA* and *Os1-CysPrxB*.

Additional file 13: Table S8. Information about predicted protein–protein interaction network depicted in Fig. 6.

Additional file 14: Table S9. Primer sequences used for RT-PCR analyses (Figs. 2 and 5).

Abbreviations

CAMK: Calcium-mediating protein kinase; CREs: cis-acting regulatory elements; DXR: 1- deoxy-D-xylulose 5-phosphate reductoisomerase; EPSPS: 5-enolpyruvoylshikimate-3-phosphate synthase; GPX3: Glutathione peroxidase 3; HSP90: Heat shock protein 90; PRXs: Peroxiredoxins; ROS: Reactive oxygen species; RURM1: Rice ubiquitin-related modifier 1; TRXh1: h-type thioredoxin 1; UBP6: Ubiquitin specific protease 6

Acknowledgments

This work was conducted with the support of the Next-Generation BioGreen 21 Program (PJ01100401 to KHJ), Rural Development Administration, Republic of Korea.

Authors' contributions

SAP, YSG, GA, SRK, and KHJ designed the research. SAP, YSG, SRK, and AKCR performed experiments. SAP, YSG, SRK, and AKCR analyzed data. SAP, YSG, SRK, and KHJ wrote the manuscript. All authors read and approved the final manuscript.

Competing interests

The authors declare that they have no competing interests.

Author details

[1]Graduate School of Biotechnology & Crop Biotech Institute, Kyung Hee University, Yongin 17104, Republic of Korea. [2]Plant Breeding, Genetics, and Biotechnology Division, International Rice Research Institute, Metro Manila, Philippines.

References

Baier M, Dietz KJ (1999) Protective function of chloroplast 2-cysteine peroxiredoxin in photosynthesis. Evidence from transgenic Arabidopsis. Plant Physiol 119: 1407–1414

Baier M, Dietz KJ (2005) Chloroplasts as source and target of cellular redox regulation: a discussion on chloroplast redox signals in the context of plant physiology. J Exp Bot 56:1449–1462

Bailey TL, Boden M, Buske FA, Frith M, Grant CE, Clementi L, Ren J, Li WW, Noble WS (2009) MEME SUITE: tools for motif discovery and searching. Nucleic Acids Res 37:W202–W208

Bhatt I, Tripathi BN (2011) Plant peroxiredoxins: catalytic mechanisms, functional significance and future perspectives. Biotechnol Adv 29:850–859

Busk PK, Pages M (1997) Protein binding to the abscisic acid-responsive element is independent of VIVIPAROUS1 in vivo. Plant Cell 9:2261–2270

Busk PK, Pages M (1998) Regulation of abscisic acid-induced transcription. Plant Mol Biol 37(3):425–435

Busk PK, Pujal J, Jessop A, Lumbreras V, Pagès M (1999) Constitutive protein-DNA interactions on theabscisic acid-responsive element before and after developmental activation of the rab28 gene. Plant Mol Biol 41:529–536.

Cao P, Jung KH, Choi D, Hwang D, Ronald PC (2012) The Rice Oligonucleotide Array database: an atlas of rice gene expression. Rice 5:17

Chandran AKN, Jung KH (2014) Resources for systems biology in rice. J Plant Biol 57:80–92

Chandran AKN, Bhatnagar N, Kim B, Jung KH (2016a) Genome-wide identification and analysis of rice genes to elucidate morphological agronomic traits. J Plant Biol 59:639–647

Chandran AKN, Jeong HY, Jung KH, Lee C (2016b) Development of functional modules based on co-expression patterns for cell-wall biosynthesis related genes in rice. J Plant Biol 59:1–15

Chandran AKN, Lee GS, Yoo YH, Yoon UH, Ahn BO, Yun DW, Kim JH, Choi HK, An G, Kim TH, Jung KH (2016c) Functional classification of rice flanking sequence tagged genes using MapMan terms and global understanding on metabolic and regulatory pathways affected by *dxr* mutant having defects in light response. Rice (N Y) 9:17

Chow CN, Zheng HQ, Wu NY, Chien CH, Huang HD, Lee TY, Chiang-Hsieh YF, Hou PF, Yang TY, Chang WC (2016) PlantPAN 2.0: an update of plant promoter analysis navigator for reconstructing transcriptional regulatory networks in plants. Nucleic Acids Res 44:D1154–D1160

Dietz KJ (2003) Plant peroxiredoxins. Annu Rev Plant Biol 54:93–107

Dietz KJ (2011) Peroxiredoxins in plants and cyanobacteria. Antioxid Redox Signal 15:1129–1160

Dietz KJ (2016) Thiol-based Peroxidases and Ascorbate Peroxidases: why plants rely on multiple Peroxidase Systems in the Photosynthesizing Chloroplast? Mol Cells 39:20–25. doi:10.14348/molcells.2016.2324

Dietz KJ, Jacob S, Oelze ML, Laxa M, Tognetti V, de Miranda SM, Baier M, Finkemeier I (2006) The function of peroxiredoxins in plant organelle redox metabolism. J Exp Bot 57:1697–1709

Ezcurra I, Wycliffe P, Nehlin L, Ellerstrom M, Rask L (2000) Transactivation of the *Brassica napus* napin promoter by ABI3 requires interaction of the conserved B2 and B3 domains of ABI3 with different cis-elements: B2 mediates activation through an ABRE, whereas B3 interacts with an RY/G-box. Plant J 24:57–66

Fehlberg V, Vieweg MF, Dohmann EM, Hohnjec N, Puhler A, Perlick AM, Kuster H (2005) The promoter of the leghaemoglobin gene VfLb29: functional analysis and identification of modules necessary for its activation in the infected cells of root nodules and in the arbuscule-containing cells of mycorrhizal roots. J Exp Bot 56:799–806

Haslekas C, Viken MK, Grini PE, Nygaard V, Nordgard SH, Meza TJ, Aalen RB (2003) Seed 1-cysteine peroxiredoxin antioxidants are not involved in dormancy, but contribute to inhibition of germination during stress. Plant Physiol 133: 1148–1157. doi:10.1104/pp.103.025916

Higgins DG, Thompson JD, Gibson TJ (1996) Using CLUSTAL for multiple sequence alignments. Methods Enzymol 266:383–402

Higo K, Ugawa Y, Iwamoto M, Higo H (1998) PLACE: a database of plant Cis-acting regulatory DNA elements. Nucleic Acids Res 26:358–359

Hong HJ, Yoo HY, Park SA, Moon S, Kim SR, An G, Jung KH (2017) Genome-wide identification and extensive analysis of rice-endosperm preferred genes using reference expression database. J Plant Biol 60:249–258

Hruz T, Laule O, Szabo G, Wessendorp F, Bleuler S, Oertle L, Widmayer P, Gruissem W, Zimmermann P (2008) Genevestigator v3: a reference expression database for the meta-analysis of transcriptomes. Adv Bioinforma 2008:420747

Itoh J, Nonomura K, Ikeda K, Yamaki S, Inukai Y, Yamagishi H, Kitano H, Nagato Y (2005) Rice plant development: from zygote to spikelet. Plant Cell Physiol 46:23–47

Jain M, Nijhawan A, Tyagi AK, Khurana JP (2006) Validation of housekeeping genes as internal control for studying gene expression in rice by quantitative real-time PCR. Biochem Biophys Res Commun 345:646–651

Jefferson RA, Kavanagh TA, Bevan MW (1987) GUS fusions: beta-glucuronidase as a sensitive and versatile gene fusion marker in higher plants. EMBO J 6:3901–3907

Jin GH, Gho HJ, Jung KH (2012) A systematic view of rice heat shock transcription factor family using phylogenomic analysis. J Plant Physiol 170:321–329

Jing LW, Chen SH, Guo XL, Zhang H, Zhao YX (2006) Overexpression of a chloroplast-located peroxiredoxin Q gene, SsPrxQ, increases the salt and low-temperature tolerance of Arabidopsis. J Integr Plant Biol 48:1244–1249

Jung KH, Lee J, Dardick C, Seo YS, Cao P, Canlas P, Phetsom J, Xu X, Ouyang S, An K, Cho YJ, Lee GC, Lee Y, An G, Ronald PC (2008) Identification and functional analysis of light-responsive unique genes and gene family members in rice. PLoS Genet 4:e1000164

Jung KH, Cao P, Seo YS, Dardick C, Ronald PC (2010) The Rice Kinase Phylogenomics database: a guide for systematic analysis of the rice kinase super-family. Trends Plant Sci 11:595–599

Jung KH, Gho HJ, Nguyen MX, Kim SR, An G (2013) Genome-wide expression analysis of HSP70 family genes in rice and identification of a cytosolic HSP70 gene highly induced under heat stress. Funct Integr Genom 13:391–402

Kamiya N, Nagasaki H, Morikami A, Sato Y, Matsuoka M (2003) Isolation and characterization of a rice WUSCHEL-type homeobox gene that is specifically expressed in the central cells of a quiescent center in the root apical meristem. Plant J 35:429–241

Kim SR, Lee DY, Yang JI, Moon S, An G (2009) Cloning vectors for rice. J Plant Biol 52(1):73–78

Lee S, Jeon JS, Jung KH, An G (1999) Binary vectors for efficient transformation of rice. J Plant Biol 42(4):310–316

Lee KO, Jang HH, Jung BG, Chi YH, Lee JY, Choi YO, Lee JR, Lim CO, Cho MJ, Lee SY (2000) Rice 1Cys-peroxiredoxin over-expressed in transgenic tobacco does not maintain dormancy but enhances antioxidant activity. FEBS Lett 486:103–106

Lu CA, Lim EK, Yu SM (1998) Sugar response sequence in the promoter of a rice alpha-amylase gene serves as a transcriptional enhancer. J Biol Chem 273: 10120–10131

Lu CA, Ho TH, Ho SL, Yu SM (2002) Three novel MYB proteins with one DNA binding repeat mediate sugar and hormone regulation of alpha-amylase gene expression. Plant Cell 14(8):1963–1980

Mitsui T, Akazawa T, Christeller JT, Tartakoff AM (1985) Biosynthesis of rice seed alpha-amylase: two pathways of amylase secretion by the scutellum. Arch Biochem Biophys 241:315–328

Moon YK, Hong JP, Cho YC, Yang SJ, An G, Kim WT (2009) Structure and expression of OsUBP6, an ubiquitin-specific protease 6 homolog in rice (*Oryza sativa* L.) Mol Cells 28:463–472

Murai H, Hara S, Ikenaka T, Goto A, Arai M, Murao S (1985) Amino acid sequence of protein alpha-amylase inhibitor from Streptomyces griseosporeus YM-25. J Biochem 97:1129–1133

Nguyen MX, Moon S, Jung KH (2013) Genome-wide expression analysis of rice aquaporin genes and development of a functional gene network mediated by aquaporin expression in roots. Planta 238:669–681

Nguyen VN, Moon S, Jung KH (2014) Genome-wide expression analysis of rice ABC transporter family across spatio-temporal samples and in response to abiotic stresses. J Plant Physiol 171:1276–1288

Nguyen QN, Lee YS, Cho LH, Jeong HJ, An G, Jung KH (2015) Genome-wide identification and analysis of *Catharanthus roseus* RLK1-like kinases in rice. Planta 241:603–613

Niogret MF, Culianez-Macia FA, Goday A, Mar Alba M, Pages M (1996) Expression and cellular localization of rab28 mRNA and Rab28 protein during maize embryogenesis. Plant J 9:549–557

Ogawa M, Hanada A, Yamauchi Y, Kuwahara A, Kamiya Y, Yamaguchi S (2003) Gibberellin biosynthesis and response during Arabidopsis seed germination. Plant Cell 15:1591–1604

Pospisil P (2009) Production of reactive oxygen species by photosystem II. BBA-bioenergetics 1787:1151–1160

Proteau PJ (2004) 1-Deoxy-D-xylulose 5-phosphate reductoisomerase: an overview. Bioorg Chem 3:483–493

Pulido P, Spínola MC, Kirchsteiger K, Guinea M, Pascual MB, Sahrawy M, Sandalio LM, Dietz KJ, González M, Cejudo FJ (2010) Functional analysis of the pathways for 2-Cys peroxiredoxin reduction in Arabidopsis thaliana chloroplasts. J Exp Bot 61:4043–4054

Romero-Puertas MC, Laxa M, Matte A, Zaninotto F, Finkemeier I, Jones AM, Perazzolli M, Vandelle E, Dietz KJ, Delledonne M (2007) S-nitrosylation of peroxiredoxin II E promotes peroxynitrite-mediated tyrosine nitration. Plant Cell 19:4120–4130

Schmittgen TD, Livak KJ (2008) Analyzing real-time PCR data by the comparative CT method. Nat Protoc 6:1101–1108

Shannon P, Markiel A, Ozier O, Baliga NS, Wang JT, Ramage D, Amin N, Schwikowski B, Ideker T (2003) Cytoscape: a software environment for integrated models of biomolecular interaction networks. Genome Res 13:2498–2504

Stougaard J, Jorgensen JE, Christensen T, Kuhle A, Marcker KA (1990) Interdependence and nodule specificity of cis-acting regulatory elements in the soybean leghemoglobin lbc3 and N23 gene promoters. Mol Gen Genet 220:353–360

Sutoh K, Yamauchi D (2003) Two cis-acting elements necessary and sufficient for gibberellin-upregulated proteinase expression in rice seeds. Plant J 34:635–645

Takase T, Yanagawa Y, Mitsuhara I, Ohashi Y, Hakagawa H, Hashimoto J (2004) Overexpression of a gene for 26S proteasome subunit RPN10 confers enhanced resistance to canavanine, an analog of arginine, in transgenic rice (*Oryza sativa* L.) Plant Biotechnol 21(3):233–236

Tamura K, Peterson D, Peterson N, Stecher G, Nei M, Kumar S (2011) MEGA5: molecular evolutionary genetics analysis using maximum likelihood, evolutionary distance, and maximum parsimony methods. Mol Biol Evol 28:2731–2739

Tsukiyama T, Teramoto S, Yasuda K, Horibata A, Mori N, Okumoto Y, Teraishi M, Saito H, Onishi A, Tamura K, Tanisaka T (2013) Loss-of-function of a ubiquitin-related modifier promotes the mobilization of the active MITE mPing. Mol Plant 6:790–801

Tuteja N, Sahoo RK, Garg B, Tuteja R (2013) OsSUV3 dual helicase functions in salinity stress tolerance by maintaining photosynthesis and antioxidant machinery in rice (*Oryza sativa* L. cv. IR64). Plant J 76(1):115–127

Umate P (2010) Genome-wide analysis of thioredoxin fold superfamily peroxiredoxins in Arabidopsis and rice. Plant Signal Behav 5:1543–1546

Vieweg MF, Fruhling M, Quandt HJ, Heim U, Baumlein H, Puhler A, Kuster H, Andreas MP (2004) The promoter of the *Vicia faba* L. leghemoglobin gene VfLb29 is specifically activated in the infected cells of root nodules and in the arbuscule-containing cells of mycorrhizal roots from different legume and nonlegume plants. Mol Plant-Microbe Interact 17:62–69

Wang W, Xia H, Yang X, Xu T, Si HJ, Cai XX, Wang F, Su J, Snow AA, Lu BR (2014) A novel 5-enolpyruvoylshikimate-3-phosphate (EPSP) synthase transgene for glyphosate resistance stimulates growth and fecundity in weedy rice (*Oryza sativa*) without herbicide. New Phytol 202:679–688

Yamamoto E, Yonemaru J, Yamamoto T, Yano M (2012) OGRO: the overview of functionally characterized genes in Rice online database. Rice 5:26

Zhang CJ, Zhao BC, Ge WN, Zhang YF, Song Y, Sun DY, Guo Y (2011) An Apoplastic H-type Thioredoxin is involved in the stress response through regulation of the Apoplastic reactive oxygen species in Rice. Plant Physiol 157(4):1884–1899

Zimmermann P, Hirsch-Hoffmann M, Hennig L, Gruissem W (2004) GENEVESTIGATOR. Arabidopsis microarray database and analysis toolbox. Plant Physiol 136:2621–2632

10

Sugary Endosperm is Modulated by *Starch Branching Enzyme IIa* in Rice (*Oryza sativa* L.)

Yunjoo Lee[1], Min-Seon Choi[2], Gileung Lee[1], Su Jang[1], Mi-Ra Yoon[3], Backki Kim[1], Rihua Piao[4], Mi-Ok Woo[1], Joong Hyoun Chin[5*] and Hee-Jong Koh[1*]

Abstract

Background: Starch biosynthesis is one of the most important pathways that determine both grain quality and yield in rice (*Oryza sativa* L.). Sugary endosperm, *sugary-1* (*sug-1*), is a mutant trait for starch biosynthesis. Rice plants carrying *sug-1* produce grains that accumulate water-soluble carbohydrates instead of starch, even after maturity. Although this trait enhances the diversity of grain quality, sugary endosperm rice has hardly been commercialized due to the severely wrinkled grains and subsequent problems in milling. This study was conducted to identify the genes responsible for the *sug-h* phenotype through a map-based cloning technology.

Results: We induced a mild sugary mutant, *sugary-h* (*sug-h*) through the chemical mutagenesis on the Korean *japonica* cultivar Hwacheong. Grains of the *sug-h* mutant were translucent and amber-colored, and the endosperm appeared less wrinkled than *sug-1*, whereas the soluble sugar content was fairly high. These characteristics confer greater marketability to the *sug-h* mutant. Genetic analyses indicated that the *sug-h* mutant phenotype was controlled by a complementary interaction of two recessive genes, *Isoamylase1* (*OsISA1*), which was reported previously, and *Starch branching enzyme IIa* (*OsBEIIa*), which was newly identified in this study. Complementation tests indicated that *OsBEIIa* regulated the properties of sugary endosperm.

Conclusions: Complementary interactions between the starch biosynthesis genes *OsISA1* and *OsBEIIa* determine the mild sugary endosperm mutant, *sugary-h*, in rice. Our finding may facilitate the breeding of sugaryendosperm rice for commercial benefit.

Keywords: Map-based cloning, *OsBEIIa*, *OsISA1*, Rice, *Sugary-h*, Sugary endosperm

Background

Rice is the staple food for more than 3 billion people globally. The endosperm is an edible part of the rice grain, and has been one of the major targets for grain geneticists and breeders to enhance grain yield and quality. Endosperm development directly regulates grain formation at the grain filling stage. Mature rice endosperm contains starch, storage proteins, lipids, and other substances. Studies on starch have been an essential focus in rice research. Starch is the primary component that makes cereal crops economically and commercially important. Starch research is also becoming increasingly relevant for industrial and manufacturing applications.

Rice starch is composed of amylose (linear α-1,4-polyglucans) and amylopectin (α-1,6-branched polyglucans). Amylopectin has a distinct fine structure called multiple cluster structure, and accounts for approximately 65–85% of storage starch (Nakamura 2002). Starch is synthesized by four enzyme classes, with multiple subunits in each class: ADP-glucose pyrophosphorylase (AGPase); starch synthase (SS); starch branching enzyme (BE); and starch debranching enzyme (DBE). Other enzymes, such as phosphorylase (Pho) and disproportionating enzyme, are thought to be involved in starch biosynthesis. BE and

* Correspondence: jhchin@sejong.ac.kr; heejkoh@snu.ac.kr
[5]Graduate School of Integrated Bioindustry, Sejong University, 209, Neungdong-ro Gwangjin-gu, Seoul, South Korea
[1]Department of Plant Science and Research Institute for Agriculture and Life Sciences, and Plant Genomics and Breeding Institute, Seoul National University, Seoul 08826, South Korea

DBE have important roles in determining amylopectin structure. BE forms the α-1,6-glycosidic bonds of amylopectin, whereas DBE trims improper branches generated by BE (Fujita 2014).

BE isoforms are classified into two groups, BEI (RBE1) and BEII. Cereals have two BEII isozymes, BEIIa (RBE4) and BEIIb (RBE3). These isoforms are classified according to the transferred amylopectin chain length. For example, BEII transfers shorter chains than BEI, and BEIIb transfers shorter chains than BEIIa, during extended incubations (Mizuno et al. 2001). The expression patterns of BE isoforms also differ. BEI and BEIIa transcripts have been localized in the endosperm and other tissues, whereas BEIIb is expressed only in the endosperm and reproductive tissues. In rice, BEIIa is expressed earlier than either BEIIb or BEI (Mizuno et al. 2001; Ohdan et al. 2005). Previous reports designated BEIIb-deficient mutants in maize and rice as *amylose-extender* (*ae*) mutants, in which the abundance of short amylopectin chains was reduced (Kim et al. 1998; Nishi et al. 2001). Other transgenic research about *BEIIb* gene was reported that the manipulation of BEIIb activity can generate various starch type rice, containing chalky and sugary endosperm (Tanaka et al. 2004). The RNA interference results demonstrated that reduced expression of BEIIa (SBEIIa) caused increase of amylose content in wheat endosperm (Regina et al. 2006). However, the specific functional role of BEIIa has not been elucidated in rice because the seed phenotypes of BEIIa-deficient mutants and wild-type plants are not significantly different (Fujita 2014).

DBEs directly hydrolyze α-1,6-glycosidic linkages of α-polyglucans. DBEs are classified into two types in higher plants, Isoamylase (ISA1, ISA2, and ISA3) and Pullulanase (PUL). According to Fujita (2014), Isoamylase1 (ISA1)-deficient mutants (*isa1*) were called as *sugary-1* mutants in rice (*sug-1*) and maize (*su1*). These mutants have a defect in the amylopectin cluster structure, which results in the accumulation of a polymeric water-soluble polysaccharide (WSP) termed phytoglycogen, and a reduction in the starch content (James et al. 2003). There are various *sug-1* mutant types, EM series, reported by Japanese group (Nakamura et al. 1997; Nakamura et al. 1996; Wong et al. 2003). The *sug-1* locus in rice is located on chromosome 8 (Fujita et al. 1999; Yano et al. 1984). In transgenic *sug-1* rice expressing the wheat *ISA1* gene, phytoglycogen synthesis is substantially replaced by starch biosynthesis in the endosperm (Kubo et al. 2005). This result implies that ISA1 is essential for amylopectin crystallinity and biosynthesis in both rice and wheat. In maize, double mutant defective in both ISA2 and SSIII generated water-soluble glucans in the mutant endosperm, although single mutants of either ISA2 or SSIII could synthesize normal amylopectin (Lin et al. 2012). By contrast, the contribution of PUL for amylopectin trimming was much smaller than that of ISA1, and PUL function partially overlaps with that of ISA1 (Fujita et al. 2009).

Since rice grains of *sug-1* mutants primarily contain water-soluble carbohydrates instead of starch even after maturity, they have the potential value in breeding programs for diversified qualities and commercial uses. However, they have not been used in practice due to the severely wrinkled grains and subsequent problems in milling. We developed a mild-type sugary mutant in rice, *sugary-h* (*sug-h*), which displayed an intermediate phenotype between the *sug-1* mutant and wild type. Grains of the *sug-h* mutant have better quality for subsequent processing and higher yield than *sug-1*. In addition, palatability, protein and amylose content which are crucial for breeding were increased in *sug-h* mutant (Yoon et al. 2009; *sug-2* was renamed as *sug-h* to avoid confusion with Nakagami et al. (2016)). Therefore, *sug-h* mutants could be valuable for practical applications and nutritional aspects. This study performed map-based cloning to identify the genes responsible for the *sug-h* phenotype. Our results not only identified a potential resource for utilizing the sugary endosperm mutant for commercial benefit, but also may provide a new insight into starch biosynthesis.

Results

Phenotypic Characterization of the *sug-h* Mutant

The *sug-h* mutant did not exhibit abnormal phenotypes during the vegetative stage of plant growth, although the heading date was delayed and plant height was reduced compared with that of wild-type plants (Table 1; Additional file 1: Figure S1). Grains of the *sug-h* mutant displayed an intermediate phenotype between the wild-type and *sug-1* mutant grains, that was medium thickness and slightly wrinkled (Fig. 1a-f). Wild type, *sug-1*, and *sug-h* mutants showed significant differences in grain dimensions and thickness. The grain length and width of *sug-h* mutant was slightly longer and shorter, respectively, than that of wild-type (Table 1).

Morphological Properties of Starch in Mutant Endosperm

Grains were stained with iodine to identify starch components. Phytoglycogen, which does not stain with iodine, had the highest abundance in *sug-1* endosperm. The *sug-h* mutant grain was partially stained in outermost endosperm layers, whereas the entire wild-type endosperm was stained (Fig. 1g-i). These results indicate that the endosperm starch components in both sugary mutants differed from those in the wild type, and starch production was at least partially restored in the *sug-h* endosperm.

Cross sections of polished rice grains were observed with scanning electron microscope (SEM). The images revealed that *sug-1* and *sug-h* mutants had loosely packed, abnormal starch granules in the cutting plane compared with the densely packed granules in wild-type

Table 1 Agronomical characters of plant and dimensions of grain in wild type (Hwacheong) and mutants

Trait	HD	CL (cm)	PN	SN	TGW (g)	GL (mm)	GW (mm)	GT (mm)	GS
WT	Aug.17	85.90 ± 1.61^{a}	14.60 ± 1.43^{b}	98.90 ± 16.97	19.88 ± 0.19^{a}	5.11 ± 0.09^{b}	3.06 ± 0.07^{b}	2.17 ± 0.03^{c}	1.67 ± 0.54^{a}
sug-1	Aug.20	87.90 ± 3.04^{b}	11.10 ± 3.57^{a}	90.30 ± 10.86	8.76 ± 0.30^{b}	4.89 ± 0.09^{a}	2.96 ± 0.09^{a}	1.07 ± 0.13^{a}	1.65 ± 0.05^{a}
sug-h	Aug.20	75.60 ± 1.31^{c}	16.40 ± 1.84^{b}	101.80 ± 9.78	14.52 ± 0.15^{c}	5.41 ± 0.13^{c}	2.91 ± 0.10^{a}	1.70 ± 0.11^{b}	1.86 ± 0.07^{b}

HD heading date, *CL* culm length, *PN* number of panicles per plant, *SN* number of spikelets per panicle, *TGW* 1000-grain weight for brown rice, *GL* grain length, *GW* grain width, *GT* grain thickness, *GS* grain shape (length/width), *WT* wild type (Hwacheong), *sug-1 sugary-1*, *sug-h sugary-h*
Different letters denote significant differences. Ten biological replicates were used to measure for each of the traits

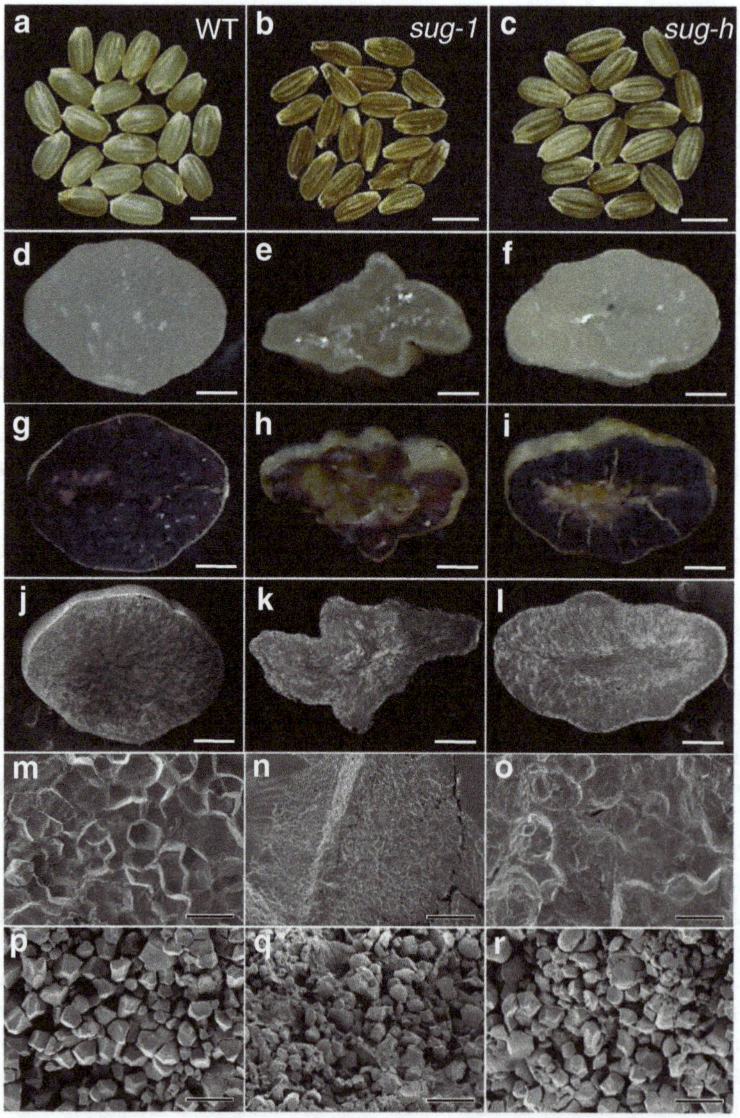

Fig. 1 Optical and scanning electron microscopy observation of grain phenotype and starch granule structure. **a-c** The *sug-h* mutant grain exhibits a phenotype that is intermediate between that of the wild type and *sug-1*. Bars = 4 mm. **d-i** Cross sections of wild-type, *sug-1*, and *sug-h* kernels at the mature stage. Sectioned seeds were stained with iodine solution. Bars = 0.5 mm. **j-o** SEM observations of wild-type, *sug-1*, and *sug-h* mutant endosperm. **j-l** Bars = 0.5 mm. (m-o) Bars = 10 μm. **p-r** SEM observations of starch granule structures in wild-type, *sug-1*, and *sug-h* mutant endosperm. Bars = 10 μm

polished grains. The *sug-h* endosperm had starch granule packing that was intermediate between that of *sug-1* and the wild type (Fig. 1j-o). Starch granules in *sug-1* and *sug-h* mutants also displayed irregular shapes with round edges, whereas those in the wild type displayed polygonal shapes with sharp edges (Fig. 1p-r). These results indicate that aberrant starch production was involved in the abnormal phenotypes of *sug-1* and *sug-h* mutants.

Starch crystallinity was investigated by performing X-ray diffraction analysis of endosperm starch from the wild type, *sug-1, sug-h,* and *Sug-1/sug-h*. Diffraction patterns of *sug-1* and *sug-h* starch displayed lower peak intensity than that of the wild type (Fig. 2). These X-ray diffraction patterns indicate that the crystallization of *sug-1* and *sug-h* starch was lower than that of wild-type starch, as reported by Yoon et al. (2009). Starch from *sug-h* mutant had an intermediate crystallization level between that from *sug-1* and the wild type. However, *Sug-1/ sug-h*, which has a normal phenotype, displayed similar starch crystallinity to that of wild-type starch. These results indicate that *sug-1* is a major effect on the starch structure and crystallization.

Amylopectin Fine Structure in Mature Mutant Endosperm

Fine structural feature of amylopectin in wild-type and mutant endosperms were compared in detail by analysis of amylopectin chain-length distribution using high performance anion–exchange chromatography and a pulsed amperometric detector (HPAEC-PAD). Both sugary mutants had the definitely increased amount of short chains in the range of DP 6–10 (Fig. 3a–d). This result was similar to the previous reports (Nakamura et al. 1997; Wong et al. 2003; Yoon et al. 2009), indicating that mutation on ISA altered the fine structure of amylopectin into phytoglycogen in rice endosperm. The difference between *sug-1* and *sug-h* on fine structure of amylopectin was relative amount of short chains. Most range of chain-length distribution of amylopectin in *sug-h* was similar with that of *sug-1*, except in the range of DP 7–8 and DP 13–17 (Fig. 3e–f). Structural feature of amylopectin suggest that the *sug-h* phenotype was determined by the relative amount of short chains.

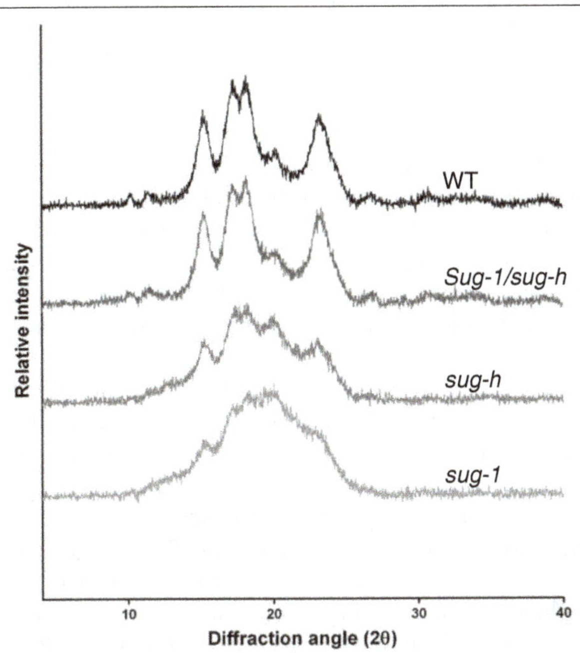

Fig. 2 X-ray diffraction analysis of purified starch granules from mature endosperm of wild-type rice (Hwacheong), *sug-1*, *sug-h*, and *Sug-1/sug-h*. The seeds possessed a normal *OsISA1* and a mutated *osbe2a* genotypes were designated as *Sug-1/sug-h*

Map-Based Cloning of Genes Related to the *sug-h* Phenotype

We performed map-based cloning to elucidate the genes related to the *sug-h* phenotype. Normal-type seeds were designated as 'N-type' and sugary-type seeds were designated as 'S-type'. S-type seeds were grouped into 'I-type' (*sugary-1* type), mixed-type (I-type and II-type), and 'II-type' (*sugary-h* type) in segregated populations. F_2 seeds derived from a cross between the *sug-h* mutant and Hwacheong were used for segregation ratio analysis. The results from 352 F_2 seeds identified 271 N-type, 57 I-type, and 24 II-type seeds, which fit the expected ratio of 12:3:1 (P = 0.45) (Additional file 2: Table S1). Expected genotypes of N-type, I-type and II-type will be *Sug-1_ Sug-h_*, *sug-1sug-1Sug-h_*, and *sug-1sug-1sug-h sug-h*, respectively. Based on this result, we hypothesized that the *sug-h* phenotype was controlled by the epistatic interaction between two genes. One of these genes is the preceding gene, which is related to the sugary endosperm phenotype, and the other is the interacting gene, which reduces the severity of the sugary abnormality in starch biosynthesis.

Preliminary mapping was initially conducted on 352 F_2 plants derived from *sug-h*/Milyang.23 (M.23, a Korean *tongil*-type cultivar), and mapped the first gene related to the sugary phenotype. Bulked DNA for the first bulked segregant analysis (BSA) was prepared to distinguish N-type, heterozygous type (N-type and S-type), and S-type plants in the F_2 population. Two of seventy-two sequence tagged site (STS) markers across all chromosomes, S08105 and S08107 on chromosome 8, were used as the flanking markers for fine mapping (Fig. 4a). Eighteen genes were identified in the fine-mapped region. The *OsISA1* gene sequence (Os08g40930) of Hwacheong and *sug-h* mutant was primarily compared, because previous studies reported that *OsISA1* affected the sugary phenotype in *sug-1* rice (Kubo et al. 1999; Nakamura et al. 1992; Nakamura et al. 1996). Our sequence analysis revealed that nucleotide residue 6179 in *OsISA1*, which was adenine (A) in the wild type, was substituted with guanine (G) in *sug-1* and *sug-h*

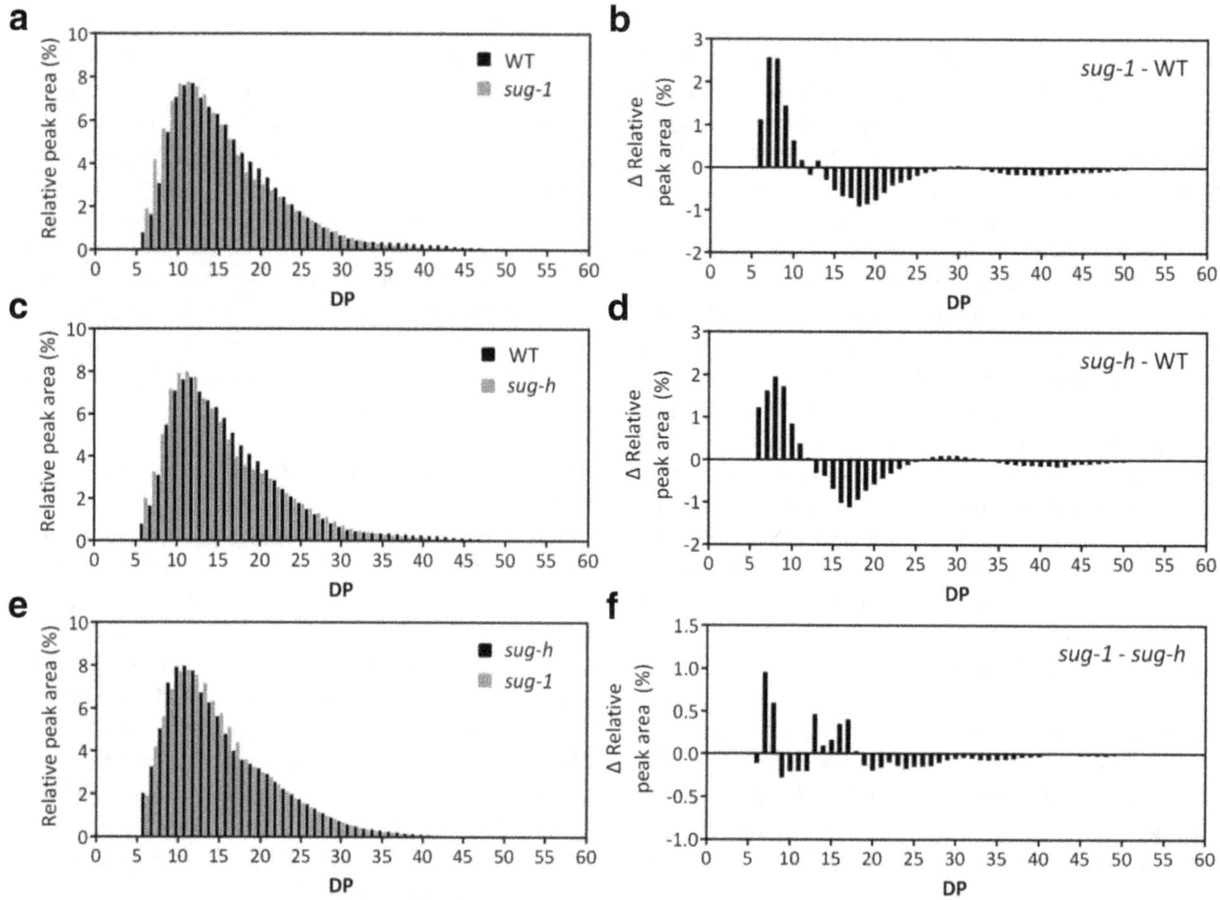

Fig. 3 Comparison of chain length distribution of amylopectin in rice endosperm between wild type (Hwacheong) and mutant. The left panels showed chain-length profiles, and the right panels showed differences in chain-length profiles. Chain-length profiles after debranching the starch fraction and differences between WT and *sug-1* mutant (**a-b**), between WT and *sug-h* mutant (**c-d**), and between *sug-1* and *sug-h* mutant (**e-f**) were shown

mutants, thereby changing the amino acid from glycine (Gly) to aspartic acid (Asp). These results indicate that the sugary endosperm phenotype could be caused by a point mutation on *OsISA1*, which was shared by both *sug-1* and *sug-h* mutants.

One of the objectives in this study was to identify the second gene controlling the thickness and wrinkling of the sugary endosperm. Therefore, 837 F_3 plants were derived from two individual F_2 plants in which the *osisa1* allele was fixed, for efficient mapping. Bulked DNA for the second BSA was prepared from I-type, mixed-type (I-type and II-type), and II-type plants selected from the F_3 population. The second BSA revealed that markers and the *sug-h* phenotype co-segregated on chromosome 4, indicating that the second gene was located on chromosome 4. For fine mapping, S04056A and S04058 were selected as flanking markers. One STS marker, S04057, and two dCAPS markers, AL731A and AL731B within the AL731641 clone, were additionally designed and applied for linkage analysis. These results reduced the size of the candidate region to 12.9 kb, which contained the following three genes: *1,4-α-glucan-branching enzyme 2*; *Ser/Thr protein phosphatase family protein*; and *histone deacetylase* (Fig. 4b). These genes were sequenced and compared in Hwacheong and *sug-h* mutant, and a point mutation was detected in the *1,4-α-glucan-branching enzyme 2* (*OsBEIIa*; *OsSBE4*) gene of the *sug-h* mutant. We found that the nucleotide residue 4732 in wild-type *OsBEIIa*, which was G, was substituted with A in the *sug-h* mutant, thereby changing the amino acid from Gly to Asp. This amino acid substitution was not found in other grain species including maize, sorghum, barley, and wheat (Additional file 3: Figure S2). These combined results indicate that the *sug-h* mutant might be caused by single point mutations resulting in amino acid substitutions in both *OsISA1* and *OsBEIIa*.

Transgenic Complementation of the *sug-h* Mutation

To confirm the function of *OsBEIIa* in the *sug-h* mutant, we generated dsRNA-mediated interference (RNAi)

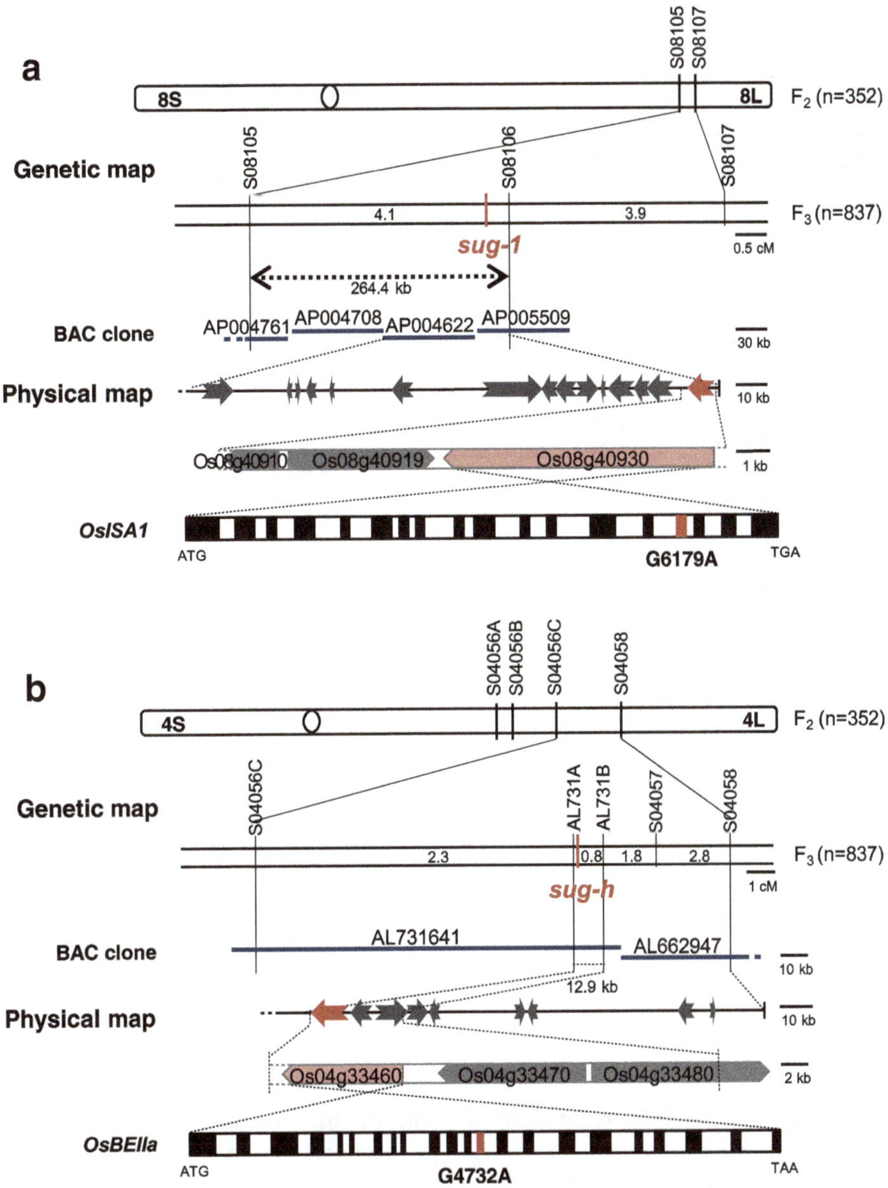

Fig. 4 Map-based cloning of the *sug-h* mutant. **a** Candidate region of the sugary gene was on the long arm of chromosome 8. Candidate gene of the first bulked segregant analysis, Os08g40930, was located on the AP005509 BAC clone and contained 18 exons (*black boxes*) and 17 introns (*white boxes*). A point mutation, G to A, occurred in the 15th exon (*red box*). **b** Schematic representation of *OsBEIIa* on the long arm of chromosome 4. Candidate region of the gene related to the *sug-h* mutant was located within the AL731641 clone. The physical distance was approximately 12.9 kb, which included three candidate genes. Schematic structure of the candidate gene, Os04g33460, contained 22 exons (*black boxes*) and 21 introns (*white boxes*). A point mutation, G to A, occurred in the 13th exon (*red box*). *Gray arrows* indicate the main genes with known functions within the candidate region. *Red arrow* indicates the candidate gene. ATG and TGA/TAA indicate the initiation and termination codons, respectively

transgenic plants. T_1 seeds of the *OsBEIIa*-RNAi transgenic line had normal phenotype, suggesting that a single mutation of *OsBEIIa* had no effect on seed phenotype (Fig. 5a). To evaluate interactions between *OsISA1* and *OsBEIIa*, artificial crossing was performed between the *OsBEIIa*-RNAi T_0 plant and the *sug-h* mutant. After phenotypic selection, F_2 seeds from the artificial cross were planted for co-segregation analysis of phenotype, genotype, and RNA expression. Phenotypes of segregated F_2 seeds showed that N-type seeds were slightly thinner than wild-type (Dongjin) seeds. By contrast, the thickness of I-type and II-type seeds was not significantly different from that of *sug-1* and *sug-h* seeds, respectively (Fig. 5b). Genotype analysis using an antibiotic resistance gene-specific primer showed that PCR bands were produced in all II-type plants, but no bands were amplified in all I-type plants (data not shown). The qRT-PCR analysis indicated that the relative *OsBEIIa*

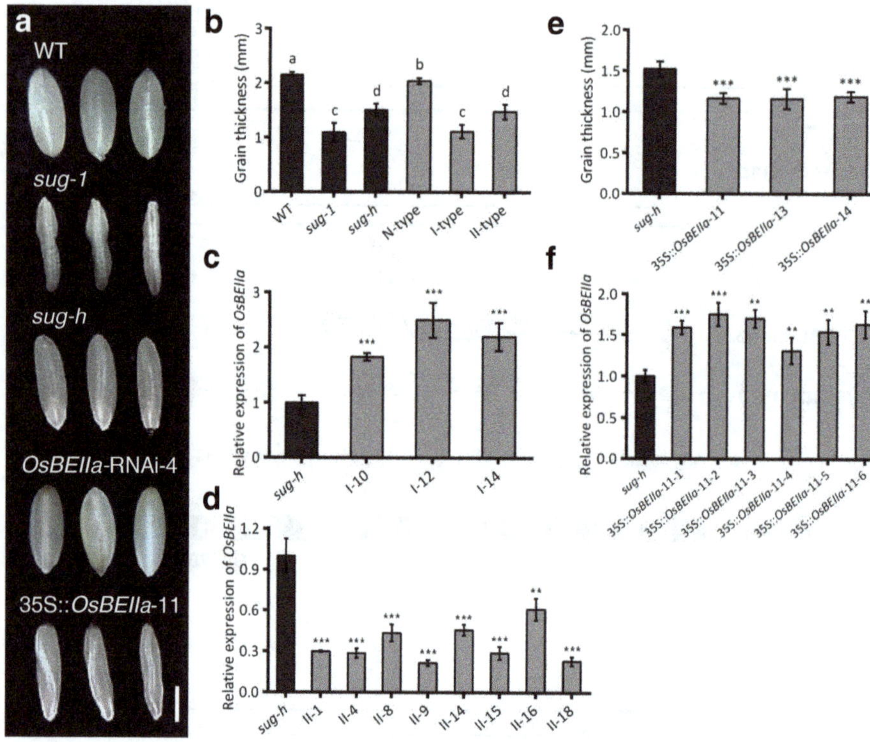

Fig. 5 Phenotype of transgenic seeds and complementation test of the *sug-h* mutation. **a** Grain morphologies of wild-type rice (Dongjin), *sug-1*, *sug-h*, and transgenic seeds. Bar = 2 mm. **b** Comparison of grain thickness in F_2 seeds derived from *OsBEIIa*-RNAi-4/*sug-h*. Different letters denote significant differences. Replicate samples were measured 20 times. **c** Relative expression level of *OsBEIIa* in each F_2 plant derived from I-type seeds using qRT-PCR. The *sug-h* mutant was used as a control. Error bars represent SD for three technical experiments. Asterisks indicate statistical significance compared with the control, as determined by Student's *t*-test (***$P < 0.001$). **d** Relative expression level of *OsBEIIa* in each F_2 plant derived from II-type seeds by qRT-PCR. Error bars represent SD for three technical experiments. Asterisks indicate statistical significance compared with the control, as determined by Student's *t*-test (**$P < 0.01$, ***$P < 0.001$). **e** Comparison of grain thickness in three lines of 35S::*OsBEIIa* and *sug-h* mutant seeds. Statistical significance was determined by Student's *t*-test (***$P < 0.001$). Replicate samples were measured 20 times. **f** qRT-PCR analysis showed increased *OsBEIIa* expression in 35S::*OsBEIIa*-11 T_1 plants. Error bars represent SD for three technical experiments. Statistical significance was determined using Student's *t*-test (***$P < 0.001$)

expression levels were higher in each F_2 plant derived from I-type seed than in the *sug-h* mutant (Fig. 5c), whereas the relative *OsBEIIa* expression levels in each F_2 plant derived from II-type seed were lower than that in the *sug-h* mutant (Fig. 5d). These results show that phenotype, genotype, and RNA expression co-segregate in the *OsBEIIa*-RNAi-4/*sug-h* F_2 population, indicating that the two genes are associated with the *sug-h* phenotype.

We also generated transgenic plants overexpressing *OsBEIIa*, in which a vector was introduced into the *sug-h* mutant to complement the phenotype. According to the mapping results that the *sug-h* mutant might be caused by single point mutations in both *OsISA1* and *OsBEIIa* genes, this complementation test was intended to reconstruct the *sug-1* phenotype from *sug-h* mutant. Most T_1 seeds were thinner and more wrinkled than *sug-h* seeds (Fig. 5a, e). The qRT-PCR analysis indicated that *OsBEIIa* transcript levels were higher in transgenic T_1 plants than in *sug-h* mutant (Fig. 5f). These complementation data suggest that *osbe2a* may improve the *sug-h* phenotype by producing grains that are less wrinkled.

Gene Expression Patterns of *OsBEIIa*

To investigate the expression patterns of *OsISA1* and *OsBEIIa* in different organs, we performed qRT-PCR analyses and β-glucuronidase (GUS) reporter gene assay. The qRT-PCR analyses showed that *OsBEIIa* was expressed primarily in leaf, stem, and seed in wild-type and *sug-h* mutant plants (Fig. 6a). *OsBEIIa* expression in 10 days after flowering (DAF) seeds was slightly down-regulated in *sug-h* mutant compared with that in the wild type, although this change was not statistically significant. By contrast, *OsISA1* expression in *sug-h* seeds was significantly lower than that of the wild type, and *OsISA1* expression was similarly low in both wild-type and mutant leaf samples (Additional file 4: Figure S3). Transcript analysis revealed that the *OsBEIIa* mutation did not show typical RNA expression patterns in

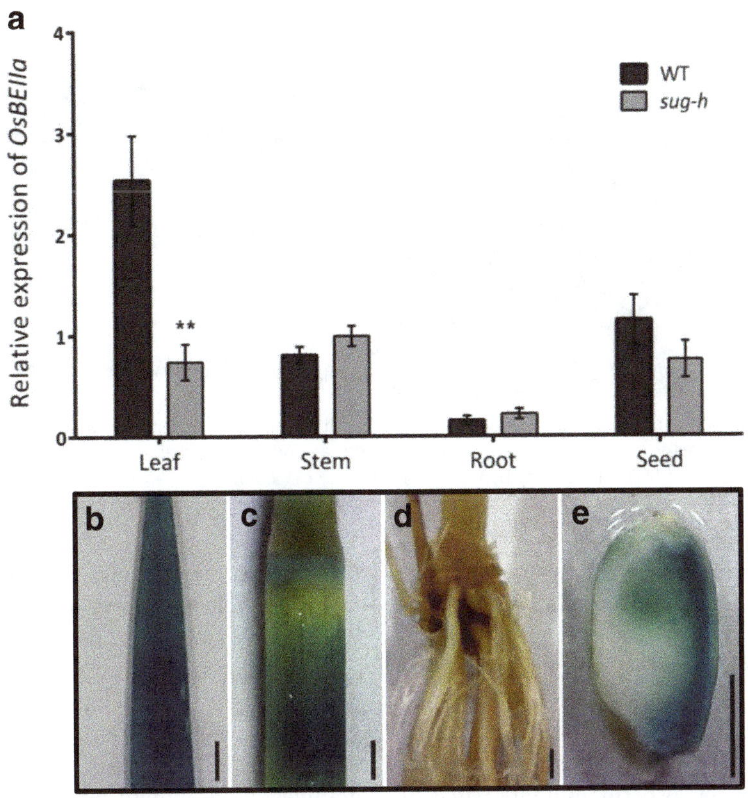

Fig. 6 Expression pattern of *OsBEIIa* gene (**a**) qRT-PCR analysis detected *OsBEIIa* transcripts in leaf, stem, root, and 10 DAF seeds. Data are mean ± SD (n = 3). Statistical significance was determined using Student's t-test (**$P < 0.01$). WT, wild-type rice (Hwacheong); DAF, days after flowering. **b-e** GUS expression was detected in leaf, root, stem base, and 20 DAF seeds in a transgenic plant expressing the *OsBEIIa* promoter::GUS reporter gene. Bars = 2 mm

different samples, implying the possibility that other factors also determine the *sug-h* phenotype.

GUS was expressed under the control of the native *OsBEIIa* promoter in the wild-type background, and the results were consistent with the qRT-PCR data. GUS expression was detected in leaf, stem base, node, and 20 DAF seeds (Fig. 6b-e). The expression levels of *OsBEIIa* were negligible in roots as determined by both qRT-PCR and GUS reporter analyses.

Debranching and Branching Enzyme Activities in Mutant Endosperm

Native-PAGE/DBE and BE activity staining was performed to determine the change of enzyme activity in the *sug-h* mutant. Debranching enzymes, ISA and PUL, were detected as blue bands on the native gel containing potato tuber amylopectin stained with an iodine solution. ISA was visualized as three major bands with low mobility. ISA activity conspicuously decreased in both sugary mutants, as the same result with mapping on chromosome 8 (Fig. 7a).

Branching enzymes, BEI and BEII, were also detected on the native gel in distinct band patterns by iodine staining. BEI and BEIIb/Pho activities were shown in both wild-type and *sug-1* mutant. However, BEIIa activity shown as reddish bands, dramatically decreased only in the *sug-h* mutant (Fig. 7b). Zymogram results imply that the difference between *sug-1* and *sug-h* mutant was caused by branching enzyme activity.

Discussion

Genetic mapping of the *sug-h* rice mutant was used to identify and isolate two recessive genes, *OsISA1* and *OsBEIIa*. The *sug-h* mutant has a mild sugary phenotype, which preferentially accumulates desirable sugar compositions of non-starch polysaccharide rather than starch, and is more commercially viable than the *sug-1* mutant because it does not display excessive wrinkling, which interferes with milling (Koh and Heu 1994; Yoon et al. 2009). Segregation ratios of the F_2 population showed that the *sug-h* phenotype was controlled by complementary interactions between *OsISA1* and *OsBEIIa*. We demonstrated that *OsISA1* and *OsBEIIa* were associated with the genetic modifications that were responsible for the sugary endosperm phenotype. Although a single mutation in *OsBEIIa* did not affect endosperm phenotype,

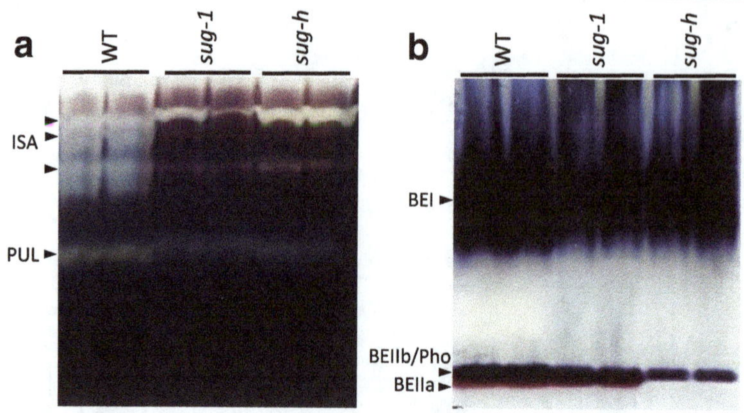

Fig. 7 Native-PAGE/activity staining of developing endosperm in wild-type and mutant seeds. **a** Native-PAGE/debranching enzyme (DBE) activity staining of rice endosperm at late-milky stage. The ISA and PUL activity bands are indicated by black arrows. **b** Native-PAGE/branching enzyme (BE) activity staining. The BEI, BEIIa and BEIIb/Pho activity bands are indicated by *black arrows*

the mutation in *OsBEIIa* moderately recovered the sugary endosperm from the severe wrinkling caused by *osisa1*. Therefore, *sug-h* mutant seed maintains a sugary phenotype, but the seed quality is superior (less wrinkled) than that of *sug-1*.

There have been several reported mutants and transgenic rice related to sugary-type endosperm. Among them, severe sugary-type endosperm mutants, such as EM914 (Nakamura et al. 1997; Wong et al. 2003), #1–1 (Tanaka et al. 2004), and *OsISA1* suppression and *OsISA2* over-expression transgenic lines (Utsumi et al. 2011), were similar to the *sug-1* mutant used in this study. Of them, phenotype of EM914 was governed by mutated *ISA1* while #1–1 and *OsISA2* over-expression line had different genes than *sug-1*. It is interesting that a mild sugary mutant, which was reported as a variation of *su-1* mutant by Nakamura et al. (1997), had the similar phenotype to *sug-h* mutant although only *sug-1* locus was involved in the *su-1* mutant. The reason for the phenotypic similarity between *su-1* mutant by Nakamura et al. (1997) and *sug-h* mutant in this study remains to be comparatively studied. Recently, a rice novel endosperm mutant, named as *sugary-2*, was reported (Nakagami et al. 2016), in which the results on the activity of BE in the *sugary-2* mutant was unlike the *sug-h* mutant, indicating that the *sugary-2* mutant was different from the *sug-h* mutant.

Nakamura (2002) reported that inhibition of BEIIa activity caused low levels of short amylopectin chains with degree of polymerization (DP) ≤10 in rice leaf sheath, in which BEIIb is not expressed; however, the BEIIa-deficient mutant does not exhibit a significant change in amylopectin chain length profile in rice endosperm. Because a direct function of the *OsBEIIa* could not have been elucidated yet, we suggest that the amino acid substitutions in *OsISA1* and *OsBEIIa* changed the protein complex or enzyme interaction involved in starch biosynthesis, and might be responsible for the *sug-h* phenotype affecting amylopectin structure. Future studies will perform enzymatic analyses to test this hypothesis.

Previous studies investigated possible interactions between ISA and other enzymes. The debranching enzyme PUL was related to the sugary phenotype (Fujita et al. 2009). Amylose content, seed morphology, and starch granules of *pul* mutant lines were essentially the same as those of wild-type plants. By contrast, double *pul* and *isa1* mutant lines contained higher levels of WSPs and had shorter amylopectin chains with DP ≤7 in the endosperm compared with the *sug-1* parents, indicating that PUL can partly compensate for starch biosynthesis. The absence of ISA activity primarily affected the sugary endosperm phenotype regardless of the presence of PUL activity. This result was very similar using the *sug-h* mutant; however, no differences in *OsPUL* sequences were identified in the Hwacheong wild type and the *sug-h* mutant.

FLOURY ENDOSPERM6 (*FLO6*) encodes a CBM48 domain-containing protein (Peng et al. 2014). FLO6 may act as a starch-binding protein interacting with ISA1, and may be a bridge between ISA1 and starch during starch biosynthesis. ISA1 may have interacting factors that mediate starch binding, and interacting enzymes that have not yet been elucidated. Previous research evaluated protein-protein interactions of starch biosynthetic enzymes. Crofts et al. (2015) reviewed that co-immunoprecipitation analysis revealed the following associations in rice: BEIIa-BEIIb, BEIIa-BEI, BEIIa-Pho1, BEIIa-SSI, and BEIIa-SSIIIa. The BEIIa-SSI interaction was also identified in wheat and maize (Tetlow et al. 2008). These results suggested that some isozymes involved in starch biosynthesis in rice formed active protein complexes. These combined results suggest a possible mechanism of BEIIa function in rice endosperm.

Phenotypic variation is a critical consideration for phenotypic analysis of sugary endosperm in *sug-h* populations because of environmental effects. Satoh et al. (2008) evaluated the effect of growth temperature on the frequency of various grain phenotypes and the extent of starch accumulation in the wild type and mutant, and reported that starch accumulation in the *phosphorylase1* (*pho1*) mutant endosperm was affected by temperature. Similarly, the seed phenotypes of *sug-1* and *sug-h* mutants differed slightly between plants grown in the field and those grown in the green house (data not shown). To reduce this phenotypic variation, all seeds from a whole main panicle of F_2 and F_3 plants grown in the field were used for genetic mapping. The mutants and wild-type seeds were grown together under the same conditions and prepared for phenotypic analysis at the same time. Future studies should assess the effects of environmental factors, especially temperature, on phenotypic variation.

In this study, we propose that mutated *OsBEIIa* plays a role in restoring the severely wrinkled sugary phenotype caused by *osisa1* in rice endosperm, although *OsBEIIa* mutation alone did not result in a significant phenotypic change. The observed complementary interaction between *OsISA1* and *OsBEIIa* provides novel insight into the roles of starch biosynthesis enzymes and their interactions. Our result can facilitate the breeding of functional rice cultivars with special nutritional qualities, and might be applicable to endosperm modification in other cereal crops.

Conclusions

Grains of *sug-h* mutant have less wrinkled feature, which make it more commercially feasible by easy polishing in processing. Here, we report the cloning of a new sugary gene in rice, which controls the thickness of sugary endosperm grains. Genetic analysis revealed that phenotype of *sug-h* mutant was controlled by a complementary interaction of two recessive genes, *OsBEIIa* and *OsISA1*. Complementation test demonstrated that sugary endosperm was modulated by *OsBEIIa*. These results may facilitate the breeding of sugary endosperm rice for commercial use and will be helpful to enlarge our understanding on starch biosynthesis in rice endosperm.

Methods

Plant Materials

The *sug-h* mutant was induced by N-methyl-N-nitrosourea (MNU) treatment on the Korean *japonica* cultivar Hwacheong (Koh and Heu 1994; The *sug-2* was renamed in this study as *sug-h* to avoid confusion with Nakagami et al. (2016)). The F_2 population was derived from a cross between the *sug-h* mutant and M.23. F_3 seeds were classified into three groups: normal, sugary, and heterozygous type. Two F_2 individuals that displayed the sugary-type in F_3 seeds were developed for the F_3 population. To calculate the segregation ratio, another F_2 population derived from a cross between the *sug-h* mutant and wild-type Hwacheong was used. The *sug-1* mutant line was selected from the *sug-h*/Hwacheong F_2 population, and isolated to the F_6 generation via self-pollination. Normal-type seeds were designated as 'N-type' and sugary-type seeds were designated as 'S-type'. S-type seeds were grouped into 'I-type' (*sugary-1* type), 'II-type' (*sugary-h* type) and mixed-type (I-type and II-type) in segregating populations. These populations were cultivated using conventional methods at the Experimental Farm of Seoul National University.

Phenotypic Analysis

All F_1, F_2, and F_3 seeds were dehulled and observed under a microscope. S-type and N-type seeds were distinguished by seed thickness and severity of wrinkling. All seeds from a whole panicle were used for phenotyping to minimize phenotypic error resulting from differences in seed maturity. Grain dimensions including length, width, and thickness were measured using digimatic calipers (Mitutoyo, Japan) and analyzed using the IBM SPSS statistics program. Each seed was stained with iodine solution [0.1% (*w*/*v*) I_2 and 1% (*w*/*v*) KI] to detect starch.

Scanning Electron Microscopy

Seeds and starch samples were visualized with a SEM according to the previously published method of Fujita et al. (2003). Gold-coated seeds and starch samples were observed using a SUPRA 55VP Scanning Electron Microscope (Carl Zeiss, Germany).

X-ray Diffraction Analysis of Starch

One of the samples for morphological analysis of starch properties, *Sug-1*/*sug-h*, possessed normal *OsISA1* and mutated *osbe2a* alleles. X-ray diffraction was used to determine the structures of starch according to the methods described previously by Kubo et al. (2005). The X-ray diffraction patterns of isolated insoluble glucans were obtained with a copper (nickel foil-filtered) K_α radiation using D8 Advanced X-ray diffractometry (Bruker, Germany) at 40 kv and 40 mA. The scanning region of the two-theta angle (2θ) ranged from 4.0 to 40.0° with a scan speed of 0.5 deg. min^{-1}.

Chain–Length Distribution of Amylopectin by HPAEC-PAD

The chain–length distributions of amylopectin from wild-type and mutants endosperm were analyzed using HPAEC-PAD as in previous reports (Hanashiro et al. 1996; Kwak et al. 2017).

DNA Extraction and PCR Amplification

Total genomic DNA was extracted from young leaves of F_2, F_3 plants and their parents according to the method of McCouch et al. (1988) with modifications. Polymerase chain reactions (PCR) were performed in a reaction volume of 20 μl containing 100 ng of template DNA, 0.1 μM each primer, 2.5 mM dNTP, 10 mM Tris-HCl (pH 8.3), 50 mM KCl, 1.5 mM $MgCl_2$, 0.01% (*w*/*v*) gelatin, and 0.5 U Taq DNA polymerase. PCR amplification was carried out in a DNA Engine Tetrad 2 and Dyad Thermal Cycler (Bio-Rad, USA) using the following reaction conditions: 5 min at 94 °C; followed by 35 cycles of 1 min at 94 °C, 30 s at 56 °C, and 40 s at 72 °C; and 10 min at 72 °C for final extension. PCR products were separated on 2.5% (*w*/*v*) agarose gels containing 0.15 μg ml^{-1} ethidium bromide (EtBr) in 0.5 × TBE buffer.

Linkage Mapping of the *sug-h* Mutant

We performed BSA to genetically map and isolate genes related to the *sug-h* mutant (Michelmore et al. 1991). Ten N-type plants, ten heterozygous type plants (N-type and S-type), and ten S-type plants were selected from the *sug-h*/M.23 F_2 population for the first BSA to identify gene distinguishing N-type and S-type. For the second BSA, ten I-type plants, 12 mixed-type plants (I-type and II-type), and 12 II-type plants were selected from the *sug-h*/M.23 F_3 population, in which the sugary allele was fixed. Three bulked samples containing randomly combined equal amounts of DNA were used for genotyping. Then, fine mapping was conducted on two chromosomes with flanking STS markers, which were developed by designing primers based on the DNA sequence differences between *indica* and *japonica* rice cultivars (Chin et al. 2007). Additional STS and derived cleaved amplified polymorphic sequence (dCAPS) primers were designed with Primer3 version 0.4.0 (http://frodo.wi.mit.edu/primer3) for additional fine mapping based on the available rice genome sequence data (http://www.gramene.org, http://www.ncbi.nlm.nih.gov). Primer sequences and other information for DNA markers designed and used in this study are listed in Additional file 5: Table S2.

Candidate Gene Analysis

To validate the candidate gene models, full-length genomic DNA of each candidate gene in Hwacheong and the *sug-h* mutant was amplified by PCR. PCR products were purified using the PCR purification kit (Inclone, Korea) and transformed into *E. coli* strain DH5α, followed by ligation of PCR amplicons into the pGEM-T Easy Vector (Promega, USA). Transformed plasmid sequences were analyzed with an ABI Prism 3730 XL DNA Analyzer (PE Applied Biosystems, USA). The PCR clones were verified by sequence alignment with the original parent using CodonCode Aligner software (CodonCode Corporation, USA). Based on the results of the sequencing analysis, multiple alignments were performed using a public database (http://www.ch.embnet.org/software/BOX_form.html).

RNA Isolation and Quantitative Real-Time PCR Analysis

Total RNA was extracted from various tissues of wild-type, mutant, and transgenic plants using MG Total RNA Extraction kit (Doctor Protein, Korea). The RNAs were converted into first-strand cDNA using M-MLV Reverse Transcriptase (Promega, USA). Quantitative RT-PCR was performed using SYBR Premix ExTaq (Takara Bio, Japan) according to the manufacturer's instructions. Gene expression levels were evaluated in leaf, stem, and root samples collected at maximum tillering stage and seed samples collected at 5, 10, and 20 DAF. The following gene-specific primer sets were used: *OsISA1*-RT, 5′-CAAATGCG CAATTTCTTTGTT-3′ (sense) and 5′-GTCCCAGCG GAAATAATTGA-3′ (antisense); *OsBEIIa*-RT, 5′-GCCAA TGCCAGGAAGATGA-3′ (sense) and 5′-GCGCAACA TAGGATGGGTTT-3′ (antisense) (Zhang et al. 2012); control *UBQ*-qPCR, 5′-GAGCCTCTGTTCGTCAAGTA-3′ and 5′-ACTCGATGGTCCATTAAACC-3′ (Tanaka et al. 2011). Quantitative RT-PCR was performed using a C1000 thermal cycler (Bio-Rad, USA).

Complementation of the *sug-h* Mutant

The RNA interference (RNAi) vector was constructed by PCR amplification of a 291 bp fragment from *OsISA1* and a 206 bp fragment from *OsBEIIa* cloned from Hwacheong cDNA. These fragments were subcloned into pDONR201 (Invitrogen, USA), and then transferred into the RNAi vector pH7GWIWG (II) using Gateway BP and LR clonase enzyme mixes (Invitrogen, USA). The full-length *OsBEIIa* cDNA was amplified from Hwacheong cDNA and used for constructing the overexpression vector. The amplified fragment was transferred into pMDC32 via pCR™ 8/GW/TOPO® TA Cloning Kit (Invitrogen, USA). The RNAi constructs were transformed into wild-type Dongjin (a *japonica* cultivar), and the overexpression construct was transformed into callus of the *sug-h* mutant. Transformation was performed using a modification of the previously published *Agrobacterium*-mediated transformation method (Nishimura et al. 2006).

Histological GUS Assay

The 1909 bp region upstream from the start codon of *OsBEIIa* was amplified and cloned into the binary vector pHGWFS7 using Gateway BP and LR clonase enzyme mixes (Invitrogen, USA). The final construct was introduced into wild-type Dongjin by *Agrobacterium*-mediated transformation. Transgenic plants containing the *OsBEIIa* promoter::GUS reporter construct were selected, and T_0 plants were used for GUS assays. GUS staining was performed as described previously (Jefferson et al. 1987). X-Gluc buffer solution was vacuum-infiltrated into several different tissue samples. The samples were incubated overnight in X-Gluc buffer solution at 37 °C, and then washed with a graded ethanol series.

Native–PAGE/Activity Staining of DBE and BE

Crude enzyme was extracted using the method described by Yamanouchi and Nakamura (1992) from ten seeds at late-milky stage. Native–PAGE/activity staining of DBE was performed using the modified method of Nakamura et al. (1997). Native slab gel was prepared with 6.5% (*w*/*v*; resolving gel) containing 0.3% (*w*/*v*) potato tuber amylopectin (Sigma, USA), and 3.3% (*w*/*v*; stacking gel) acrylamide. Electrophoresis was carried out at 4 °C at constant current of 20 mA for 2.5 h. For detection of the isoamylase activity, the gel was rinsed with 20 ml of 50 mM Na-acetate (pH 5.4), 50 mM 2-mercaptoethanol and 50 mM $CaCl_2$ at room temperature and then incubated at 37 °C for 3 h with 20 ml of the same buffer solution. Native–PAGE/activity staining of BE was carried out according to the modified method by Yamanouchi and Nakamura (1992). A slab gel prepared with 5% (*w*/*v*; resolving gel) and 3.3% (*w*/*v*; stacking gel) acrylamide. Electrophoresis was carried out at 4 °C at constant current of 20 mA. After electrophoresis, the gel was imbibed with 20 mL of a solution containing 50 mM HEPES-NaOH buffer (pH 7.4) and 20% (*v*/v) glycerol for 15 min on ice. For detection of BE, the gel was incubated in 20 ml of the reaction mixture, which consisted of 50 mM HEPES-NaOH buffer (pH 7.4), 50 mM Glc-1-P (Sigma, USA), 2.5 mM AMP (Sigma, USA), 10% (*v*/v) glycerol, and rabbit muscle phosphorylase a (about 30 units; Sigma, USA) for 6 h at 30 °C with gentle shaking. Iodine solution [0.1% (*w*/*v*) I_2 and 1% (*w*/*v*) KI] was used for staining both gels.

Additional files

Additional file 1: Figure S1. Plant and panicle morphology of wild-type rice (Hwacheong) and mutants. (a-b) Plant phenotype of wild-type and mutant plants 54 days after transplanting (a) and at the milky stage (b). (c) Panicle length of wild-type and mutant plants at the yellow ripe stage.

Additional file 2: Table S1. Genetic analysis of the *sug-h* mutant using F_3 seeds

Additional file 3: Figure S2. Multiple alignments of cereal OsBEIIa proteins. Protein of the *sug-h* mutant (top line) was aligned with that of wild-type rice (Hwacheong) and four cereal plants (barley, maize, sorghum, and wheat). Black boxes indicate identical residues; gray boxes indicate similar residues. Mutated region is marked with an asterisk. Color bars indicate the domains; E_set_GBE_euk_N (green), AmyAc_bac_euk_BE (blue), and Alpha-amylase_C (orange).

Additional file 4: Figure S3. *OsISA1* expression patterns in different organs and at different stages of seed development using qRT-PCR analysis. (a) Transcript levels decreased in seed (10 DAF) of the *sug-h* mutant. (b) *OsISA1* expression in 10 and 20 DAF seeds decreased in the *sug-h* mutant. All data are mean ± SD (n = 3). Statistical significance was determined using Student's *t*-test (*$P < 0.05$, **$P < 0.01$). WT, wild-type rice (Hwacheong); DAF, days after flowering.

Additional file 5: Table S2. Molecular markers used for fine mapping of the *sug-h* mutant.

Abbreviations

BEIIa: Branching enzyme IIa; BSA: Bulked segregant analysis; dCAPS: Derived cleaved amplified polymorphic sequence; ISA1: Isoamylase1; PCR: Polymerase chain reaction; STS: Sequence tagged site; sug-1: sugary-1; sug-h: sugary-h; WSP: Water-soluble polysaccharide

Acknowledgments

Authors express sincere thanks to Prof. Fujita in Akita Prefectural University, Japan, for her kind advice on native-PAGE/BE staining. Authors also thank Dr. Joohyun Lee in Konkuk University and Dr. Soon-Wook Kwon in Pusan National University for supporting material development and helpful discussion. This study was supported by a grant from the Next-Generation BioGreen 21 Program (no. PJ01102401) of the Rural Development Administration, Korea.

Authors' Contributions

HJK and JHC supervised the work and improved the manuscript. YL designed and performed experiments, and wrote the manuscript. MSC developed the plant materials, and performed experiments. GL, SJ, BK, RP provided advice on the experiments. MRY supported experimental data. MOW improved the manuscript. All authors reviewed the manuscript. All authors read and approved the final manuscript.

Competing Interests

The authors declare that they have no potential competing interests.

Author details

[1]Department of Plant Science and Research Institute for Agriculture and Life Sciences, and Plant Genomics and Breeding Institute, Seoul National University, Seoul 08826, South Korea. [2]Vegetable Crop Division, National Institute of Horticultural and Herbal Science, Rural Development Administration, Muan 534-833, South Korea. [3]Department of Central Area Crop Science, National Institute of Crop Science (NICS), RDA, Suwon 16429, South Korea. [4]Rice Research Institute, Jilin Academy of Agricultural Sciences, Gongzhuling, Jilin 136100, China. [5]Graduate School of Integrated Bioindustry, Sejong University, 209, Neungdong-ro Gwangjin-gu, Seoul, South Korea.

References

Chin J-H, Kim J-H, Jiang W, Chu S-H, Woo M-O, Han L, Brar D, Koh H-J (2007) Identification of subspecies-specific STS markers and their association with segregation distortion in rice (*Oryza sativa* L.) J Crop Sci Biotec 10:175–184

Crofts N, Abe N, Oitome NF, Matsushima R, Hayashi M, Tetlow IJ, Emes MJ, Nakamura Y, Fujita N (2015) Amylopectin biosynthetic enzymes from developing rice seed form enzymatically active protein complexes. J Exp Bot 66:4469–4482

Fujita N (2014) Starch biosynthesis in Rice endosperm. AGBM 4:1–18

Fujita N, Kubo A, Francisco PB Jr, Nakakita M, Harada K, Minaka N, Nakamura Y (1999) Purification, characterization, and cDNA structure of isoamylase from developing endosperm of rice. Planta 208:283–293

Fujita N, Kubo A, Suh DS, Wong KS, Jane JL, Ozawa K, Takaiwa F, Inaba Y, Nakamura Y (2003) Antisense inhibition of isoamylase alters the structure of amylopectin and the physicochemical properties of starch in rice endosperm. Plant Cell Physiol 44:607–618

Fujita N, Toyosawa Y, Utsumi Y, Higuchi T, Hanashiro I, Ikegami A, Akuzawa S, Yoshida M, Mori A, Inomata K, Itoh R, Miyao A, Hirochika H, Satoh H, Nakamura Y (2009) Characterization of pullulanase (PUL)-deficient mutants of rice (*Oryza sativa* L.) and the function of PUL on starch biosynthesis in the developing rice endosperm. J Exp Bot 60:1009–1023

Hanashiro I, J-i A, Hizukuri S (1996) A periodic distribution of the chain length of amylopectin as revealed by high-performance anion-exchange chromatography. Carbohydr Res 283:151–159

James MG, Denyer K, Myers AM (2003) Starch synthesis in the cereal endosperm. Curr Opin Plant Biol 6:215–222

Jefferson RA, Kavanagh TA, Bevan MW (1987) GUS fusions: beta-glucuronidase as a sensitive and versatile gene fusion marker in higher plants. EMBO J 6:3901–3907

Kim K-N, Fisher DK, Gao M, Guiltinan MJ (1998) Molecular cloning and characterization of the Amylose-extender gene encoding starch branching enzyme IIB in maize. Plant Mol Biol 38:945–956

Koh HJ, Heu MH (1994) Physicochemical properties of sugary-endosperm mutants in Rice. Korean J Crop Sci 39:1–6

Kubo A, Fujita N, Harada K, Matsuda T, Satoh H, Nakamura Y (1999) The starch-debranching enzymes isoamylase and pullulanase are both involved in amylopectin biosynthesis in rice endosperm. Plant Physiol 121:399–410

Kubo A, Rahman S, Utsumi Y, Li Z, Mukai Y, Yamamoto M, Ugaki M, Harada K, Satoh H, Konik-Rose C, Morell M, Nakamura Y (2005) Complementation of sugary-1 phenotype in rice endosperm with the wheat isoamylase1 gene supports a direct role for isoamylase1 in amylopectin biosynthesis. Plant Physiol 137:43–56

Kwak J, Yoon M-R, Lee J-S, Lee J-H, Ko S, Tai TH, Won Y-J (2017) Morphological and starch characteristics of the japonica rice mutant variety Seolgaeng for dry-milled flour. Food Sci and Biotech 26:43–48

Lin Q, Huang B, Zhang M, Zhang X, Rivenbark J, Lappe RL, James MG, Myers AM, Hennen-Bierwagen TA (2012) Functional interactions between starch synthase III and isoamylase-type starch-debranching enzyme in maize endosperm. Plant Physiol 158:679–692

Mccouch SR, Kochert G, Yu ZH, Wang ZY, Khush GS, Coffman WR, Tanksley SD (1988) Molecular mapping of Rice chromosomes. Theor Appl Genet 76:815–829

Michelmore RW, Paran I, Kesseli RV (1991) Identification of markers linked to disease-resistance genes by bulked segregant analysis: a rapid method to detect markers in specific genomic regions by using segregating populations. Proc Natl Acad Sci U S A 88:9828–9832

Mizuno K, Kobayashi E, Tachibana M, Kawasaki T, Fujimura T, Funane K, Kobayashi M, Baba T (2001) Characterization of an Isoform of Rice starch branching enzyme, RBE4, in developing seeds. Plant Cell Physiol 42:349–357

Nakagami T, Yoshihara H, Nakamura T, Utsumi Y, Sawada T, Fujita N, Satoh H, Nakamura Y (2016) Biochemical analysis of new type mutants of japonica rice that accumulate water-soluble α-glucans in the endosperm but retain full starch debranching enzyme activities. Starch-Stärke 68:1–11

Nakamura Y (2002) Towards a better understanding of the metabolic system for amylopectin biosynthesis in plants: rice endosperm as a model tissue. Plant Cell Physiol 43:718–725

Nakamura Y, Kubo A, Shimamune T, Matsuda T, Harada K, Satoh H (1997) Correlation between activities of starch debranching enzyme and alpha-polyglucan structure in endosperms of sugary-1 mutants of rice. Plant J 12:143–153

Nakamura Y, Umemoto T, Takahata Y, Amano E (1992) Characteristics and roles of key enzymes associated with starch biosynthesis in rice endosperm. Gamma Field Symposia 31:25–44

Nakamura Y, Umemoto T, Takahata Y, Komae K, Amano E, Satoh H (1996) Changes in structure of starch and enzyme activities affected by sugary mutations in developing rice endosperm. Possible role of starch debranching enzyme (R-enzyme) in amylopectin biosynthesis. Physiol plantarum 97:491–498

Nishi A, Nakamura Y, Tanaka N, Satoh H (2001) Biochemical and genetic analysis of the effects of Amylose-extender mutation in Rice endosperm. Plant Physiol 127:459–472

Nishimura A, Aichi I, Matsuoka M (2006) A protocol for Agrobacterium-mediated transformation in rice. Nat Protoc 1:2796–2802

Ohdan T, Francisco PB, Sawada T, Hirose T, Terao T, Satoh H, Nakamura Y (2005) Expression profiling of genes involved in starch synthesis in sink and source organs of rice. J Exp Bot 56:3229–3244

Peng C, Wang Y, Liu F, Ren Y, Zhou K, Lv J, Zheng M, Zhao S, Zhang L, Wang C, Jiang L, Zhang X, Guo X, Bao Y, Wan J (2014) FLOURY ENDOSPERM6 encodes a CBM48 domain-containing protein involved in compound granule formation and starch synthesis in rice ENDOSPERM. Plant J 77:917–930

Regina A, Bird A, Topping D, Bowden S, Freeman J, Barsby T, Kosar-Hashemi B, Li Z, Rahman S, Morell M (2006) High-amylose wheat generated by RNA interference improves indices of large-bowel health in rats. Proc Natl Acad Sci U S A 103:3546–3551

Satoh H, Shibahara K, Tokunaga T, Nishi A, Tasaki M, Hwang SK, Okita TW, Kaneko N, Fujita N, Yoshida M, Hosaka Y, Sato A, Utsumi Y, Ohdan T, Nakamura Y (2008) Mutation of the plastidial alpha-glucan phosphorylase gene in rice affects the synthesis and structure of starch in the endosperm. Plant Cell 20:1833–1849

Tanaka N, Fujita N, Nishi A, Satoh H, Hosaka Y, Ugaki M, Kawasaki S, Nakamura Y (2004) The structure of starch can be manipulated by changing the expression levels of starch branching enzyme IIb in rice endosperm. Plant Biotechnol J 2:507–516

Tanaka N, Itoh H, Sentoku N, Kojima M, Sakakibara H, Izawa T, Itoh J, Nagato Y (2011) The COP1 ortholog PPS regulates the juvenile-adult and vegetative-reproductive phase changes in rice. Plant Cell 23:2143–2154

Tetlow IJ, Beisel KG, Cameron S, Makhmoudova A, Liu F, Bresolin NS, Wait R, Morell MK, Emes MJ (2008) Analysis of protein complexes in wheat amyloplasts reveals functional interactions among starch biosynthetic enzymes. Plant Physiol 146:1878–1891

Utsumi Y, Utsumi C, Sawada T, Fujita N, Nakamura Y (2011) Functional diversity of isoamylase oligomers: the ISA1 homo-oligomer is essential for amylopectin biosynthesis in rice endosperm. Plant Physiol 156:61–77

Wong K-S, Kubo A, J-l J, Harada K, Satoh H, Nakamura Y (2003) Structures and properties of amylopectin and phytoglycogen in the endosperm of sugary-1 mutants of rice. J Cereal Sci 37:139–149

Yamanouchi H, Nakamura Y (1992) Organ specificity of isoforms of starch branching enzyme (Q-enzyme) in rice. Plant Cell Physiol 33:985–991

Yano M, Isono Y, Satoh H, Omura T (1984) Gene analysis of sugary and shrunken mutants of rice, *Oryza sativa* L. Jpn J Breed 34:43–49

Yoon M-R, Koh H-J, Lee S-C, Kang M-Y (2009) Comparative study of the physicochemical properties of Rice endosperm components expressing sugary-2 mutant in different genetic backgrounds. J Korean Soc Appl Biol Chem 52:509–515

Zhang M-Z, Fang J-H, Yan X, Liu J, Bao J-S, Fransson G, Andersson R, Jansson C, Åman P, Sun C (2012) Molecular insights into how a deficiency of amylose affects carbon allocation-carbohydrate and oil analyses and gene expression profiling in the seeds of a rice waxy mutant. BMC Plant Biol 12:1–18

11

Mesocotyl Elongation is Essential for Seedling Emergence Under Deep-Seeding Condition in Rice

Hyun-Sook Lee[1,2], Kazuhiro Sasaki[2,3], Ju-Won Kang[1], Tadashi Sato[2,4], Won-Yong Song[5] and Sang-Nag Ahn[1*]

Abstract

Background: Direct-seeding cultivation by deep-seeding of seeds (drill seeding) is becoming popular due to the scarcity of land and labor. However, poor emergence and inadequate seedling establishment can lead to yield loss in direct-seeding cultivation by deep-sowing. In rice, mesocotyl and coleoptile are primarily responsible for seedling emergence from deeper levels of soil.

Results: Quantitative trait loci (QTLs) for mesocotyl and coleoptile length at 5-cm seeding depth were detected using 98 backcross inbred lines from a cross between Kasalath and Nipponbare. Three QTLs *qMel-1*, *qMel-3*, and *qMel-6* for mesocotyl length were identified on chromosomes 1, 3, and 6, respectively, in two independent replicates. At two QTLs, *qMel-1* and *qMel-3*, the Kasalath alleles increased mesocotyl length, whereas Nipponbare allele increased at *qMel-6*. The Nipponbare alleles at two QTLs (*qCol-3* and *qCol-5*) increased the coleoptile length. Further, seeds of 54 chromosome segment substitution lines (CSSLs) from the cross between Kasalath and Nipponbare sown at 5 cm soil depth showed a significant positive correlation between seedling emergence and mesocotyl elongation ($r > 0.6$, $P < 0.0001$), but not with coleoptile elongation ($r = 0.05$, $P = 0.7$). Seedling emergence of Nipponbare, Kasalath, and the 3 of the 54 CSSLs rapidly decreased with increasing sowing depth. Seedling emergence at seeding depths of 7 and 10 cm was faster in Kasalath and CSSL-5 that harbored the Kasalath alleles across the *qMel-1* and *qMel-3* regions than in the other two CSSLs that contained a single QTL and Nipponbare alleles. CSSL-5 showed the longest mesocotyl among the 3 CSSLs, but no difference in coleoptile length was observed among the 3 CSSLs at seeding depths of 7 and 10 cm.

Conclusion: Variation of mesocotyl elongation was found to be associated with seedling emergence at the seeding depth of 5 cm. To our knowledge, this is the first study performed using CSSLs to detect QTLs for mesocotyl or coleoptile elongation and to determine the effect of mesocotyl elongation on seedling emergence in rice. Our findings provides a foundation for developing rice cultivars that show higher seedling emergence after direct seeding by introgressing QTLs for mesocotyl elongation in rice breeding.

Keywords: Rice, Mesocotyl, Coleoptile, Seedling emergence, QTL, CSSLs (chromosome segment substitution lines)

* Correspondence: ahnsn@cnu.ac.kr
[1]College of Agriculture and Life Sciences, Chungnam National University, Daejeon 305-764, South Korea
Full list of author information is available at the end of the article

Background

In Asia, two rice planting methods are used: transplanting and direct seeding. In direct seeding, seeds are sown directly in wet/puddled soil, unpuddled soil, or standing water (Kumar and Ladha 2011). In Asia, approximately 21% of the total rice area is used for direct seeding, and this is expected to increase owing to the scarcity of land, water, and labor (Pandey and Velasco 2005). Rice varieties suitable for direct seeding should include high germination ability, seedling vigor, fast root growth, early tillering ability, and lodging resistance (Lee et al. 2002; Tang 2002). Faster and uniform germination and seedling emergence resulted in more vigorous seedling growth and increasing yield in direct-seeding (Farooq et al. 2006). In direct seeding, deep seeding (drill seeding) is known to reduce damages from wildlife and improve lodging tolerance. However, when seeds are sown deep, the seedlings need to elongate their organs to ensure that the plumule reaches the soil surface. Seedling emergence is an important criterion for determining the actual yield during direct seeding cultivation of rice. The coleoptile (the protective sheath that covers the emerging shoot) and mesocotyl (the structure between the scutellar and coleoptilar nodes in an embryo) of rice are primarily responsible for the emergence of seedlings from deeper soil layers (Turner et al. 1982; Dilday et al. 1990). Seeding depth is also an important factor for dry direct-seeding rice cultivation. Seedling growth for temperate *japonica* cultivars was adversely affected when seeds were sown deeper than the optimum seeding depth of 2–3 cm (Lee et al. 2002). However, shallow seeding can increase the incidences of damage by birds, and plants might suffer lodging after the heading stage. Thus, obtaining information on the optimum seeding depth for direct seeding for diverse germplasm is necessary (Lee et al. 2002).

Previous studies reported that mesocotyl and coleoptile elongation are governed by many genetic and environmental factors (Takahasi 1978; Takahasi 1984). Their elongation shows significant variation across genotypes. Mesocotyl elongation is greater in *indica* rice than in *japonica* rice (Suge 1972; Takahashi 1978; Lee et al. 2012), whereas the coleoptile is longer in *japonica* cultivars than in *indica* cultivars under submerged condition (Takahashi 1978). Among *japonica* cultivars, upland rice generally produced shorter coleoptiles and longer mesocotyls than lowland type (Chang and Vergara 1975).

Quantitative trait loci (QTLs) associated with mesocotyl and coleoptile elongation in rice have been reported. QTLs for mesocotyl elongation were detected using various segregating populations from interspecific or intrasubspecific crosses (Katsuta-Seki et al. 1996; Cai and Morishima 2002; Cao et al. 2002; Huang et al. 2010; Lee et al. 2012). Katsuta-Seki et al. (1996) reported 3 QTLs for mesocotyl elongation on chromosomes 3, 6, and 11 by using F3 populations from the cross between 2 *indica* cultivars grown in glass tubes. Cai and Morishima (2002) detected 11 QTLs for mesocotyl length by using 125 recombinant inbred lines (RILs) derived from a cross between an *indica* cultivar and a strain of wild rice. Further, 5 QTLs for mesocotyl length were detectedusing an RIL population from the cross between 2 *japonica* cultivars by using the filter paper method (Huang et al. 2010). Cao et al. (2002) detected 8 QTLs on chromosomes 1, 3, 6, 7, 8, and 12 under moderate and low temperate conditions in a doubled haploid population derived from a cross between a *japonica* cultivar and an *indica* cultivar. In a previous study, we detected 5 QTLs for mesocotyl length on chromosomes 1, 3, 7, 9, and 12 by using agar media and the same backcross inbred line (BIL) populations used in this study (Lee et al. 2012).

QTLs for improving seedling vigor or seedling establishment, including traits for seedling emergence, seed germination, or shoot length, and coleoptile length have been detected (Redoña and Mackill 1996; Zhou et al. 2007; Xie et al. 2014). Redoña and Mackill (1996) detected 2 QTLs for coleoptile length and 5 QTLs for mesocotyl length by using an F2:F3 population derived from a cross between *japonica* cultivar Labelle and *indica* cultivar Black Gora. Further, 5 QTLs for seed vigor traits such as coleoptile length and radicle length and 3 QTLs for seed germination under low and normal temperature conditions were identified using populations derived from a cross between two *indica* cultivars (Xie et al. 2014). In addition, 9 QTLs for seedling vigor traits, including coleoptile emergence, were detected under two field conditions—drained and flooded soil—and 2 QTLs for coleoptile emergence were identified on chromosomes 1 and 3 under flooded soil condition (Zhou et al. 2007).

In this study we aimed to show that mesocotyl elongation is essential to seedling emergence under deep-seeding using chromosome segment substitution lines (CSSLs). Although several studies identified QTLs for mesocotyl or coleoptile elongation, the locations and effects of the detected QTLs varied according to the test methods (Redoña and Mackill 1996; Zhou et al. 2007; Huang et al. 2010; Xie et al. 2014). Furthermore, no study tried to characterize and detect QTLs for mesocotyl under deep-seeding condition in rice. Also, study to analyze the association between mesocotyl elongation and seedling emergence under various soil depth conditions is rare in crop including rice (Chung 2010; Alibu et al. 2011; Lu et al. 2016).

This study attempted to (1) identify QTLs associated with mesocotyl and coleoptile elongation under the deep-seeding soil condition, and (2) to evaluate the effect of mesocotyl elongation on seedling emergence using deep-seeded CSSLs.

Results

Phenotypic variation of mesocotyl and coleoptile elongation

Nipponbare and Kasalath were selected to determine the seeding depth for detecting QTLs controlling mesocotyl and coleoptile lengths. At first, Nipponbare and Kasalath seeds were placed at 5 cm and 7 cm sowing depths based on previous reports (Chung 2010, Alibu et al. 2011). Although Nipponbare and Kasalath seeds completely germinated, some seedlings showed poor growth at 7 cm depth (Additional file 1: Figure S1). Two soil depth conditions, 3 cm (shallower soil depth than 5 cm) and 10 cm (deeper soil depth than 7 cm) were added to measure mesocotyl and coleoptile length at 14 days after sowing (Fig. 1). Kasalath showed exceptionally long mesocotyls than those in Nipponbare at all soil depths. Such difference in mesocotyl length between these two varieties has also been observed using agar media (Lee et al. 2012). In contrast, the coleoptile length was significantly longer in Nipponbare than in Kasalath at all soil depths ($P < 0.05$). Both mesocotyl and coleoptile lengths increased with sowing depth in both the cultivars, but the difference was not significant between 7 and 10 cm sowing depths. The mesocotyl lengths were 24.9 mm (seeding depth, 5 cm), 33.7 mm (7 cm), and 36.5 mm (10 cm) in Kasalath, whereas the coleoptile lengths were 40.8 mm (5 cm), 44.2 mm (7 cm), and 43.2 mm (10 cm) in Nipponbare. These results suggested that the mesocotyl of Kasalath maximally elongates to about 35 mm, and the coleoptile in Nipponbare maximally elongates to about 43 mm, under the experimental conditions used in this study. Therefore, 7 cm seeding depth was considered suitable for QTL analysis of mesocotyl and coleoptile lengths in the BILs. Nevertheless, 5 cm seeding depth was selected for QTL analysis. Because, some seed grown insufficiently after germination hindered the accurate phenotypic evaluation of the traits when planted in 7 cm and 10 cm soil depths (Additional file 1: Figure S1). Mesocotyl and coleoptile length of Nipponbare and Kasalath seedling were measured except insufficiently grown seedling in 7 cm and 10 cm soil depth (Fig. 3c and d).

Fifty-seven accessions from a rice diversity research set of germplasm (RDRS) were evaluated for mesocotyl and coleoptile length to know the association between two traits and also to clarify whether 5 cm seeding depth was a proper condition for analyzing phenotypic variation. This collection carries 91% of the alleles identified in the 332 original rice varieties and covers most of the range of variation in several agro-morphological traits from original varieties (Kojima et al. 2005). Phenotypic variation in mesocotyl and coleoptile lengths in 57 RDRS accessions is shown in Fig. 2 and Additional file 2: Table S1. The rice accessions showed large variations for both mesocotyl and coleoptile lengths. The mesocotyl length of the 57 accessions ranged from 0.3 mm to 37 mm, whereas the coleoptile length ranged from 17.3 mm to 49.2 mm (Fig. 2a). The mesocotyl length showed a significant negative correlation with the coleoptile length ($r = -0.73$, $P < 0.001$). Moreover, comparison of mesocotyl and coleoptile elongation among the four groups—*japonica*, tropical *japonica*, *indica*-I, and *indica*-II, classified by Kojima et al. (2005) (Uga et al. 2009; http://www.gene.affrc.go.jp/databases-core_collections_wr.php)—showed that the difference among the groups was highly significant for both the traits ($P < 0.001$; Fig. 2b). The *indica*-I group (including Kasalath) and *japonica* group (including Nipponbare) showed the longest mesocotyl (27.2 mm) and coleoptile (41.9 mm), respectively (Fig 2b). Among the

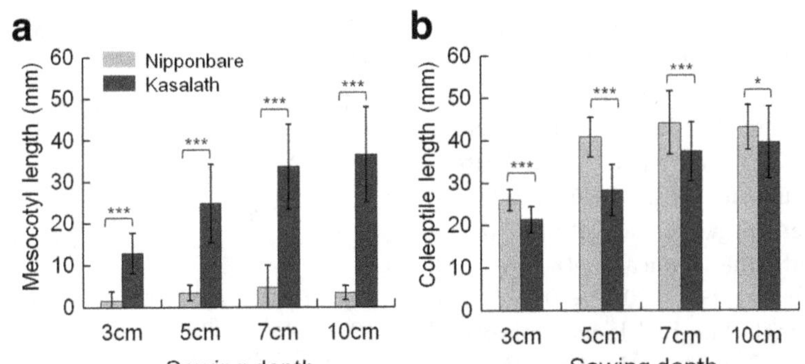

Fig. 1 Phenotypic variation of mesocotyl and coleoptile elongation at varying soil depths. **a** Mesocotyl and **b** coleoptile length of Nipponbare and Kasalath under different burial depth in soil; 12 Seeds of Nipponbare and Kasalath were sown at 3 cm, 5 cm, 7 cm, and 10 cm soil depth and incubated at alternate temperatures of 30 °C and 26 °C (14 h/10 h). At 14 days after sowing, the seedling were excavated and length were measured. Bars represent mean of length with SD ($n = 3$). Comparisons of the varieties were made with the ANOVA test. * $P < 0.05$, *** $P < 0.001$

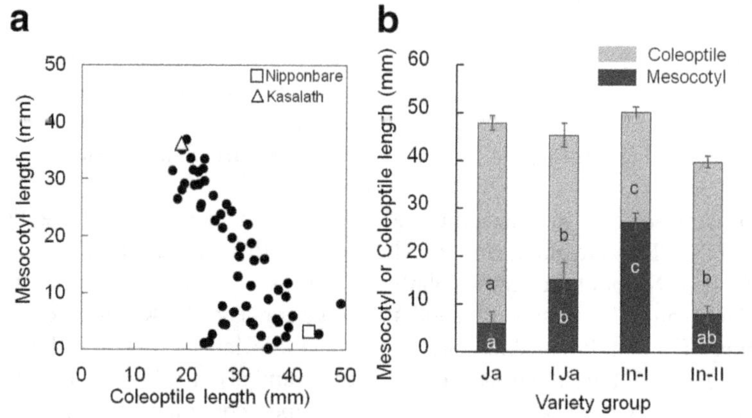

Fig. 2 Mesocotyl and coleoptile length for 57 accessions of RDRS (**a**), Comparison of mesocotyl and coleoptile length in 4 rice variety groups (**b**) Japonica (Ja), Tropical Japonica (TJa), Indica-I (In-I), and Indica-II (In-II); Seeds were sown at 5 cm soil depth and incubated at alternate temperatures of 30 °C and 26 °C (14 h/10 h). At 14 days after sowing, the seedling were excavated and length were measured. **b** Means with different letters in the same row indicate significant differences according to the Duncan multiple range test ($P < 0.001$). *Bars* represent mean of length with SE

accessions, Nipponbare and Kasalath displayed huge difference in mesocotyl and coleoptile length indicating that the BILs from a cross between Nipponbare and Kasalath were suitable for QTL analysis. Also, the BILs were used for mapping QTLs for a wide variety of traits (Lin et al. 1998; Murai et al. 2002; Yamaya et al. 2002). Therefore, 5 cm seeding depth was used to analyze the genetic variation in mesocotyl and coleoptile elongation among the 98 BILs from a cross between Kasalath and Nipponbare.

Two independent measurements of mesocotyl and coleoptile lengths of the 98 BILs were obtained at the 5 cm seeding depth (Fig. 3). The mesocotyl and coleoptile lengths were not significantly different between the two independent experiments (r = 0.79 for mesocotyl, r = 0.66 for coleoptile; $P < 0.001$). The mesocotyl lengths of Nipponbare and Kasalath were 3.3 mm and 25.1 mm, respectively, in one replicate (Fig. 3a). In the 98 BILs, the mesocotyl length ranged from 0.7 mm to 19.3 mm, with an average of 6.5 mm. The coleoptile lengths of Nipponbare and Kasalath were 33.5 mm and 28.8 mm, respectively, in one replicate (Fig. 3b.), and the average coleoptile length was 37.1 mm and ranged from 27.6 mm to 47.2 mm in the BILs. Mesocotyl length showed a highly significant negative correlation with coleoptiles length in both the replicates (r = −0.37, $P < 0.001$ in 1st replicate and r = −0.20, $P < 0.05$ in 2nd replicate). In BILs, no single line had longer mesocotyl than Kasalath whereas some transgressive lines showing longer coleoptile length than Nipponbare were observed. Lack of transgressive lines for mesocotyl length is possibly due to the finding that Kasalath has major mesocotyl increasing QTLs, *qMel-1* and *qMel-3* whereas Nipponbare has a minor QTL whose effect was

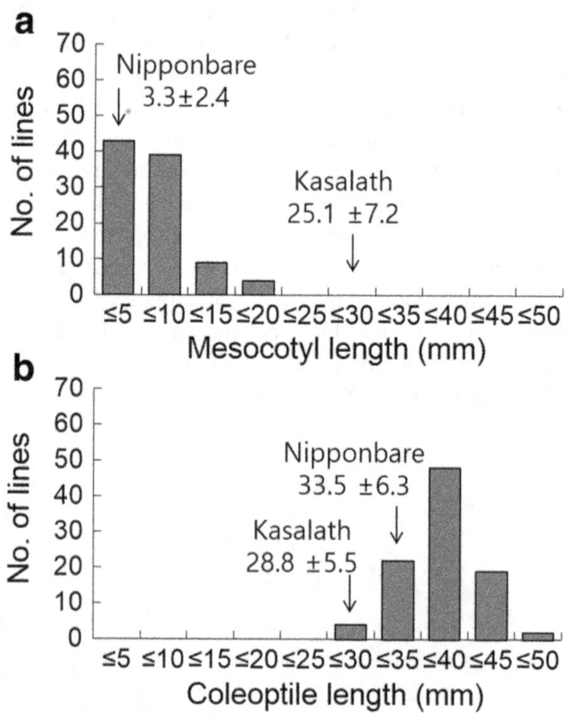

Fig. 3 Frequency distribution of the mesocotyl length (**a**) and coleoptile length (**b**) in 98 backcross inbred lines (BILs) derived from Nipponbare/Kasalath at 5 cm soil depth; *Arrowheads* indicate mean values for Nipponbare and Kasalath. 12 seeds of 98 BILs, Nipponbare and Kasalath were sown at 5 cm soil depth and incubated at alternate temperatures of 30 °C and 26 °C (14 h/10 h). At 14 days after sowing, length were measured

negligible in BILs. For coleoptile length, it is possible that Kasalath might possess a few QTL with positive effects on coleoptile elongation although two minor QTLs were detected with increasing effects from Nipponbare due to small population size of BIL, and gene interactions. This result is consistent with that of the previous studies (Redoña and Mackill 1996; Xie et al. 2014).

QTL analysis for deep-seeding tolerance-related traits

QTL was detected based on logarithm of odds ratio (LOD) thresholds after the permutation test. In all, 5 QTLs, including 3 for mesocotyl length and 2 for coleoptile length, were detected at the 5 cm seeding depth (Table 1, Additional file 3: Figure S2). The three QTLs (qMel-*1*, *qMel-3*, and *qMel-6*) for mesocotyl length were mapped near the markers—R2414, R1618, and R2123 on chromosomes 1, 3, and 6, respectively. The *qMel-3* QTL was detected in both the replicates (Rep1 and 2), and the Kasalath alleles at two QTL loci (qMel-*1* and *qMel-3*) contributed to an increase in mesocotyl length. The QTL *qMel-1* accounted for 13.89% and *qMel-3* for 37.56% and 26.97% of phenotypic variance. These QTLs were detected even when the varieties were cultured in agar (Lee et al. 2012). Two QTLs, *qCol-3* and *qCol-5* for coleoptile length were mapped near the markers, C25 and C597 on chromosomes 3 and 5, respectively, in one replicate (Rep 1). In the other replicate (Rep 2), no QTLs for coleoptile length were detected. The two QTLs, *qCol-3* and *qCol-5* accounted for 11.8 and 12.0% of phenotypic variance, respectively. Nipponbare alleles at the two QTL loci contributed to an increase in the coleoptile length. Interestingly, *qMel-3* and *qCol-3* showed loose linkage suggesting a possibility of combining two QTLs via marker-assisted selection.

Effect of mesocotyl elongation on seedling emergence

In all, 54 CSSLs were used for evaluating the effect of QTLs for mesocotyl or coleoptile length on seedling emergence at the 5 cm seeding depth (Fig. 4). Each CSSL carries a single or a few Kasalath segments in the near-isogenic background of Nipponbare. Seedlings started to emerge at 6 days after sowing, when the analysis was conducted.

A positive correlation was found between mesocotyl length and seedling emergence at 6 days after sowing ($r = 0.8$, $P < 0.0001$), whereas no significant correlation was noted between coleoptile length and seedling emergence ($P > 0.5$). This result suggests that seedling emergence is dependent on mesocotyl elongation.

All lines except 5 lines were not significantly different from Nipponbare in mesocotyl and coleoptile lengths (Fig. 4). Four lines, CSSL-5, CSSL-15, CSSL33 and CSSL-54 showed significantly longer mesocotyl length than Nipponbare and the other CSSLs (Fig. 4b). CSSL-5 showed the highest emergence percentage (70%) among CSSLs followed by CSSL-33 (17%) and CSSL-15 (12.5%). (Fig. 4a). CSSL-5 possessed Kasalath segments across *qMel-1* and *qMel-3* and across only *qMel-3* in CSSL-15. On the other hand, CSSL-33 possessed the Kasalath segments across *qMel-3* and *qMel-7*. The *qMel-7* QTL was not detected in this study whereas it was observed in the agar condition (Lee et al. 2012). Coleoptile length of Kasalath and CSSL-53 was significantly shorter than Nipponbare and the other CSSLs, but the emergence percentage of Kasalath (87.5%) was the highest (Fig. 4a and C). This result suggests that seedling emergence is dependent on mesocotyl elongation.

In addition, we tried to confirm the effect of mesocotyl length on seedling emergence using 3 CSSLs (Fig. 5). Lengths of mesocotyl and coleoptile and seedling emergence were investigated in 3 CSSLs, possessing the Kasalath segments across the *qMel-1* (CSSL-6) or *qMel-3* (CSSL-15) regions and both (CSSL-5). Seedlings of all genotypes emerged over 90% at 3 cm and 5 cm seeding depths from 8 to 9 days after sowing (Fig. 5a). At 7-cm seeding depth, Kasalath and CSSL-5 showed faster emergence than the other genotypes. At 10-cm seeding

Table 1 Characteristics of QTLs for mesocotyl and coleoptile length in 98 BILs

Locus [a]	Chr.	Nearest marker	Position	Replication [b]	LOD score [c]	R2 (%) [d]	Additive effect	Positive allele [e]
qMel-1	1	R2414	39.6- Mb	Rep 2	3.44	13.89	4.34	K
qMel-3	3	R1618	30.7-Mb	Rep 1	12.09	37.56	6.41	K
		R1618	30.7-Mb	Rep 2	7.18	26.97	6.46	K
qMel-6	6	R2123	11.68-Mb	Rep 2	2.81	8.28	3.89	N
qCol-3	3	C25	3.97-Mb	Rep 1	3.06	11.8	2.96	N
qCol-5	5	C597	0.26-Mb	Rep 1	3.20	12.0	2.70	N

[a] QTLs were designated as "*qMel* chromosome number" and "*qCol* chromosome number"
[b] QTLs were detected under soil conditions in two replications, respectively
[c] Maximum LOD score over threshold significance level at $P < 0.05$
[d] Proportion of the phenotypic variation explained by the nearest marker of QTL
[e] K and N respectively represent the positive effects of QTL contributed by Kasalath and Nipponbare alleles

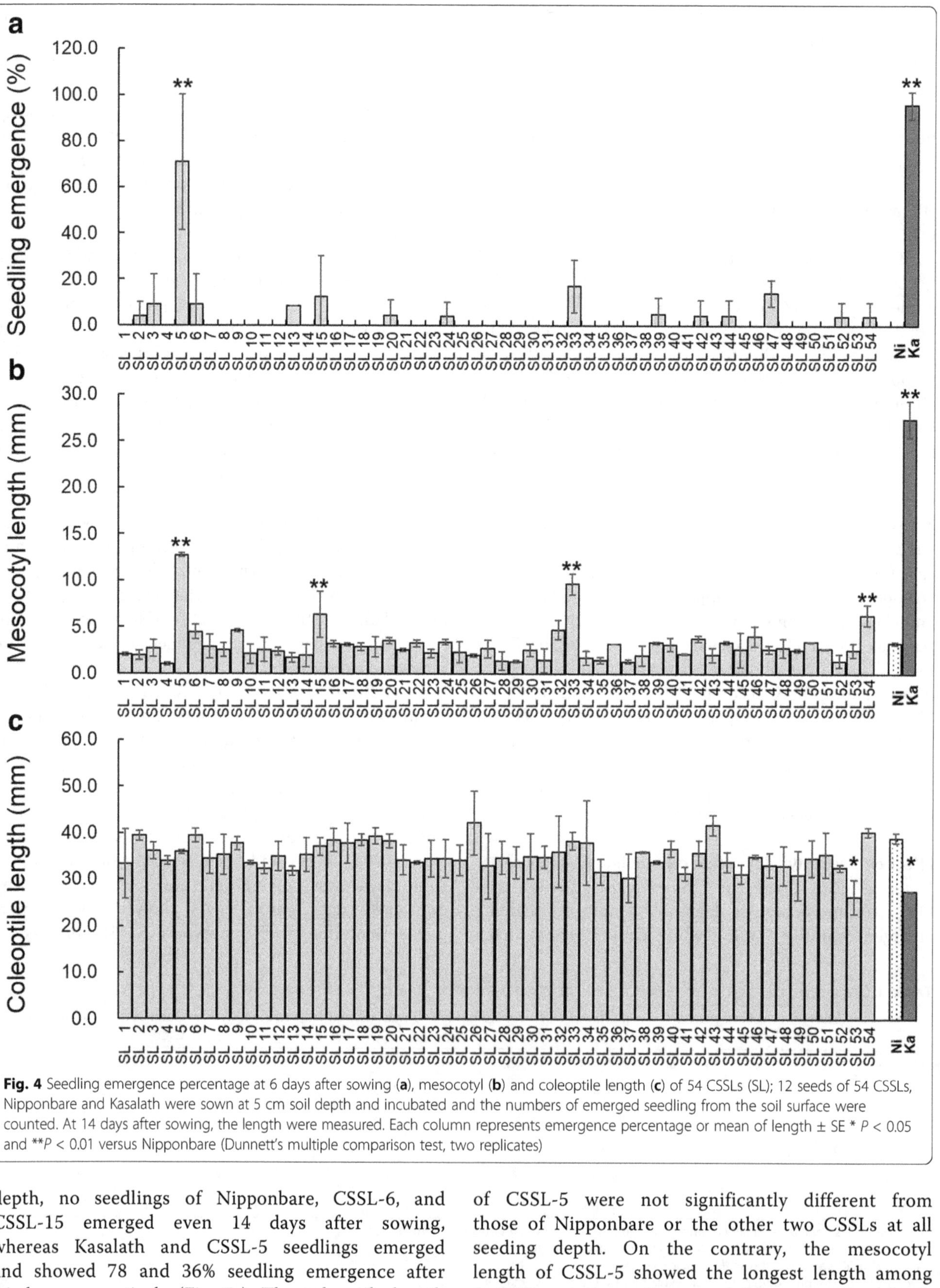

Fig. 4 Seedling emergence percentage at 6 days after sowing (**a**), mesocotyl (**b**) and coleoptile length (**c**) of 54 CSSLs (SL); 12 seeds of 54 CSSLs, Nipponbare and Kasalath were sown at 5 cm soil depth and incubated and the numbers of emerged seedling from the soil surface were counted. At 14 days after sowing, the length were measured. Each column represents emergence percentage or mean of length ± SE * $P < 0.05$ and ** $P < 0.01$ versus Nipponbare (Dunnett's multiple comparison test, two replicates)

depth, no seedlings of Nipponbare, CSSL-6, and CSSL-15 emerged even 14 days after sowing, whereas Kasalath and CSSL-5 seedlings emerged and showed 78 and 36% seedling emergence after 14 days, respectively (Fig. 5a). The coleoptile length of CSSL-5 were not significantly different from those of Nipponbare or the other two CSSLs at all seeding depth. On the contrary, the mesocotyl length of CSSL-5 showed the longest length among 3 CSSLs at 5-cm, 7-cm and 10-cm seeding depth

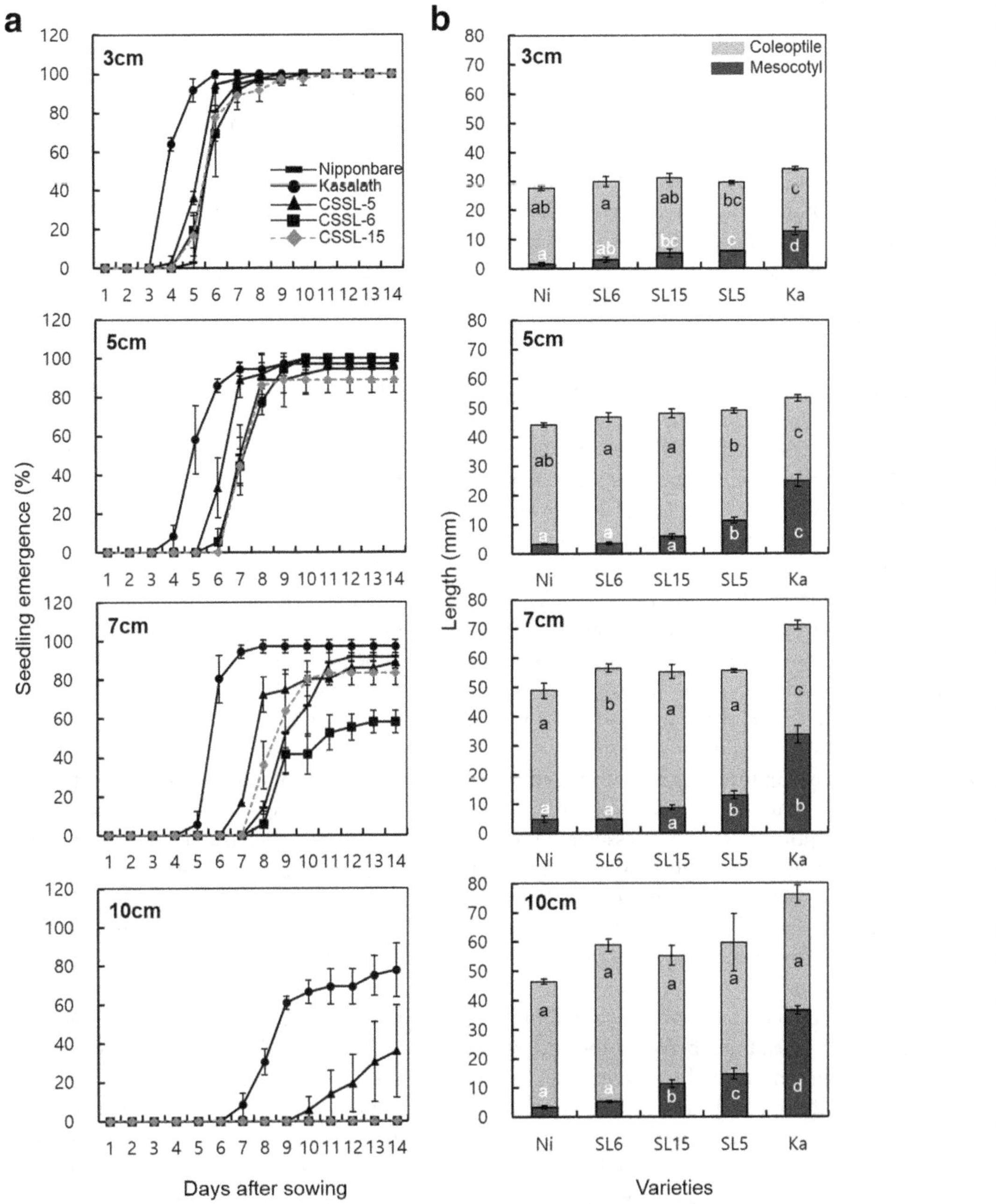

Fig. 5 Time course of seedling emergence (**a**) and mesocotyl and coleoptile length (**b**) of 3 CSSLs, Nipponbare and Kasalath in four soil burial depth; (**a**) Lines and bars represent the mean with SE of 5 lines, Nipponbare(Ni), Kasalath (Ka), CSSL-6 (SL6), CSSL-15 (SL15), and CSSL-5 (SL5). Seeds were sown at 3 cm, 5 cm, 7 cm, and 10 cm soil depth and the number of emerged seedling form soil surface were counted daily up to 14 days after sowing. **b** At 14 days after sowing, length of mesocotyl and coleoptile were measured. Means of length ± SE (n = 3) with different letters in the same row indicate significant differences according to the Duncan multiple range test ($P < 0.05$)

(Fig. 5b). These results indicate that a significant relationship exists between mesocotyl elongation and seedling emergence in deep-seeded plants.

Additionally, these results also show that the 2 QTLs, *qMel-1* and *qMel-3* act additively in distinct or complementary pathways in controlling mesocotyl elongation under this study condition.

Discussion

In rice, seeding depth for direct seeding is an important factor for ensuring seedling establishment. Lee et al. (2002) recommended a 2–3-cm sowing depth for *japonica* rice. When the seeding depth is >3 cm, seedling emergence was markedly delayed (Kawatei et al. 1963; Murai et al. 1995; Luo et al. 2007; Chung

2010; Alibu et al. 2012). Moreover, when seeds were sown deeper than 5 cm, the first leaf developed under the soil surface, and seedling establishment decreased remarkably (Chung 2010). Similar result was obtained in this study: seedling emergence speed and percentage at 7 cm and 10 cm soil depths were lower than those at the seeding depth of 3 cm in all the genotypes (Fig. 5). Although 3-cm seeding depth might be optimal for direct seeding based on these results, many studies have used seeding depths of less than 3 cm, which caused seed desiccation and floating and seedling lodging (Lee et al. 2002, Kumar and Ladha 2011). Genotype differences for seedling emergence and coleoptile length were more pronounced at 5 cm than at 3 cm seeding depth (Figs. 1 and 5). The two parents and the 3 CSSLs showed >90% seedling emergence within 7 days when the seeding depth was 3 cm, whereas Kasalath and CSSL-5 showed >90% emergence and the other three lines reached a plateau at 80% emergence at 5 cm seeding depth (Fig. 5). These results indicate that soil depth of 3–5 cm is appropriate for direct seeding of cultivars that show good mesocotyl elongation without affecting seedling establishment. Based on these findings, we selected 5 cm soil depth for screening seedling emergence traits under the deep-seeding condition.

Mesocotyl and coleoptile lengths are directly related with seedling emergence in deep seeding and enhanced mesocotyl or coleoptile elongation is associated with better seedling emergence and establishment (Turner et al. 1982; Murai et al. 1995; Luo et al. 2007; Chung 2010; Alibu et al. 2012). Murai et al. (1995) reported that a relationship exists between seedling emergence ability and lengths of leaf, mesocotyl, coleoptile, and leaf internode using dwarf lines under 7 cm soil depth. Chung (2010) reported positive correlations between the rate of seedling emergence and mesocotyl or coleoptile length at 5 cm seeding depth by using 116 Korean weedy rice accessions. Alibu et al. (2012) suggested that upland rice accessions with long mesocotyl showed better seedling emergence than those with long coleoptile under the dry direct-seeding condition. Lu et al. (2016) showed that long mesocotyl rice accessions had higher emergence rate than that of the short one at 2 and 5-cm sowing depth. In addition to the seeding depth, in dry seeding, the lengths of mesocotyl and coleoptile can be affected by moisture content. Under submergence, coleoptile growth was stimulated, although no increase in mesocotyl length was observed (Takahashi 1978; Alibu et al. 2011). This suggests that mesocotyl and coleoptile elongation is important for seedling emergence depending on the environment. In our study, a positive correlation between mesocotyl length and seedling emergence was observed (Figs. 4 and 5). Seedlings of Kasalath and CSSL-5 that have long mesocotyls emerged considerably faster than those of the other genotypes that have short mesocotyl at the 7 and 10 cm seeding depths (Figs. 4 and 5). At 7 cm seeding depth, seedlings of Kasalath and CSSL-5 emerged over 70% in 7-8 days after sowing, whereas those of the other genotypes and Nipponbare showed 70% emergence at 10–11 days after sowing. This delayed seedling emergence of up to 3–4 days might be unfavorable for seedling establishment and competition with weeds.

Evaluation of RDRS showed genotypic variation of mesocotyl and coleoptile length (Fig. 2). *Japonica* accessions tend to have shorter mesocotyls than *indica*, whereas *japonica* showed the longest coleoptiles among four groups. In this study, *indica*-I accessions had the longest mesocotyls among the four groups, followed by the tropical *japonica* group (Fig. 2b). These results are consistent with the previous findings (Suge 1972; Takahashi 1978, Sato 1987).

QTL analysis was conducted for mesocotyl and coleoptile length of seedlings sown at 5 cm soil depth. Measuring mesocotyl and coleoptile traits in the soil is difficult mainly due to the environmental influence and experimental errors. Several studies attempted to map QTLs for mesocotyl and coleoptile length by using the glass tube method (Katsuta-Seki et al. 1996), slant-board test (Redoña and Mackill 1996), filter paper with distilled water (Huang et al. 2010), agar medium (Lee et al. 2012), and filter paper on agar medium (Xie et al. 2014). However, QTL analysis for mesocotyl or coleoptile elongation at the 5 cm seeding depth by using CSSLs in rice has not yet been performed. We detected 5 QTLs under field condition in two replicates. Interestingly, QTLs for mesocotyl elongation were commonly mapped to chromosomes 1 and 3 for different mapping populations and under different experimental conditions. The QTL *qMel-3* was located on the long arm of chromosome 3, and this QTL interval overlapped with regions of QTLs reported in previous studies (Katsuta-Seki et al. 1996; Redoña and Mackill 1996; Cai and Morishima 2002; Cao et al. 2002; Huang et al. 2010). Using genome-wide association study (GWAS) association mapping, Wu et al. (2015) and Lu et al. (2016) identified some candidate genes controlling mesocotyl elongation on chromosome 1 and 3. Coleoptile QTLs were detected on chromosomes 3 and 6 by Redoña and Mackill (1996) and on chromosomes 5, 8, and 11 by Xie et al. (2014). On chromosomes 1 and 3, two QTLs for coleoptile length were identified under flooded soil condition (Zhou et al. 2007). Interestingly, *qCol-3* and *qCol-5* that were detected in the present study did not overlap with the coleoptile length QTLs that was reported previously; this could be attributed to the different genetic materials and test methods used.

Among 54 CSSLs, two lines, CSSL-5 and CSSL-33 showed significantly longer mesocotyl than Nipponbare. CSSL-5 showed the highest emergence ratio (70%) followed by CSSL-33 (17%) (Fig. 4a and b). CSSL-5 and CSSL-33 shared *qMel-3* suggesting that *qMel-3* is a major QTL. In contrast, *qMel-7* is minor and affected by testing methods based on the finding that the *qMel-7* QTL was detected in the agar condition (Lee et al. 2012) but not in the soil condition in this study. However, CSSL-5 seedlings were shorter in mesocotyl length and lower in emergence rate than Kasalath. The results from this study indicate that the *qMel-3* QTL with a high additive effect has the potential to enhance early seedling emergence in combination with the *qMel-1, qMel-7* and other QTLs not detected. Also, information on the interaction among these QTLs will be necessary to understand the genetic control of mesocotyl length. These findings will contribute to our understanding of the genetic control of mesocotyl and coleoptile elongation, and seedling emergence and provides information on QTL regions associated with early seedling emergence for rice breeding programs.

Conclusions

We showed that the variation in mesocotyl elongation is associated with seedling emergence under the deep-seeding condition. Genetic analysis using BIL and CSSL plants confirmed that two major QTLs, *qMel-1* and *qMel-3* for mesocotyl elongation have the potential to enhance early seedling emergence at 5 cm seeding depth. This is the first report to show the effect of two QTLs for mesocotyl elongation detected both in agar media and deep-seeding soil on seedling emergence using CSSLs using deep-seeding soil condition. (Lee et al. 2012). High-resolution mapping and cloning of these QTLs should be performed to understand the genetic control of mesocotyl elongation to ensure higher seedling emergence and to develop deep-seeding-tolerant rice cultivars for direct seeding by using marker-assisted selection.

Methods

Plant materials

Fifty-seven rice accessions selected from the RDRS of germplasm collection were used to detect variation in mesocotyl and coleoptile elongation (*Oryza sativa* L.) (Kojima et al. 2005; Lee et al. 2012; Additional file 2: Table S1). QTLs for mesocotyl and coleoptile elongation were detected and the effect of mesocotyl elongation on seedling emergence was determined by using 98 BILs and 54 CSSLs developed from the backcross of the Nipponbare/Kasalath (Lin et al. 1998). Seeds of BILs and CSSLs were provided by Rice Genome Resource Center (RGRC), Japan (http://www.rgrc.dna.affrc.go.jp/stock.html).

The effect of QTLs for mesocotyl elongation *qMel-1* and *qMel-3* on seedling emergence was determined by selecting 3 CSSLs—CSSL-6, CSSL-15, and CSSL-5 (Lee et al. 2012). The genotype data of the CSSLs were referred from a database of RGRC (http://www.rgrc.dna.affrc.go.jp/ineNKCSSL54.html). CSSL-6 and CSSL-15 carried the QTLs *qMel-1* and *qMel-3*, respectively. CSSL-6 contained a 34-cM Kasalath segment flanked by restriction fragment length polymorphism (RFLP) markers C86-C112 carrying *qMel-1* on chromosome 1. CSSL-15 contained a 61-cM Kasalath segment flanked by RFLP markers R19–R1925 carrying *qMel-3* on chromosome 3 in the Nipponbare background. CSSL-5 carried both QTLs *qMel-1* and *qMel-3* in the Nipponbare background. In CSSL-5, 2 Kasalath segments on qMel-1 and qMel-3 were introgressed, a 34-cM Kasalath segment near the RFLP makers C1370–2018, including *qMel-1*, 43-cM Kasalath segment near RFLP markers C136–R1925, in the Nipponbare background.

RDRS, BIL, and CSSL plants were grown in the experimental lowland field of the Graduate School of Life Sciences, Tohoku University, Kashimadai, Osaki, Miyagi Prefecture, Japan. The panicles of these plants were harvested at 40–50 days after heading and then dried in a well-ventilated space for 3 months. Dried seeds were collected from the panicle and placed in paper envelopes. All of the seeds were completely germinated in 30 °C imbibitions for 4 days. For maintaining the germinability, the seed envelopes were packed in plastic bags sealed with silica gel and stored at 4 °C in a refrigerator until use.

Measurement of mesocotyl and coleoptile elongation

Mesocotyl and coleoptile elongation was measured and seedling emergence under the soil condition was evaluated by sowing 12 good-quality seeds from each RDRS accession, BILs, and CSSLs at various soil depths. Seeds of Nipponbare, Kasalath, and 3 CSSLs were buried at a depth of 3, 5, 7, and 10 cm in plastic pots (diameter, 11 cm; height, 15 cm) in soil containing chemical fertilizer (Nursery culture soil No. 3; Mitsui-Toatsu, Japan). For determining phenotype variation in mesocotyl and coleoptile length, seeds of RDRS, 98 BILs and 54 CSSLs were buried in 5 cm soil depth. The plastic pots were kept in plant growth cabinets at alternate temperatures of 30 °C and 26 °C (14 h/10 h). The water level of the soil was maintained 2 cm from the bottom of the pots to maintain adequate soil moisture for germination and seedling growth. At 14 days after sowing, the seedlings were carefully excavated and washed, and their lengths were measured. Seedlings insufficiently grown were excluded from measurement. In this study, measurement was made on seedlings showing 3rd-4th leaves expansions at 14 days after sowing. Mesocotyl (the distance from the basal part of the seminal root to the coleoptilar node) and coleoptile (the distance from the coleoptilar node to the tip of the coleoptile) lengths of each seedling were measured using a ruler. Seedlings that germinated but grew insufficiently were excluded from the length measurements. The numbers of emerged seedlings from the soil surface were counted daily up to 14 days when seedling emergence was thought to be complete. The experiment was performed in a completely randomized design with two or three replicates of each line.

Genotype data of BILs and QTL analysis

The QTLs controlling mesocotyl and coleoptile elongation in BILs were mapped by using genotype data generated using 245 RFLP markers (http://www.rgrc.dna.affrc.go.jp/ineNKBIL98.html). Linkage analysis of the genotypic data was performed using the Kosambi function of Mapmaker/EXP 3.0 software (Lander et al. 1987). QTL analysis was performed using composite interval mapping (CIM) by using the QTL Cartographer version 2.5 software (Wang et al. 2012). CIM analysis was performed using forward-backward stepwise regression model 6 with a 10-cM window size. The LOD threshold significance level ($P < 0.05$) was determined by computing 1000 permutations. The QTL positions were assigned to the points of maximum LOD score in the target regions. The percentage of total phenotypic variance noted for each QTL was estimated on the basis of the R^2 value.

Additional files

Additional file 1: Figure S1. Seedling of Nipponbare and Kasalath growing at 5 cm and 7 cm soil depth condition; 50 seeds of Nipponbare and Kasalath were sown at 5 cm and 7 cm soil depth and incubated at alternate temperatures of 30 °C and 26 °C (14 h/10 h). At 21 days after sowing, the seedling were excavated and the length of coleoptile (Col) and mesocotyl (Me) were measured. Arrows indicate mesocotyl.

Additional file 2: Table S1. Variation of the mesocotyl and coleoptile length for 57 rice accessions from RDRS collection.

Additional file 3: Figure S2. QTL cartographer LOD peak for mesocotyl and coleoptile length at 5 cm soil depth; A QTL Cartographer plot obtained following composite interval mapping (CIM) using 2 replicates (Rep1 and 2). Significance of QTL is indicated by LOD score above the threshold values determined by permutation analysis at a significant level of $P < 0.05$. The graph below shows the additive effects for each of the QTL identified. Marker designations are given at the bottom and the genetic distances (cM) are given above the horizontal line.

Acknowledgements

This work was carried out with the support of and Basic Science Research Program through the National Research Foundation of Korea (NRF) funded by the Ministry of Education (2015R1A6A3A01060388) and "Cooperative Research Program for Agriculture Science & Technology Development (Project No. PJ 011048)", Rural Development Administration, Republic of Korea.

Authors' contributions

HL conceived of the study and carried out molecular genetic analysis and statistical analysis, and wrote manuscript. KS helped to design research and performed statistical analysis. JK and SW advised to draft the manuscript. TS participated in the design of the study. SA designed research and helped to draft the manuscript. All authors read and approved the final manuscript.

Competing interests

The authors declare that they have no competing interests.

Author details

[1]College of Agriculture and Life Sciences, Chungnam National University, Daejeon 305-764, South Korea. [2]Graduate School of Life Sciences, Tohoku University, 2-1-1 Katahira, Aoba-ku, Sendai 980-8577, Japan. [3]Institute for Sustainable Agro-ecosystem Services, Graduate School of Agricultural and Life Sciences, The University of Tokyo, 1-1-1 Midoricho, Nishitokyo, Tokyo 188-0002, Japan. [4]Graduate School of Agricultural Science, Tohoku University, 1-1 Amamiya-machi, Tsutsumidori, Aoba-ku, Sendai, Miyagi 980-8555, Japan. [5]POTECH-UZH Cooperative Laboratory, Department of Integrative Bioscience and Biotechnology, Pohang University of Science and Technology, Pohang 790-784, South Korea.

References

Alibu S, Saito Y, Shiwachi H, Irie K (2011) Relationship between coleoptile and mesocotyl elongation of upland rice (*Oryza sativa* L.) seedlings under submergence and soil-sand culture. Afr J Agric Res 6(31):6463–6472

Alibu S, Saito Y, Shiwachi H, Irie K (2012) Genotypic variation in coleoptile or mesocotyl lengths of upland rice (*Oryza sativa* L.) and seedling emergence in deep sowing. Afr J Agric Res 7(47):6239–6248

Cai HW, Morishima H (2002) QTL clusters reflect character associations in wild and cultivated rice. Theor Appl Genet 104:1217–1228

Cao L, Zhu J, Yan Q, He L, Wei X, Cheng S (2002) Mapping QTLs with epistasis for mesocotyl length in a DH population from indica-japonica cross of rice (*Oryza sativa*). Chinese J Rice Sci 16(3):221–224 (in Chinese with English abstract)

Chang T, Vergara BS (1975) Varietal diversity and morpho-agronomic characteristics of upland rice. In: Major research in upland rice, International Rice Research Institute, Philippines

Chung NJ (2010) Elongation habit of mesocotyls and coleoptiles in weedy rice with high emergence ability in direct-seeding on dry paddy fields. Crop Pasture Sci 61(11):911–917

Dilday RH, Mgonja MA, Amonsilpa SA, Collins FC, Wells BR (1990) Plant height vs. mesocotyl and coleoptile elongation in rice: Linkage or pleiotropism? Crop Sci 30:815–818

Farooq M, Barsa S, Wahid A (2006) Priming of field-sown rice seed enhances germination, seedling establishment, allometry and yield. Plant Growth Regul 49:285–294

Huang C, Jiang S-K, Feng L-L, Xu Z-J, Chen W-F (2010) Analysis of QTLs for mesocotyl length in rice (*Oryza sativa* L.) Acta Agron Sin 36(7):1108–1113

Katsuta-Seki M, Ebana K, Okuno K (1996) QTL analysis for mesocotyl elongation in rice. Rice Genetics Newsletter 13:126

Kawatei K, Hoshikawa K, Takashima Y (1963) On germination and subsequent early growth of seedling in direct seeding culture of paddy rice on upland field. Jpn J Crop Sci 31(3):267–271

Kojima Y, Ebana K, Fukuoka S, Nagamine T, Kawase M (2005) Development of an RFLP-based rice diversity research set of germplasm. Breed Sci 55:431–440

Kumar V, Ladha JK (2011) Direct Seeding of Rice: Recent developments and future research needs. Adv Agron 111:297–413

Lander ES, Green P, Abrahamson J, Barlow A, Daly MJ, Lincoln SE, Newburg L (1987) MAPMAKER: An interactive computer package for constructing primary genetic linkage maps of experimental and natural populations. Genomics 1:174–181

Lee M-H, Kim JK, Kim SS, Park ST (2002) Status of dry-seeding technologies for rice in Korea. In: Pandey S, Mortimer M, Wade L, Tuong TP, Lopez K, Hardy B (eds) Direct seeding: Research strategies and opportunities. International Rice Research Institute, Philippines, pp 161–176

Lee HS, Sasaki K, Higashitani A, Ahn SN, Sato T (2012) Mapping and characterization of quantitative trait loci for mesocotyl elongation in rice (*Oryza sativa* L.). Rice 5: 13

Lin SY, Sasaki T, Yano M (1998) Mapping quantitative trait loci controlling seed dormancy and heading date in rice, *Oryza sativa* L., using backcross inbred lines. Theor Appl Genet 96:997–1003

Lu Q, Zhang M, Niu X, Wang C, Xu Q, Feng Y, Wang S, Yuan X, Yu H, Wang Y, Wei X (2016) Uncovering novel loci for mesocotyl elongation and shoot length in indica rice through genome wide association mapping. Planta 243:645–657

Luo J, Tang S, Hu P, Aleman L, Jiao G, Tang J (2007) Analysis on factors affecting seedling establishment in rice. Rice Sci 14(1):27–32

Murai M, Sato S, Ise K, Semihira Y (1995) Effects of the dwarfing gene from dee-geo-woo-gen and others on emergence ability under deep seeding condition in rice. Breed Sci 45:163–171

Murai K, Lin S, Yano M, Nagamine T (2002) Mapping quantitative trait loci controlling seed loggevity in rice (*Oryza sativa* L.) Theor Appl Genet 104(6):981–986

Pandey S (2005) Velasco L (2005) Trends in crop establishment methods in Asia and research issues. In: Toriyama K, Heong KL, Hardy B (eds) Rice Is Life: Scientific perspectives for the 21st Century, p.178–181. Proceedings of the world rice research conference, Tsukuba

Redoña ED, Mackill DJ (1996) Mapping quantitative trait loci for seedling vigor in rice using RFLPs. Theor Appl Genet 92:395–402

Sato YI (1987) Character association patterns among Japonica rice varieties of Asian origin. In: Hsieh SC (ed) Crop exploration and utilization of genetic resources. Taichung District Agricultural Improvement Station, Taiwan

Suge H (1972) Mesocotyl elongation in *japonica* rice: Effect of high temperature pre-treatment and ethylene. Plant Cell Physiol 13:401–405

Takahashi N (1978) Adaptive importance of mesocotyl and coleoptile growth in rice under different moisture regimes. Aust J Plant Physiol 5:511–517

Takahashi N (1984) Seed germination and seedling growth. In: Tsunoda S, Takahashi N (eds) Biology of rice. Japan Science Society press Elsevier, Japan, pp 71–88

Tang S (2002) Seedling broadcasting in china: and overview. In: Pandey S, Mortimer M, Wade L, Tuong TP, Lopez K, Hardy B (eds) Direct seeding: Research strategies and opportunities. International Rice Research Institute, Philippines, pp 177–184

Turner FT, Chen CC, Bollich CN (1982) Coleoptile and mesocotyl lengths in semidwarf rice seedlings. Crop Sci 22:43–46

Uga Y, Ebana K, Abe J, Morita S, Okuno K, Yano M (2009) Variation in root morphology and anatomy among accessions of cultivated rice (*Oryza sativa* L.) with different genetic backgrounds. Breed Sci 59:87–93

Wang S, Basten CJ, Zeng ZB (2012) Windows QTL Cartographer 2.5. Department of Statistics, North Carolina State University, Raleigh, NC. http://statgen.ncsu.edu/qtlcart/WQTLCart.htm.

Wu J, Feng F, Lian X, Teng X, Wei H, Yu H, Xie W, Yan M, Fan P, Li Y, Ma X, Liu H, Yu S, Wang G, Zhou F, Luo L, Mei H (2015) Genome-wide association study (GWAS) of mesocotyl elongation based on re-sequencing approach in rice. BMC Plant Biol 15:218

Xie L, Tan Z, Zhou Y, Xu R, Feng L, Xing Y, Qi X (2014) Identification and fine mapping of quantitative trait loci for seed vigor in germination and seedling establishment in rice. J Integr Plant Biol 56:749–759

Yamaya T, Obara M, Nakajima H, Sasaki S, Hayakawa T, Sato T (2002) Genetic manipulation and quantitative-trait loci mapping for nigtogen recycling in rice. J Exp Bot 53(370):917–925

Zhou L, Wang J-K, Yi Q, Wang Y-Z, Zhu Y-G, Zhang Z-H (2007) Quantitative trait loci for seedling vigor in rice under field conditions. Field Crop Res 100:294–301

12

Identification and Validation of a Novel Major QTL for Harvest Index in Rice (*Oryza sativa* L.)

Shaohong Zhang[1,2†], Xiuying He[1,2†], Junliang Zhao[1,2], Yongsheng Cheng[1,2], Zhimei Xie[3,4], Yuehan Chen[1,2], Tifeng Yang[1,2], Jingfang Dong[1,2], Xiaofei Wang[1,2], Qing Liu[1,2], Wei Liu[1,2], Xingxue Mao[1,2], Hua Fu[1,2], Zhaoming Chen[1,2], Yaoping Liao[1,2] and Bin Liu[1,2*]

Abstract

Background: Harvest index (HI) in rice is defined as the ratio of grain yield (GY) to biomass (BM). Although it has been demonstrated that HI is significantly related to yield and is considered as one of the most important traits in high-yielding rice breeding, HI-based high-yielding rice breeding is difficult due to its polygenic nature and insufficient knowledge on the genetic basis of HI. Therefore, searching for rice varieties with high HI and mapping genes associated with high HI can facilitate marker-assisted breeding for high HI in rice.

Results: Yuexiangzhan, a popular *indica* cultivar with good reputation of high HI was crossed with Shengbasimiao, an *indica* cultivar with lower HI to develop a recombinant inbred line population, and QTL mapping for HI and its component traits was conducted. In total, five QTLs for HI, three QTLs for GY, and six QTLs for BM were detected in two-year experiments. Among the three GY QTLs, one co-located with the HI QTL on chromosome 8, while the other two co-located with the two tightly-linked BM QTLs on chromosome 3. The co-located QTLs in each of the chromosomal regions produced additive effects in the same direction. Particularly, the HI QTL on chromosome 8, *qHI-8*, could be detected across two years and explained 42.8% and 44.5% of the phenotypic variation, respectively. The existence of *qHI-8* was confirmed by the evaluation of the near isogenic lines derived from a residual heterozygous line, and this QTL was delimitated to a 1070 kb interval by substitution mapping.

Conclusion: In the present study, the detected GY QTLs overlapped with both HI QTL and BM QTL, suggesting a positive relationship between GY and HI or BM, respectively. With an understanding of the genetic basis for grain yield, harvest index and biomass, it is possible to achieve higher yield through enhancing HI and BM by pyramiding the favorable alleles for the two traits via marker-assisted selection (MAS). As *qHI-8* has a large phenotypic effect on HI and expresses stably in different environments, it provides a promising target for further genetic characterization of HI and MAS of high HI in rice breeding.

Keywords: Rice (*Oryza sativa* L.), Harvest index, Grain yield, Biomass, Quantitative trait locus (QTL), Mapping, Validation

* Correspondence: lbgz1009@163.com
[†]Equal contributors
[1]Rice Research Institute, Guangdong Academy of Agricultural Sciences, Guangzhou 510640, China
[2]Guangdong Key Laboratory of New Technology in Rice Breeding, Guangzhou 510640, China
Full list of author information is available at the end of the article

Background

Global food demand is increasing with the growth of the world population. Rice is the most important food crop, serving as the staple food for more than half of the world's population. By 2035, a 26% increase (176 million tons) in rice production will be necessary to feed the growing population (Khush 2013). However, the area available for rice planting in major production countries has been decreasing because of acceleration of urbanization and industrialization. Therefore, increasing the yield per unit area on existing land is the only way to increase rice production.

Harvest index (HI) is an important trait associated with yield. HI is defined as the ratio of grain yield (GY) to biomass (BM), and is taken as a measure of biological success in partitioning photosynthate to the harvestable product (Donald and Hamblin 1976; Sinclair 1998). In the past decades, dramatic increases in crop yield have been largely attributed to improvement in HI (Sinclair 1998). Before the Green Revolution, rice varieties were tall and leafy with weak stems, and had a HI of approximately 0.3. The incorporation of the dwarfing gene *sd1* resulted in increasing HI to 0.5, leading to dramatic increases in rice yield (Khush 2001; Mann 1999). Higher HI is considered to be one of the most important traits in the breeding program for new plant type at the International Rice Research Institute (IRRI) (Khush 1995) and for super hybrid rice in China (Yuan 2001). Although HI is an emphasized trait, rice breeding for high HI is difficult due to the trait's polygenic nature and insufficient knowledge of its genetic basis. Since the advent of molecular marker technology, some quantitative trait loci (QTLs) for HI have been identified and mapped in rice. Using a double-haploid (DH) population of an *indica-japonica* cross, Hittalmani et al. (2003) detected eleven QTLs associated with HI on chromosomes 1, 3, 4, 7 and 8 at nine locations in Asia. Among these QTLs, seven were detected in two to four locations, while the others were only detected in one location. Zhang et al. (2004) identified four QTLs for HI on chromosomes 1, 4, 8 and 11 using a DH population of another *indica-japonica* cross. Marri et al. (2005) detected one QTL for HI on chromosome 2 in an interspecific BC_2 testcross progeny. Sabouri et al. (2009) mapped five QTLs for HI on chromosomes 2, 3, 4 and 5 in the F_2 population of an *indica-indica* cross. Through association mapping, Li et al. (2012) identified nine markers associated with HI on chromosomes 1, 2, 4, 8 and 9 in Arkansas and Texas, USA, but no common associated marker was found at two locations. Although some QTLs associated with HI have been identified, successful marker-assisted breeding for high HI in rice has not been reported. This issue may be attributed to the rice varieties for studies and the complexity of the trait. In previous studies, the rice varieties used to develop mapping populations have a HI of less than 0.5, and were thus not the ideal sources for genetic studies and high HI breeding. Furthermore, HI is a quantitative trait controlled by multiple genes (QTLs) and the expression of QTL for HI is affected by environments (Hittalmani et al. 2003; Li et al. 2012). Therefore, screening additional rice varieties with higher HI and identifying stably expressed QTLs with large effect on HI will facilitate marker-assisted breeding for high HI in rice.

Yuexiangzhan (YXZ), an *indica* cultivar developed by Rice Research Institute of Guangdong Academy of Agricultural Sciences, China, is an elite variety with high-yield and good quality. Particularly, this variety possesses a HI of 0.616 (Chen et al. 1999). To dissect the genetic basis of HI and its component traits, a recombinant inbred line (RIL) population was developed by crossing YXZ with Shengbasimiao (SBSM), an *indica* cultivar with lower HI, and QTL analysis was conducted to identify and map QTLs for HI, GY and BM in the present study. A novel major QTL for HI, *qHI-8*, could be detected in two-year experiments, and explained 42.8% and 44.5% of the phenotypic variations, respectively. The existence of *qHI-8* and its phenotypic effect on HI was validated and quantified using the near isogenic lines (NILs) derived from a residual heterozygous line (RHL). The results enhance our understanding on the genetic basis of HI in rice. In particular, identification of *qHI-8*, the stably expressed and large effect QTL, provides a promising target for further genetic characterization and marker-assisted breeding for high HI in rice.

Results

Harvest indexes of the two parents and their derived RIL population

The two parents, YXZ and SBSM, exhibited significant differences in HI. The HI of YXZ was 0.555 and 0.618 in the years 2010 and 2011, respectively (Table 1). Large HI variations in the RIL population were observed, ranging from 0.486 to 0.568 in 2010, and from 0.471 to 0.642 in 2011 (Table 1). Wider HI variation was observed in 2011 compared to that in 2010 (Fig. 1). Some RI lines (3.2% in 2010 and 1.6% in 2011) had a HI higher than that of YXZ in both years, while some RI lines (17.2%) had a HI lower than that of SBSM in 2011, indicating different transgressive segregation in two-year experiments. The RIL population displayed a continuous and normal distribution for HI (Table 1 and Fig. 1), suggestive of quantitative inheritance and the involvement of multiple genes for the trait. Furthermore, a significantly positive correlation was observed between the HI measured in 2010 and 2011 in the RIL population ($r = 0.6031, p < 0.0001$).

Table 1 The harvest index of YXZ, SBSM and their derived RIL population in the two-year experiments

Year	Parent		RIL population			
	YXZ (Mean ± SD)	SBSM (Mean ± SD)	Mean ± SD	Range	Skewness	Kurtosis
2010	0.555 ± 0.001****	0.472 ± 0.001	0.515 ± 0.020	0.486–0.568	0.768	−0.437
2011	0.618 ± 0.009***	0.531 ± 0.007	0.563 ± 0.032	0.471–0.642	−0.625	0.089

*** and ****represent the significance of difference between the two parents at $p < 0.001$ and $p < 0.0001$, respectively based on *t*-test

QTLs for harvest index, grain yield and biomass in RIL population

For harvest index, a total of five QTLs were detected and mapped on chromosomes 1, 3, 7, 8 and 12 in two years (Table 2 and Fig. 2). Among the five QTLs, *qHI-7* was detected in 2010 and explained 7.3% of the phenotypic variation, while *qHI-1*, *qHI-3* and *qHI-12* were detected in 2011 and could explain phenotypic variations ranging from 4.6% to 7.4%. It is noteworthy that the QTL on chromosome 8, *qHI-8*, could be detected across both years and could explain 42.8% and 44.5% of the phenotypic variation, with a high LOD of 25.1 and 29.2, in 2010 and 2011, respectively. All positive alleles contributing to HI came from the high HI parent YXZ.

For grain yield and biomass, three and six QTLs were detected, respectively (Table 2 and Fig. 2). Among the three GY QTLs, *qGY-3b* was detected in 2010, while *qGY-3a* and *qGY-8* were detected in 2011. The positive allele at *qGY-8* came from YXZ, while the positive allele at *qGY-3a* and *qGY-3b* came from SBSM. Among the six BM QTLs, *qBM-3b*, *qBM-3c* and *qBM-12* were detected in 2010, while *qBM-2*, *qBM-3a* and *qBM-3d* were detected in 2011. The positive alleles contributing to BM at these QTLs came from SBSM.

Comparing chromosomal locations of the QTLs for GY, HI and BM identified in this study, we found that one GY QTL, *qGY-8*, co-located with the major HI QTL, *qHI-8*, in the interval RM447-RM6845 on chromosome 8; they produced their additive effects in the same direction. The other two tightly linked GY QTLs, *qGY-3a* and *qGY-3b*, co-located with two tightly linked BM QTLs, *qBM-3a* and *qBM-3b*, on chromosome 3. These co-located GY and BM QTLs also produced additive effects in the same direction.

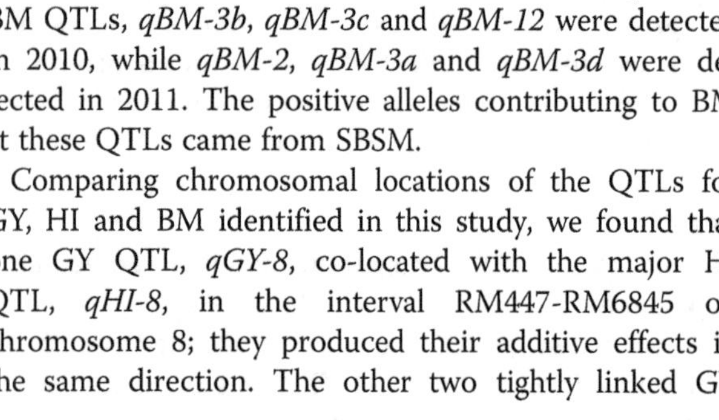

Fig. 1 Distribution of harvest index in RIL population derived from YXZ and SBSM in two years (2010 and 2011). A and B indicate the harvest index of YXZ and SBSM, respectively

No significant digenic epistatic interaction between the QTLs, nor between the QTLs and other loci, were found ($p > 0.05$) in the present study based on analysis using QTLNetwork-2.0.

Validation and delimitation of *qHI-8* using NILs derived from a residual heterozygous line

The mapping results reveal that *qHI-8* is a stably expressing and large-effect QTL for harvest index. Additionally, *qHI-8* co-located with *qGY-8*, the QTL for grain yield and they produce their additive effects in the same direction, suggesting its positive effect on grain yield. We believe that *qHI-8* has great potential in high-yielding rice breeding. To validate *qHI-8*, NILs were developed from a residual heterozygous line (RHL). Based on the results of QTL mapping, we searched for RHLs from the 186 RI lines using the markers defining the QTL region of *qHI-8*. One RI line exhibiting heterozygous in the region of *qHI-8* was found. A single plant (RHL15) from this RI line was selected; it is heterozygous in the region harboring *qHI-8* (ID08.26-RM477), but homozygous in other genomic regions based on the genotypes determined by the 142 polymorphic SSR and InDel markers distributed across the 12 rice chromosomes (Additional file 1: Figure S1). RHL15 selfed to produce progenies segregating at the target region in a near isogenic background (equivalent to F_2, designated as NIL-F_2). The NIL-F_2 plants were screened with the polymorphic markers ID08.26, RM502, RM447, RM6845, and RM477 to identify the homozygotes and recombinants within the target region. Based on the genotypes determined by the five markers, three plants (15–2, 15–27 and 15–183) with homozygous YXZ genotypes and three plants (15–193, 15–307 and 15–398) with homozygous SBSM genotypes were selected. They were grouped as two sets of contrasting NILs with and without *qHI-8*, and designated as NIL15-YXZ and NIL15-SBSM, respectively (Table 3 and Fig. 3). Then, NIL15-YXZ, NIL15-SBSM, and the identified recombinant plants further selfed to generate their corresponding F_3 families. Ten informative homozygous recombinants were identified by genotyping F_3 individuals with the five

Table 2 The QTLs for harvest index, grain yield and biomass identified in RIL population derived from YXZ and SBSM in two years

Trait[a]	QTL	Chr	2010					2011				
			Interval	Interval distance (cM)	LOD	Additive effect[b]	Variation explained (%)	Interval	Interval distance (cM)	LOD	Additive effect	Variation explained (%)
HI	*qHI-1*	1						RM488-RM237	10.4 (6.0)[c]	3.8	0.0069	4.6
	qHI-3	3						RM15857-RM520	7.7 (6.0)	5.8	0.0089	7.4
	qHI-7	7	RM427-RM7217	31.6 (10.0)	3.2	0.0053	7.3					
	qHI-8	8	RM447-RM6845	2.6 (0)	25.1	0.0129	42.8	RM447-RM6845	2.6 (0)	29.2	0.0216	44.5
	qHI-12	12						RM28172-RM463	19.4 (10.0)	4.0	0.0082	6.3
GY	*qGY-3a*	3						RM282-RM6676	15.0 (8.0)	11.6	−2.4719	28.0
	qGY-3b	3	RM6676-RM6283	11.4 (2.0)	5.8	−1.3457	12.9					
	qGY-8	8						RM447-RM6845	2.6 (2.0)	6.5	1.5962	12.1
BM	*qBM-2*	2						RM290-RM7413	16.7 (7.4)	3.5	−2.2656	9.2
	qBM-3a	3						RM282-RM6676	15.0 (8.0)	11.0	−4.1157	28.9
	qBM-3b	3	RM6676-RM6283	11.4 (8.0)	4.6	−2.4995	10.9					
	qBM-3c	3	RM15857-RM520	7.7 (6.0)	3.2	−2.1424	7.2					
	qBM-3d	3						RM3867-RM571	9.1 (0)	3.5	−1.9126	6.3
	qBM-12	12	RM7102-RM28172	4.4 (2.0)	3.0	−1.8650	6.0					

[a] *HI* Harvest index, *GY* Grain yield, *BM* Biomass
[b] Positive values indicate effects of allele from YXZ, negative values indicate effects of allele from SBSM
[c] Numbers in parenthesis indicate genetic distance from the LOD peak position to the left marker on the interval

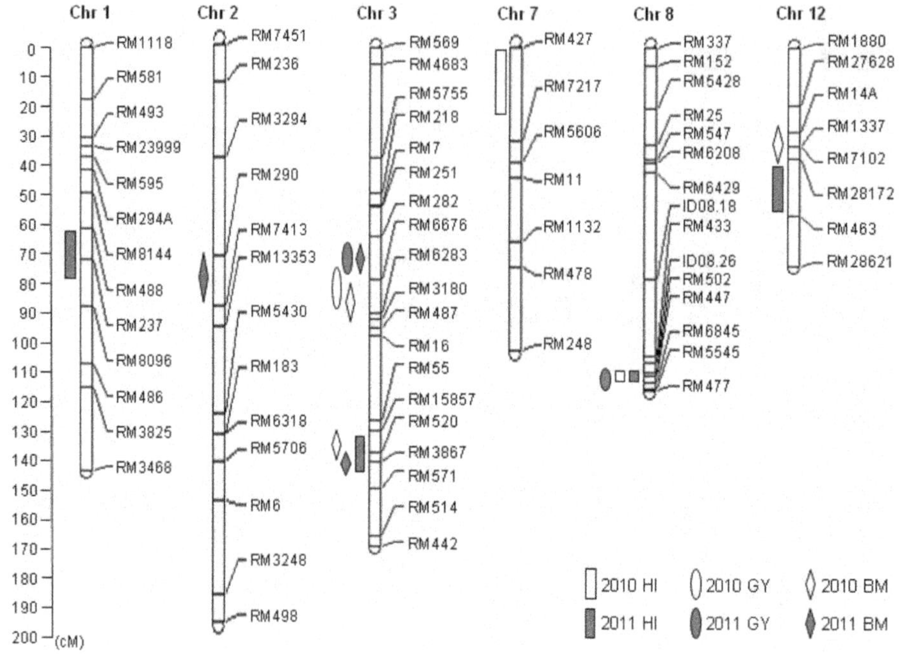

Fig. 2 Chromosomal locations of QTLs for harvest index, grain yield and biomass identified in two years (2010 and 2011)

markers. These recombinants can be grouped into six genotypes (G1 to G6) according to their positions of recombinant breakpoints and allelic composition (Fig. 3). They have the same genetic background as NIL15-YXZ and NIL15-SBSM except for the target regions, and NIL15-YXZ and NIL15-SBSM were used as controls (CK1 and CK2) to measure HI, respectively. The t-test revealed that no significant differences in HI were observed within lines in NIL15-YXZ or NIL15-SBSM ($p > 0.05$), but highly significant difference in mean HI was observed between NIL15-YXZ and NIL15-SBSM ($p < 0.0001$). The lines carrying *qHI-8* showed consistently higher HI, an average of 0.570, while the lines without *qHI-8* showed consistently lower HI, an average of 0.499 (Table 3). Through HI comparison between NIL15-YXZ and NIL15-SBSM, we could confirm the existence of *qHI-8*. The presence of *qHI-8* resulted in a 0.071 increase in HI.

To further delimitate the *qHI-8* region, differences in HI among recombinant lines and controls (CK1 and CK2) were analyzed using Duncan's multiple range test. All recombinant lines in G1, G2, and G6 were significantly different from CK2 ($p < 0.0001$), but not significantly different from CK1 ($p > 0.05$), suggesting that *qHI-8* is located within the interval of ID08.26-RM6845. Furthermore, all recombinant lines in G3, G4, and G5 were significantly different from CK1 ($p < 0.0001$), but not significantly different from CK2 ($p > 0.05$), suggesting that *qHI-8* is not located in the intervals of ID08.26-RM502 and RM6845-RM477. Thus, we can delimitate *qHI-8* to a 1070 kb region flanked by RM502 and RM6845 on chromosome 8 (Fig. 3).

Table 3 Estimation of the relative effect of *qHI-8* on harvest index using two sets of contrasting NILs

NILs	Lines	*qHI-8*[a]	Harvest index	Average of harvest index	Difference[b]
NIL15-YXZ	15–2	+	0.565 ± 0.006		
	15–27	+	0.572 ± 0.007	0.570	
	15–183	+	0.574 ± 0.006		
					0.071****
NIL15-SBSM	15–193	–	0.494 ± 0.010		
	15–307	–	0.508 ± 0.004	0.499	
	15–398	–	0.494 ± 0.005		

[a]+ and - indicate the presence and absence of *qHI-8* in the lines, respectively
[b]Difference in the average HI between NIL15-YXZ and NIL15-SBSM
****represents the significance of difference in the average HI between NIL15-YXZ and NIL15-SBSM at $p < 0.0001$ based on *t*-test

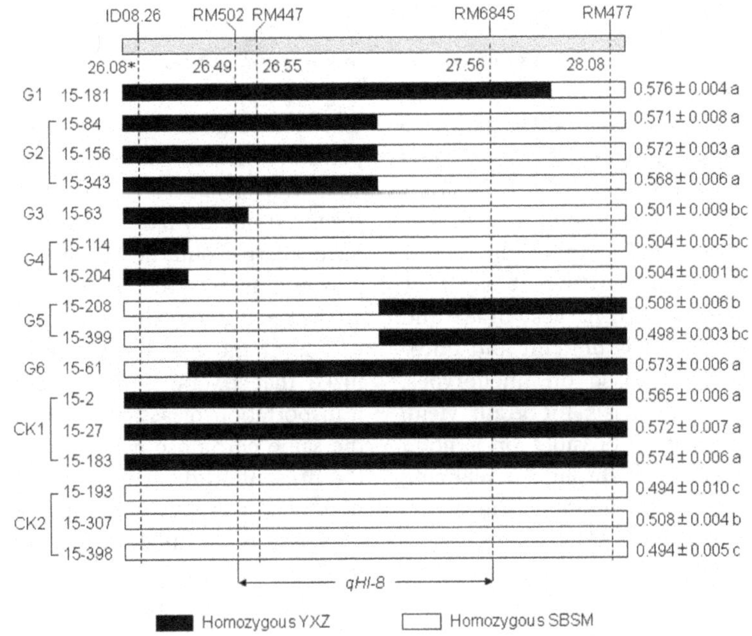

Fig. 3 Delimitation of *qHI-8* using the homozygous recombinants. The *qHI-8* was delimited to a 1070 kb interval defined by markers RM502 and RM6845. G1 to G6 indicate the recombinants used in this study, consist of six genotypes according to their positions of recombinant breakpoints and allelic composition. Values in the right side are the HI phenotypes of each recombinant and the controls, CK1 and CK2 (mean ± SD). The values with the same letter indicate no significant difference in HI at $p = 0.05$ based on Duncan's multiple range test. * The physical position (Mb) of markers on the chromosome

Discussion

Rice harvest index is an environment-sensitive quantitative trait

Although HI is considered as an important trait closely related to yield in high-yielding rice breeding (Khush, 1995; Yuan, 2001), its genetics is poorly studied compared to other yield-related traits. To better understand the genetic basis of HI in rice, we evaluated the HI of the RIL population and mapped the QTLs for HI in two-year experiments. Our results showed that the HI exhibited continuous and normal distribution in the RIL population, and a total of five QTLs for HI were detected. These results suggest that HI is a quantitative trait controlled by multiple genes (QTLs). Although a good correlation was observed between the HI measured in 2010 and 2011 in the RIL population (r = 0.6031, p < 0.0001), wider variation of HI among RI lines and more QTLs for HI were detected in 2011compared to that in 2010 (Fig. 1 and Table 2). These differences are mainly attributed to different environments in 2010 and 2011, suggesting that the expression of QTL for HI is affected by the environment. Furthermore, the HI for most of the RI lines fall within the range of their parental values; however, some of RI lines (17.2% in 2011) had lower HI than the low HI parent SBSM, and a few lines (3.2% in 2010 and 1.6% in 2011) had higher HI than the high HI parent YXZ. The recombination of the genetic factors associated with HI in the two parents is one of the main reasons causing transgressive segregation. Even though SBSM has a low HI, there may be undetected alleles with smaller phenotypic effects that are responsible for its HI. Thus, it is possible that some lines carry more favorable combinations of alleles from their parents and have higher HI compared to the high HI parent YXZ, while some lines carry less favorable combinations of alleles and have lower HI compared to the low HI parent SBSM. Transgressive segregation was also observed in the previous studies on yield and its component traits in rice (Hittalmani et al. 2003; Zhang et al. 2004; Liu et al. 2010). Since HI is controlled by multiple genes and the gene expression is affected by the environments, it is necessary to identify the major-effect and stably expressed QTL in order to achieve successful MAS of HI in rice breeding.

Both harvest index and biomass contribute to grain yield

In the present study, the mapping results showed that the three QTLs for GY, *qGY-8* co-located with the major QTL for HI, *qHI-8* on chromosome 8, the other two tightly linked QTLs, *qGY-3a* and *qGY-3b*, co-located with the two tightly linked QTLs for BM, *qBM-3a* and *qBM-3b* on chromosome 3. The co-located QTLs in each of the chromosomal regions produced their additive effects in the same direction, which was in agreement with significantly positive correlation between GY and HI (r = 0.4084, p < 0.0001), and between

GY and BM (r = 0.8859, p < 0.0001) in the RIL population (Additional file 2: Figure S2). The results suggest that grain yield increase can be achieved by either harvest index improvement or biomass enhancement. That equal attention should be given to harvest index and biomass in cereal crop breeding was proposed by Donald and Hamblin in 1976. It is common attempt to increase rice productivity by enhancing both harvest index and biomass in super-high yield rice breeding programs (Peng et al. 1999; Ying et al. 1998). However, as both harvest index and biomass are quantitative traits, it is difficult and unreliable to select the two traits based on phenotypes. With knowledge of the genetic basis for grain yield, harvest index and biomass, it is possible or a good strategy to achieve higher yield through enhancing harvest index and biomass by pyramiding the favorable alleles for the two traits via MAS.

The *qHI-8* is a novel QTL for HI and provides a promising target in molecular rice breeding for high HI

In total, five QTLs for HI from YXZ, *qHI-1*, *qHI-3*, *qHI-7*, *qHI-8* and *qHI-12* were detected in the two-year experiments in the present study. Compared with the chromosomal locations of the QTLs for HI identified in previous studies, *qHI-1* and *qHI-8* are reported for the first time in the present study. Our results also showed that different sets of QTLs for HI were detected in the experiments in 2010 and 2011 (Table 2). Similarly, Hittalmani et al. (2003) detected a total of eleven QTLs for HI in rice at nine locations of Asia representing different environments. Among these QTLs, seven were detected in two to four locations, the others were only detected in one location. Through association mapping, Li et al. (2012) identified seven and two SSR markers significantly associated with HI in Arkansas and Texas, USA, respectively, but no common associated marker was found at two locations. Together with our results, this suggests that most of the QTLs for HI in rice are environment-specific. It is notable that the *qHI-8* identified in the present study could be consistently detected in the two-year experiments representing two different environments, with a high LOD score of 25.1 and 29.2, and explaining 42.8% and 44.5% of the phenotypic variation in 2010 and 2011, respectively. Through the evaluation of HI of the contrasting NILs with and without *qHI-8*, we were able to confirm the existence of *qHI-8* and estimate its effect on HI. Furthermore, through substitution mapping, *qHI-8* was delimited to a 1070 kb region flanked by RM502 and RM6845 on chromosome 8 (Fig. 3). The stable expression and large effect of *qHI-8* on HI makes it to be a promising target in molecular breeding for high HI in rice.

The function of *qHI-8* on HI may be associated with its promotion in sink formation or transportation of assimilation product

HI in rice reflects the capacity of photosynthate translocation from leaves to panicles during the grain filling period. Therefore, it is related to the allocation of assimilation products and the development of grain yield related traits. To have a higher HI, the rice plant should have a larger sink and higher efficiency in transporting photosynthate to the grains. Therefore, the potential role of the gene underlying the QTLs for HI may be related to one or more of these traits or processes. We recognized that several QTLs related to sink formation and transportation of assimilation product were mapped on the same chromosomal region as the *qHI-8* identified in the present study. These include the QTLs controlling number of primary rachis branches (Nagata et al. 2002; Xu et al. 2004) and number of vascular bundles (Cui et al. 2003; Zhang et al. 2002). Furthermore, *OsSPL16*, a functional gene of the QTL for grain width (Wang et al. 2012) also resides in the region of *qHI-8*. These QTLs/genes located in the same chromosomal region may be the same QTL/gene with pleiotropic effects, crosstalk in the metabolic pathways of sink formation and assimilate transportation or may be the clustering of unrelated genes at this locus. To elucidate if the role of *qHI-8* on increasing HI in rice is due to its positive effect on these traits, we are pursuing fine mapping and cloning of the genes underlying *qHI-8*.

Conclusion

In the present study, a RIL population derived from a cross betweenYXZ, a popular *indica* cultivar with good reputation of high HI, and SBSM, an *indica* cultivar with lower HI, was used for genetic analysis of HI and its component traits. Our two-year evaluation of HI of the RI lines suggests that the HI in rice is a quantitative trait and affected by the environment. In total, five QTLs for HI, three QTLs for GY, and six QTLs for BM were detected. The detected GY QTLs overlapped with the HI QTL and the BM QTL, suggesting the positive relationships between GY and HI and between GY and BM, respectively. With knowledge of the genetic basis for grain yield, harvest index and biomass, it is possible to achieve higher yield through enhancing HI and BM by pyramiding the favorable alleles for the two traits via MAS. Importantly, the major QTL for HI, *qHI-8*, is reported for the first time in the present study. This QTL explained more than 40% of the phenotypic variation and stably expressed in the two-year experiments. The existence of *qHI-8* was confirmed using NILs, and this QTL was delimitated to a 1070 kb interval flanked by RM502 and RM6845. We believe that the identification of the stably expressed and large-effect *qHI-8* provides a promising target for further genetic characterization of HI and MAS of high HI in rice breeding.

Methods

Plant materials

YXZ (high HI) was crossed with SBSM (low HI) and advanced from F_2 to F_8 by single seed descent. The resulting RIL population consisting of 186 RI lines was used for QTL analysis in the present study. NILs derived from a residual heterozygous line (RHL15) were used to validate and estimate the relative phenotypic effect of the major QTL for HI, *qHI-8*.

Trait measurements

The RIL population and their parents were planted in the paddy field equipped with bird-net at the experimental base of Guangdong Academy of Agricultural Sciences in Guangdong, China, in the second cropping season in 2010 and 2011. The experiments were arranged in a randomized complete block design with three replicates. Seeds were sown in the seedling nursery and twelve 20-day-old seedlings of each line were transplanted into one row in the field with a space of 20 cm × 20 cm. The field management, including irrigation, fertilization, and disease and pest control, followed the conventional practice in rice production. At complete maturity, ten plants in the middle of each row were harvested individually at ground level. After drying to constant weight, grain yield (GY; filled grain weight per plant, g) and biomass (BM; filled grains, unfilled grains and straw, g) were measured. HI was calculated as the ratio of GY to BM. Trait measurements averaged over the three replicates in each year were used as the data in QTL analysis.

To validate the existence of the major QTL for HI, *qHI-8*, and estimate its phenotypic effect, HI of the 10 homozygous recombinants and two sets of contrasting NILs, NIL15-YXZ and NIL15-SBSM, were assessed at the same experimental base in the first cropping season in 2014. The experiments were arranged in a randomized complete block design with three replicates. Seeds were sown in the seedling nursery and ten 35-day-old seedlings of each line were transplanted into one row in the field with a space of 20 cm × 20 cm. Eight plants in the middle of each row were harvested individually to evaluate HI as described above.

DNA extraction and marker analysis

Rice genomic DNA from young seedling was extracted using a modified method described by Dellaporta et al. (1983). For genotyping, simple sequence repeat (SSR) markers evenly distributed over the 12 chromosomes Temnykh et al. 2001; McCouch et al. 2002; International Rice Genome Sequencing Project 2005) and InDel markers designed based on the insertion-deletion polymorphism between YXZ and SBSM were used. PCR reaction was performed in a total volume of 15 ul, containing 15 ng genomic DNA, 0.25 uM each of SSR or InDel primer, 200 uM dNTP mix, 1× PCR buffer, and 0.5 U of Taq polymerase (TaKaRa, Japan) using a Thermal Cycler (DNA Engine, Bio-Rad Laboratories, Inc., USA). The PCR protocol consisted of an initial denaturation at 94 °C for 4 min, followed by 35 cycles of denaturation at 94 °C for 30 s, annealing at 55 °C (50 °C or 61 °C depending on primer sequence) for 30 s, extension at 72 °C for 1 min, and a final extension at 72 °C for 8 min. The PCR products were separated on nondenaturing 8% polyacrylamide gels and stained with Goldview. The gels were documented by a molecular imager (Gel Doc XR, Bio-Rad Laboratories, Inc., USA) for band scoring.

Construction of linkage map

The linkage map based on the RIL population of 186 lines was constructed using Mapmaker/Exp. 3.0 (Lincoln et al. 1993) with the Kosambi function. A genetic map consisting of 142 markers, covering 12 rice chromosomes and spanning 1427 cM (with the average distance between two markers of 10 cM) was used for QTL analysis.

Data analysis

The *t*-test, correlation analysis and Duncan's multiple range test were conducted using SAS program (SAS Institute, 2000).

Composite interval mapping (CIM) was conducted to map QTL by using Windows QTL Cartographer 2.5 (Wang et al. 2007). A LOD threshold of 3.0 was used to declare main-effect QTL. The 1-LOD confidence intervals for the positions of the QTL were also defined based on the CIM results. Digenic epistatic interactions were analyzed using QTLNetwork-2.0 (Yang et al. 2005). The nomenclature of QTL was as described by McCouch et al. (1997).

Abbreviations

BM: Biomass; cM: centi-Morgan; GY: Grain yield; HI: Harvest index; InDel: Insertion-deletion; MAS: Marker-assisted selection; NIL: Near isogenic line; QTLs: Quantitative trait loci; RHL: Residual heterozygous line; RIL: Recombinant inbred line; SBSM: Shengbasimiao; SSR: Simple sequence repeat; YXZ: Yuexiangzhan

Acknowledgments

We appreciate Dr. Hei Leung at the International Rice Research Institute for critical review on this manuscript. This study was partially supported by Guangdong Provincial Natural Science Foundation (9251064001000001), Guangdong Provincial Scientific and Technological Plan (2016A020210111, 2015B020231002) and the team project of Guangdong Agricultural Department (2016LM2148).

Authors' contributions

SHZ, BL, XYH and YPL conceived and designed the experiment. SHZ conducted the experiments, performed data analysis and wrote the manuscript. XYH, JLZ, YSC, ZMX, YHC, TFY, JFD, XFW, QL, WL, XXM, HF and ZMC participated in material development, molecular maker analysis and measurement of harvest index. BL drafted the proposal and corrected the manuscript. All authors read and approved the final manuscript.

Competing interests

The authors declare that they have no competing interests.

Author details

[1]Rice Research Institute, Guangdong Academy of Agricultural Sciences, Guangzhou 510640, China. [2]Guangdong Key Laboratory of New Technology in Rice Breeding, Guangzhou 510640, China. [3]Rice Research Institute, Hunan Agricultural University, Changsha 410128, China. [4]Present address: Yahua Seed Science Research Institute, Longping High-tech, Changsha 410116, China.

References

Chen Z, Liao Y, Chen S, He X, Chen Y (1999) Yuexiangzhan: a new indica rice variety with high harvest index and good grain quality. Chinese J Rice Sci 13:61

Cui K, Peng S, Xing Y, Yu S, Xu C, Zhang Q (2003) Molecular dissection of the genetic relationships of source, sink and transport tissue with yield traits in rice. Theor Appl Genet 106:649–658

Dellaporta SL, Wood J, Hicks JB (1983) A plant DNA minipreparation: version II. Plant Mol Biol Rep 1:19–21

Donald CM, Hamblin J (1976) The biological yield and harvest index of cereals as agronomic and plant breeding criteria. Adv Agron 28:361–405

Hittalmani S, Huang N, Courtois B, Venuprasad R, Shashidhar HE, Zhuang JY, Zheng KL, Liu GF, Wang GC, Sidhu JS, Srivantaneeyakul S, Singh VP, Bagali PG, Prasanna HC, McLaren G, Khush GS (2003) Identification of QTL for growth-and grain yield-related traits in rice across nine locations of Asia. Theor Appl Genet 107:679–690

International Rice Genome Sequencing Project (2005) The map-based sequence of the rice genome. Nature 436:793–800.

Khush GS (1995) Breaking the yield frontier of rice. GeoJournal 35:329–332

Khush GS (2001) Green revolution: the way forward. Nat Rev Genet 2:815–822

Khush GS (2013) Strategies for increasing the yield potential of cereals: case of rice as an example. Plant Breed 132:433–436

Li X, Yan W, Agrama H, Jia L, Jackson A, Moldenhauer K, Yeater K, McClung A, Wu D (2012) Unraveling the complex trait of harvest index with association mapping in rice (*Oryza sativa* L.) PLoS One 7:e29350. https://doi.org/10.1371/journal.pone.0029350

Lincoln SE, Daly MJ, Lander ES (1993) Mapmaker/Exp. 3.0 and Mapmaker/QTL 1.1, tutorial and reference manual. Whitehead Institute, 9 Cambridge Center, Cambridge MA 02142, USA

Liu T, Zhang Y, Xue W, Xu C, Li X, Xing Y (2010) Comparison of quantitative trait loci for 1,000-grain weight and spikelets per panicle across three connected rice populations. Euphytica 175:383–394

Mann CC (1999) Crop scientists seek a new revolution. Science 283:310–314

Marri PR, Sarla N, Reddy LV, Siddiq EA (2005) Identification and mapping of yield and yield related QTLs from an Indian accession of Oryza rufipogon. BMC Genet 6:33

McCouch S, Cho Y, Yano M, Paul E, Blinstrub M, Morishima H, Kinoshita T (1997) Report on QTL nomenclature. Rice Genet Newslett 14:11–13

McCouch S, Teytelman L, Xu Y, Lobos K, Clare K, Walton M, Fu B, Maghirang R, Li Z, Xing Y, Zhang Q, Kono I, Yano M, Fjellstrom R, DeClerck G, Schneider D, Cartinhour S, Ware D, Stein L (2002) Development and mapping of 2240 new SSR markers for rice (Oryza sativa L.). DNA Res 9:199–207.

Nagata K, Fukuta Y, Shimizu H, Yagi T, Terao T (2002) Quantitative trait loci for sink size and ripening traits in rice (*Oryza sativa* L.) Breeding Sci 52:259–273

Peng S, Cassman KG, Virmani SS, Sheehy J, Khush GS (1999) Yield potential trends of tropical rice since the release of IR8 and the challenge of increasing rice yield potential. Crop Sci 39:1552–1559

Sabouri H, Sabouri A, Dadras AR (2009) Genetic dissection of biomass production, harvest index and panicle characteristics in indica-indica crosses of Iranian rice (Oryza sativa L.) cultivars. Aust J Crop Sci 3:155–166

SAS Institute (2000) SAS/STAT user's guide, version 8.1. SAS Institute Inc, Cary, North Carolina, USA

Sinclair TR (1998) Historical changes in harvest index and crop nitrogen accumulation. Crop Sci 38:638–643

Temnykh S, DeClerck G, Lukashova A, Lipovich L, Cartinhour S, McCouch S (2001) Computational and experimental analysis of microsatellites in rice (Oryza sativa L.): frequency, length variation, transposon associations, and genetic marker potential. Genome Res 11:1441–1452.

Wang S, Basten C, Zeng Z (2007) Windows QTL Cartographer 2.5. Department of Statistics, North Carolina State University, Raleigh

Wang S, Wu K, Yuan Q, Liu X, Liu Z, Lin X, Zeng R, Zhu H, Dong G, Qian Q, Zhang G, Fu X (2012) Control of grain size, shape and quality by *OsSPL16* in rice. Nat Genet 44:950–954

Xu JL, Yu SB, Luo LJ, Zhong DB, Mei HW, Li ZK (2004) Molecular dissection of the primary sink size and its related traits in rice. Plant Breed 123:43–50

Yang J, Hu C, Ye X, Zhu J (2005) QTLNetwork-2.0. Institute of Bioinformatics, Zhejiang University, Hangzhou, China (http://ibi.zju.edu.cn/software/qtlnetwork/)

Ying J, Peng S, He Q, Yang H, Yang C, Visperas RM, Cassman KG (1998) Comparison of high-yield rice in tropical and subtropical environments. I. Determinants of grain and dry matter yields. Field Crops Res 57:71–84

Yuan L (2001) Breeding of super hybrid rice. In: Peng S and Hardy B (eds) Rice Research for Food Security and Poverty Alleviation. International Rice Research Institute, Los Baños, Philippines, pp143–149

Zhang ZH, Li P, Wang LX, Hu ZL, Zhu LH, Zhu YG (2004) Genetic dissection of the relationships of biomass production and partitioning with yield and yield related traits in rice. Plant Sci 167:1–8

Zhang ZH, Li P, Wang LX, Tan CJ, Hu ZL, Zhu YG, Zhu LH (2002) Identification of quantitative trait loci (QTLs) for the characters of vascular bundles in peduncle related to *indica-japonica* differentiation in rice (*Oryza sativa* L.) Euphytica 128:279–284

13

Genetic Evidence for the Role of a Rice Vacuolar Invertase as a Molecular Sink Strength Determinant

Shamitha Rao Morey[1], Tatsuro Hirose[1,2], Yoichi Hashida[1,4], Akio Miyao[3], Hirohiko Hirochika[3], Ryu Ohsugi[1], Junko Yamagishi[1] and Naohiro Aoki[1*]

Abstract

Background: Rice is a major crop feeding the majority of the global population, and increasing its sink strength is one of the modes to alleviate the declining availability of food for the rapidly growing world population. We demonstrate a role for an important rice vacuolar invertase isoform, OsINV3, in sink strength determination.

Results: *OsINV3* mutants showed shorter panicles with lighter and smaller grains, owing to a smaller cell size on the outer and inner surfaces of the palea and lemma as observed by scanning electron microscopy. Further, strong *promoter::GUS* expression was observed in the palea, lemma and the rachis branches in the young elongating panicles, which supported the role of *OsINV3* in cell expansion and thus, in spikelet size and panicle length determination. Size of the spikelet was found to directly influence the grain weight, which was confirmed by the lack of differences in weights of hulled grain for differently segregated alleles in the heterozygous lines. Assessment of field grown mutants not only revealed a drastic reduction in the percentage of ripened grain, 1000-grain weight and final yield, but also significantly reduced partitioning of assimilates to the panicles, whereby the total dry weight remained unaffected. Determination of the non-structural carbohydrate contents revealed a lower hexose-to-sucrose ratio in the panicles of the mutants from panicle initiation to 10 days after heading, a stage that identifies as the critical pre-storage phase of grain filling, whereas the starch contents were not affected. In addition, strong *promoter::GUS* expression was observed in the dorsal end of ovary during the pre-storage phase until 6 days after flowering, highlighting a function for OsINV3 in monitoring the initial grain filling stage.

Conclusions: OsINV3 was found to regulate spikelet size by playing a key role in cell expansion, driving the movement of assimilates for grain filling by modulating the hexose-to-sucrose ratio, contributing in grain weight determination and thus, the grain yield.

Keywords: Vacuolar invertase, Grain size, Cell expansion, Hexose-to-sucrose ratio, Grain weight, Sink strength, Assimilate partitioning

Background

Carbon that is assimilated by plants through photosynthesis is partitioned among various tissues based on their competitive ability to import assimilates. This ability is commonly termed as sink strength and has been previously described to be dependent on physical as well as physiological capacity of the sinks (Ho 1988; Marcelis 1996; Herbers and Sonnewald 1998; Bihmidine et al. 2013). While sink size determines the physical capacity, sink activity involving assimilate transport, utilization and storage by the sink cells, determines the physiological capacity (Ho 1988; Herbers and Sonnewald 1998). An increase in sink strength directly translates to an increase in yield, mainly due to higher assimilate partitioning into the sinks co-ordinated by an increased sink size. With respect to rice (*Oryza sativa* L.), an important cereal crop that feeds a majority of the global population, there is a pressing need to increase the sink strength in order to

* Correspondence: aaokin@mail.ecc.u-tokyo.ac.jp
[1]Graduate School of Agricultural and Life Sciences, The University of Tokyo, 1-1-1 Yayoi, Bunkyo-ku, Tokyo 113-8657, Japan
Full list of author information is available at the end of the article

achieve global food security and enable sustenance by ensuring food for the rapidly growing population.

Cleavage of sucrose, a major photoassimilate in plants, plays an important role in carbon allocation, not only by osmotically controlling the flux, but also by generating hexose-based signals to regulate genes involved in carbon metabolism in the sink organs (Herbers and Sonnewald 1998; Smeekens 2000; Koch 2004). The enzymes that thus, catalyze sucrose cleavage are considered to be potential molecular contributors to sink strength (Herbers and Sonnewald 1998). One such class of enzymes are the invertases (EC 3.2.1.26) that catalyze the irreversible cleavage of sucrose into glucose and fructose. Based on their sub-cellular localization, solubility and optimum pH of function, they are classified into insoluble acid (cell wall), soluble acid (vacuolar) and soluble neutral (cytosolic) invertases. The entry of sucrose into different utilization pathways is regulated by various isoforms of invertases (Sturm 1999).

In rice, eight neutral invertases (NINs), nine cell wall invertases (CWINs) and two vacuolar invertases (VINs) have been identified (Ji et al. 2005). Till date, CWINs have been found to play a primary role in assimilate partitioning, thus regulating the grain weight in crops including rice (Cheng et al. 1996; Hirose et al. 2002; Wang et al. 2008; Tang et al. 2017; Li et al. 2013). For example, two rice CWIN isogenes, *OsCIN1* and *OsCIN2*, play a role in determination of sink strength by regulation of assimilate partitioning, where *OsCIN1* was found to play a role during the early grain filling stages, and *OsCIN2* in the regulation of grain size and weights (Hirose et al. 2002; Wang et al. 2008).

In the past, vacuolar invertases (VINs) have been attributed to roles in cell elongation of seedling hypocotyls in Arabidopsis (Sergeeva et al. 2006), fiber cell elongation in cotton (Wang et al. 2010), and rapidly expanding tissues in carrot taproot (Tang et al. 1999) and sugar beet petioles (González et al. 2005). It has been suggested that VIN regulates the sink size by driving cell expansion, owing to a turgor generated by influx of osmotic solutes in response to an increase in hexoses (Koch 2004; Sergeeva et al. 2006). It has also been considered that VINs drive cell expansion based on availability of carbohydrates (Koch 2004). Further, role of VINs in determination of hexose-to-sucrose ratio in potato (Zrenner et al. 1996) and tomato (Klann et al. 1996) suggests the importance of sink sugar composition in regulation of the sink size (Klann et al. 1996).

Roles of CWINs as contributors to sink strength, exhibiting a role in phloem unloading, have been extensively studied (Miller and Chourey 1992; Weber et al. 1995), and a CWIN *OsCIN2 (GIF1)* has been previously reported to play a key role in grain filling by affecting sucrose import by sinks (Wang et al. 2008). However, roles of vacuolar invertases in terms of sucrose import by sinks has not been a research focus in recent times. The present study examines the physiological role of a VIN isogene, *OsINV3*, in rice, in terms of sink size and assimilate partitioning, in concert with analysis of cell size and sink sugar composition, thus, establishing evidence for its role as a molecular sink strength determinant.

Results

Isolation of the Gene-Disruption Mutant of *OsINV3* and Phenotypic Analysis with Respect to Panicle Height, Grain Size and Grain Weight

Following screening of approximately 40,000 lines with *Tos17* retro-transposon insertions, a mutant line NG6441 bearing a *Tos17* insertion within the *OsINV3* gene was isolated. Sequence analysis revealed that the *Tos17* retrotransposon in NG6411 was introduced in the second exon (Fig. 1a). The gene-disruption mutants segregated from NG6441 and the WT were grown under controlled conditions to assess major phenotypic differences. Differences in the observed traits were reconfirmed through complementation. The size of the amplicon indicating the allele carrying WT gene was 350 bp, and for that of mutant was 315 bp. Genotyping of the WT, mutant and the complementation lines C3, C4 and C13 confirmed a homozygous insertion of the *Tos17* retrotransposon in the second exon of the *OsINV3* gene in the mutants, and subsequent successful incorporation of *OsINV3* gene in the mutant lines, enabling an effective complementation (Fig. 1c). The mRNA transcript levels revealed through real-time PCR showed an overexpression in the C4 line (Fig. 1d). The presence of aberrant transcript in the mutants was reconfirmed by semi-quantitative PCR using primers F5 and R5 spanning the *Tos17* insertion, that showed an absence of amplicon in the mutants (Additional file 1: Figure S1A), while primers F6 and R6 targeting the region downstream of the *Tos17* insertion showed a presence of amplicon, however, lighter in intensity (Additional file 1: Figure S1B), coinciding with the real-time data (Fig. 1d). The enzyme activity in the mature leaves of the mutant lines was lower by 28.8% relative to the WT, suggesting a reduction in soluble acid invertase activity in the mutants, with observed recovery in the C4 line (Fig. 1e). It was observed that the panicles of the mutants were shorter than that of the WT, with displayed recovery of the trait upon complementation (Fig. 1f). Grain length and width of unhulled grain was also smaller in the mutant when compared with the WT, with significant recovery observed for both the traits in C4 (Fig. 1g). Similar results were obtained for the area of both filled and unfilled spikelets, where smaller area of the mutants was significantly recovered in the C4 lines (Fig. 1h). In addition to smaller grain size and panicle height, average single grain weight was also lower in the mutant lines, displaying an absolute recovery for the trait in C4 (Fig. 1i). These results indicate a significant role for

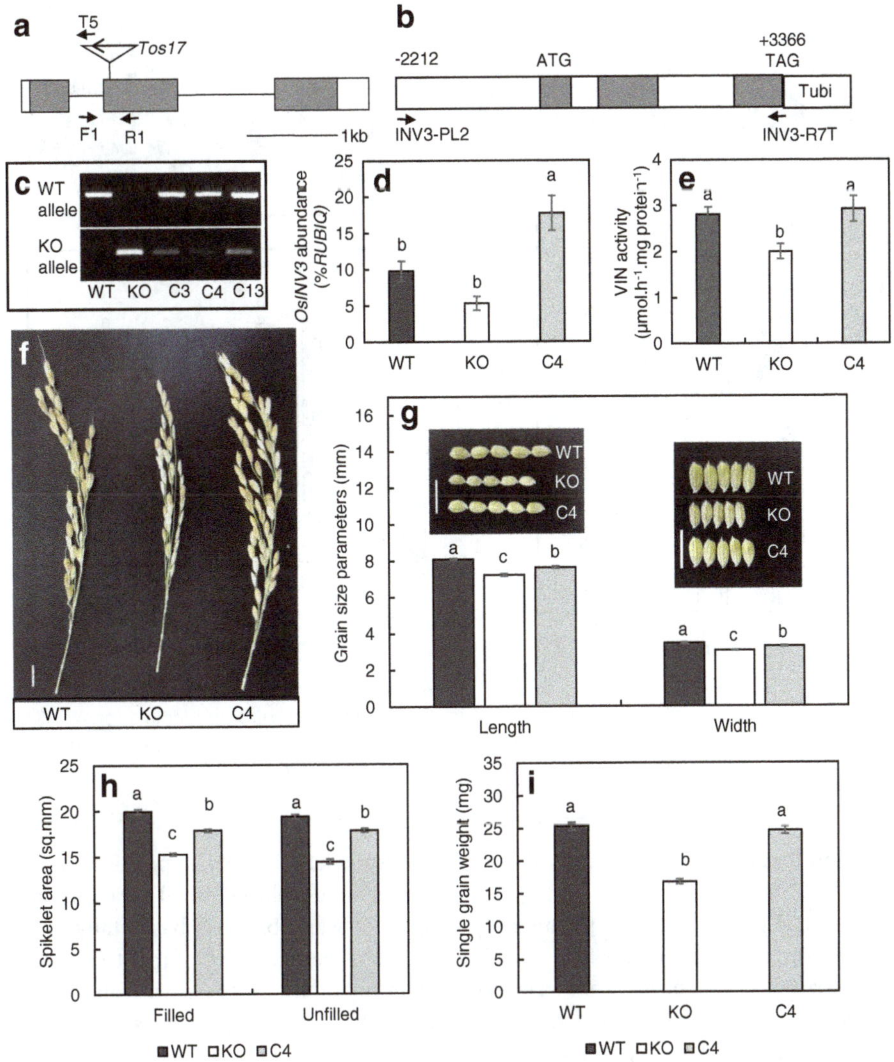

Fig. 1 Characterization and phenotypic analysis of the mutants (KO). **a** *OsINV3* gene in 5'-3' direction with the *Tos17* insertion in the second exon in reverse orientation. Grey boxes denote exons separated by lines that denote introns, white boxes indicate 5'(left) and 3'(right) UTR, and arrows indicate primers F1, R1 and T5, **b** The construct used for complementation test containing the ORF of *OsINV3* along with 2212 bp upstream of the transcription start site, grey boxes indicating the exons, and arrows indicate primers INV3-PL2 and INV3-R7T, **c** Gel data showing the presence of WT and KO alleles in WT, KO and complement lines, **d** mRNA transcript level of *OsINV3* relative to *RUBIQ* ($n = 3$) and **e** Vacuolar invertase activity expressed as Vmax ($n = 5$), in mature leaf of WT, KO and C4, **f** Panicles from WT, KO and C4 lines, **g** Grain length and width parameters for WT, KO and C4 ($n = 12$), **h** Spikelet area for filled and unfilled spikelets from WT, KO and C4 lines ($n = 12$), **i** average weight per grain, for WT, KO and C4 lines ($n =$ whole grain set from 3 plants). Data represent the mean ± SE. Same letters indicate insignificant differences between the lines tested using ANOVA (Tukey's test) with significance level of $p < 0.05$. Vertical white bars indicate 1 cm

OsINV3 in the regulation of panicle height, grain size and grain weight.

In order to isolate the role of *OsINV3* in spikelet cell expansion, spikelets at 2 days before heading (DBH) from WT, mutant and the C4 lines were analyzed using scanning electron microscopy (SEM). The cell density in mutant on the outer surface of both palea and lemma were significantly higher than the WT, with observed recovery in the C4 line (Fig. 2a–c). The cells were stacked closer in the mutants when compared to the WT and the C4 lines. For an observed 23.5% decrease in spikelet area (Fig. 1h), a 24.3% increase in the cell density (Fig. 2c) was established suggesting a possible lack of difference in cell number. However, cell size on the outer surface of the spikelets of mutant was considerably smaller (Fig. 2a and b). The epidermal cells on the inner surface of palea and lemma were analyzed, with decreased cell area, cell width and cell length in the mutant, and recovery for these traits in the C4 lines (Fig. 2d and e). Although absolute recovery in the cell length, and cell area in the lemma of C4 was observed; cell width and cell area in the palea, although significant, were only partially recovered. Thus, the role of *OsINV3* in cell expansion in regulation of spikelet size was established.

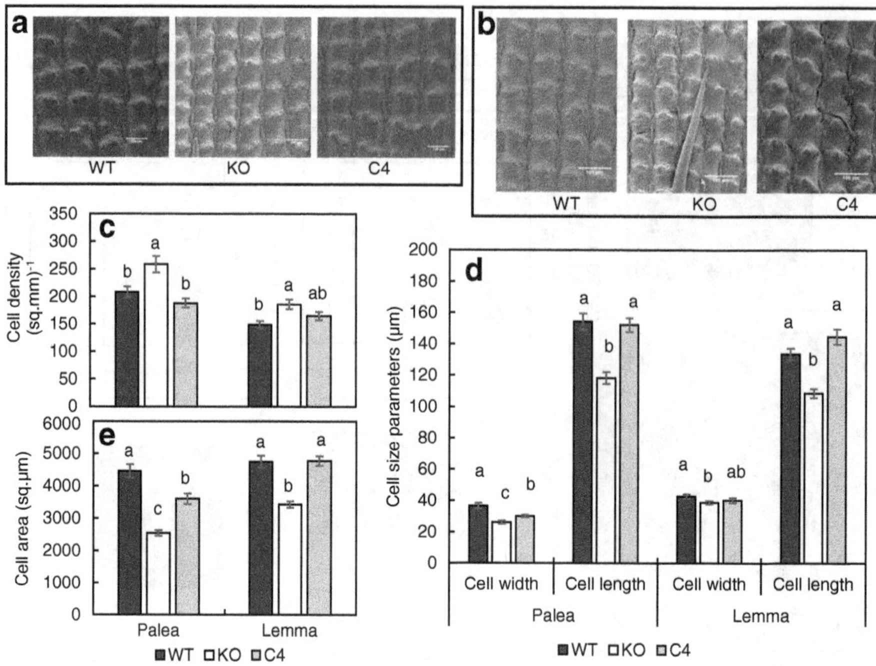

Fig. 2 Spikelet cell number and size analysis using SEM. Cell density for a fixed area in the outer surface of **a** palea and **b** lemma for WT, mutant (KO) and C4. **c** Cell density expressed as number of cells/sq.mm of the outer surface of palea and lemma for WT, KO and C4. **d** Cell size parameters, depicting cell width and cell length and **e** Cell area in sq.µm, for the inner surface of palea and lemma. Data represent the mean ± SE (n = 9). Same letters indicate insignificant differences between the lines tested using ANOVA (Tukey's test) with significance level of $p < 0.05$. Horizontal bars represent 100 µm

Grain Weight Distribution for Segregated Alleles in the Heterozygous Lines

To explore whether the small grain phenotype is caused by disruption of *OsINV3* in the maternal or the filial tissues, seeds borne on the heterozygous (+/−) plants were analyzed, and corresponding weights of hulled grain were determined for the segregated alleles, namely INV3 +/−, INV3 +/+ and INV3 −/−. A lower grain weight as observed in the mutant (−/−) plant was not the case in INV3 −/− grain borne on the heterozygous lines (Table 1). Lack of differences in the grain weight of seeds on INV3 +/− lines irrespective of the genotype indicates that the small grain phenotype of the mutant (−/−) plant is not solely caused by the filial factor (highlighting a maternal influence on grain weight).

Expression of *OsINV3* During Various Stages of Panicle Development

OsINV3 promoter:GUS assay for panicles at progressing stages of panicle development were studied to isolate the temporal and spatial expression pattern in the panicles. Early panicle development stages of young panicles, 0.5 cm – 1 cm in length, showed no GUS activity except for in the bract hair (Additional file 1: Figure S2A). Expression of *OsINV3* was first observed in panicles ~ 3 cm in length, in the rudimentary glume, pedicel and the empty glumes in the spikelets in upper position (Fig. 3a and a'), following a similar expression pattern in the spikelets in lower position along the progressing stages of development. The expression in rachilla and spikelet was observed basipetally along the progressing stages of panicle development (Fig. 3b-f, 3b'-f'). A strong staining in the anthers was also observed in all stages (Fig. 3b'-f'), showing expression in the epidermal cells as

Table 1 Distribution of grain weights for the WT and mutant (KO) alleles segregated in the heterozygous lines

Genotype of plant	Plant Number	Genotype of seeds	N	Single grain weight (mg)
Heterozygous (+/−)	1	INV3 +/+	11	22.0 ± 2.3
		INV3 +/−	15	22.9 ± 2.7
		INV3 −/−	6	21.6 ± 1.3
	2	INV3 +/+	9	22.7 ± 2.0
		INV3 +/−	16	22.4 ± 1.5
		INV3 −/−	7	22.0 ± 2.0
	3	INV3 +/+	8	21.7 ± 2.0
		INV3 +/−	17	21.4 ± 1.7
		INV3 −/−	7	21.3 ± 2.0
WT (+/+)		INV3 +/+	32	22.1 ± 1.6
KO (−/−)		INV3 −/−	32	14.0 ± 1.5

Seeds borne on heterozygous (+/−) plants were individually weighed and genotyped to determine the filial regulation of grain weights. Note that grain weight differences were absent between the segregated alleles on +/− plants, while were found to be significant between +/+ and −/−. Data represent the mean ± standard deviation

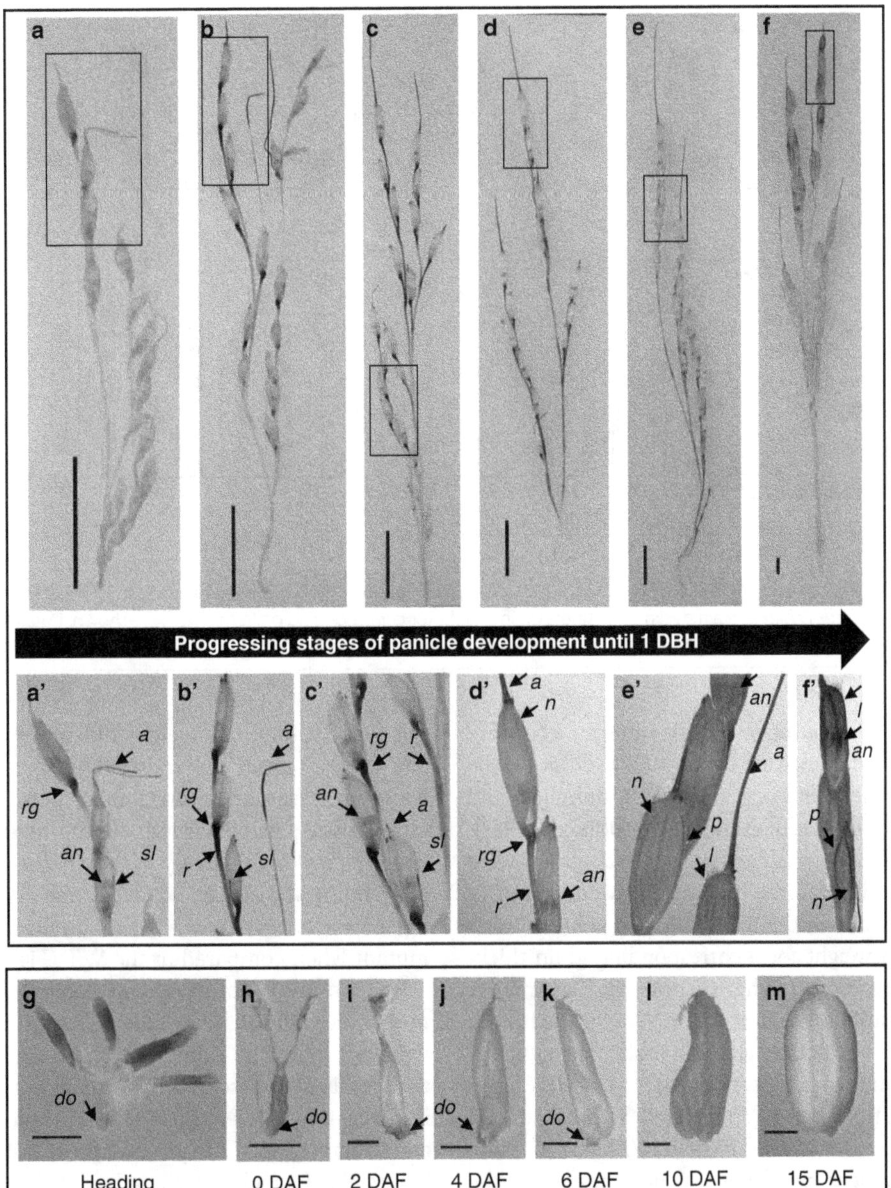

Fig. 3 Spatio-temporal expression of *OsINV3* in the panicle using *promoter::GUS* lines. **a-f** GUS stained panicles at different progressing stages of panicle development until 1 DBH; **a'-f'** magnified images of the area marked in **a-f**. **g-m** *Promoter::GUS* expression in the developing endosperm at heading, 0 DAF, 2 DAF, 4 DAF, 6 DAF, 10 DAF and 15 DAF respectively. Vertical bars indicate 1 cm and horizontal bars indicate 1 mm. Arrows show the areas stained, isolating a vascular trace of *OsINV3* expression in rachis branches (*r*), palea (*p*), lemma (*l*), nerves (*n*), anther (*an*), sterile lemmas (*sl*), rudimentary glume (*rg*), awn (*a*) and the dorsal end of the ovary (*do*)

well as the pollen (Additional file 1: Figure S2D). Strong expression was observed in the dorsal end of the ovary following fertilization until 6 days after flowering (DAF) (Fig. 3g-k). However, no expression in the developing grain was observed following this stage (Fig. 3l and m).

Assessment of Agronomic Traits of the Field Grown Mutants

Field trials for the *OsINV3* WT and the mutant were performed to test their agronomic and growth characteristics. Not only were the grains of the mutant smaller (Fig. 4b), the yield, grain weight and the percentage of ripened grain were also significantly lower in the mutant when compared to the WT, with 17.8%, 30.3% and 47.9% reduction in the percentage of ripened grain, 1000-grain weight and yield respectively (Table 2).

The field grown WT and the mutants for *OsINV3* reconfirmed the phenotype of shorter panicles and smaller grains as observed in controlled conditions. The mutant panicles were significantly shorter (Fig. 4a), however, with

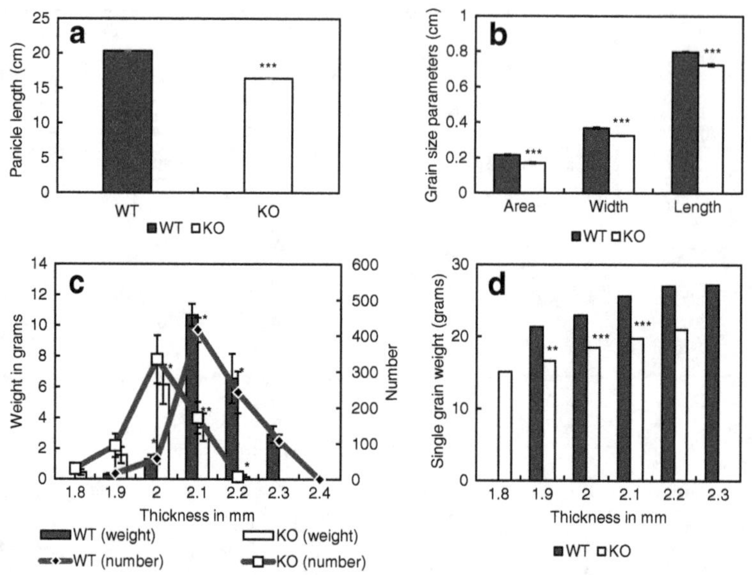

Fig. 4 Panicle and grain size analysis for field grown mutants (KO). **a** Panicle length ($n = 29$), **b** grain size parameters of unhulled grain ($n = 30$), **c** grain number and weight for corresponding grain thickness, **d** single grain weight for corresponding grain thickness are shown. Data represent the mean ± SE ($n = 3$ plants). Asterisks indicate statistical significance of difference using Student's T-statistic with *-$p < 0.05$, **-$p < 0.01$ and ***-$p < 0.001$

no differences in the number of primary or secondary rachis branches when compared to the WT (Additional file 1: Figure S3A). The mutant grains, both unhulled and hulled were smaller in size than the WT (Additional file 1: Figure S3B). The length, width and area of filled grains were all significantly smaller for the mutant when compared to the WT (Fig. 4b). Further, the distribution of grain number (Fig. 4c) and grain weight for a corresponding grain thickness revealed a significantly shorter peak and a smaller area under the distribution curve for the mutant, with the single grain weight for a corresponding grain thickness significantly smaller in the mutants (Fig. 4d).

Dry weight partitioning is considered to be the end result of assimilate partitioning in plants (Marcelis 1996). Hence, total dry weight, per plant was determined for the WT and mutant at different stages from panicle initiation to harvest (Fig. 5a) in order to estimate possible differences in assimilate partitioning. The differences that existed at panicle initiation (with a lower plant dry weight for the mutant), ceased to exist at all other stages. Given, the lower grain weight of the mutant (Table 1) at harvest,

Table 2 Yield and its components for WT and mutant (KO)

	Number of productive panicles	Number of spikelet per panicle	Percentage of ripened grain	Grain weight (g 1000 grains^{-1})	Yield (g m^{-2})
WT	9.3 ± 0.4	112.4 ± 2.1	76.6 ± 2.8	27.7 ± 0.1	491.8 ± 25.2
KO	9.2 ± 0.4	107.8 ± 2.8	63 ± 2.2***	19.3 ± 0.1***	256.2 ± 18.7***
% reduction	n.s.	n.s.	17.75	30.32	47.91

Data represent the mean ± SE ($n = 12$ plants). Asterisks indicate statistical significance of difference using Student's T-statistic with ***-$p < 0.001$. Results of the 2016 field experiment indicated, with similar results obtained in 2015

this lack of total dry weight difference between the WT and the mutant at the grain filling stages indicates differences in assimilate partitioning between panicles and other tissues. As conceived, the dry matter partitioning to panicles was significantly lower in the mutant at all the grain filling stages (Fig. 5c), while the dry matter partitioning to vegetative tissues was significantly higher in the mutant when compared to the WT (Fig. 5b).

Further, the photosynthetic capacity of the WT and mutant at panicle initiation, heading and late ripening were not different (Additional file 1: Figure S4), displaying a maintained source ability. Thus, it could be conceptualized that for a given amount of carbon fixed by the WT and mutant plants, the percentage of assimilate partitioned into sinks (panicles) was indeed lower in the mutants, suggestive of a lower sink ability to import photoassimilates in the vacuolar invertase mutants.

Analyses of Non-Structural Carbohydrate (NSC) Contents During the Period of Panicle Development and Grain Filling

The amounts of NSCs in the whole panicles of field grown plants were estimated at stages from panicle initiation to maturity (Fig. 6). While levels of starch in both the WT and the mutant were not different (Fig. 6h), significant differences in the sugar levels were determined till 20 days after heading (DAH) (Fig. 6a-e), where the mutant showed higher sucrose and lower hexose levels until 10 DAH (Fig. 6a-d). At 20 DAH, sucrose and hexoses were found to be lower in the mutant than in the WT (Fig. 6e). After this stage, no differences between the WT and the mutant were observed until maturity (Fig. 6f and g). The differences in

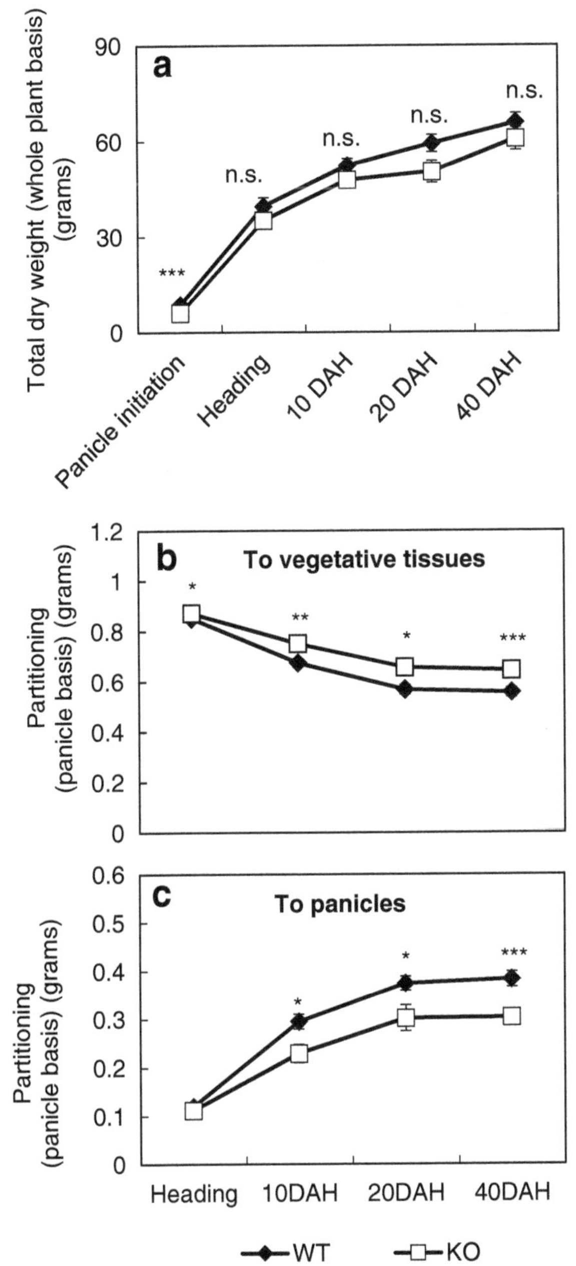

Fig. 5 Dry matter weight differences between the WT and the mutants (KO). **a** Total dry weights measured on whole plant basis, **b** Dry matter partitioning to the vegetative tissues and, **c** panicles, on per panicle basis for *OsINV3* WT and KO. Data represent the mean ± SE ($n = 9$). Asterisks indicate statistical significance of difference using Student's T-statistic with *- $p < 0.05$, **-$p < 0.01$ and ***-$p < 0.001$; and n.s. – not significant. X-axis – representative of stage, and not scaled to number of days

the sugar composition were reflected in the hexose-to-sucrose ratios, where, although the trend was maintained from heading to harvest for both WT and mutants, the ratio was significantly lower for the mutants between heading to 10 DAH, which signify the pre-storage phase of grain filling (Fig. 6i). However, a stark difference in hexose-to-sucrose ratio between the WT and the mutant was observed at panicle initiation stage (Fig. 6i) usually characterized by high hexose levels, owing to a higher sucrose and negligible hexose levels in the mutants (Fig. 6a). The vacuolar invertase activity levels were however, not different between the WT and mutant for young panicles at this stage (Additional file 1: Figure S5A), nor were the neutral (Additional file 1: Figure S5B) and cell wall invertase activities (Additional file 1: Figure S5C). Absence of differences in vacuolar invertase activity between the WT and mutant could be attributed to functional redundancy exhibited by *OsINV2*, the transcript levels of which were significantly higher in the mutants when compared to the WT (Additional file 1: Figure S5D).

Discussion

The present study characterizes one of the isoforms of vacuolar invertase genes, *OsINV3* in rice and determines its physiological roles in terms of sink expansion and partitioning into the sinks.

Key Role of OsINV3 in Determination of Spikelet Size and Its Influence on Grain Weight

The *OsINV3* mutants demonstrated smaller sinks by displaying shorter panicle and smaller grain phenotypes, owing to smaller cell size in both the inner and outer surfaces of palea and lemma of the mutant spikelets as demonstrated by the SEM analysis (Fig. 2). Role of vacuolar invertases in cell expansion by generation of a turgor in the cell has previously been suggested in Arabidopsis roots (Sergeeva et al. 2006), developing petioles and primary roots of carrot (Sturm et al. 1995) and has been widely accepted (Tang et al. 1999; Roitsch and González 2004; González et al. 2005; Ruan et al. 2010). The expression of *OsINV3* in the spikelet as observed in the *promoter::GUS* lines in the palea, lemma and the nerves during the panicle development stages (Fig. 3d-f, 3d'-f') and the early grain filling stages (Additional file 1: Figure S2B), is consistent with the role of *OsINV3* in cell expansion in the spikelet. Size of the spikelet plays a crucial role in determination of final grain weight (Wang et al. 2008; Song et al. 2015), mainly by modifying the grain filling rate, eventually affecting the grain weight (Wang et al. 2008). Absence in individual grain weight differences for grains borne on the heterozygous lines (Table 1), further strengthens evidence for the importance of spikelet size in determination of grain weight.

OsINV3 Regulates Assimilate Partitioning Into the Sinks by Modulating the Hexose-to-Sucrose Ratio

The lower grain yield for the mutants, was primarily owing to the lower percentage of grain ripening and lower grain weight (Table 2). A lower grain ripening could be attributed to failure in normal development of anthers and pollen, and the process of flowering. This was consistent with the

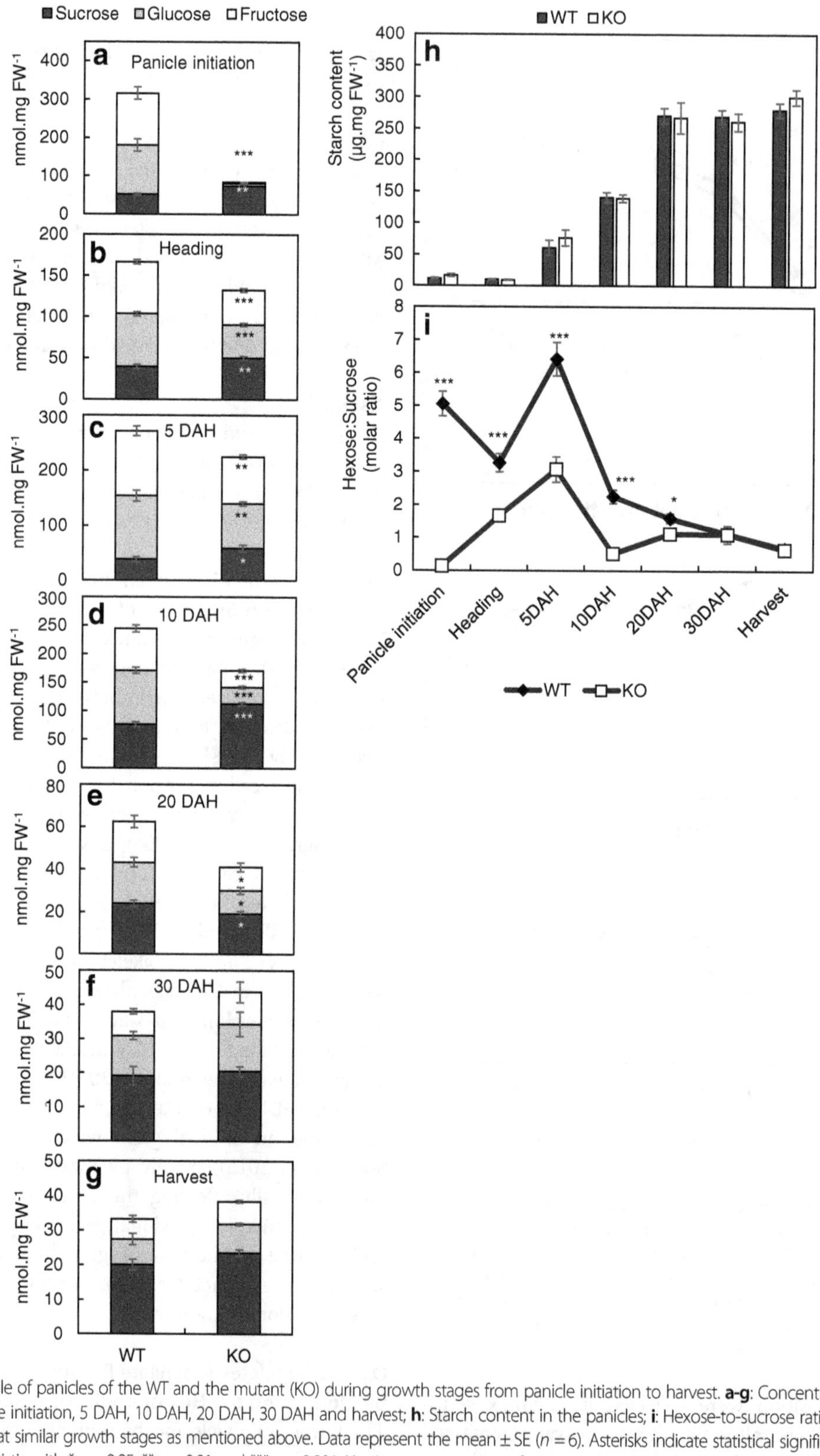

Fig. 6 a-i: NSC profile of panicles of the WT and the mutant (KO) during growth stages from panicle initiation to harvest. **a-g**: Concentration of sugars in the panicle at panicle initiation, 5 DAH, 10 DAH, 20 DAH, 30 DAH and harvest; **h**: Starch content in the panicles; **i**: Hexose-to-sucrose ratio in the panicles of WT and the mutant at similar growth stages as mentioned above. Data represent the mean ± SE ($n = 6$). Asterisks indicate statistical significance of difference using Student's T-statistic with *-$p < 0.05$, **-$p < 0.01$ and ***-$p < 0.001$. X-axis – representative of stage, and not scaled to number of days

observation in *promoter::GUS* lines beginning from the early panicle development stages until after anthesis, where a strong expression of *OsINV3* in anthers (Fig. 3a-f, and Additional file 1: Figure S2D, left) and pollen (Additional file 1:Figure S1D, right), and the lodicules (Additional file 1: Figure S2C) was found. Although this prospect is co-incident with recent findings (Wang and Ruan 2016, Goetz et al. 2017), further studies are needed to delve deeper into specific role of OsINV3 in reproductive structures. While the size restriction of the grain could be stated as one of the reasons for smaller grain weight, a lower assimilate partitioning to grain was also observed (Fig. 5c), indicating an impaired transport of sucrose from source to sink tissues. In rice, sucrose transport from the source to maternal sinks along the phloem is largely believed to be symplasmic, driven by a sucrose gradient owing to the osmotic potential at the unloading site from the sieve element/companion cell complex (Aoki et al. 2012). Modifications in the sugar composition, thus the osmotic potential at the site of unloading could lead to altered capacities of sucrose import by sinks.

Hexose-to-sucrose ratio has long been considered to play a major role in modulating the osmotic gradient, thus, regulating the sink strength during the period of grain filling (Herbers and Sonnewald 1998). This ratio is found to increase during the pre-storage phase, followed by its drastic decrease, characterizing the beginning of the active storage phase of grain filling (Herbers and Sonnewald 1998). In our study, although the ratio from heading to harvest was maintained between the WT and the mutants, the ratio tended to be significantly lower in the mutants during the pre-storage phase of grain filling (Fig. 6i), with an increased sucrose concentration, which diminished the sucrose gradient along the site of unloading, facilitating a lower sucrose influx thus, a lower availability of sucrose for unloading into the filial tissues. The *promoter::GUS* studies revealed a strong *OsINV3* expression at the dorsal ovary immediately after fertilization until 6 DAF (Fig. 3g-m), serving as the critical point for directing sucrose import.

Functional Redundancy of Vacuolar Invertase

Key biochemical functions are redundantly encoded in various isoforms to ensure normal development upon failure of one of the isoforms. Rice consists of two isoforms of vacuolar invertase, *OsINV2* and *OsINV3* (Ji et al. 2005). In the developing panicles of *OsINV3* mutants, although the development of sinks was not defective, they were reduced in size (Fig. 1f and g), owing to the lower ratios of hexose to sucrose in comparison to the WT (Fig. 6i). Lack of differences in the VIN activity between the WT and mutants (Additional file 1: Figure S5A) failed to account for the reduced hexose levels in the young panicles of the mutant (Fig. 6a). This could be attributed to the functional redundancy of VINs, where *OsINV2* was overexpressed under the absence of *OsINV3*, owing to a greater demand of hexoses in the rapidly expanding tissues; which was confirmed by an increase in mRNA transcript levels of *OsINV2* in the tissues of the *OsINV3* mutants (Additional file 1:Figure S5D). Despite this overexpression, *OsINV2* failed to substitute for the key physiological roles of *OsINV3* in maintaining the hexose-to-sucrose ratios (Fig. 6i). We suggest two possible rationale to explain this finding. First, the higher Vmax of 510 nkat mg^{-1} for OsINV2 when compared to 20.5 nkat mg^{-1} for OsINV3 (Ji et al. 2007), outlines a stronger possibility of the former being regulated post-translationally by inhibitors, as reported in CWINs (French et al. 2014 (in rice); Jin et al. 2009 (in tomato)). However, lack of studies on mode of regulation by VIN inhibitors renders it difficult to replicate this in in vitro assays as observed in our study (Additional file 1: Figure S5A). Second, differences in cellular or sub-cellular localization of the two isoforms could explain the inadequate substitution by OsINV2 for physiological roles of OsINV3. However, further studies need to be carried out to establish this proposition.

Conclusion

We demonstrate a key role for a vacuolar invertase isoform, *OsINV3* in sink strength determination during the reproductive phase, mainly by regulation of grain size and assimilate partitioning to grain. *OsINV3* exercises a role in spikelet size determination mainly by regulating cell expansion, and in assimilate partitioning to grain by modulation of hexose-to-sucrose ratios during the initial phase of grain filling. In terms of assimilate partitioning and grain weight determination in cereals, by far, CWINs have been found to be key contributors. Our study attributes a similar role to VIN, mainly enabled by its regulation of spikelet size and hexose-to-sucrose ratios, and a direct influence of spikelet size on grain weights. In terms of achieving global food sustainability, attaining higher yield has been of utmost priority and our findings demonstrate a key candidate for generation of high yielding cultivars through molecular breeding studies.

Methods

Plant Material and Genotyping

To elucidate the physiological function of a vacuolar invertase isoform, *OsINV3*, the gene disruption lines were screened from the population of mutants carrying insertions of the *Tos17* retro-transposon, which were obtained from the National Institute of Agrobiological Sciences, Ibaraki, Japan. Detailed procedure for generation of the insertion mutants are as previously described (Miyao et al. 2003). PCR-screening was conducted using DNA pools prepared with three-dimensional sampling method from approximately 40,000 plants bearing *Tos17* insertion

(Agrawal et al. 2005). Nested PCR was carried out with PrimeStar GXL DNA polymerase (Takara Bio Inc., Shiga, Japan) according to manufacturer's instructions using combinations of primers specific to *Tos17* and *OsINV3* sequences as listed in Additional file 1:Table S1. Positive products were sequenced after gel-purification to identify the location of insertions in the rice genome. An insertion line NG6411 was isolated with a *Tos17* insertion in the second exon of *OsINV3* (Os02g0106100). The WT and mutant were segregated from the same *Tos17* insertion line NG6441 following genotyping using the primers F1 and R1 for WT allele, and F1 and T5 for the mutant allele (Additional file 1:Table S1). Genomic DNA isolation from leaves was performed using methods similar to those previously described (Okamura et al. 2013). One μl of the genomic DNA template was used for PCR with PrimeStar GXL DNA polymerase (Takara Bio Inc.) according to manufacturer's instructions.

Growth Conditions and Grain Size Analyses for Controlled Experiments

Seeds were sown in nursery soil in a plastic tray following chemical sterilization with 2.5% sodium hypochlorite for 30 min, an elaborate wash and imbibition with water at 30 ° C for 3 days. They were placed in a growth chamber at 27 ° C/22 °C, 14 h-light/10 h-dark cycle, 65% relative humidity and a light intensity of 900 $\mu mol.m^{-2}.s^{-1}$. At around 20 days after sowing, seedlings were transplanted to bigger plastic pots and grown under the same conditions.

Grain length, width and area for ten respective grains from each of the three replicate plants from WT, mutant and the T1 complement (C4) lines were determined using a digital microscope (VHX-6000, Keyence, Osaka, Japan) in concert with ImageJ 1.46r (Schneider et al. 2012).

Generation of OsINV3 Complement Lines

The full-length ORF of *OsINV3* along with 2212 bp upstream of the putative translation start site was isolated by PCR using a high-fidelity DNA polymerase (PrimeStar GXL, Takara Bio Inc.) and primers PL2 and R7T (Additional file 1: Table S1) with genomic DNA from a rice cultivar, Nipponbare, as the template (Fig. 1b). The resultant DNA fragment was cloned into a binary vector, pZH2B (Kuroda et al. 2010) using In-Fusion HD cloning kit (Takara Bio Inc.) according to manufacturer's instructions. The complement lines for *OsINV3* were generated by incorporation of this vector into the homozygous *Tos17* mutant using Agrobacterium (EHA105) mediated transformation (Toki 1997). Successful complementation was assessed by genotyping using F1, R1 and T5 primers (Additional file 1: Table S1). Out of the 16 complementation lines isolated, C3 and C4 were selected and propagated to T1 generation, solely based on availability of grain. Complementation tests were carried out with T2 grain from C3 and C4 showing recovery in grain size and weight characteristics, however, results are demonstrated for only the C4 line in the current study.

SEM Analysis

Spikelet hulls for WT, mutant and C4, 2 DBH were sampled and processed as previously described (Li et al. 2011) followed by dehydration in isoamyl alcohol. The samples were dried with a critical-point drier (JEOL JCPD-5, Tokyo, Japan) and coated with Pt/Pd using a sputter coater (Hitachi E-1030, Tokyo, Japan). The SEM (Hitachi S-4800, Tokyo, Japan) was operated at 2 kV, with an aperture of 15 mm, and magnification of 120X and 200X for lemma and palea respectively. Cell density was determined as number of cells/unit area for the outer surface, and the cell area, width and height were determined for the cells on the inner surface of palea and lemma using ImageJ. Mean and SE were determined for 9 spikelet hulls for each line, with cell size analyses for 10 characteristic cells from palea and lemma of each of the 9 spikelets.

Promoter: GUS Assay

A genomic DNA fragment that covers from − 1930 to + 30 nucleotides of the translation start site of *OsINV3* was amplified by PCR with primers PINV3-L2H and -R2X, having restriction sites HindIII and XbaI, respectively, at the 5'end of each primer (Additional file 1: Table S1). Following the double restriction digestion, the genomic DNA fragment was cloned into a binary vector, pZH2B-GUS (Kuroda et al. 2010), with a pre-existing insertion of the beta-glucoronidase (GUS) gene. Eventually, a plasmid construct, with the GUS gene flanked by 30 nucleotides downstream of the translation start site of *OsINV3* and an interconnecting XbaI (six nucleotides) site, expected to be driven by the *OsINV3* promoter was obtained (Additional file 1: Figure S1E). The GUS lines of *OsINV3* were generated by incorporation of this vector into the WT lines segregated from NG6441 using transformation techniques mentioned above. Following genotyping using PINV3-L2H and PINV3-R2X (Additional file 1: Table S1), 3 lines were isolated and propagated to T1 generation. The T1 plants that tested positive for staining were isolated and their grain used for further analyses.

Panicles at different progressive stages of panicle initiation, and through stages from heading to harvest were sampled in ice cold 90% acetone (*v*/*v*), on ice. Samples were incubated at room temperature for 20 min, followed by wash with 50 mM phosphate buffer (pH 7.0) which was repeated thrice to eliminate the acetone completely. Following this, the GUS staining was performed similar to the methods previously described (Hirose et al. 2014). Observation was carried out using a stereo microscope (SMZ745T, Nikon, Tokyo, Japan).

Field Trials and Determination of Yield and its Components

Plants were grown in an experimental paddy field at the Institute for Sustainable Agro-ecosystem Services, The University of Tokyo (35.73°N, 139.53°E, 58 m above sea level), in Tokyo, Japan. Cultivation was carried out between the months of May and October 2015 and 2016, where one-month old seedlings were transplanted to the field on May 28, 2015 and May 25, 2016. Hills were spaced at 0.3 m × 0.15 m with one seedling per hill, leading to a planting density of 22.2 hills m^{-2}. Basal dressing of fertilizer was applied at 50 $g.m^{-2}$ with a $N:P_2O_5:K_2O$ of 12:16:18. Three sub-plots were maintained for each of the lines, their positions alternating with one another. Five biological replicates from each of the three sub-plots for both WT and mutant were used for assessment of yield and its components. The yield component data were obtained using methods previously described (Hirose et al. 2013).

Dry Weight and Photosynthesis Rate Measurements

Leaves, stem, panicles and dead plant tissue from each plant were sampled separately and dried at 80 °C for at least 1 week before weighing them. Five biological replicates from each of the three sub-plots for each line were used for dry weight estimation at panicle initiation (28 DBH), heading, 10 DAH, 20 DAH and harvest (40 DAH). Total dry weight was measured on a whole plant basis by determining the sum total weights of each plant part. Dry matter partitioning to panicles or vegetative tissues was estimated on per panicle basis, as the ratio of total dry weight of panicles or vegetative tissues to the total dry weight of the plant.

Photosynthesis rates for the flag leaves of the main stem of *OsINV3* WT and the mutant (KO) were measured at panicle initiation, heading and late ripening stages using CIRAS-3 Photosynthesis Systems (PP systems, Amesbury, MA, USA) between 1100 and 1200 h. The leaf chamber was set to 30 °C, photosynthetic photon flux density of 1500 $\mu mol.m^{-2}.s^{-1}$ and CO_2 concentration of 390 $ml.L^{-1}$.

Panicle and Grain Size Analyses for Field Grown Panicles

The panicles from WT and mutant lines were harvested post-maturity (40 DAH) and air-dried for 2 weeks before analyses. The length of the panicles was measured from the tip to the base for all panicles in three replicate plants for both the lines, and the number of primary and secondary rachis branches were determined for the same sample set.

Grain length, width and area were determined for ten corresponding grains from each of the three replicate plants for both the WT and the mutant using ImageJ. For the grain thickness distribution curve, the whole grain set for three replicate plants were analyzed. The grains were made to pass through a series of sieves sized from 1.8 mm to 2.4 mm, and the grains trapped in each of these sieves were counted and weighed. The average weight per grain was obtained by the ratio of the total weight to total number of the grains corresponding the particular grain thickness.

Quantitative Real-Time PCR

Young panicles (~ 3–4 cm in length) from WT, mutant and C4 were sampled in liquid nitrogen and pulverized cryogenically using a Multi-beads shocker (Yasui Kikai, Osaka, Japan) at 2000 rpm for 15 s. RNA isolation, cDNA synthesis and real-time PCR were performed similar to the methods previously described (Hashida et al. 2016). Primers specific to *OsINV2, OsINV3* and polyubiquitin gene *RUBIQ1* (Wang et al. 2000) are listed in Additional file 1:Table S1.

Plant Sampling and Pulverization

Whole panicles from the field grown plants with two biological replicates from each of the three sub-plots for both WT and mutant, were sampled on dry ice at panicle initiation (28 DBH), heading, 5 DAH, 10 DAH, 20 DAH, 30 DAH and harvest (40 DAH). Panicle initiation was identified by the appearance of first internode at 28 DBH, where young panicles (~ 3–4 cm in length) at the base of the stem were isolated and sampled. Samples were then pulverized cryogenically as mentioned above, and weighed separately for NSC determination and the enzyme assay.

Determination of Starch and Soluble Sugar Contents, and Invertase Enzyme Assay

NSC (Starch, sucrose, glucose and fructose) contents were determined similarly to the methods previously described (Okamura et al. 2013). Preparation of crude extract, and the soluble and insoluble enzyme assays were performed similarly to the methods previously described (Ishimaru et al. 2005), with the following modifications. The reaction buffer for acid invertase contained 100 mM of sodium acetate (pH 4.5) and 20 mM sucrose, to which 50 μl aliquot of the crude extract was added to initiate the reaction. The reaction volume was maintained at 200 μl. The reaction progressed at 37 °C for 30 min and was stopped by the addition of 20 μl of 1 M HEPES-NaOH (pH 7.5) and boiling at 98 °C for 2 min. Neutral invertase assay was performed same as above, except that the reaction buffer used contained 100 mM HEPES-NaOH (pH 7.5) instead of the acid buffer. The enzyme activity was determined as Vmax in terms of glucose generated by the reaction and was measured as described in the determination of sugar content.

Statistical Analyses

Unpaired student's t-test statistic (Microsoft excel 2016) was used to determine significance in differences between the WT and the mutants from minimum 3 independent biological replicates. For analyses with the complementation line, one-way ANOVA with a post-hoc analysis (Tukey's test) was used (SPSS 13.0 for Windows). *P*-values ≤0.05, ≤ 0.01 and ≤0.001 were indicated by *, ** and *** respectively, and considered significant.

The GenBank database accession numbers for *OsINV2* and *OsINV3* used in this article are AF276703.1 and AF276704.1 respectively.

Abbreviations

C: Complementation; CWIN: Cell wall invertase; DAF: Days after flowering; DAH: Days after heading; DBH: Days before heading; NIN: Neutral invertase; NSC: Non-structural carbohydrates; SEM: Scanning electron microscopy; VIN: Vacuolar invertase; WT: Wild type

Acknowledgements

The authors would like to thank the technical support staff of the Institute for Sustainable Agro-ecosystem Services, The University of Tokyo for their cultivation management of rice, and Dr. Fumiko Ishizuna for her assistance with the SEM data acquisition. We also thank Dr. Tomohiro Kurai for his invaluable suggestions during the course of the study, and Dr. Masaki Okamura for providing direction with the grain size analyses. A word of gratitude to Mr. Akira Murakami for his assistance with preliminary experiments, and the members of the Laboratory of Crop Science, The University of Tokyo, for their continued assistance with the field samplings.

Funding

This work was supported by the JSPS KAKENHI [grant number 15H02428] to TH, RO and NA.

Author's Contributions

T.H. and N.A. conceived the original screening and research plan; T.H., A.M. and H.H. generated the *Tos17* insertion mutants and performed screening for isolation of the mutants used in this study, N.A., R.O. and J.Y. supervised the experiments; S.R.M. designed and performed most of the experiments; Y.H. provided technical assistance to S.R.M.; S.R.M. analyzed the data and wrote the article with contributions from all the authors; N.A. supervised and complemented the writing. All authors read and approved the final manuscript.

Competing Interests

The authors declare that they have no competing interests.

Author details

[1]Graduate School of Agricultural and Life Sciences, The University of Tokyo, 1-1-1 Yayoi, Bunkyo-ku, Tokyo 113-8657, Japan. [2]Central Region Agricultural Research Center, NARO, 1-2-1 Inada, Joetsu, Niigata 943-0193, Japan. [3]Advanced Genomics Breeding Section, Institute of Crop Science, NARO, 2-1-2, Kannondai, Tsukuba, Ibaraki 305-8518, Japan. [4]Present addresses: Research Institute for Food and Agriculture, Ryukoku University, 1-5 Yokotani, Seta Oe-cho, Otsu, Shiga 520-2194, Japan.

References

Agrawal GK, Abe K, Yamazaki M, Miyao A, Hirochika H (2005) Conservation of the E-function for floral organ identity in rice revealed by the analysis of tissue culture-induced loss-of-function mutants of the OsMADS1 gene. Plant Mol Biol 59:125–135

Aoki N, Hirose T, Furbank RT (2012) Sucrose transport in higher plants: from source to sink. In: Eaton-Rye JJ, Tripathy BC, Sharkey TD (eds) Photosynthesis: plastid biology, energy conversion and carbon assimilation, advances in photosynthesis and respiration, vol 34, pp 703–729

Bihmidine S, Hunter CT, Johns CE, Koch KE, Braun DM (2013) Regulation of assimilate import into sink organs: update on molecular drivers of sink strength. Front Plant Sci 4:1–15

Cheng W, Taliercio E, Chourey P (1996) The miniature1 seed locus of maize encodes a cell wall invertase required for normal development of endosperm and maternal cells in the pedicel. Plant Cell 8:971–983

French S, Abu-Zaitoon Y, Uddin M, Bennett K, Nonhebel H (2014) Auxin and cell wall invertase related signaling during rice grain development. Plants 3:95–112

Goetz M, Guivarćh A, Hirsche J, Bauerfeind MA, González M-C, Hyun TK, Eom SH, Chriqui D, Engelke T, Großkinsky DK, Roitsch T (2017) Metabolic control of tobacco pollination by sugars and invertases. Plant Physiol 173:984–997

González MC, Roitsch T, Cejudo FJ (2005) Circadian and developmental regulation of vacuolar invertase expression in petioles of sugar beet plants. Planta 222:386–395

Hashida Y, Hirose T, Okamura M, Hibara KI, Ohsugi R, Aoki N (2016) A reduction of sucrose phosphate synthase (SPS) activity affects sucrose/starch ratio in leaves but does not inhibit normal plant growth in rice. Plant Sci 253:40–49

Herbers K, Sonnewald U (1998) Molecular determinants of sink strength. Curr Opin Plant Biol 1:207–216

Hirose T, Aoki N, Harada Y, Okamura M, Hashida Y, Ohsugi R, Akio M, Hirochika H, Terao T (2013) Disruption of a rice gene for α-glucan water dikinase, OsGWD1, leads to hyperaccumulation of starch in leaves but exhibits limited effects on growth. Front Plant Sci 4:1–9

Hirose T, Hashida Y, Aoki N, Okamura M, Yonekura M, Ohto C, Terao T, Ohsugi R (2014) Analysis of gene-disruption mutants of a sucrose phosphate synthase gene in rice, OsSPS1, shows the importance of sucrose synthesis in pollen germination. Plant Sci 225:102–106

Hirose T, Takano M, Terao T (2002) Cell wall invertase in developing rice caryopsis: molecular cloning of OsCIN1 and analysis of its expression in relation to its role in grain filling. Plant Cell Physiol 43:452–459

Ho LC (1988) Metabolism and compartmentation of imported sugars in sink organs in relation to sink strength. Annu Rev Plant Physiol Plant Mol Biol 39:355–378

Ishimaru T, Hirose T, Matsuda T, Goto A, Takahashi K, Sasaki H, Terao T, Ishii RI, Ohsugi R, Yamagishi T (2005) Expression patterns of genes encoding carbohydrate-metabolizing enzymes and their relationship to grain filling in rice (Oryza Sativa L.): comparison of caryopses located at different positions in a panicle. Plant Cell Physiol 46:620–628

Ji X, Van Den Ende W, Schroeven L, Clerens S, Geuten K, Cheng S, Bennett J (2007) The rice genome encodes two vacuolar invertases with fructan exohydrolase activity but lacks the related fructan biosynthesis genes of the Pooideae. New Phytol 173:50–62

Ji X, Van Den Ende W, Van Laere A, Cheng S, Bennett J (2005) Structure, evolution, and expression of the two invertase gene families of rice. J Mol Evol 60:615–634

Jin Y, Ni DA, Ruan YL (2009) Posttranslational elevation of cell wall invertase activity by silencing its inhibitor in tomato delays leaf senescence and increases seed weight and fruit hexose level. Plant Cell 21: 2072-2089

Klann EM, Hall B, Bennett AB (1996) Antisense acid invertase (TIV1) gene alters soluble sugar composition and size in transgenic tomato fruit. Plant Physiol 112:1321–1330

Koch K (2004) Sucrose metabolism: regulatory mechanisms and pivotal roles in sugar sensing and plant development. Curr Opin Plant Biol 7:235–246

Kuroda M, Kimizu M, Mikami C (2010) A simple set of plasmids for the production of transgenic plants. Biosci Biotechnol Biochem 74:2348–2351

Li B, Liu H, Zhang Y, Kang T, Zhang L, Tong J, Xiao L, Zhang H (2013) Constitutive expression of cell wall invertase genes increases grain yield and starch content in maize. Plant Biotechnol J 11:1080–1091

Li Y, Fan C, Xing Y, Jiang Y, Luo L, Sun L, Shao D, Xu C, Li X, Xiao J, He Y, Zhang Q (2011) Natural variation in GS5 plays an important role in regulating grain size and yield in rice. Nat Genet 43:1266–1269

Marcelis LM (1996) Sink strength as a determinant of dry matter partitioning in the whole plant. J Exp Bot 47:1281–1291

Miller M, Chourey P (1992) The maize invertase-deficient miniature-1 seed mutation is associated with aberrant pedicel and endosperm development. Plant Cell 4:297–305

Miyao A, Tanaka K, Murata K, Sawaki H, Takeda S, Abe K, Shinozuka Y, Onosato K, Hirochika H (2003) Target site specificity of the Tos17 retrotransposon shows a preference for insertion within genes and against insertion in retrotransposon-rich regions of the genome. Plant Cell 15: 1771-1780

Okamura M, Hirose T, Hashida Y, Yamagishi T, Ohsugi R, Aoki N (2013) Starch reduction in rice stems due to a lack of OsAGPL1 or OsAPL3 decreases grain yield under low irradiance during ripening and modifies plant architecture. Funct Plant Biol 40:1137–1146

Roitsch T, González MC (2004) Function and regulation of plant invertases: sweet sensations. Trends Plant Sci 9:606–613

Ruan YL, Jin Y, Yang YJ, Li GJ, Boyer JS (2010) Sugar input, metabolism, and signaling mediated by invertase: roles in development, yield potential, and response to drought and heat. Mol Plant 3:942–955

Schneider Ca, Rasband WS, Eliceiri KW (2012) NIH image to ImageJ: 25 years of image analysis. Nat Methods 9:671–675

Sergeeva LI, Keurentjes JJB, Bentsink L, Vonk J, Van der Plas LHW, Koornneef M, Vreugdenhil D (2006) Vacuolar invertase regulates elongation of Arabidopsis Thaliana roots as revealed by QTL and mutant analysis. Proc Natl Acad Sci 103:2994–2999

Smeekens S (2000) Sugar-induced signal transduction in plants. Annual reviews of plant physiology and. Plant Mol Biol 51:49–81

Song XJ, Kuroha T, Ayano M, Furuta T, Nagai K, Komeda N, Segami S, Miura K, Ogawa D, Kamura T, Suzuki T, Higashiyama T, Yamasaki M, Mori H, Inukai Y, Wu J, Kitano H, Sakakibara H, Jacobsen SE, Ashikari M (2015) Rare allele of a previously unidentified histone H4 acetyltransferase enhances grain weight, yield, and plant biomass in rice. Proc Natl Acad Sci 112:76–81

Sturm A (1999) Invertases. Primary structures, functions, and roles in plant development and sucrose partitioning. Plant Physiol 121:1–7

Sturm A, Šebková V, Lorenz K, Hardegger M, Lienhard S, Unger C (1995) Development- and organ-specific expression of the genes for sucrose synthase and three isoenzymes of acid β-fructofuranosidase in carrot. Planta 195:601–610

Tang GQ, Lüscher M and Sturm A (1999) Antisense repression of vacuolar and cell wall invertase in transgenic carrot alters early plant development and sucrose partitioning. Plant Cell 11:177–190

Tang X, Su T, Han M, Wei L, Wang W, Yu Z, Xue Y, Wei H, Du Y, Greiner S, Rausch T, Liu L (2017) Suppression of extracellular invertase inhibitor gene expression improves seed weight in soybean (Glycine max). J Exp Bot 68:469–482

Toki S (1997) Rapid and efficient Agrobacterium-mediated transformation in rice. Plant Mol Biol Report 15:16–21

Wang E, Wang J, Zhu X, Hao W, Wang L, Li Q, Zhang L, He W, Lu B, Lin H, Ma H, Zhang G, He Z (2008) Control of rice grain-filling and yield by a gene with a potential signature of domestication. Nat Genet 40:1370–1374

Wang J, Jiang J, Oard JH (2000) Structure, expression and promoter activity of two polyubiquitin genes from rice (Oryza Sativa L). Plant Sci 156:201–211

Wang L, Li XR, Lian H, Da N, He Y, Chen XY, Ruan YL (2010) Evidence that high activity of vacuolar invertase is required for cotton fiber and Arabidopsis root elongation through osmotic dependent and independent pathways, respectively. Plant Physiol 154:744–756

Wang L, Ruan Y-L (2016) Critical roles of vacuolar invertase in floral organ development and male and female fertilities are revealed through characterization of GhVIN1-RNAi cotton plants. Plant Physiol 171:405–423

Weber H, Borisjuk L, Heim U, Buchner P, Wobus U (1995) Seed coat-associated invertases of fava bean control both unloading and storage functions: cloning of cDNAs and cell type-specific expression. Plant Cell 7:1835–1846

Zrenner R, Schüler K, Sonnewald U (1996) Soluble acid invertase determines the hexose-to-sucrose ratio in cold-stored potato tubers. Planta 198:246–252

14

Identification and Fine Mapping of *Bph33*, a New Brown Planthopper Resistance Gene in Rice (*Oryza sativa* L.)

Jie Hu[1,2], Xingyuan Chang[1,2], Ling Zou[1,2], Weiqi Tang[1,2] and Weiren Wu[1,2*]

Abstract

Background: Host-plant resistance is the most desirable and economic way to overcome BPH damage to rice. As single-gene resistance is easily lost due to the evolution of new BPH biotypes, it is urgent to explore and identify new BPH resistance genes.

Results: In this study, using $F_{2:3}$ populations and near-isogenic lines (NILs) derived from crosses between two BPH-resistant Sri Lankan rice cultivars (KOLAYAL and POLIYAL) and a BPH-susceptible cultivar 9311, a new resistance gene *Bph33* was fine mapped to a 60-kb region ranging 0.91–0.97 Mb on the short arm of chromosome 4 (4S), which was at least 4 Mb distant from those genes/QTLs (*Bph12, Bph15, Bph3, Bph20, QBph4* and *QBph4.2*) reported before. Seven genes were predicted in this region. Based on sequence and expression analyses, a Leucine Rich Repeat (LRR) family gene (LOC_Os04g02520) was identified as the most possible candidate of *Bph33*. The gene exhibited continuous and stable resistance from seedling stage to tillering stage, showing both antixenosis and antibiosis effects on BPH.

Conclusion: The results of this study will facilitate map-based cloning and marker-assisted selection of the gene.

Background

Brown planthopper (BPH; *Nilaparvata lugens* Stål) is one of the most destructive insect pests of rice in Asia-Pacific region, often causing hopper burn and severe yield loss due to its monophagy and migration ability (Normile 2008). Presently, spraying insecticides (e.g. imidacloprid) is the chief way to control BPH, which is costly and hazardous to health and environment. In addition, it often makes BPH develop resistance to insecticides and therefore leads to resurrection of BPH population (Tanaka et al. 2000). Hence, using rice resistance to BPH should be the most economic and effective approach for the management of BPH (Jena et al. 2006).

Rice resistance to BPH has been studied in the last half century. To date, more than 30 BPH-resistance genes or quantitative trait loci (QTLs) have been identified from ssp. *indica* and wild relatives of rice (Prahalada et al. 2017; Guo et al. 2018). Of them, 22 genes or QTLs (*Bph1, Bph2/Bph26, Bph3, Bph6, bph7, Bph9, Bph12, Bph14, Bph15, Bph17, Bph18, Bph19, Bph20, Bph21, Bph27, Bph27(t), Bph28(t), bph29, QBph3, QBph4, QBph4.2, Bph31* and *Bph32*) have been fine-mapped (Cha et al. 2008; Murai et al. 2001; Jairin et al. 2007; Qiu et al. 2010, 2012, 2014; Zhao et al. 2016; Du et al. 2009; Lv et al. 2014; Liu et al. 2015; Jena et al. 2006; Chen et al. 2006; Rahman et al. 2009; Huang et al. 2013; He et al. 2013; Wu et al. 2014; Wang et al. 2015; Hu et al. 2015a,b; Prahalada et al. 2017; Ren et al. 2016). However, only 8 genes (*Bph14, Bph26, Bph3, bph29, Bph32, Bph18, Bph9* and *Bph6*) have been isolated by map-based cloning (Du et al. 2009; Tamura et al. 2014; Liu et al. 2015; Wang et al. 2015; Ren et al. 2016; Ji et al. 2016; Zhao et al. 2016; Guo et al. 2018). *Bph14, Bph26* (or *Bph2*), *Bph18* and *Bph9* are found to encode coiled-coil, nucleotide-binding, and leucine-rich repeat (CC-NB-LRR) proteins, and the latter three (*Bph26, Bph18* and *Bph9*) are allelic. Four haplotypes in *Bph9* locus exhibit different resistance to different BPH biotypes, indicating that allelic variation is an important strategy for rice to resist BPH (Zhao et al. 2016).

Since 1970s, many BPH resistance genes have been utilized in rice breeding and a series of cultivars with

* Correspondence: wuwr@fafu.edu.cn
[1]Key Laboratory of Genetics, Breeding and Multiple Utilization of Crops, Ministry of Education, Fujian Agriculture and Forestry University, Fuzhou 350002, China
[2]Fujian Key laboratory of Crop Breeding by Design, Fujian Agriculture and Forestry University, Fuzhou 350002, China

improved resistance (e.g. IR26, IR36 and IR50) have been developed and released. However, because of the evolution of new BPH biotypes, the improved cultivars are easy to lose their BPH resistance controlled by a single resistant gene (Jena and Kim 2010). Moreover, only a few resistance genes (*Bph3, Bph6, Bph9* and *Bph31*) show high resistance to more than one BPH biotypes/populations. Therefore, it is urgent to discover new broad-spectrum BPH resistance genes and integrate them into rice cultivars.

Two Sri Lankan *indica* rice cultivars, KOLAYAL and POLIYAL, were previously found to be highly resistant to the most dominated BPH strain in Wuhan (seedling resistance score RS = 1.3 and 2.2 in two tests) and that in Fuzhou (RS = 1.5 and 1.6 in two tests), China, suggesting that they could be useful donors of BPH resistance genes in rice breeding. The present study aimed to understand the genetic basis and possible mechanism of the BPH resistance in KOLAYAL and POLIYAL, so as to facilitate the utilization of these two varieties as BPH resistance donors in rice breeding.

Results

Segregation of BPH resistance in $F_{2:3}$ populations

To understand the genetic basis of BPH resistance in KOLAYAL and POLIYAL, we investigated the RS segregation at seedling stage in the two $F_{2:3}$ populations derived from crosses KOLAYAL × 9311 and POLIYAL × 9311. In both $F_{2:3}$ populations, the RS at 12 DAI showed a bimodal distribution, with the valley bottom approximately at the point between 6.00 and 6.99 (Fig. 1). According to the valley bottom, the $F_{2:3}$ lines could be divided into two classes, the resistant (RS ≤ 6.99) and the susceptible (RS ≥ 7.00), which was consistent with the scoring criterion of BPH resistance (Qiu et al. 2010). Thus, the segregation ratios of resistant lines to susceptible lines in the two populations were found to be 71:22 and 69:20, respectively, both of which were in agreement with the theoretical 3:1 ratio (chi-squares = 0.76 and 0.58, $P = 0.313$ and 0.446). These results indicated that the BPH resistance of KOLAYAL and POLIYAL were both controlled by a major gene.

Mapping of resistance gene

To map the genes in the two resistant parents, we performed BSA-seq based on the two $F_{2:3}$ populations. A total of 161.5 K SNP and short InDel markers were obtained for gene mapping analysis. The results showed that there was only a common peak of allele frequency difference in the two populations ranging 0–1.5 Mb on the short arm of chromosome 4 (4S) (Fig. 2), suggesting that the resistance genes of KOLAYAL and POLIYAL were possibly the same. We named it *Bph33*, following the two (*Bph31* and *Bph32*) reported by Prahalada et al. (2017) and Ren et al. (2016), respectively.

Verification of *Bph33*

To verify the location of *Bph33*, we examined its effect in both $BC_1F_{1:2}$ and BC_2F_1 populations of the two crosses. Based on the BSA-seq results, two flanking InDel markers, H25 and D17 (Additional file 1: Table S1), which delimited an interval of ~ 1.5 Mb, were developed to survey *Bph33* genotype of the $BC_1F_{1:2}$ lines and BC_2F_1 plants. At the same time, the resistance levels of $BC_1F_{1:2}$ lines and BC_2F_1 plants were evaluated. T-test showed that the difference of resistance level was very significant between the homozygous resistant genotype and the homozygous susceptible genotype in the two $BC_1F_{1:2}$ populations (Fig. 3a, b) and between the heterozygous resistant genotype and the homozygous susceptible genotype in the two BC_2F_1 populations (Fig. 3c, d). These results strongly confirmed the existence of *Bph33* on the region between markers H25 and D17.

Fine mapping of *Bph33*

Since *Bph33* was present in both KOLAYLA and POLIYAL, we only used the BC_3F_2 plants from KOLAYLA × 9311 to fine map the gene. A total of 3200 BC_3F_2 plants segregating at the *Bph33* locus were genotyped with H25 and D17, from which 184 recombinants between the two markers were identified. Based on these recombinants, a local map of the H25-D17 region containing additional nine markers inside was constructed (Fig. 4b). The 184 recombinants were classified into different groups according to their marker genotypes in this region and 3–4 recombinants was randomly selected from each group for subsequent analysis. Each recombinant selected was selfed and the homozygous recombinants in its progeny line were identified using markers and tested for BPH resistance. According to the genotypes and resistance phenotypes of the homozygous recombinants, it could be inferred that lines 80, 140 and 124, 117 delimited the left and right margins of the *Bph33* position, respectively (Fig. 4c). Thus, *Bph33* was located between InDel markers H14 and H84 (Additional file 1: Table S1), within an interval of 470 kb in the Nipponbare genome sequence.

To narrow down the *Bph33* interval, we further screened the genotypes of 6600 BC_3F_3 seedlings and obtained 128 recombinants between H14 and H84, from which the progeny homozygous recombinants were genotyped using 10 new indel markers evenly distributed between H14 and H84. Thus, a high resolution map of *Bph33* locus was constructed (Fig. 4d). By testing the resistance of the homozygous recombinants obtained through the same procedure as described above, we found that the *Bph33* position could be confined between markers H99 and H101 ((Additional file 1: Table S1)), which spanned a distance of 60 kb (Fig. 4d, e), covering part of the two BAC clones OSJNBa0073L13 and OSJNBa0050O03 of Nipponbare each.

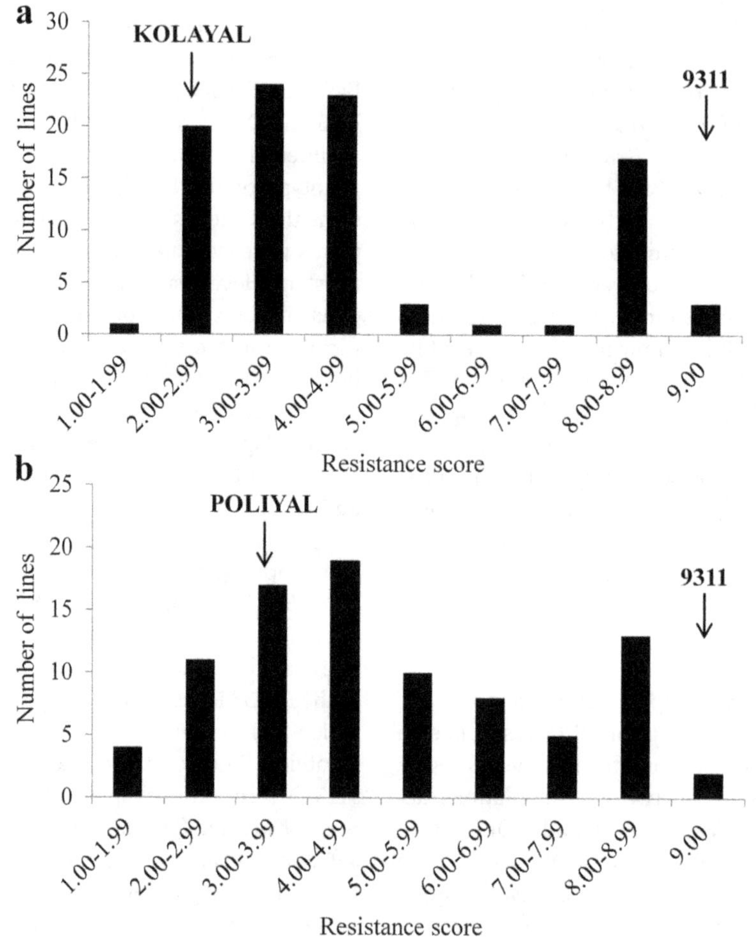

Fig. 1 Frequency distribution of BPH resistance score in two $F_{2:3}$ populations. **a** Population from 9311 × KOLAYAL; **b** Population from 9311 × POLIYAL. The resistance scores of individuals were measured at the time of 12 days after infestation (DAI)

Candidates of *Bph33*

According the reference genome of Nipponbare, seven genes are predicted in the 60 kb region harboring *Bph33* (Fig. 4f). Among them, two genes encode a zinc finger protein (LOC_Os04g02510) and a Leucine Rich Repeat (LRR) family protein (LOC_Os04g02520), respectively, while the other five genes (LOC_Os04g02530, LOC_Os04g02550, LOC_Os04g02560, LOC_Os04g02570 and LOC_Os04g02580) all encode expressed proteins. To identify the possible candidate gene(s), using the BSA-seq data, we compared the allele sequences of these genes from the resistant parents KOLAYAL and POLIYAL with those from the susceptible parent 9311 and Nipponbare (Additional file 2: Table S3). LOC_Os04g02510 and LOC_Os04g02520 both contained many variations between the resistant and the susceptible varieties, and their alleles from the resistant varieties appeared to be functional. So, they both were potential to be *Bph33*. Among the five expressed-protein genes, LOC_Os04g02530 and LOC_Os04g02570 showed loss-of-function (start codon lost and frameshift) mutations in KOLAYAL and POLIYAL; LOC_Os04g02580 had no exon variations; LOC_Os04g02550 only contained a SNP in 3′-UTR; and LOC_Os04g02560 had a SNP causing non-synonymous change of a codon. Obviously, the first two genes were impossible to be *Bph33* because resistant was dominant to susceptible. The middle two were also unlikely to be *Bph33*. The possibility for the last one to be *Bph33* could not be eliminated, but it should be very small compared with those of LOC_Os04g02510 and LOC_Os04g02520. Hence, we had good reasons to exclude the five expressed-protein genes as the candidates of *Bph33*.

To further examine the possibility of candidate genes to be Bph33, we analyzed their expression before and after infestation in NIL(+) (near-isogenic lines carrying Bph33 allele from the resistant parent) and NIL(−) (near-isogenic lines carrying Bph33 allele from the susceptible parent), respectively. The results (Additional file 3: Figure S1) indicated that at the time of 0 h (before infestation), five genes, LOC_Os04g02510, LOC_Os04g02520, LOC_Os04g02530, LOC_Os04g02550 and LOC_Os04g02570 showed significantly lower expression level in NIL (+) than those of NIL

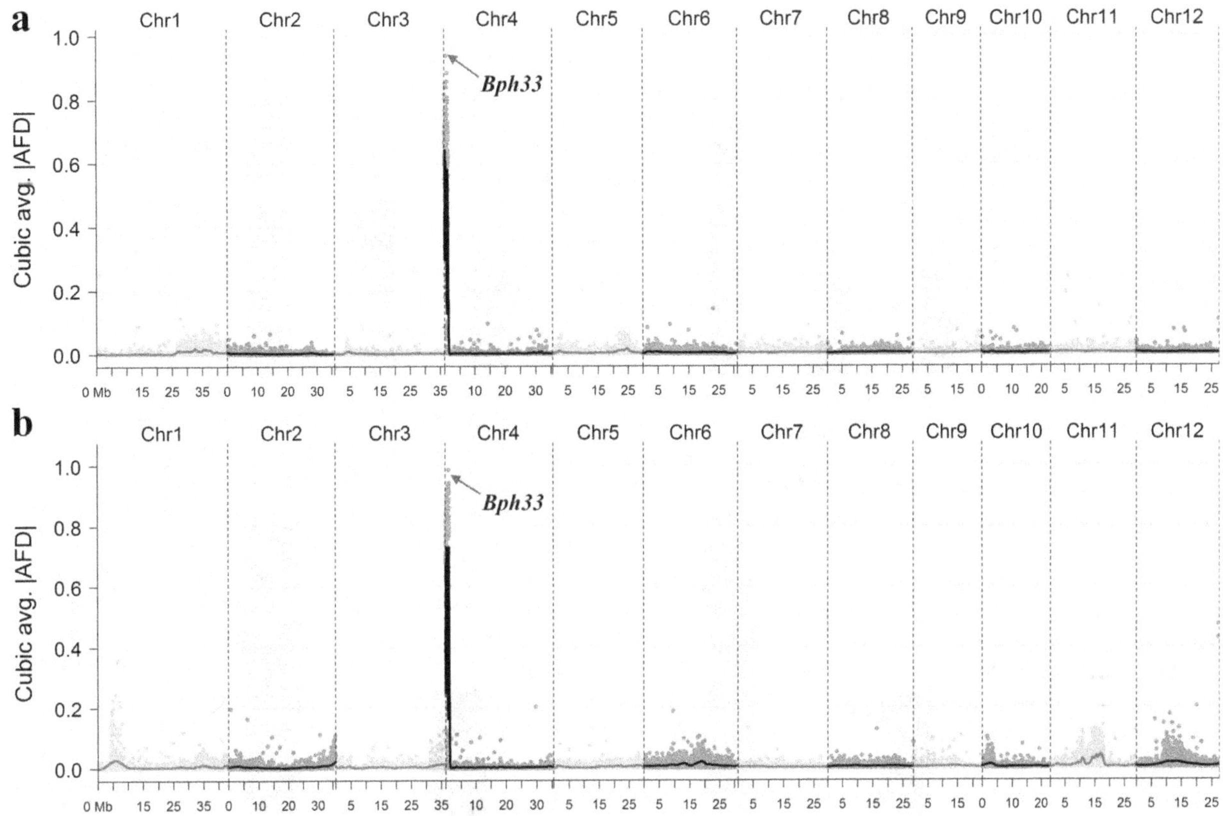

Fig. 2 Mapping of *Bph33* using the method of BSA-seq based on the $F_{2:3}$ populations from 9311 × KOLAYAL (**a**) and 9311 × POLIYAL (**b**). AFD, allele frequency difference of individual marker between the two opposite DNA pools. The arrows indicate the peaks where *Bph33* was located

(–). At the time of 6 h and 48 h (after infestation), four and two genes had lower expression level in NIL (+) that those of NIL (–). However, the expression levels of none of the genes in NIL (+) were significantly up-regulated and significantly higher than those in NIL (–) almost at both of time points after BPH infestation. These results suggested that the structure, not expression variation might be related to the function of *Bph33*.

The expression level of LOC_Os04g02510 was dramatically reduced at 48 h after infestation in both NIL (+) and NIL (–), but LOC_Os04g02520 did not show significant expression changes. In addition, both genes had a lower expression level in NIL (+) than in NIL(–), but the difference was much greater for LOC_Os04g02510. Considering that resistant was dominant to susceptible as mentioned above, these results suggested that LOC_Os04g02510 should be less likely to be *Bph33*. So, we took LOC_Os04g02520 as the most possible candidate of *Bph33*. There were six SNPs and two InDels in the second and the third exons of LOC_Os04g02520 between the two resistant and the two susceptible varieties, causing six amino acid substitutions and one and two amino acid insertions in POLIYAL and KOLAYAL, respectively (Fig. 4f). The possibility for LOC_Os04g02520 to be *Bph33* was also indirectly supported by the findings that LRR domain exists in some BPH-resistance genes (Du et al. 2009; Tamura et al. 2014; Ji et al. 2016; Zhao et al. 2016).

Resistance effect of *Bph33*

To better understand the resistance effect of *Bph33*, we examined the resistance of NIL(+) and NIL(–) together with the three parents. The NIL(+) and NIL(–) showed similar resistance levels to those of the resistant and the susceptible parents, respectively, exhibiting significant difference between them at both seedling stage (Fig. 5a, c) and tillering stage (Fig. 5b). These results suggested that *Bph33* had continuous resistance effect during the plant growth.

Antixenosis effect of *Bph33*

In the host selection experiment, BPHs dispersed inside a bucket soon after release, with most landing on soil and the edge of the bucket. As shown in Fig. 6a, 3h after release, about half of the BPHs were found on plants, with ~ 15 nymphs on NIL (+), KOLAYAL or POLIYAL, and ~ 35 nymphs on NIL (–) or 9311. Afterwards, the number of BPHs on the resistant plants decreased slightly, while that on the susceptible plants increased

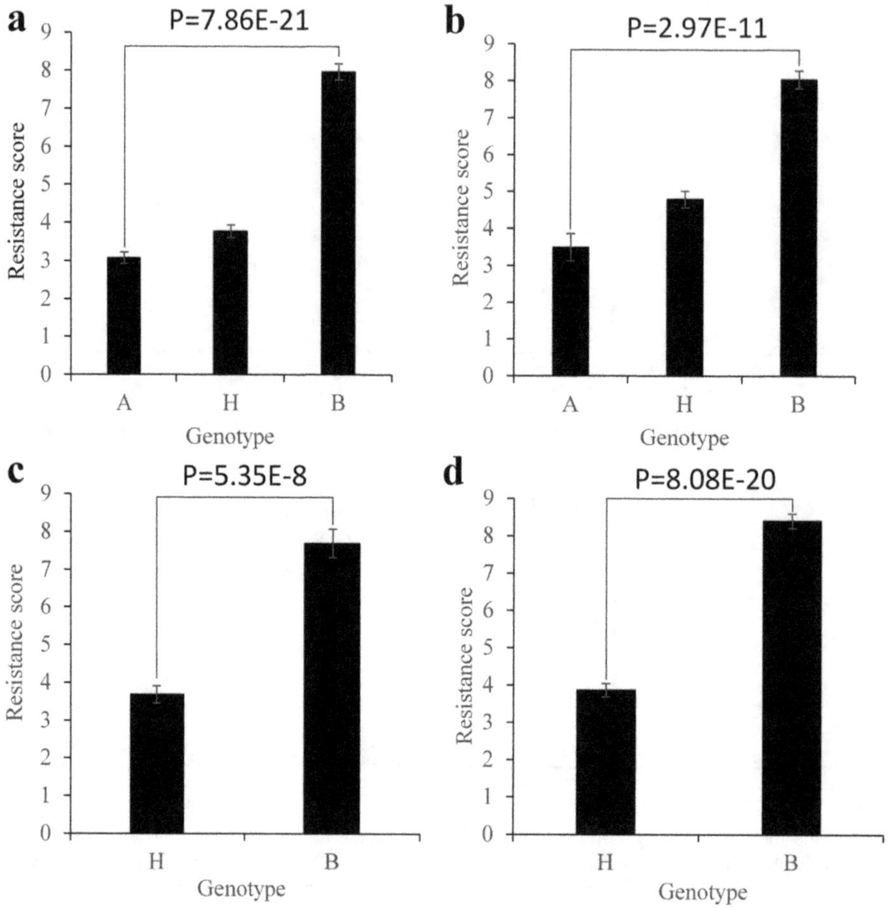

Fig. 3 Comparison of resistance levels among different *Bph33* genotypes in four populations. **a** and **b** In the $BC_1F_{2:3}$ populations from 9311 × KOLAYLA (**a**) and 9311 × POLIYAL (**b**); **c** and **d** In the BC_2F_1 populations from 9311 × KOLAYLA (**c**) and 9311 × POLIYAL (**d**). The *Bph33* genotype was identified according to its flanking markers H25 and D17. A, KOLAYLA or POLIYAL genotype; B, 9311 genotype; H, heterozygote genotype. *P*-values are shown for the comparison between A and B (in **a** and **b**) or between B and H (in **c** and **d**)

rapidly, reaching the maximum value of ~ 65 at 24 h and then maintaining approximately stable with only slight decrease at 48 h. The results suggested that after one day the BPHs had basically settled down on the plants for food. The difference in the number of settled BPHs on plants was very great and significant between the resistant and the susceptible genotypes throughout the period of experiment, indicating that antixenosis was a factor in the BPH resistance conferred by *Bph33*.

Antibiosis effect of *Bph33*

The antibiosis experiment showed that the weight of honeydew secreted by BPH on NIL (+) (1.14 mg/female) was significantly lower than that on NIL (-) (5.84 mg/female; Fig. 6b). Likewise, the honeydew weight on the resistant parents KOLAYLA (1.28 mg/female) and POLIYAL (0.95 mg/female) were significantly lower than that on the susceptible parent 9311 (6.84 mg/female; Fig. 6b). Consistently, the area of honeydew stains generated by BPH on NIL (+) was much smaller than that on NIL (–) (Fig. 6c). Moreover, BPH survival rates on NIL (+) and the two resistant parents were significantly lower than those on NIL (–) and 9311 from 3 to 7 DAI, and the difference between the resistant genotype and the susceptible genotype increased rapidly with the time (Fig. 6d). At 7 DAI, only 25.75% BPH survived on NIL (+), while 87.78% were still alive on NIL (–). These results indicated that antibiosis was also a factor in the BPH resistance conferred by *Bph33*.

Discussion

Up to now, at least 30 BPH resistance genes/QTLs have been found in rice, distributed on chromosomes 3, 4, 6, 11 and 12. Interestingly, multiple BPH resistance genes cluster together on chromosomes 4, 6 and 12. Six genes/QTLs (*Bph12, Bph3, Bph15, Bph20, QBph4 and QBph4.2*) are located in the region 5–9 Mb on chromosome 4S (Qiu et al. 2012; Lv et al. 2014; Liu et al. 2015; Rahman et al. 2009; Hu et al. 2015a,b). In the present study, *Bph33* was fine mapped in a region of ~ 60 kb (0.91–0.97 Mb) on chromosome 4S, which was at least 4 Mb distant from

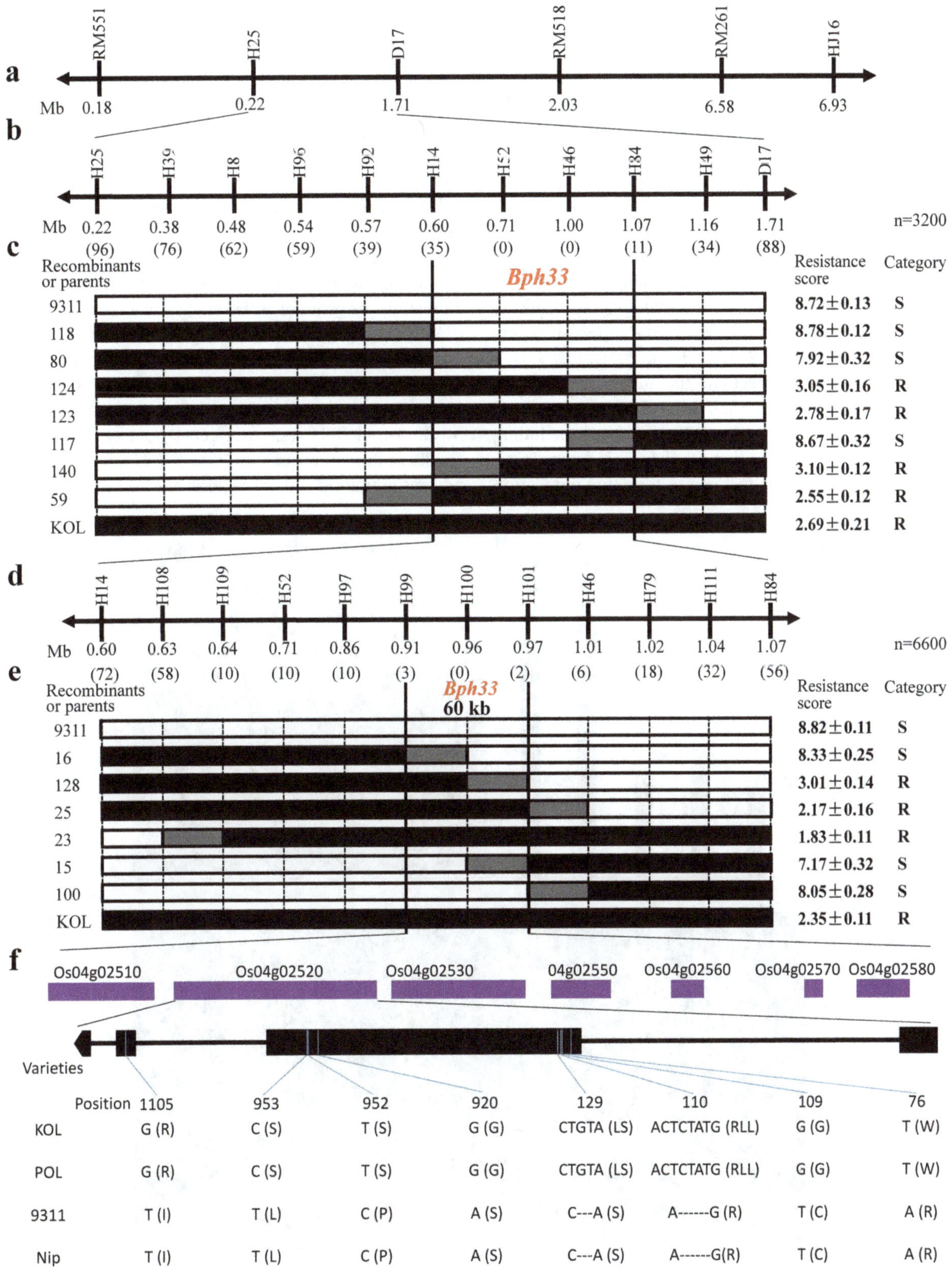

Fig. 4 (See legend on next page.)

(See figure on previous page.)
Fig. 4 Fine mapping of *Bph33*. **a** Physical map of part of chromosome 4S containing *Bph33*. **b, d** Physical map of marker interval H25-D17 (**b**) or H14-H84 harboring *Bph33*. The numbers in brackets show the times of recombination occurred between the marker loci and *Bph33* among 3200 (**b**) or 6600 (**d**) individuals. **c, e** Graphical genotypes and resistance phenotypes of the recombinants. Black, white and grey bars denote KOLAYLA, 9311 and recombined segments, respectively. R, resistant; S, susceptible. **f** Annotated genes within the final 60-kb interval. The nucleotide sequence differences and resultant amino acid changes (in brackets) among the parents and Nipponbare in the CDS of candidate gene LOC Os04g02520 are shown

those genes/QTLs reported before (Fig. 7). Recently, a new major QTL, *Qbph4.3*, was mapped between markers RM551 (0.177 Mb) and RM335 (0.688 Mb) on chromosome 4S (Mohanty et al. 2017). The position of *Bph33* was outside this interval (Fig. 7). Therefore, *Bph33* should be different from *Qbph4.3*. There is another QTL, *Qbph4.4*, claimed beside *Qbph4.3*, of which the interval covers *Bph33* (Mohanty et al. 2017). However, *Qbph4.4* is a minor QTL with small effect and its interval is very large, ranging from 0.688 Mb to 13.07 Mb, which covers a total of seven BPH resistance genes/QTLs reported (including *Bph33*; Fig. 7). This means that *Qbph4.4* is indeterminate. In addition, as *Qbph4.4* is mapped in the same population with *Qbph4.3*, it is likely to be a false QTL because two QTLs located in adjacent marker intervals are indistinguishable according to the statistical theory of QTL mapping (Zeng 1994). So, it is hard to say that *Qbph4.4* is a new QTL. Hence, we have reason to believe that *Bph33* is a novel major gene for BPH resistance.

At the individual level, three general resistance mechanisms have been recognized to defend plants against BPH, including antixenosis, antibiosis and tolerance (Cohen et al. 1997). Among them, antibiosis is the major one conferred by almost all BPH resistance genes (Qiu et al. 2010). The results of this study indicated that *Bph33* also exhibited significant antibiosis and antixenosis effects, with its resistant

Fig. 5 Test of resistance effect of *Bph33* at seedling stage and tillering stage. **a** Seedling resistance of NILs and parents. Asterisks indicate significant difference between the resistant and the susceptible genotypes at 0.01 level. Error bar indicates standard error. **b** NIL(+) and NIL(−) plants at 15 DAI. **c** Seedlings of NILs and parents at 12 DAI

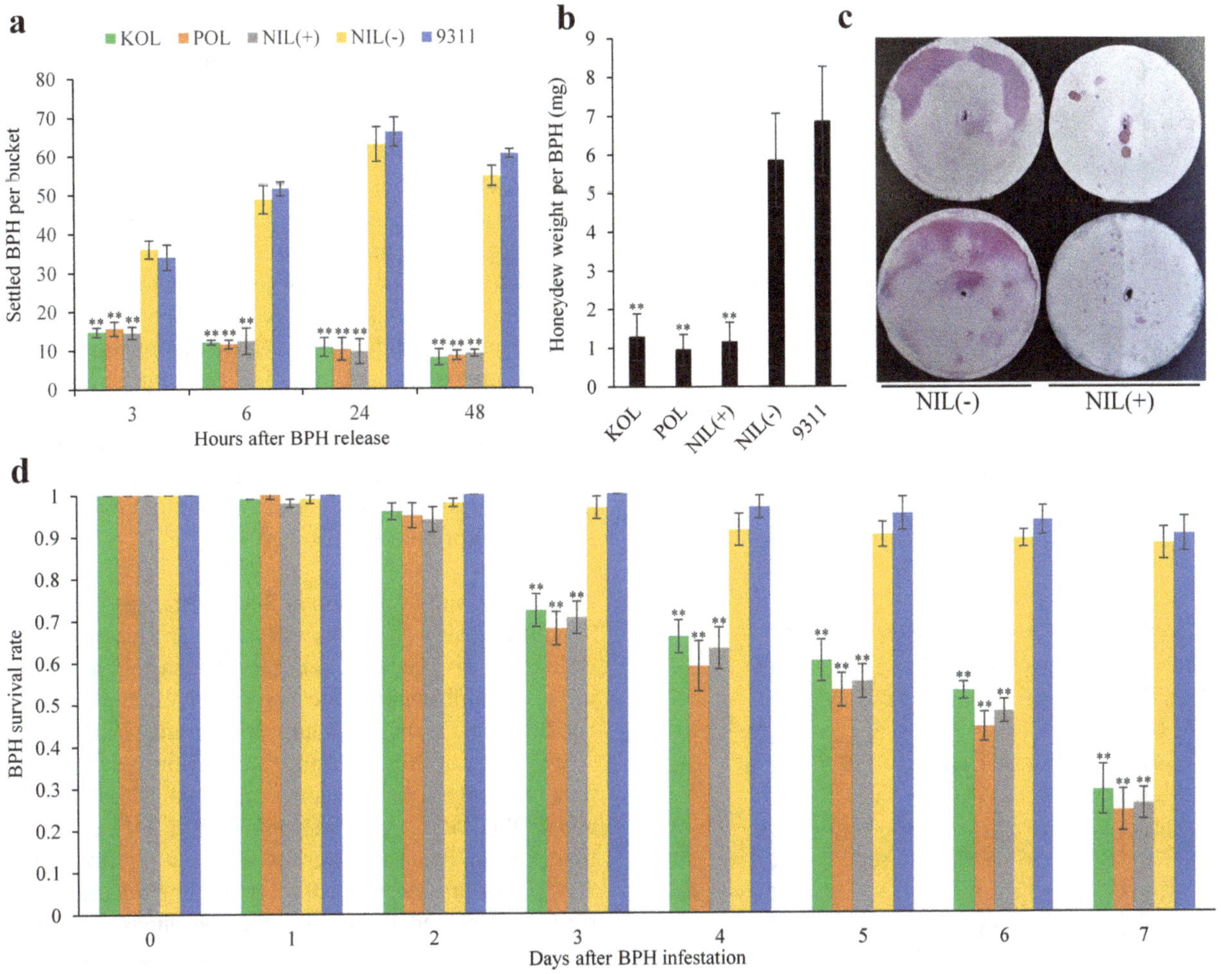

Fig. 6 Tests of antixenosis and antibiosis conferred by *Bph33*. **a** BPH settlement preference on plants of different *Bph33* genotypes. **b** Weights of honeydew secreted by BPH on NILs and parents. **c** Areas of honeydew secreted by BPH feeding on NIL(+) and NIL(−) plants. **d** Survival rates of BPH on NILs and parents. The column colors have the same meanings as in **a**. KOL, KOLAYLA; POL, POLIYAL. Asterisks indicate significant difference between the resistant and the susceptible genotypes at 0.01 level. Error bar indicates standard error

allele acting to strengthen the antibiosis and antixenosis (Fig. 6).

It has been found that a large proportion of rice cultivars from Sri Lanka are probably resistant to different biotypes of BPH (Hu et al. 2016). Rathu Heenati, a famous Sri Lankan cultivar, is found to be resistant to all the four biotypes of BPH (Sun et al. 2005; Jairin et al. 2007). In this study, the two resistant cultivars, KOLAYLA and POLIYAL, were also originated from Sri Lanka. This implies that they (or *Bph33* gene) have the potential to resist to multiple biotypes of BPH. In fact, as mentioned above, these two cultivars showed high resistance to BPH from Wuhan (~ 30 N, 114E) and Fuzhou (~ 26 N, 119E). The two cities are far apart to each other (the straight line distance between them is ~ 900 km) and have different climates. So, it is possible that the BPH populations from these two places are divergent in genetic structure with different constitution of biotypes or different predominant biotypes. In addition, the resistance mechanisms of *Bph33* discussed above and its feature of continuous resistance during plant growth (Fig. 5) are all similar to those found in broad-spectrum BPH resistance genes such as *Bph6* and *Bph9* (Zhao et al. 2016; Guo et al. 2018). All these imply that *Bph33* could be a potential resistance gene of wide spectrum.

Advances in sequencing technology and functional genomics have significantly facilitated the mapping and cloning of BPH resistance genes. The first BPH resistance gene *Bph14* was cloned in 2009, while seven other genes, *Bph26*, *Bph3*, *bph29*, *Bph32*, *Bph18*, *Bph9* and *Bph6*, were cloned only within recent two years (Du et al. 2009; Tamura et al. 2014; Liu et al. 2015; Wang et al. 2015; Ren et al. 2016; Ji et al. 2016; Zhao et al. 2016; Guo et al. 2018). BSA-seq is very efficient method for gene/QTL mapping (Tang et al. 2018). In this study, we used BSA-seq to rapidly map *Bph33* in a 1.5 Mb region and further fine mapped the gene using

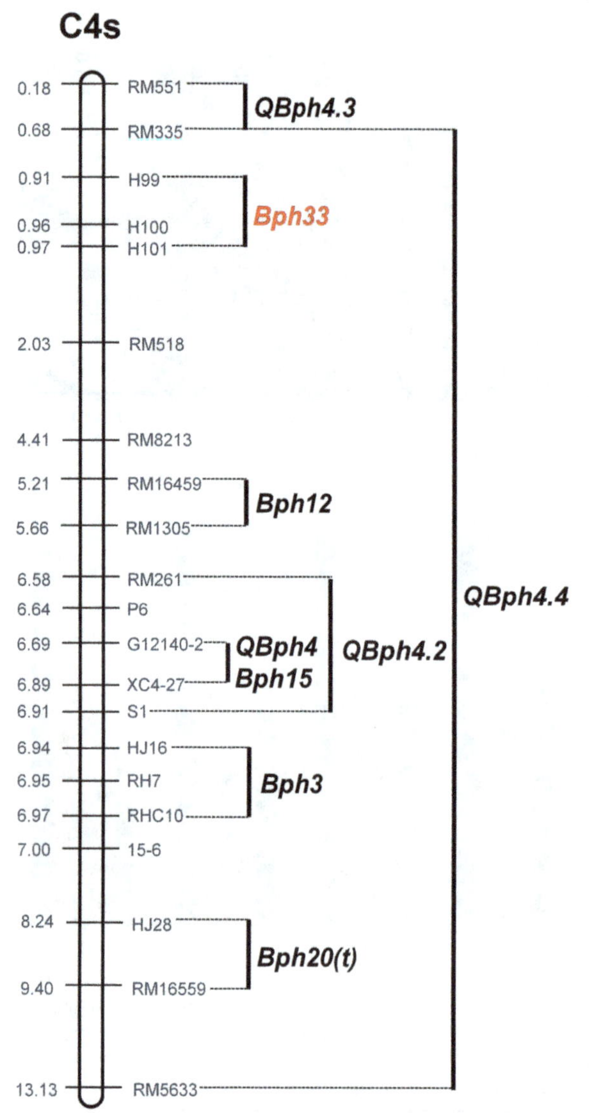

Fig. 7 Locations of all BPH genes or QTLs around *Bph33* on the short arm of chromosome 4. Data from Qiu et al. (2012), Lv et al. (2014), Hu et al. (2015a, b), Mohanty et al. (2017), Liu et al. (2015) and Rahman et al. (2009)

closely linked markers developed according to the deep sequencing data, demonstrating the efficiency of BSA-seq mapping. Our results will facilitate map-based cloning of *Bph33* and adopting marker-assisted backcrossing (MABC) strategy in BPH resistance breeding.

Conclusion

A new major BPH resistance gene *Bph33* was fine mapped within an interval of 60 kb on rice chromosome 4S. Antixenosis and antibiosis were two factors in the BPH resistance conferred by *Bph33*. Our results will facilitate map-based cloning and marker-assisted selection of the gene.

Methods

Plant materials and mapping populations

A BPH-susceptible Chinese *indica* rice cultivar 9311 was crossed with two BPH-resistant Sri Lankan *indica* rice cultivars, KOLAYAL (IRGC 36295) and POLIYAL (IRGC 36352), which were kindly provided by the International Rice Research Institute (IRRI), to develop two $F_{2:3}$ populations for genetic analysis and gene mapping. A few highly-resistant $F_{2:3}$ lines were selected and backcrossed to 9311 to develop several backcross populations for validating the resistance gene. Secondary segregating populations, in which genetic segregation only existed in a few genomic segments including the target region, and isogenic lines (NILs) with 9311 background were further derived from the backcross populations for fine mapping the resistance gene and analyzing its resistance mechanism. The detailed procedure of creating the experimental populations or lines is illustrated in Fig. 8.

BPH insects and evaluation of BPH resistance

The BPH insects used for infestation in this study were collected from rice fields in Fuzhou, China, and maintained on a susceptible *indica* rice variety (TN1) under natural conditions at Fujian Agriculture and Forestry University.

Evaluation of BPH resistance was performed by seedling bulk test as described (Huang et al. 2001), with KOLAYAL and POLIYAL as resistant controls and 9311 as susceptible control. Rice seeds were sown in plastic trays of 50 × 30 × 10 cm^3 in size, with 20 seeds per line (including control) in a row and 12 lines per tray. Seedlings were grown in a greenhouse at 24–28 °C under natural light. At the three-leaf stage (~ 12 days after sowing), seedlings were thinned out to the density of 10 plants per line, and then infestation was performed with second-instar nymphs in a number of 10 insects per seedling on average. At ~ 12 days after infestation (DAI) when all seedlings of 9311 died, all other lines were examined. The resistance level of a seedling was indicated with a resistance score (RS = 1, 3, 5, 7 or 9). A smaller RS value represented a higher resistance level. The resistance level of a line was indicated by the average RS of all the seedlings it contained.

For evaluation of resistance in tillering stage, two plants (45–50 days old) of NIL (+) or NIL (–), or 9311 (as control) were transplanted into a plastic cup of 17 cm in height and 8 cm in diameter. Then, 400 nymphs of 2–3 instar were released in two plastic cups (one was NIL (+) or NIL (–), the other was 9311) covered by a transparent gauze. At 15 DAI, the plants were evaluated relative to the degree of susceptibility of the control and rated as resistant and susceptible. The experiments were repeated for six times.

Antixenosis test

To test antixenosis, ten 30-day-old rice plants, with five of NIL (+) and NIL (–) each, or of KOLAYAL and 9311

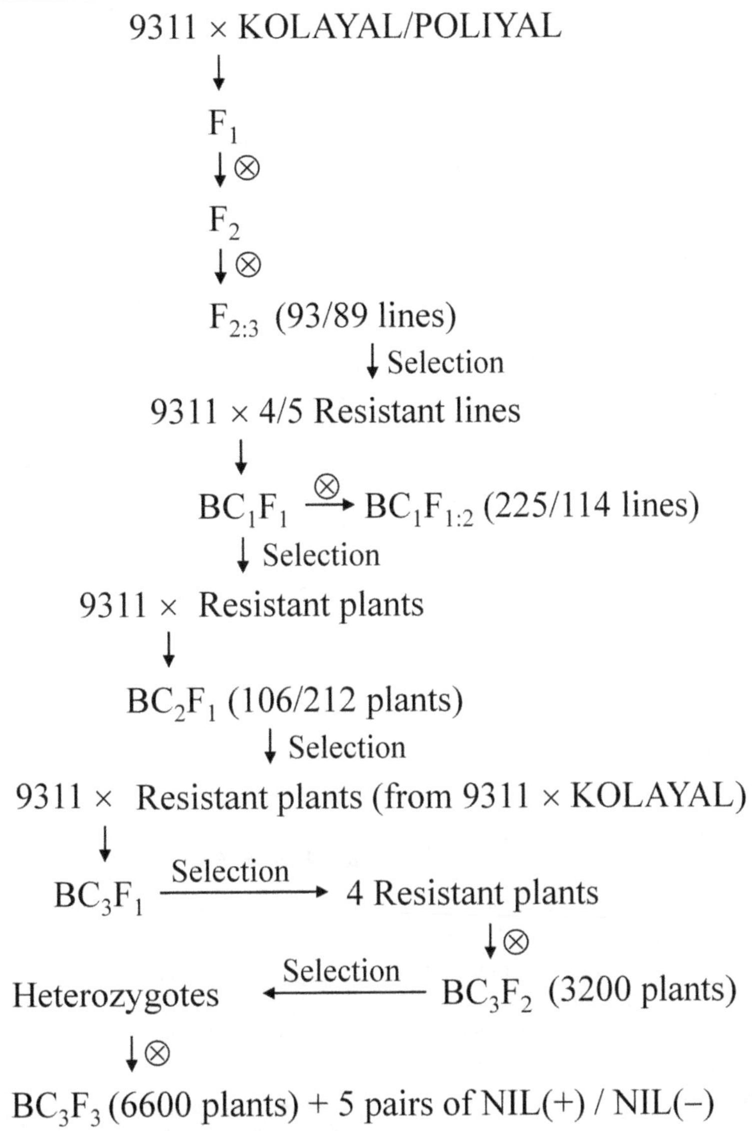

Fig. 8 Procedure of creating experimental populations or lines used in this study. NIL(+)/NIL(−), near-isogenic lines with 9311 background carrying the BPH resistant/susceptible allele. Both target gene and genetic background were selected. Target gene was selected according to resistance phenotype or flanking markers. Background was selected according to phenotypes of morphological traits

each, or of POLIYAL and 9311 each, were randomly planted in a circle within a plastic bucket of 20 cm in diameter. Plants of the same genotype were marked with small labels. Then, 100 nymphs of 2–3 instar were released in each bucket covered by a transparent gauze. The number of BPHs settling on each plant and the total number of nymphs on 5 plants of the same genotype were recorded at 3 h, 6 h, 24 h and 48 h after release. Six replicates were set, each had 3 buckets in a treatment.

Antibiosis test

To test antibiosis, the amount (weight and area) of honeydew excreted by BPH and the survival rate of BPH were measured. Honeydew collection was performed using the method of Pathak et al. (1982) with slight modification. Newly-emerged fifth-instar female nymphs that had starved for 2 h were enclosed in Parafilm sachets (two nymphs per sachet) and attached to the main stems of rice plants (35 days old) for 24 h. The honeydew in each sachet was then weighed using an electronic balance of 0.1-mg precision. In total, 30 sachets were tested on NIL (+) and NIL (-) each, with three sachets on one plant.

Honeydew area measurement was performed following the method of Du et al. (2009) with modification. A 35-day-old rice plant, of which the leaves and tillers were removed and only the main stem remained, was transplanted into a transparent plastic cup of 17 cm in height and 8 cm in diameter. The cup was covered mouth to mouth by another cup of the same type. The two cups were separated by a piece of filter paper. There was a small hole at the center of the filter paper, through which the main stem stretched into the upper cup. Several small holes were made at the bottom of the upper cup for air exchange. Five fifth-instar BPH nymphs that had starved for 2 h were put into the upper cup. After 2 days, the filter paper was dried under 60 °C for 30 min and then treated with 0.1% solution of ninhydrin in acetone to visualize honeydew stains, which appeared as violet or purple areas on the filter paper. Ten replicates were tested.

BPH survival rate experiment was performed in a similar way to the honeydew area test. Each 35-day-old plant was infested with 15 first-instar nymphs, and the number of survival nymphs in the cup was recorded every day for a period of 7 days. Again, all experiments were conducted in ten replicates.

Marker sources and genotyping

Simple sequence repeats (SSR) markers were obtained from the database Gramene (http://www.gramene.org), and insertion-deletion (InDel) markers (Additional file 1: Table S1) were developed based on the genome sequences of 1,479 rice cultivars (http://ricevarmap.ncpgr.cn) as well as the genome sequencing data of the three parents (see below). Genomic DNA was extracted from fresh leaves using the CTAB method with modification. The PCR system for SSR/InDel assay contained 10.0 μL of 2 × buffer, 2.0 μL of dNTP (2 mM), 0.3 μL of each primer (50 ng/μL), 5.3 μL of ddH_2O, 0.1 μL of rTaq E (5 U/μL), and 2.0 μL of DNA template (50–100 ng/μL). The PCR program used was: 94 °C for 4 min; 32 cycles of 94 °C for 30 s, 55 °C for 30 s, and 72 °C for 30 s; and 72 °C for 5 min. PCR products were separated on 4% denaturing polyacrylamide gels and visualized by silver staining.

Gene mapping

The next-generation sequencing-based bulked segregant analysis method (BSA-seq) was used to map the BPH resistance gene. A pair of bulks of individuals with opposite extreme phenotypes (extremely-resistant bulk consisting of 15 or 17 lines vs. extremely-susceptible bulk consisting of 16 or 18 lines) was selected from each of the two $F_{2:3}$ populations, and genomic DNA pools of the four bulks were prepared. The DNA pools as well as the DNA of the three parents were deeply (~ 20×) sequenced on the platform of Hiseq X Ten. Single nucleotide polymorphism (SNP) and short-InDel markers were identified based on the sequencing data, and gene mapping was conducted using the markers.

Analysis of gene expression response to BPH infestation

Quantitative real-time PCR (qRT-PCR) was conducted to analyze the expression of genes in *Bph33* region in NIL (+) and NIL (−) before and after BPH infestation, using the primers described by Du et al. (2009) and Zhao et al. (2016) (Additional file 4: Table S2). A thirty-day-old plant with all leaves removed was transplanted into a transparent plastic cup and covered by another transparent plastic cup as described above. Total RNA was isolated from the stems before BPH infestation and 6 h and 48 h after infestation, respectively, using TRIzol reagent (Invitrogen) following the manufacturer's instructions, and then converted into first-strand cDNA. Total RNA was also extracted from non-infested NIL (−) plants at every time point. qRT-PCR was performed using SYBR Green PCR Master Mix (Applied Bio systems) and a CFX96 Real-Time System (Bio-Rad) following the manufacturer's instructions. Three biological replicates were set for each treatment. Each replicate was a mixture of 10 independent NILs of the same genotype in BPH resistance (either resistant or susceptible), and each NIL consisted of 5 plants.

Statistical analysis

Student's t-test was used to examine the difference between treatment and control. One-way ANOVA and Duncan tests were used to compare multiple samples. Statistical tests were conducted using the software SPSS.

Additional files

Additional file 1: Table S1. Indel markers for fine mapping of *Bph33*.

Additional file 2: Table S3. Structural variations in candidate genes of *Bph33*.

Additional file 3: Figure S1. Expression analysis of *Bph33* candidate genes in the two NILs with contrary *Bph33* genotypes. Letter 'a' or 'b' indicates significant difference between NIL(+) and NIL(−) at the same time point after BPH infestation at 0.01 or 0.05 level. Asterisks indicate significant difference between the none-infested (0 h) and the infested plants at 0.01 (**) or 0.05 (*) level. Error bar indicates standard error

Additional file 4: Table S2. List of primers used for analyzing the expression of genes in *Bph33* region.

Acknowledgments

We are very grateful to IRRI for providing seeds of resistant parents.

Funding

This work was supported in part by National Key R&D Program of China (2017YFD0100103), Fujian Provincial Natural Science Foundation (2017 J01437), National Natural Science Foundation for the Youth of China (31401355).

Authors' contributions

WW and JH conceived the idea and wrote the manuscript. JH, XC and LZ conducted the research. WT analyzed the data in BSA-seq experiment. All authors read and approved of the final manuscript.

Competing interests

The authors declare that they have no competing interests.

References

Cha YS, Ji H, Yun DW et al (2008) Fine mapping of the rice *Bph1* gene, which confers resistance to the brown planthopper (*Nilaparvata lugens*) and development of STS markers for marker assisted selection. Mol Cells 26:146–151

Chen JW, Wang L, Pang F (2006) Genetic analysis and fine mapping of a rice brown planthopper (*Nilaparvata lugens* Stål) resistance gene bph19(t). Mol Gen Genomics 275:321–329

Cohen MB, Alam SN, Medina EB, Bernal CC (1997) Brown planthopper, *Nilaparvata lugens*, resistance in rice cultivar IR64: mechanism and role in successful N. *lugens* management in Central Luzon, Philippines. Entomol Exp Appl 85:221–229

Du B, Zhang WL, Liu BF et al (2009) Identification and characterization of *Bph14*, a gene conferring resistance to brown planthopper in rice. P Natl Acad Sci USA 106:22163–22168

Guo JP, Xu CX, Wu D et al (2018) Bph6 encodes an exocyst-localized protein and confers broad resistance to planthoppers in rice. Nat Genet. https://doi.org/10.1038/s41588-018-0039-6

He J, Liu YQ, Liu YL et al (2013) High-resolution mapping of brown planthopper (BPH) resistance gene *Bph27(t)* in rice (*Oryza sativa* L.). Mol Breeding 31:549–557

Hu J, Xiao C, Cheng MX et al (2015a) Fine mapping and pyramiding of brown planthopper resistance genes *QBph3* and *QBph4* in an introgression line from wild rice *O. officinalis*. Mol Breeding 35:3. https://doi.org/10.1007/s11032-015-0228-2

Hu J, Xiao C, Cheng MX et al (2015b) A new finely mapped *Oryza australiensis*-derived QTL in rice confers resistance to brown planthopper. Gene 561:132–137

Hu J, Xiao C, He YQ (2016) Recent progress on the genetics and molecular breeding of brown planthopper resistance in rice. Rice 9:30. https://doi.org/10.1186/s12284-016-0099-0

Huang D, Qiu Y, Zhang Y et al (2013) Fine mapping and characterization of *BPH27*, a brown planthopper resistance gene from wild rice (*Oryza rufipogon* Griff.). Theor Appl Genet 126:219–229

Huang Z, He GC, Shu LH, Li XH, Zhang QF (2001) Identification and mapping of two brown planthopper resistance genes in rice. Theor Appl Genet 102:929–934

Jairin J, Phengrat K, Teangdeerith S et al (2007) Mapping of a broad-spectrum brown planthopper resistance gene, *Bph3*, on rice chromosome 6. Mol Breeding 19:35–44

Jena KK, Jeung JU, Lee JH et al (2006) High-resolution mapping of a new brown planthopper (BPH) resistance gene, *Bph18(t)*, and marker-assisted selection for BPH resistance in rice (*Oryza sativa* L.). Theor Appl Genet 112:288–297

Jena KK, Kim SM (2010) Current status of Brown Planthopper (BPH) resistance and genetics. Rice 3:161–171

Ji H, Kim SR, Kim YH et al (2016) Map-based cloning and characterization of the *Bph18* gene from wild rice conferring resistance to brown planthopper (BPH) insect pest. Sci Rep 6:34376

Liu YQ, Wu H, Chen H et al (2015) A gene cluster encoding lectin receptor kinases confers broadspectrum and durable insect resistance in rice. Nat Biotechnol 33:301–305

Lv WT, Du B, Shangguan XX et al (2014) BAC and RNA sequencing reveal the brown planthopper resistance gene *BPH15* in a recombination cold spot that mediates a unique defense mechanism. BMC Genomics 15:674–589

Mohanty SK, Panda RS, Mohapatra SL et al (2017) Identification of novel quantitative trait loci associated with brown planthopper resistance in the rice landrace Salkathi. Euphytica 213:38. https://doi.org/10.1007/s10681-017-1835-2

Murai H, Hashimoto Z, Sharma P et al (2001) Construction of a high resolution linkage map of a rice brown planthopper (*Nilaparvata lugens* Stål) resistance gene bph2. Theor Appl Genet 103:526–532

Normile D (2008) Reinventing rice to feed the world. Science 321:330–333

Pathak PK, Saxena RC, Heinrichs EA (1982) Parafilm sachet for measuring honeydew excretion by *Nilaparvata lugens* on rice. J Econ Entomol 75:194–195

Prahalada GD, Shivakumar N, Lohithaswa HC et al (2017) Identification and fine mapping of a new gene, BPH31 conferring resistance to brown planthopper biotype 4 of India to improve rice, Oryza sativa L. Rice 10:41. https://doi.org/10.1186/s12284-017-0178-x

Qiu Y, Guo J, Jing S et al (2010) High-resolution mapping of the brown planthopper resistance gene *Bph6* in rice and characterizing its resistance in the 93-11 and *Nipponbare* near isogenic backgrounds. Theor Appl Genet 121:1601–1611

Qiu Y, Guo J, Jing S et al (2012) Development and characterization of japonica rice lines carrying the brown planthopper-resistance genes *BPH12* and *BPH6*. Theor Appl Genet 124:485–494

Qiu YF, Guo JP, Jing SL et al (2014) Fine mapping of the rice brown planthopper resistance gene *BPH7* and characterization of its resistance in the 93-11background. Euphytica 198:369–379

Rahman ML, Jiang W, Chu SH et al (2009) High-resolution mapping of two rice brown planthopper resistance genes, *Bph20(t)* and *Bph21(t)*, originating from *Oryza minuta*. Theor Appl Genet 119:1237–1246

Ren J, Gao F, Wu X, Zeng L, Lv J, Su X, Luo H, Ren G (2016) *BPH32*, a novel gene encoding an unknown SCR domain-containing protein, confers resistance against the brown planthopper in rice. Sci Rep 6:37645. https://doi.org/10.1038/srep37645.8

Sun L, Su C, Wang C et al (2005) Mapping of a major resistance gene to brown planthopper in the rice cultivar Rathu Heenati. Breed Sci 55:391–396

Tamura Y, Hattori M, Yoshioka H et al (2014) Map-based cloning and characterization of a brown planthopper resistance gene *BPH26* from *Oryza sativa* L. ssp. *indica* cultivar ADR52. Sci Rep 4:5872

Tanaka K, Endo S, Kazana H (2000) Toxicity of insecticides to predators of rice planthoppers: spiders, the mirid bug and the dryinid wasp. Appl Entomol Zool 35:177–187

Tang WQ, Huang LK, Bu SH, Zhang XZ, Wu WR (2018) Estimation of QTL heritability based on pooled sequencing data. Bioinformatics 34:978–984

Wang Y, Cao LM, Zhang YX et al (2015) Map-based cloning and characterization of *BPH29*, a B3 domain-containing recessive gene conferring brown planthopper resistance in rice. J Exp Bot 66:6035–6045

Wu H, Liu YQ, He J et al (2014) Fine mapping of brown planthopper (*Nilaparvata lugens* Stål) resistance gene *Bph28(t)* in rice (*Oryza sativa* L.). Mol Breeding 33:909–918

Zeng ZB (1994) Precision mapping of quantitative trait loci. Genetics 136:1457–1468

Zhao Y, Huang J, Wang ZZ et al (2016) Allelic diversity in an NLR gene *BPH9* enables rice to combat planthopper variation. P Natl Acad Sci USA 113:12850–12855

15

Genome-wide Association Study and Candidate Gene Analysis of Rice Cadmium Accumulation in Grain in a Diverse Rice Collection

Junliang Zhao[1,2†], Wu Yang[1,2†], Shaohong Zhang[1,2], Tifeng Yang[1,2], Qin Liu[1,2], Jingfang Dong[1,2], Hua Fu[1,2], Xingxue Mao[1,2] and Bin Liu[1,2*]

Abstract

Background: Cadmium (Cd) accumulation in rice followed by transfer to the food chain causes severe health problems in humans. Breeding of low Cd accumulation varieties is one of the most economical ways to solve the problem. However, information on the identity of rice germplasm with low Cd accumulation is limited, particularly in *indica*, and the genetic basis of Cd accumulation in rice is not well understood.

Results: Screening of 312 diverse rice accessions revealed that the grain Cd concentrations of these rice accessions ranged from 0.12 to 1.23 mg/kg, with 24 accessions less than 0.20 mg/kg. Three of the 24 accessions belong to *indica*. *Japonica* accumulated significantly less Cd than *indica* ($p < 0.001$), while tropical *japonica* accumulated significantly less Cd than temperate *japonica* ($p < 0.01$). GWAS in all accessions identified 14 QTLs for Cd accumulation, with 7 identified in *indica* and 7 identified in *japonica* subpopulations. No common QTL was identified between *indica* and *japonica*. The previously identified genes (*OsHMA3, OsNRAMP1,* and *OsNRAMP5*) from *japonica* were colocalized with QTLs identified in *japonica* instead of *indica*. Expression analysis of *OsNRAMP2*, the candidate gene of the novel QTL (*qCd3–2*) identified in the present study, demonstrated that *OsNRAMP2* was mainly induced in the shoots of high Cd accumulation accessions after Cd treatment. Four amino acid differences were found in the open reading frame of *OsNRAMP2* between high and low Cd accumulation accessions. The allele from low Cd accumulation accessions significantly increased the Cd sensitivity and accumulation in yeast. Subcellular localization analysis demonstrated OsNRAMP2 expressed in the tonoplast of rice protoplast.

Conclusion: The results suggest that grain Cd concentrations are significantly different among subgroups, with Cd concentrations decreasing from *indica* to temperate *japonica* to tropical *japonica*. However, considerable variations exist within subgroups. The fact that no common QTL was identified between *indica* and *japonica* implies that there is a different genetic basis for determining Cd accumulation between *indica* and *japonica*, or that some QTLs for Cd accumulation in rice are subspecies-specific. Through further integrated analysis, it is speculated that *OsNRAMP2* could be a novel functional gene associated with Cd accumulation in rice.

Keywords: Rice (*Oryza sativa* L.), Cadmium accumulation, Quantitative trait locus (QTL), Genome-wide association study (GWAS)

* Correspondence: lbgz1009@163.com
†Junliang Zhao and Wu Yang contributed equally to this work.
[1]Guangdong Key Laboratory of New Technology in Rice Breeding, Guangzhou 510640, China
[2]Rice Research Institute, Guangdong Academy of Agricultural Sciences, Guangzhou 510640, China

Background

Cadmium (Cd) is one of the most mobile and toxic heavy metals. Due to rapid industrialization and environmental pollution, Cd pollution has become a major problem in paddy fields. It has been reported that approximately 20 million hectares of cultivated lands in China, or 20% of the total, were contaminated by Cd (Hu et al. 2016). Cd can be readily absorbed by the roots of crops and transferred to aboveground organs, then enter the human body through the food chain (Li et al. 2017). Rice is a staple food for nearly half of the world's population, and compared to other cereal crops, rice tends to accumulate more Cd (Hu et al. 2016). A recent survey showed that 10% of rice grain with Cd exceeded the allowable concentration of 0.2 mg/kg as stipulated by the National Food Hygiene Standard of China (Xie et al. 2017). As rice is the primary source of dietary Cd intake (Clemens and Ma 2016; Uraguchi and Fujiwara 2013), Cd contamination in rice is becoming a severe public health problem.

Several techniques have been proposed to minimize Cd contamination in rice, such as phytoremediation and soil washing. Murakami et al. (2009) demonstrated that phytoextraction with the *indica* rice Chokoukoku, grown for 2 years without irrigation after drainage, removed 883 g Cd ha^{-1}, reducing the total soil Cd content by 38%. Makino et al. (2008) tested the feasibility of extracting Cd from the soil with iron, manganese, and zinc salts. Their results revealed that manganese, zinc, and iron salts extracted 4–41%, 8–44%, and 24–66% of the total Cd, respectively. Although these techniques are effective, they are time-consuming, expensive, or impractical in some situations. Many studies have revealed that Cd accumulation in rice varied considerably among cultivars and suggested that the breeding of low Cd-accumulating rice cultivars is a feasible strategy (Arao and Ae 2003; Cao et al. 2014; Yu et al. 2006). To date, breeding and use of rice cultivars with low Cd concentration in grain has been considered a promising solution to reducing Cd contamination in rice.

Obtaining rice germplasm with low Cd accumulation and identifying the genes responsible for Cd intake and transport are the prerequisites for the effective breeding of low-Cd-accumulation rice. In the past decade, much effort has been dedicated to the identification of rice germplasm with low Cd accumulation and of QTLs/genes controlling Cd accumulation (Ishikawa et al. 2005; Miyadate et al. 2011; Shimo et al. 2011). Ishikawa et al. (2005) conducted the first QTL analysis of Cd accumulation in rice, using 39 chromosome segment substitution lines derived from "Kasalath" and "Koshihikari". They identified QTLs controlling the Cd concentration on chromosomes 3, 6, and 8 in brown rice. Since then, many QTLs for Cd accumulation in rice have been identified and mapped (Ueno et al. 2009; Xue et al. 2009; Norton et al. 2010; Ishikawa et al. 2010; Abe et al. 2013). Several genes have been cloned, and their functions have been confirmed regarding intake and transport of Cd in rice (Clemens and Ma 2016; Uraguchi and Fujiwara 2013). The P_{1B}-type Heavy Metal ATPase 3 (*OsHMA3*) is the first cloned gene responsible for Cd accumulation in rice (Miyadate et al. 2011; Ueno et al. 2010). *OsHMA3* is expressed in the tonoplast of root cells, and limits translocation of Cd from the roots to the aboveground tissues by selectively sequestrating Cd into the root vacuoles (Miyadate et al. 2011; Ueno et al. 2010). Some *indica* and temperate *japonica* cultivars contain the non-functional allele of *OsHMA3*, which is related to higher Cd accumulation in these cultivars (Ueno et al. 2010; Yan et al. 2016). *OsHMA2* is another HMA family gene confirmed to be responsible for Cd accumulation in rice. Studies on OsHMA2 suggest that it plays a role in Cd loading to the xylem and participates in root-to-shoot translocation of Cd (Satoh-Nagasawa et al. 2012; Takahashi et al. 2012). Another study by Yamaji et.al (2013) found that OsHMA2 also plays an important role in preferential distribution of Cd through the phloem to the developing tissues. Additionally, *OsNRAMP1* and *OsNRAMP5*, belonging to the natural resistance-associated macrophage protein (NRAMP) family, were also identified as Cd transporters in rice (Sasaki et al. 2012; Takahashi et al. 2011). OsNRAMP1 participates in cellular Cd uptake and Cd transport within plants, and the higher expression of OsNRAMP1 in the roots could lead to an increase in Cd accumulation in the shoots (Takahashi et al. 2011). OsNRAMP5 is a plasma membrane-localized protein polarly localized at the distal side of both exodermis and endodermis cells. Functional analysis revealed that OsNRAMP5 is responsible for the transport of Mn and Cd from the external solution to root cells (Ishikawa et al. 2012; Sasaki et al. 2012). Despite significant progress being made in the identification of rice germplasm and associated QTLs/genes with low Cd accumulation, identification of both the *indica* germplasm with low accumulation and functional genes associated with low Cd accumulation from *indica* are limited.

Fortunately, rice germplasm is very rich, providing a good condition in searching for rice cultivars with low Cd accumulation. Genome-wide association study (GWAS) has emerged as a powerful approach for identifying the genes underlying complex traits at an unprecedented rate (Huang et al. 2010). This approach was first applied to the genetic analysis of human diseases (Hirschhorn and Daly 2005), and then used extensively for the genetic dissection of complex traits in plants (Li et al. 2016; Weng et al. 2011; Yang et al. 2014). With the rapid development in next-generation sequencing technology, GWAS with high-density molecular markers and more diverse resources is possible. Recently, McCouch et al. (2016) established an open-access resources for

GWAS in rice, which included a collection consisting of 1568 diverse inbred rice varieties and their genotypes determined by 700,000 single-nucleotide polymorphisms (SNPs), providing a good resource for the genetic dissection of complex traits in rice through GWAS.

To identify rice germplasms with low Cd accumulation and the responsible genes, 312 diverse rice accessions, selected from the 1568 rice accessions (McCouch et al. 2016), were subjected to evaluation of Cd accumulation; GWAS was performed to map QTLs associated with low Cd accumulation in the present study. Twenty-four rice accessions were identified, including three *indica* varieties, with grain Cd accumulation less than 0.2 mg/kg. In total, 14, 7, and 7 QTLs for low Cd accumulation in grain were identified by GWAS, using all 312 rice accessions, in the *indica* subpopulation, and in the *japonica* subpopulation, respectively. The previously identified functional genes, including *OsHMA3, OsNRAMP1, OsNRAMP5,* and *OsLCD,* co-localized with the QTLs identified in the present study, suggesting the reliability of the GWAS results. Through differential expression analysis, genomic sequence analysis, yeast expression experiments, and subcellular localization analysis, *Os03g0208500,* a member of the rice NRAMP gene family, was identified as the possible functional gene of a novel QTL *(qCd3–2)* identified in the present study. The results provide a good basis for cloning of the gene and molecular breeding for low Cd accumulation in rice.

Results

Phylogenetic analysis of 312 rice accessions

A total of 312 rice accessions, which are from 53 countries worldwide, were selected from 1568 rice accessions (McCouch et al. 2016) according to their diversity of genotypes, origins, and subpopulations. Phylogenetic tree (Fig. S1) of the selected 312 rice accessions was constructed using SNP data from the previous study (McCouch et al. 2016). The phylogenetic analysis showed that 312 rice accessions could be divided into two subpopulations, namely *indica* population and *japonica* population, according to their genetic similarities. The *indica* population consists of 213 rice accessions, while the *japonica* population consists of 99 rice accessions. The *japonica* population can be further divided into two groups (temperate *japonica* group and tropical *japonica* group) according to the phylogenetic analysis and information provided by the previous study (McCouch et al. 2016).

Variations in cd accumulation in grain among 312 rice accessions

Substantial variation and normal distribution in grain Cd accumulation were observed in 312 rice accessions (Fig. 1A). The *indica* cultivar "BOL ZO" from Korea Republic had the lowest Cd accumulation in grain (0.12 mg/kg), while the *indica* cultivar "Jariyu" from India had the highest Cd accumulation in grain (1.23 mg/kg). In total, 24 rice accessions had Cd accumulation in grain lower than 0.20 mg/kg (Table 1). Three out of those 24 rice accessions belong to *indica* and the others belong to tropical *japonica*. The comparisons in grain Cd accumulation of subspecies and subgroups revealed that the mean grain Cd accumulation of *indica* was significantly higher than that of *japonica* ($p < 0.001$) (Fig. 1B), while the mean grain Cd accumulation of temperate *japonica* was markedly higher than that of tropical *japonica* ($p < 0.01$) (Fig. 1C).

Mapping of QTLs for cd accumulation in grain by GWAS

Based on the criteria of having less than 15% missing data and minor allele frequency (MAF) larger than 5% in the selected population, 183,884 SNPs were selected for GWAS from the SNP dataset in the present study (McCouch et al. 2016). The QQ plot is shown in Fig. 2A. According to the LD decay rate of 312 rice accessions, a significant LD decay was observed after 200 kb (Fig. 2B). Therefore, a region was considered as one QTL where it had more than two SNPs with $P < 0.0001$ within a 200-kb genomic window. The Manhattan plot of the GWAS results is shown in Fig. 2C. In total, 14 QTLs with 49 SNPs were significantly associated with grain Cd accumulation in the 312 rice accessions (Table 2). These QTLs (designate as *qCd* hereafter) distributed on chromosomes 1, 2, 3, 4, 7, 8 and 11, with detailed information of the 14 QTLs listed in Table 2 of this report. It is noted that three QTLs identified in the present study co-localized with the four previously cloned genes associated with Cd accumulation in rice (*OsLCD, OsHMA3, OsNRAMP1, OsNRAMP5*) (Miyadate et al. 2011; Sasaki et al. 2012; Shimo et al. 2011; Takahashi et al. 2011) (Fig. 2 D and 2E) (Table 2).

GWAS was further conducted using the *indica* and *japonica* subpopulations of the 312 rice accessions and compared with all the identified QTLs (Fig. 3). Since fewer SNPs were used for GWAS in *indica* and *japonica* subpopulations according to the criteria of MAF > 0.05, the *p*-value was reduced for selecting significant SNPs to $p < 0.001$. In total, 7 and 7 QTLs were identified in the *indica* and *japonica* subpopulations, respectively (Table 2). Comparisons of the QTLs identified in different populations indicated that *qCd1, qCd2–1, qCd2–2, qCd4* and *qCd8–1,* which were identified in the whole population, overlapped with *qCd_ind-1, qCd_ind-2-1, qCd_ind-2-2, qCd_ind-4* and *qCd_ind-8,* which were identified in the *indica* subpopulation. The *qCd2–3, qCd3–2, qCd7–1, qCd7–2* identified in the whole population overlapped with *qCd_jap-2, qCd_jap-3, qCd_jap-7-1* and *qCd_jap-7-2* from the *japonica* subpopulation. The three cloned genes functioning in rice Cd accumulation (*OsHMA3, OsNRAMP1, OsNRAMP5*) were also located within the interval of *qCd_jap-7-1* and *qCd_jap-7-2,* respectively (Table 2). Interestingly, no common QTL was detected between the *indica* and *japonica* subpopulations (Fig. 3, Table 2).

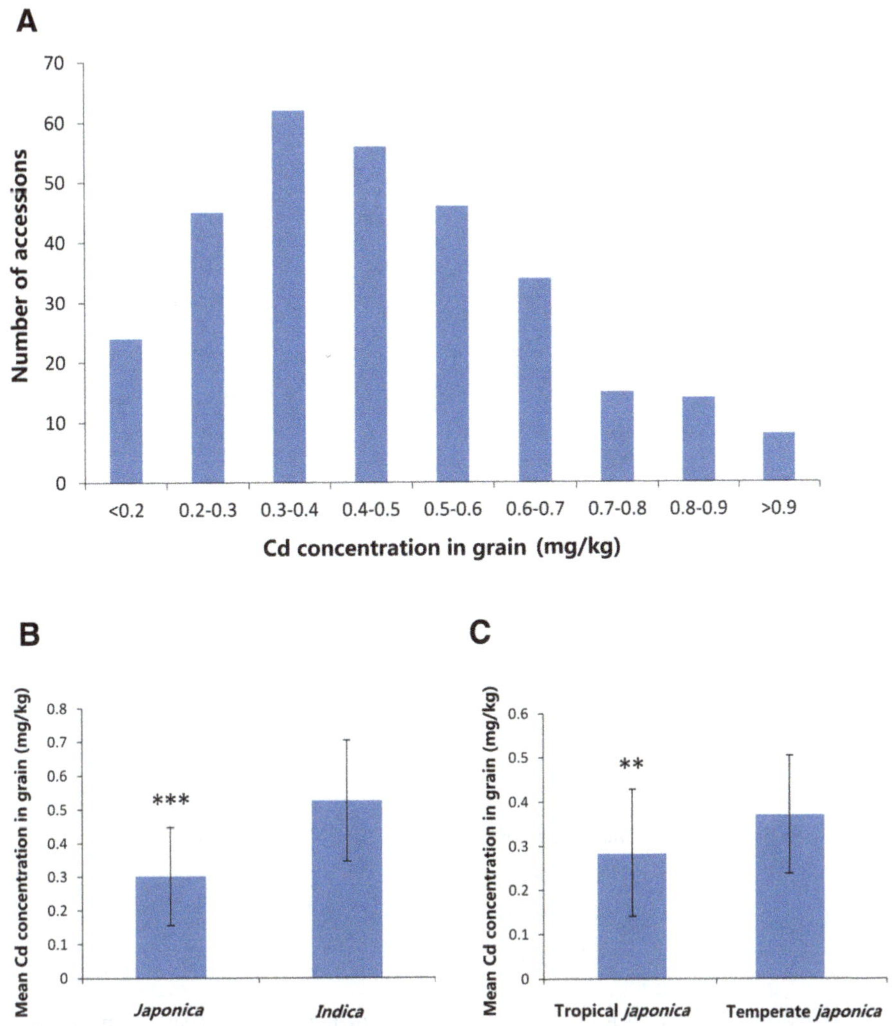

Fig. 1 Distribution of Cd concentration in grain in 312 rice accessions and comparisons of Cd concentration in grain between subspecies and between subgroups. **a**, Distribution of Cd concentration in grain of 312 rice accessions. **b**, Comparison of Cd concentration in grain between *indica* and *japonica* species. **c**, Comparison of Cd concentration in grain between Temperate and Tropical *japonica*. *** and ** represent significant difference at $p < 0.001$,and $p < 0.01$ in *t*-test

Identification of candidate genes related to cd accumulation in grain

To identify candidate genes related to Cd accumulation in grain, annotations of the genes were analyzed within the identified QTLs (200 kb up and down stream of the most significant SNP of the QTL) based on the Rice Annotation Project (RAP) (Kawahara et al. 2013). Based on its annotation, a potential functional gene (*Os03g0208500*) belongs to the rice NRAMP gene family and is located 14 kb downstream from the most significant SNP in *qCd3–2* (Fig. 4A). The gene symbol synonym of *Os03g0208500* in RAP is *OsNRAMP2*. Since other two members, *OsNRAMP1* and *OsNRAMP5*, have been reported to function in Cd transportation in rice (Sasaki et al. 2012; Takahashi et al. 2011), it was reasoned that *OsNRAMP2* might be the functional gene of *qCd3–2* associated with grain Cd accumulation in rice.

To confirm the function of *OsNRAMP2* in Cd accumulation in rice, the gene structure of *OsNRAMP2* and its expression patterns in both high and low Cd accumulation rice accessions were analyzed. The haplotypes associated with high and low Cd accumulation were determined by analyzing SNPs in the *qCd3–2* region and grain Cd accumulation of the tested rice accessions. In total, 15 rice accessions with low Cd accumulation phenotype and low Cd accumulation haplotype, and 15 accessions with the contrary phenotype and haplotype were selected. The genomic sequences of *OsNRAMP2* in the selected 30 rice accessions were amplified by PCR and individually sequenced. Sequence analysis revealed four SNPs and one Indel in the coding sequence (CDS) of *OsNRAMP2* among the 30 rice accessions (Fig. 4B). Among these differences, there were one Indel and three SNPs between the 15 high Cd accumulation accessions and

Table 1 The rice accessions with Cd concentration lower than 0.2 mg/kg in grain

Accession name	Subpopulation	Origin	Cd concentration in grain (mg/kg)
BOL ZO	*Indica*	Korea Rep	0.12
IAC 164	Tropical *Japonica*	Brazil	0.13
CHERIVIRUPPU	*Indica*	India	0.14
LEVANTE HOMEM	Tropical *Japonica*	Brazil	0.15
PACHOLINHA	Tropical *Japonica*	Brazil	0.15
ARAGUAIA	Tropical *Japonica*	Brazil	0.15
KWADWO AMOA	Tropical *Japonica*	Ghana	0.15
GBANTE	Tropical *Japonica*	Ivory Coast	0.16
MUT IAC 25–44-807	Tropical *Japonica*	Guyana	0.16
TOANG	Tropical *Japonica*	Indonesia	0.16
KETAN MERAH	Tropical *Japonica*	Indonesia	0.16
PATO DE GALLINAZO	Tropical *Japonica*	Australia	0.16
BOTRA FOTSY	Tropical *Japonica*	Madagascar	0.16
VARY MAINTY	Tropical *Japonica*	Madagascar	0.16
VARY SOMOTRA SIHANAKA	Tropical *Japonica*	Madagascar	0.17
MANGAVAVA FOTSILANSTSIKA	Tropical *Japonica*	Madagascar	0.17
BAKAW	Tropical *Japonica*	Philippines	0.18
HONDURAS	Tropical *Japonica*	Spain	0.18
IRAT 364	Tropical *Japonica*	Nicaragua	0.18
ARC 18294	*Indica*	India	0.19
RXAR RGUE:	Tropical *Japonica*	United States	0.19
WAB 56–125	Tropical *Japonica*	Ivory Coast	0.19
1–52-6	Tropical *Japonica*	Brazil	0.19
BOMALASANG	Tropical *Japonica*	Philippines	0.19

the 15 low Cd accumulation accessions except for the rice accession 194 from this study. The sequences of allele from low Cd accumulation accessions (designated as *OsNRAMP2-L*) are exactly the same as that from Nipponbare. The allele from high Cd accumulation accessions (designated as *OsNRAMP2-H*) has two amino acid deletions in the 24th and 25th position, as well as an amino acid mutation at the 235th position, with Glu being substituted by Asp, and the 508th amino acid Arg being substituted by Gln. None of these amino acids are within the putative conserved domains previously identified (Cellier et al. 1995; Mani and Sankaranarayanan 2018). However, according to the alignment results with all rice NRAMP genes (Mani and Sankaranarayanan 2018), the Arg in the 508th position is a conserved amino acid across all 7 rice NRAMP genes. This transition may have some functional implications.

To further validate the function of *OsNRAMP2* in rice Cd accumulation, the expression pattern of *OsNRAMP2* in shoot and roots was investigated. The four rice accessions with high Cd accumulation, as well as four accessions with low Cd accumulation, were selected from the 30 rice accessions above. Results demonstrated that the expression levels of *OsNRAMP2* did not significantly change in the roots of any of the eight rice accessions after Cd treatment, compared to their respective controls ($p > 0.05$) (Fig. 4C). However, the expression levels of *OsNRAMP2* in the shoots of the high Cd accumulation rice accessions increased after Cd treatment, while no significant change was observed in low Cd accumulation rice accessions (Fig. 4D).

To investigate Cd transport function of the two alleles of *OsNRAMP2*, expression vectors (pYES2) containing *OsNRAMP2-H* and *OsNRAMP2-L* were used to transform the yeast INVSc1 strain and *Δycf1* mutant strain. The *Δycf1* mutant strain is sensitive to Cd treatment. There was no difference in Cd sensitivity between yeast strains expressing vector control or different alleles of *OsNRAMP2* in the presence of glucose in both the INVSc1 strain and *Δycf1* mutant strain (Fig. 5A, B). In the presence of galactose, which induced gene expression, significant increase in Cd sensitivity was observed in yeast strain *Δycf1* under 5 μM of Cd treatment, and in yeast strain INVSc1 under 10 μM of Cd treatment, when expressing *OsNRAMP2-L*. However, *OsNRAMP2-H* only slightly enhanced yeast Cd sensitivity compared to the

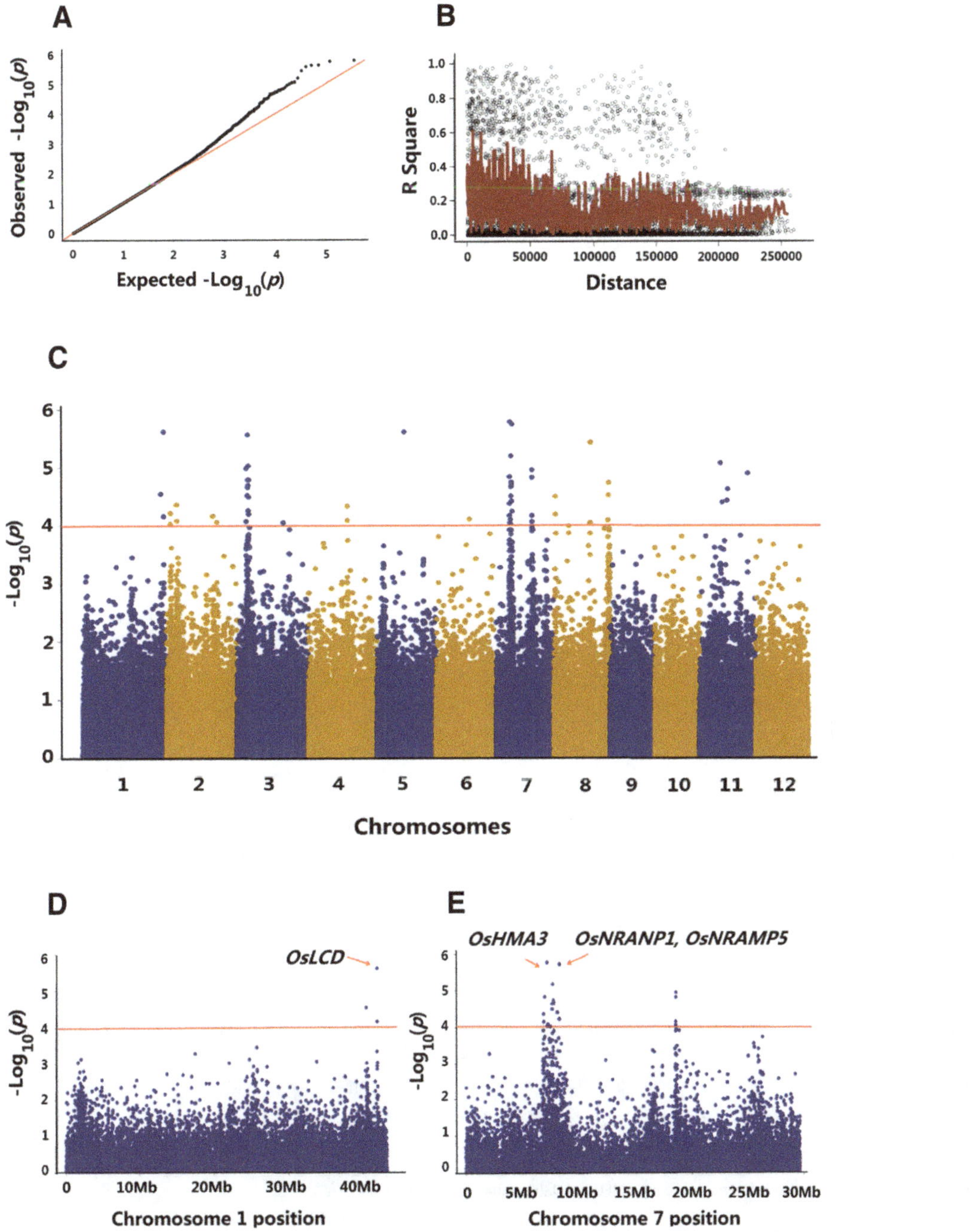

Fig. 2 GWAS of grain Cd accumulation using the 312 rice accessions. **a**, QQ-plot for GWAS of Cd concentration in grain. **b**, LD decay of the whole population. **c**, Manhattan plots of GWAS of grain Cd accumulation in 12 chromosomes. **d**, Colocalization of *OsLCD* with *qCd-1*. **e**, Co-localization of *OsHMA3*, *OsNRAMP1* and *OsNRAMP5* with *qCd3–1* and *qCd3–2*

empty vector (Fig. 5C, D). Further analysis of the Cd accumulation in *Δycf1* strain expressing different alleles demonstrated that yeast expressing *OsNRAMP2-L* accumulated more Cd than those expressing *OsNRAMP2-H* or the empty vector (Fig. 5E). These results indicate that the OsNRAMP2-L has Cd transport function in yeast, but the Cd transport function of OsNRAMP2-H is reduced due to the change in amino acid sequence.

Subcellular localization of OsNRAMP2 protein was further investigated in rice protoplasts. Enhanced green fluorescent protein (EGFP) or OsNRAMP2-L/OsNRAMP2-H fused with EGFP at the C-terminal (designate as OsNRAMP2-L::EGFP and OsNRAMP2-H::EGFP) were transiently expressed in protoplasts from rice seedlings (Zhang et al. 2011). Under confocal microscopy, the green fluorescent signal of both the OsNRAMP2-L::EGFP and

Table 2 QTL associated with Cd accumulation identified by GWAS using different populations

QTLs	Chromosome	Linked SNP position [a]	*P*-value	Phenotype contribution (%)	Candidate gene
Composite population					
qCd1	1	41,982,531	2.43E-06	4.66	*OsLCD*
qCd2–1	2	2,060,980	6.01E-05	2.40	
qCd2–2	2	5,523,027	4.38E-05	3.47	
qCd2–3	2	23,732,572	6.88E-05	2.71	
qCd3–1	3	4,848,498	1.04E-05	4.06	
qCd3–2	3	5,635,117	2.78E-06	4.61	
qCd4	4	20,074,216	4.67E-05	2.53	
qCd7–1	7	7,186,204	1.69E-06	4.82	*OsHMA3*
qCd7–2	7	8,263,482	1.86E-06	4.78	*OsNRAMP1*, *OsNRAMP5*
qCd7–3	7	18,553,958	1.14E-05	4.02	
qCd8–1	8	902,085	3.25E-05	3.01	
qCd8–2	8	18,365,933	3.82E-06	4.48	
qCd8–3	8	27,463,194	3.12E-05	2.60	
qCd11	11	14,891,373	2.48E-05		
Indica					
qCd_ind_1	1	41,982,531	8.89E-05	6.85	*OsLCD*
qCd_ind_2–1	2	2,060,980	7.66E-04	5.00	
qCd_ind_2–2	2	4,986,931	2.27E-04	6.04	
qCd_ind_3	3	27,195,551	3.06E-04	5.78	
qCd_ind_4	4	20,368,258	1.34E-04	5.93	
qCd_ind_5	5	23,588,056	4.53E-04	5.45	
qCd_ind_8	8	899,377	1.93E-05	8.20	
Japonica					
qCd_jap-2	2	23,666,135	5.23E-05	12.74	
qCd_jap-3	3	5,507,369	3.81E-04	9.63	
qCd_jap-7-1	7	7,451,789	3.52E-04	9.75	*OsHMA3*
qCd_jap-7-2	7	8,467,983	1.56E-04	11.01	*OsNRAMP1*, *OsNRAMP5*
qCd_jap-7-3	7	24,677,317	3.71E-05	13.29	
qCd_jap-7-4	7	27,037,840	6.24E-04	8.88	
qCd_jap-11	11	25,263,712	4.09E-04	9.52	

[a]Position of SNP is base on rice reference sequence MSU V 7.0. (Kawahara et al. 2013)

OsNRAMP2-H::EGFP was observed outside of the chloroplasts (the chloroplasts showed red signal) (Fig. 6). These results suggest that both proteins are localized to the tonoplast.

Discussion

Cd accumulation in grain is diverse in rice germplasm, and is significantly different among subspecies and subgroups

Cd is a highly toxic and mobile element. Consequently, it can easily enter the food chain and become detrimental to human and animal health. To minimize the influx of Cd into the human food chain through rice consumption, Yu et al. (2006) proposed the concept of pollution-safe cultivars (PSCs), i.e. cultivars whose edible parts accumulate a specific pollutant at a level low enough for safe consumption, even when grown in contaminated soil. Selecting or breeding low Cd accumulation cultivars is an attractive and economical method for minimizing the Cd contamination in rice, as the benefit will persist in the seed and reduces the need for other management techniques such as fertilization management or crop rotations.

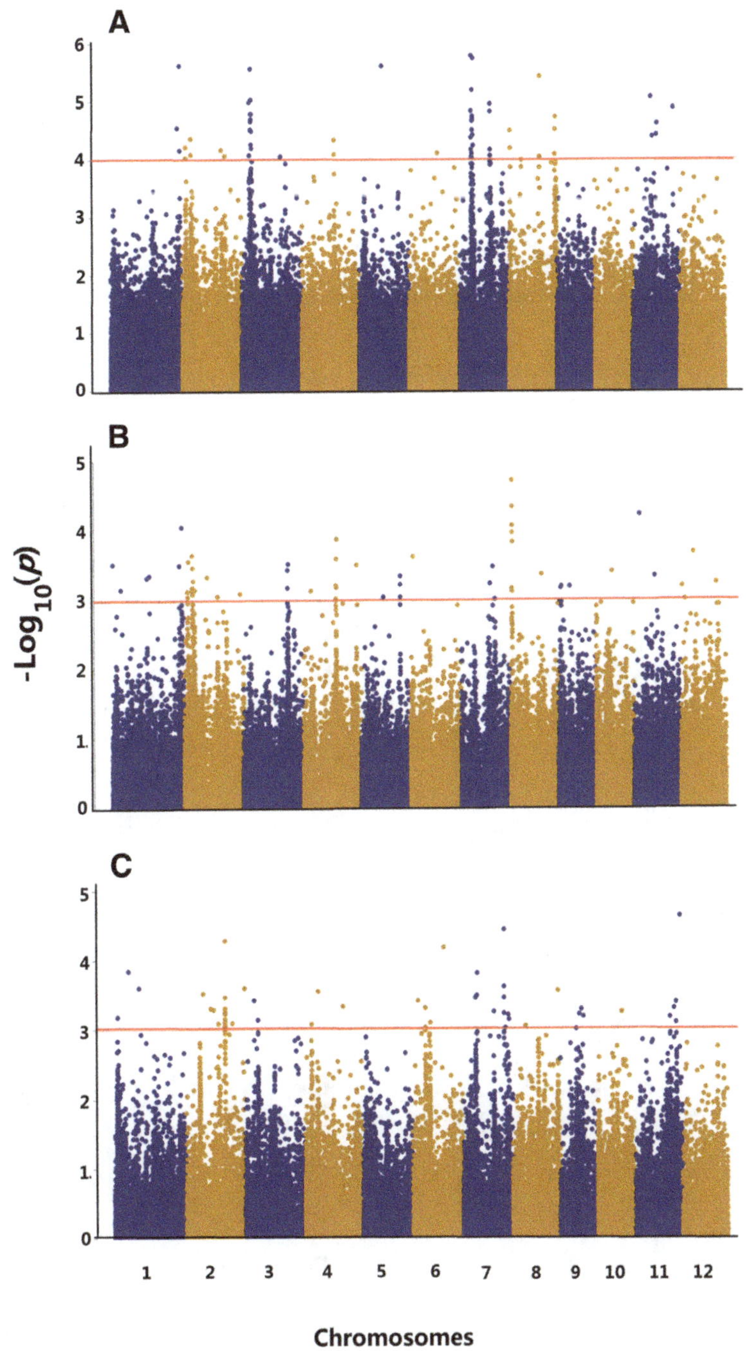

Fig. 3 Manhattan plots of GWAS of grain Cd accumulation using different populations. **a**, Composite population. **b**, *Indica* subpopulation. **c**, *Japonica* subpopulation

To search for rice resources with low Cd accumulation for breeding, a diverse rice collection consisting of 312 rice accessions was subjected to evaluation of Cd accumulation. The results showed that Cd concentration in grains among the 312 rice accessions varied considerably, ranging from 0.12 mg/kg to 1.23 mg/kg when these rice accessions were grown in Cd-contaminated paddy fields (Fig. 1A). Similar to previous studies (Arao and Ae 2003, Liu et al. 2005; Pinson et al. 2015; Zhou et al. 2017), the results indicate that the mean grain Cd accumulation in *japonica* accessions is significantly lower than that in *indica* accessions ($p < 0.001$) (Fig. 1B). Furthermore, the mean grain Cd concentration in tropical *japonica* accessions is significantly lower than that in temperate *japonica* accessions (Fig. 1C). This result contradicts previous findings claiming that temperate *japonica* has lower Cd accumulation compared to tropical *japonica* (Yan

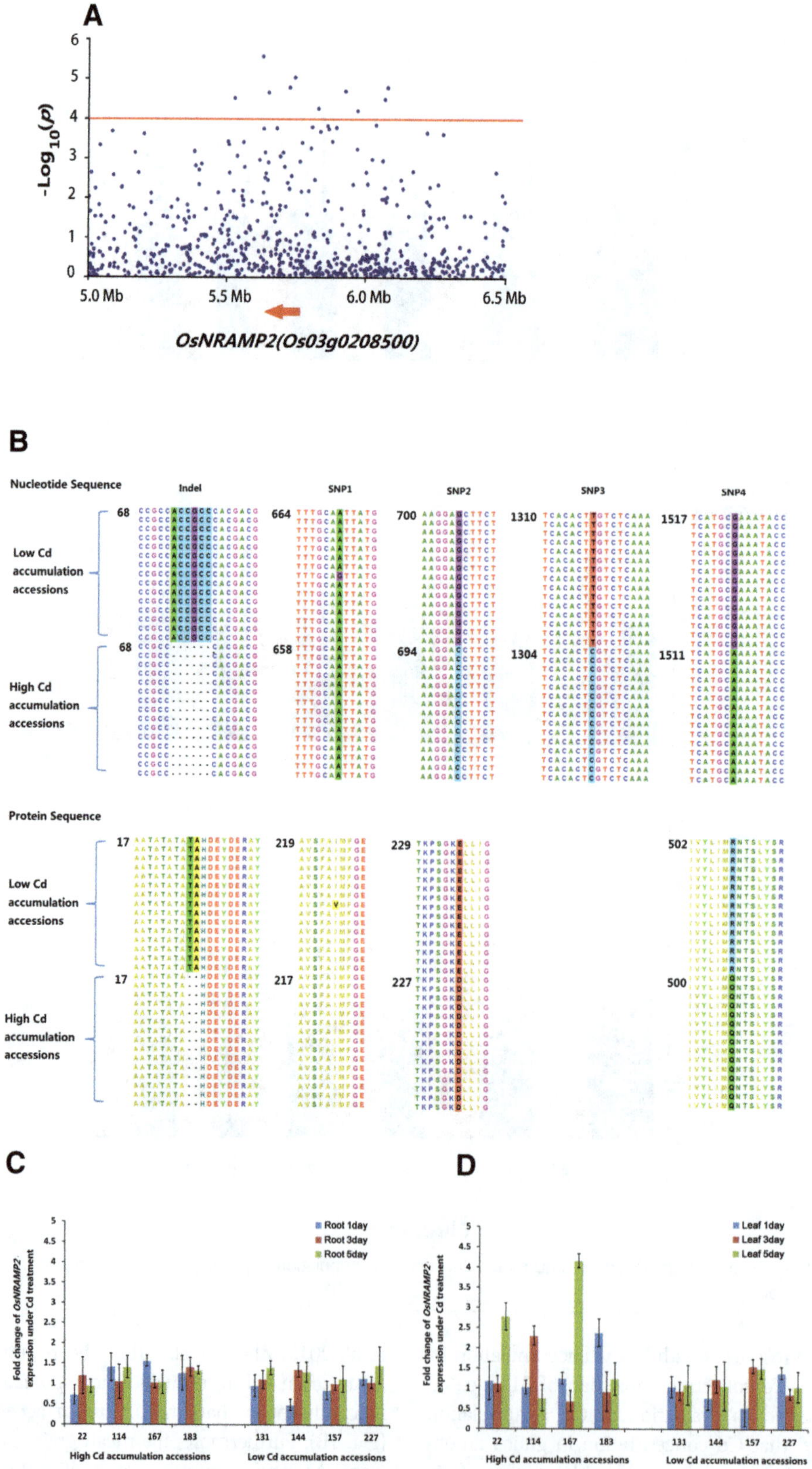

Fig. 4 (See legend on next page.)

(See figure on previous page.)
Fig. 4 The location of *OsNRAMP2* on chromosome 3 and sequence differences and expression analysis of *OsNRAMP2*. **a**, Colocalization of *OsNRAMP2* with *qCd3–2*. The red arrow indicates the location and direction of *OsNRAMP2*. b, Sequence differences of OsNRAMP2 protein and the respective causal nucleotide differences between high and low Cd accumulation accessions. Sequence differences between high and low Cd accumulation rice accessions are indicated with different background colors. The numbers on the left of each local alignment indicate the position of the first nucleotide/amino acid for the alignment. **c**, Expression changes of *OsNRAMP2* in roots after Cd treatment between high and low Cd accumulation rice accessions. **d**, Expression changes of *OsNRAMP2* in shoot after Cd treatment between high and low Cd accumulation rice accessions

et al. 2010). Although *indica* cultivars accumulate more Cd, three *indica* accessions with grain Cd concentrations lower than 0.20 mg/kg were still identified. Interestingly, the *indica* "BOL ZO" from the Korean Republic had the lowest grain Cd concentration (0.12 mg/kg) among the 312 rice accessions (Table 1). Together, these results suggest that Cd accumulation in grain is diverse in rice germplasm. However, the Cd accumulation in grain is significant different among subgroups, with accumulation in *indica* > temperate *japonica* > tropical *japonica*. Regardless of subspecies or subgroups, it is possible to screen for rice germplasm with low Cd accumulation, but there is a higher likelihood of identifying them in tropical *japonica*.

There is a different genetic basis for cd accumulation between *indica* and *japonica* rice

In the present study, 14 QTLs related to Cd accumulation were identified through GWAS using a population consisting of 312 rice accessions. Chromosomal position comparisons reveal that the previously identified genes

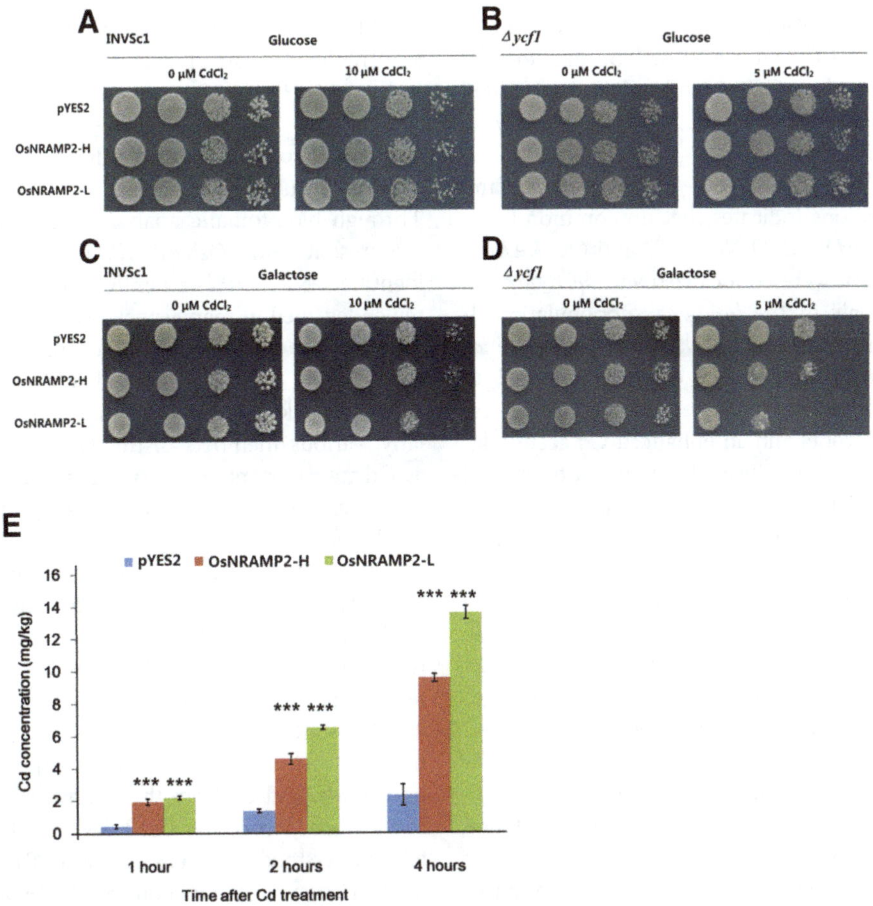

Fig. 5 Functional assay of *OsNRAMP2* by heterologous expression in yeast. Yeast INVSc1 strain and Δ*ycf1* mutant strain expressing the empty vector (pYES2), *OsNRAMP2-H* and *OsNRAMP2-L* were grown in SD-U medium containing concentrations of 0 and 5 μM/10 μM Cd in the presence of glucose (**a**, **b**) to suppress the expression of transformed gene, or galactose (**c**, **d**) to induce the expression of transformed gene. E, Cd accumulation in Δ*ycf1* strain expressing empty vector and *OsNRAMP2-H*, *OsNRAMP2-L* after exposure to 5 μM Cd for 1, 2 and 4 h. Data are means ± SD of three biological replicates. *** $p < 0.001$ in *t*-test

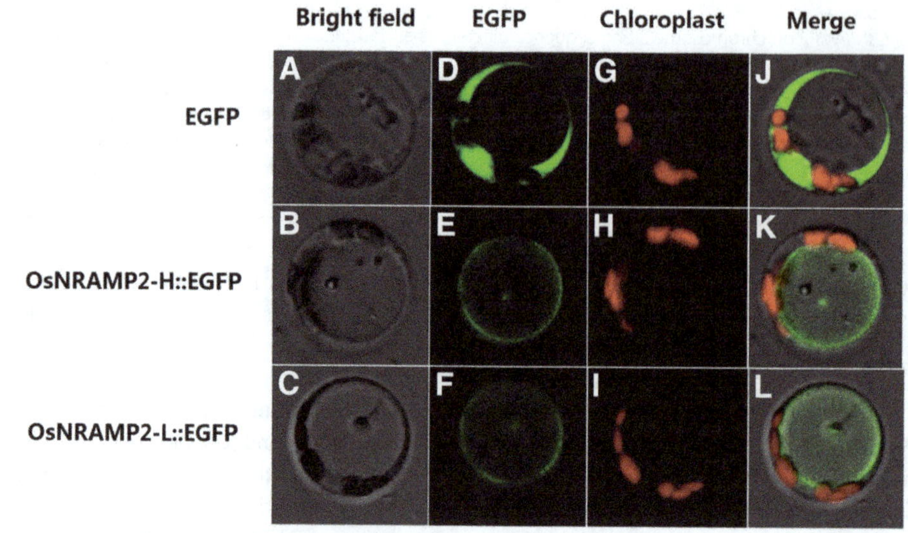

Fig. 6 Subcellular localization of OsNRAMP2. Images of bright field (**a-c**), EGFP-derived green fluorescence (**d-f**), chloroplast auto-fluorescence (**g-i**) and merge images (**j-l**) of protoplasts expressing EGFP, OsNRAMP2-H::EGFP and OsNRAMP2-L::EGFP

related to Cd transportation and accumulation, *OsLCD, OsHMA3, OsNRAMP1* and *OsNRAMP5*, are located within the QTLs *qCd1*, *qCd7–1* and *qCd7–2* identified in the present study, indicating the reliability of GWAS results in this study.

Comparison of QTLs identified by GWAS in *indica* and *japonica* subpopulations indicates that no common QTL. The three genes, *OsHMA3, OsNRAMP1* and *OsNRAMP5* cloned from *japonica*, were co-localized with *qCd_jap-7-1* and *qCd_jap-7-2* only in the *japonica* subpopulation. It was reported that some *indica* cultivars and temperate *japonica* cultivars contained a loss-of-function allele of *OsHMA3*, in which results in weakened vacuolar sequestration of Cd in the roots and an enhanced Cd accumulation, while many *japonica* cultivars contained a functional allele of *OsHMA3* (Ueno et al. 2010; Yan et al. 2016). Similarly, a significant difference was observed in the expression levels of *OsNRAMP1* between the *indica* and *japonica* subspecies, which may also explain the difference in Cd accumulation between these subspecies (Zhou et al. 2017). Our findings, together with the results from previous studies, suggest that there may be a different genetic basis of Cd accumulation in grain between *indica* and *japonica*, and the genes/QTLs responsible for Cd transportation and accumulation may be subspecies-specific.

Since *indica* rice accumulates more Cd and the cloned genes, so far, are from *japonica* rice, there is an urgent need to identify the genes related to Cd accumulation in *indica* rice and understand their molecular mechanisms. In the present study, three *indica* accessions with very low Cd concentrations in grain (Table 1) and 7 QTLs associated with Cd accumulation in *indica* have been identified by GWAS (Table 2), providing a good basis for the identification of genes responsible for Cd accumulation and molecular breeding for low Cd accumulation in *indica* rice.

OsNRAMP2 could be a novel functional gene controlling grain cd accumulation in rice

Through bioinfomatic analysis of the QTLs identified in the present study, *OsNRAMP2* in the *qCd3–2* region was identified. NRAMP is a family of integral membrane proteins identified in many species (Cellier et al. 1995). Previous studies demonstrated that many members of this family could function as metal transporters in yeast and plants (Supek et al. 1996; Thomine et al. 2000). Particularly, various members of this family had been identified as Cd transporters in plants (Clemens and Ma 2016). For example, AtNRAMP6 is a Cd transporter that functions inside the cell either by mobilizing cadmium from its storage compartment or by bringing cadmium to the cellular compartment where it is toxic (Cailliatte et al. 2009).

Altogether, seven *NRAMP* genes have been identified in rice, and five of them (*OsNRAMP1, 3, 4, 5,* and *6*) have been functionally characterized (Mani and Sankaranarayanan 2018). OsNRAMP1 and OsNRAMP5 were identified as Cd transporters in rice. OsNRAMP5 is responsible for the transport of Mn and Cd from the external solution to root cells, but the function of the OsNRAMP1 Cd transporter is not fully understood (Clemens and Ma 2016; Takahashi et al. 2011). However, the function of OsNRAMP2 as a Cd transporter had not been reported previously.

OsNRAMP2 localizes within a 200 kb region of *qCd3–2* as identified in the present study. There were four amino acid differences in the gene between high and low Cd accumulation accessions. The heterologous assay in yeast (Fig. 5) showed that OsNRAMP2-L can increase Cd sensitivity and accumulation in yeast, while OsNRAMP2-H only slightly increased Cd sensitivity and accumulation in yeast, suggesting that OsNRAMP2 could function as a Cd transporter and OsNRAMP2-L is its functional form. Furthermore, the subcellular localization analysis demonstrated that both OsNRAMP2-L and OsNRAMP2-H are localized at the tonoplast (Fig. 6). Base on previous studies, OsNRAMP1 and OsNRAMP5 are localized to the plasma membrane and mediate metal uptake into the cytosol (Sasaki et al. 2012; Takahashi et al. 2011). In the present study, the OsNRAMP2 protein was localized to the tonoplast, suggesting that it might mediate vacuole-to-cytosol transport of Cd as like AtNRAMP3 and AtNRAMP4, which influence metal accumulation by mobilizing vacuolar metal pool to the cytosol (Lanquar et al. 2005; Thomine et al. 2000; Thomine, et al. 2003). Further physiological, cellular and molecular studies are underway to elucidate the actual function and mechanism of OsNRAMP2 in Cd transport in rice.

Conclusion

In the present study, Cd accumulation in grain from a diverse rice collection was screened, and QTLs related to rice Cd accumulation were identified through GWAS. The results suggest that the Cd accumulation in grain is significantly different among subgroups (*indica* > temperate *japonica* > tropical *japonica*). Substantial variation also exists within subgroups. It is possible to screen for *indica* germplasm with low Cd accumulation as long as diverse *indica* accessions are investigated, but more success could be found in tropical *japonica*. The fact that no common QTL was identified between the *indica* group and *japonica* group in the present study imply that there may be a different genetic basis for Cd accumulation between *indica* and *japonica*, or that many QTLs for Cd accumulation in rice are subspecies-specific. Through an integrated analysis using GWAS, gene annotation, and functional analysis, it is speculated that OsNRAMP2 could be a functional Cd transporter, and is considered as a novel candidate functional gene associated with Cd accumulation in rice. This study provides new insights into the genetic basis of Cd accumulation, as well as a novel candidate functional gene associated with rice Cd accumulation. These results lay a strong foundation for gene cloning and molecular breeding for low Cd accumulation in rice.

Methods

Materials

A total of 312 rice accessions from 53 countries were selected according to the 1568 diverse rice accessions based on their 700,000 SNP genotypes and their origins (McCouch et al. 2016). Seeds of all 312 lines are provided by the International Rice Research Institute (IRRI).

Field experiment and cd treatment

The field experiments were conducted in 2016 in Renhua, Guangdong, China. The fields are rice-growing paddy fields belonging to light loam soils. Paddy water management, fertilizer applications, and crop protection followed the local farming practices. The 312 rice accessions were planted in a randomized complete design with three replicates. Ten plants per accession were planted in each replicate. The Cd concentration of the field is 1.4 mg/kg on average, which is higher than the second criteria level of the National Environmental Quality Standard for Soil (GB 15618–1995) (People's Republic of China). The concentrations of other heavy metals (Pb, Cr, Hg, As) are lower than the national standard level 1, indicating that the soil is not polluted by these heavy metals. The pH value of the field was 5.5.

Sampling and cd detection

To determine the grain Cd concentrations of the 312 rice accessions, 8 plants in the middle of the 10 plants in each replicate were harvested. The grains of eight plants in each replicate were pooled and dried in oven at 70 °C for 24 h, then de-husked. The de-husked grains were grounded into powder and digested with an acid mixture of HNO_3-$HClO_4$. The Cd concentration was determined using a Inductively Coupled Plasma Optical Emission Spectrometry (ICP-OES, iCAP6000,Thermo Scientific, USA).

GWAS, QTL delimitation and identification of candidate gene

GAPIT version 2 was used for GWAS analysis (Tang et al. 2016). SNPs were selected for GWAS analysis from the 700 K assay of a previous study by the criteria of having less than 15% missing data and minor allele frequency (MAF) > 0.05 (McCouch et al. 2016). GWAS was conducted using the mix liner model with kinship matrix, and PC was set to 2 in GAPIT. Manhattan and QQ plots were produced using R package qqman.

Rice genome sequence version of MSU V7.0 was used as reference for analysis (Kawahara et al. 2013). After GWAS analysis, we follow the criteria of having one associated locus between any two significant SNPs within a 200 kb interval. After determining the QTL of each GWAS analysis, the candidate genes were searched from 200 kb upstream and downstream of the most significant SNP in each QTL.

Sequencing of *OsNRAMP2* in the rice accessions with low and high cd accumulation

The low and high Cd accumulation rice accessions with 15 accessions each were selected according to the haplotype associated with Cd accumulation within *qCd3–2* and Cd accumulation in the grains of 312 rice accessions. Seedlings of selected rice accessions were used for DNA extraction by the DNeasy Plant Mini Kit (Qiagen, Germany). The DNA samples of the selected rice accessions were used as templates to amplify the full-length genomic sequence of *OsNRAMP2* by KOD-FX polymerase (Toyobo, Japan) using the following primers, forward primer: acaaccactcctagagtccagaga, reverse primer: ggcgacatctcctgaagataacct. After PCR amplification, PCR products were sequenced by Sangon Biotech Co., Ltd. (Shanghai, China). Sequences of *OsNRAMP2* from the 30 accessions were aligned and analyzed by MEGA 7.0 (Kumar et al. 2016).

Determining the expression patterns of *OsNRAMP2* in rice after cd treatment

Four rice accessions with high Cd accumulation and four rice accessions with low Cd accumulation were selected from the 30 rice accessions mentioned above to analyze the expression patterns of *OsNRAMP2* after Cd treatment. Seeds of selected rice accessions were germinated and grown hydroponically in half-strength Kimura B solution for 2 weeks, with pH adjusted to 5.6 for the duration. The solution was renewed every two days. Plants were grown in an RXZ-1000C growth chamber (Ningbo Jiangnan Instrument Factory, NingBo, China) under a light intensity of 359 μmol m^{-2} s^{-1} with a 12 h day length and a relative humidity of 80 ± 5%. The temperature was maintained at 25 °C. After two weeks of growing, plants were treated with 5 μM $CdCl_2$ for various days. After Cd treatment for 1d, 3d and 5d, plants were separated into root and shoot, and RNA was extracted by TRIzol (Thermo Fisher, USA) according to the manufacturer's instructions. Rice plants without Cd treatment were used as control. All assays were conducted in three biological replicates. The RNA samples were reversely transcribed using PrimeScript Reverse Transcriptase (Takara, Japan). Gene expression was determined by qRT-PCR using Biorad CFX96 (Biorad, USA) and SYBR Premix Ex Taq (Takara, Japan).

Functional analysis of *OsNRAMP2* alleles in yeast

Saccharomyces cerevisiae strain INVSc1 (Yan et al. 2016) and Cd-sensitive mutant strain *Δycf1* (Ueno et al. 2010) were used for heterologous expression of *OsNRAMP2-H* and *OsNRAMP2-L*. The ORF of *OsNRAMP2-H* and *OsNRAMP2-L* were amplified by KOD FX (Toyobo, Japan) from the cDNA of high and low Cd accumulation accessions as described above, using the following primers: forward primer, aaaagaattcaaaaaaatgtctatggcgtcgcgcgacctcg, reverse primer, aaaactcgagatcatgtgctctttgtcattgct. PCR products were cloned into expression vector pYES2 (Thermo Fisher, USA). Transformation and selection of empty vector and vectors with *OsNRAMP2* alleles were conducted according to a previous study (Ueno et al. 2010). Positive clones were cultured in SD medium without uracil (SD-U) liquid media with 2% glucose to the early log phase. Five μL of the cell suspension with an initial OD value of 0.1 and three serial 1:10 dilutions were spotted on SD-U plates containing 0 and 10 μM $CdCl_2$ for the INVSc1 strain, and 0 and 5 μM $CdCl_2$ for the *Δycf1* strain, in the presence of 2% glucose or galactose. The plates were incubated at 30 °C for 3d before the growth phenotypes were evaluated. Cd accumulation in yeast *Δycf1* strains containing pYES2 empty vector and vectors expressing *OsNRAMP2* alleles was determined according to a previous study (Yan et al. 2016). The transformed yeast strains were cultured in SD-U liquid media containing 2% glucose to the early log phase, then enriched by centrifugation and washed with sterile water three times. Cells were then adjusted to the same density (OD600 value 0.2) in SD-U liquid media containing 2% galactose. After incubation for 3 h to induce protein expression, the media was amended with 5 μM $CdCl_2$ and the cells were sampled after culturing for 1, 2 and 4 h. The sampled cells were washed with a cold (4 °C) EDTA solution (10 μM, pH 5.0) twice, and then with deionized water twice, and freeze-dried. The cells were weighed, and Cd concentration was determined as described above.

Subcellular localization of OsNRAMP2 protein

The CDS sequences of *OsNRAMP2-L and OsNRAMP2-H* were amplified using primers as following: forward primer, aaaagaattcatggcgtcgcgcgacctcg, reverse primer, aaaagtcgactgtgctctttgtcattgctgag. The CDS sequences were inserted into the 35S::EGFP vector to produce the OsNRAMP2-L::EGFP and OsNRAMP2-H::EGFP fusion proteins. The fusion plasmids and the empty 35S::EGFP plasmid were transferred into rice protoplasts according to a previous study (Zhang et al. 2011). Laser confocal microscopy (Zeiss, LSM710, Germany) was used to detect fluorescence after incubation for 24 h.

Data analysis

Phylogenetic tree of all 312 lines was constructed by MEGA 7.0 (Kumar et al. 2016) using SNP data mentioned above. A t-test was conducted using SAS (SAS Institute, 2000) to detect the differences in Cd accumulation in grain between *indica* rice and *japonica* rice, and between tropical *japonica* rice and temperate *japonica* rice.

Abbreviations

Cd: Cadmium; CDS: Coding sequence; EGFP: Enhanced green fluorescent protein; GWAS: Genome-wide association study; InDel: Insertion-deletion; MAF: Minor allele frequency; NRAMP: Natural resistance-associated macrophage protein; PSC: Pollution-safe cultivar; QTL: Quantitative trait loci; SNP: Single-nucleotide polymorphism

Acknowledgments

The seeds of the 312 rice accessions in the present study were provided by International Rice Research Institute. This study was partially supported by Guangdong Provincial Scientific and Technological Plan (2015B020231002, 2015B020201001), Science and Technology Program of Guangzhou

(201510010279), the Team Project of Guangdong Agricultural Department (2017LM2148) and the Special Fund of Central Government Guided Local Scientific Development.

Funding

This study was partially supported by Guangdong Provincial Scientific and Technological Plan (2015B020231002, 2015B020201001), Science and Technology Program of Guangzhou (201510010279), the Team Project of Guangdong Agricultural Department (2017LM2148) and the Special Fund of Central Government Guided Local Scientific Development.

Authors' contributions

JL Z, B L, and SH Z conceived and designed the experiment. JL Z, W Y, SH Z conducted the experiments, performed data analysis and wrote the manuscript. TF Y, W Y, SH Z, Q L, JF D, H F and XX M participated in material development, sample preparation and data analysis. B L and JL Z drafted proposals and corrected the manuscript. All authors read and approved the final manuscript.

Authors' information

B L, Ph D in Plant Molecular Genetics, Professor and Head of the Rice Molecular Breeding Lab, Rice Research Institute, Guangdong Academy of Agricultural Sciences, China;
JL Z, Ph D in Plant Molecular Biology, Associate Professor of the Rice Molecular Breeding Lab, Rice Research Institute, Guangdong Academy of Agricultural Sciences, China;
W Y, Ph D in Plant Molecular Biology, Assistant Professor of the Rice Molecular Breeding Lab, Rice Research Institute, Guangdong Academy of Agricultural Sciences, China;
SH Z, M.S. in Plant Physiology and Biochemistry, Professor of the Rice Molecular Breeding Lab, Rice.
Research Institute, Guangdong Academy of Agricultural Sciences, China;
TF Yang, Ph D in Plant Genetics and Breeding, Associate Professor of the Rice Molecular Breeding Lab, Rice Research Institute, Guangdong Academy of Agricultural Sciences, China;
Q L, Ph D in Plant Molecular Biology, Associate Professor of the Rice Molecular Breeding Lab, Rice Research Institute, Guangdong Academy of Agricultural Sciences, China;
JF D, Ph D in Plant Molecular Biology, Assistant Professor of the Rice Molecular Breeding Lab, Rice Research Institute, Guangdong Academy of Agricultural Sciences, China;
H F, B.S in Agronomy, Assistant Professor of the Rice Molecular Breeding Lab, Rice Research Institute, Guangdong Academy of Agricultural Sciences, China;
XX M, Ph D in Plant Molecular Biology, Associate Professor of the Rice Molecular Breeding Lab, Rice Research Institute, Guangdong Academy of Agricultural Sciences, China;

Competing interests

The authors declare that they have no competing interests.

References

Abe T, Nonoue Y, Ono N, Omoteno M, Kuramata M, Fukuoka S, Yamamoto T, Yano M, Ishikawa S (2013) Detection of QTLs to reduce cadmium content in rice grains using LAC23/Koshihikari chromosome segment substitution lines. Breed Sci 63:284–291

Arao T, Ae N (2003) Genotypic variations in cadmium levels of rice grain. Soil Sci Plant Nutr 49:473–479

Cailliatte R, Lapeyre B, Briat JF, Mari S, Curie C (2009) The NRAMP6 metal transporter contributes to cadmium toxicity. Biochem J 422:217–228

Cao F, Wang R, Cheng W, Zeng F, Mosaddek Ahmed I, Hu X, Zhang G, Wu F (2014) Genotypic and environmental variation in cadmium, chromium, lead and copper in rice and approaches for reducing the accumulation. Sci Total Environ 496:275–281

Cellier M, Prive G, Belouchi A, Kwan T, Rodrigues V, Chia W, Gros P (1995) Nramp defines a family of membrane proteins. Proc Natl Acad Sci U S A 92:10089–10093

Clemens S, Ma JF (2016) Toxic heavy metal and metalloid accumulation in crop plants and foods. Annu Rev Plant Biol 67(1):489–512

Hirschhorn J, Daly M (2005) Genome-wide association studies for common diseases and complex traits. Nat Rev Genet 6:95–108

Hu Y, Cheng H, Tao S (2016) The challenges and solutions for cadmium-contaminated rice in China: a critical review. Environ Int 92:515–532

Huang X, Wei X, Sang T, Zhao Q, Feng Q, Zhao Y, Li C, Zhu C, Lu T, Zhang Z, Li M, Fan D, Guo Y, Wang A, Wang L, Deng L, Li W, Lu y WQ, Liu K, Huang T, Zhou T, Jing Y, Li W, Lin Z, Buckler E, Qian Q, Zhang Q, Li J, Han B (2010) Genome-wide association studies of 14 agronomic traits in rice landraces. Nat Genet 42:961–967

Ishikawa S, Abe T, Kuramata M, Yamaguchi M, Ando T, Yamamoto T, Yano M (2010) A major quantitative trait locus for increasing cadmium-specific concentration in rice grain is located on the short arm of chromosome 7. J Exp Bot 61:923–934

Ishikawa S, Ae N, Yano M (2005) Chromosomal regions with quantitative trait loci controlling cadmium concentration in brown rice (Oryza sativa). New Phytol 168:345–350

Ishikawa S, Ishimaru Y, Igura M, Kuramata M, Abe T, Senoura T, Hase Y, Arao T, Nishizawa NK, Nakanishi H (2012) Ion-beam irradiation, gene identification, and marker-assisted breeding in the development of low-cadmium rice. Proc Natl Acad Sci U S A 109:19166–19171

Kawahara Y, Bastide M, Hamilton J, Kanamori H, McCombie W, Ouyang S, Schwartz D, Tanaka T, Wu J, Zhou S, Childs K, Davidson R, Lin H, Quesada-Ocampo L, Vaillancourt B, Sakai H, Lee S, Kim J, Numa H, Itoh T, Buell C, Matsumoto T (2013) Improvement of the Oryza sativa Nipponbare reference genome using next generation sequence and optical map data. Rice 6:4

Kumar S, Stecher G, Tamura K (2016) MEGA7: molecular evolutionary genetics analysis version 7.0 for bigger datasets. Mol Biol Evol 33:1870–1874

Lanquar V, Lelièvre F, Bolte S, Hamès C, Alcon C, Neumann D, Vansuyt G, Curie C, Schröder A, Krämer U, Barbier-Brygoo H, Thomine S (2005) Mobilization of vacuolar iron by AtNRAMP3 and AtNRAMP4 is essential for seed germination on low iron. EMBO J 24:4041–4051

Li F, Chen B, Xu K, Gao G, Yan G, Qiao J, Li J, Li H, Li L, Xiao X, Zhang T, Nishio T, Wu X (2016) A genome-wide association study of plant height and primary branch number in rapeseed (Brassica napus). Plant Sci 242:169–177

Li K, Yu H, Li T, Chen G, Huang F (2017) Cadmium accumulation characteristics of low-cadmium rice (Oryza sativa L.) line and F1 hybrids grown in cadmium-contaminated soils. Environ Sci and Pollut Res 24:17566–17576

Liu J, Zhu Q, Zhang Z, Xu J, Yang J, Wong M (2005) Variations in cadmium accumulation among rice cultivars and types and the selection of cultivars for reducing cadmium in the diet. J Sci Food Agric 85:147–153

Makino T, Takano H, Kamiya T, Itou T, Sekiya N, Inahara M, Sakurai Y (2008) Restoration of cadmium-contaminated paddy soils by washing with ferric chloride: cd extraction mechanism and bench-scale verification. Chemosphere 70:1035–1043

Mani A, Sankaranarayanan K (2018) In silico analysis of natural resistance-associated macrophage protein (NRAMP) family of transporters in rice. Protein J 37:237–247

McCouch S, Wright M, Tung C, Maron L, McNally K, Fitzgerald M, Singh N, DeClerck G, Agosto-Perez F, Korniliev GA, Naredo MS, Harrington S, Shi Y, Branchini D, Kuser-Falcao P, Leung H, Ebana K, Yano M, Eizenga G, McClung A, Mezey J (2016) Open access resources for genome-wide association mapping in rice. Nat Commun 7:10532

Miyadate H, Adachi S, Hiraizumi A, Tezuka K, Nakazawa N, Kawamoto T, Katou K, Kodama I, Sakurai K, Takahashi H, Satoh-Nagasawa N, Watanabe A, Fujimura T, Akagi H (2011) OsHMA3, a P1B-type of ATPase affects root-to-shoot cadmium translocation in rice by mediating efflux into vacuoles. New Phytol 189:190–199

Murakami M, Nakagawa F, Ae N, Ito M, Arao T (2009) Phytoextraction by rice capable of accumulating cd at high levels: reduction of cd content of rice grain. Environ Sci Technol 43:5878–5883

Norton GJ, Deacon CM, Xiong L, Huang S, Meharg AA, Price AH (2010) Genetic mapping of the rice ionome in leaves and grain: identification of qtls for 17 elements including arsenic, cadmium, iron and selenium. Plant Soil 329:139–153

Pinson SRM, Tarpley L, Yan W, Yeater K, Lahner B, Yakubova E, Huang XY, Zhang HM, Guerinot ML, Salt DE (2015) Worldwide genetic diversity for mineral element concentrations in rice grain. Crop Sci 55:294–311

SAS Institute (2000) SAS/STAT user's guide, version 8.1. SAS Institute Inc, Cary, North Carolina,USA

Sasaki A, Yamaji N, Yokosho K, Ma J (2012) Nramp5 is a major transporter responsible for manganese and cadmium uptake in rice. Plant Cell 24:2155–2167

Satoh-Nagasawa N, Mori M, Nakazawa N, Kawamoto T, Nagato Y, Sakurai K, Takahashi H, Watanabe A, Akagi H (2012) Mutations in rice (Oryza sativa) heavy metal ATPase 2 (OsHMA2) restrict the translocation of zinc and cadmium. Plant Cell Physiol 53:213–224

Shimo H, Ishimaru Y, An G, Yamakawa T, Nakanishi H, Nishizawa N (2011) *Low cadmium (LCD)*, a novel gene related to cadmium tolerance and accumulation in rice. J Exp Bot 62:5727–5734

Supek F, Supekova L, Nelson H, Nelson N (1996) A yeast manganese transporter related to the macrophage protein involved in conferring resistance to mycobacteria. Proc Natl Acad Sci U S A 93:5105–5110

Takahashi R, Ishimaru Y, Senoura T, Shimo H, Ishikawa S, Arao T, Nakanishi H, Nishizawa N (2011) The OsNRAMP1 iron transporter is involved in cd accumulation in rice. J Exp Bot 62:4843–4850

Takahashi R, Ishimaru Y, Shimo H, Ogo Y, Senoura T, Nishizawa N, Nakanishi H (2012) The OsHMA2 transporter is involved in root-to-shoot translocation of Zn and cd in rice. Plant Cell Environ 35:1948–1957

Tang Y, Liu X, Wang J, Li M, Wang Q, Tian F, Su Z, Pan Y, Liu D, Lipkai A, Buckler E, Zhang Z (2016) GAPIT version 2: an enhanced integrated tool for genomic association and prediction. The Plant Genome 9:2

Thomine S, Lelièvre F, Debarbieux E, Schroeder JI, Barbier-Brygoo H (2003) AtNRAMP3, a multispecific vacuolar metal transporter involved in plant responses to iron deficiency. Plant J 34:685–695

Thomine S, Wang R, Ward JM, Crawford NM, Schroeder JI (2000) Cadmium and Iron transport by members of a plant metal transporter family in Arabidopsis with homology to Nramp genes. Proc Natl Acad Sci U S A 97:4991–4996

Ueno D, Kono I, Yokosho K, Ando T, Yano M, Ma JF (2009) A major quantitative trait locus controlling cadmium translocation in rice (Oryza sativa). New Phytol 182:644–653

Ueno D, Yamaji N, Kono I, Huang C, Ando T, Yano M, Ma J (2010) Gene limiting cadmium accumulation in rice. Proc Natl Acad Sci U S A 107:16500–16505

Uraguchi S, Fujiwara T (2013) Rice breaks ground for cadmium-free cereals. Curr Opin Plant Biol 16:328–334

Weng J, Xie C, Hao Z, Wang J, Liu C, Li M, Zhang D, Bai L, Zhang S, Li X (2011) Genome-wide association study identifies candidate genes that affect plant height in Chinese elite maize (Zea mays L.) inbred lines. PLoS One 6:e29229

Xie L, Tang S, Wei X, Shao G, Jiao G, Sheng Z, Luo J, Hu P (2017) The cadmium and lead content of the grain produced by leading Chinese rice cultivars. Food Chem 217:217–224

Xue D, Chen M, Zhang G (2009) Mapping of QTLs associated with cadmium tolerance and accumulation during seedling stage in rice (Oryza sativa, L.). Euphytica 165:587–596

Yamaji N, Xia J, Mitaniueno N, Yokosho K, Ma JF (2013) Preferential delivery of zinc to developing tissues in rice is mediated by P-type heavy metal atpase OsHMA2. Plant Physiol 162:927–939

Yan J, Wang P, Wang P, Yang M, Lian X, Tang Z, Huang C, Salt D, Zhao F (2016) A loss-of-function allele of OsHMA3 associated with high cadmium accumulation in shoots and grain of japonica rice cultivars. Plant Cell Environ 39:1941–1954

Yan Y, Choi D, Kim D, Lee B (2010) Genotypic variation of cadmium accumulation and distribution in rice. J Crop Sci Biotech 13:69–73

Yang W, Guo Z, Huang C, Duan L, Chen G, Jiang N, Fang W, Feng H, Xie W, Lian X, Wang G, Luo Q, Zhang Q, Liu Q, Xiong L (2014) Combining high-throughput phenotyping and genome-wide association studies to reveal natural genetic variation in rice. Nat Commun 5:5087

Yu H, Wang J, Fang W, Yuan J, Yang Z (2006) Cadmium accumulation in different rice cultivars and screening for pollution-safe cultivars of rice. Sci Total Environ 370:302–309

Zhang Y, Su J, Duan S, Ao Y, Dai J, Liu J, Wang P, Li Y, Liu B, Feng D, Wang J, Wang H (2011) A highly efficient rice green tissue protoplast system for transient gene expression and studying light/chloroplast-related processes. Plant Methods 7:30

Zhou Q, Shao G, Zhang Y, Dong Q, Wang H, Cheng S, Cao L, Shen X (2017) The difference of cadmium accumulation between the indica and japonica subspecies and the mechanism of it. Plant Growth Regul 81:523–532

16

Kinetic Analysis of Zinc/Cadmium Reciprocal Competitions Suggests a Possible Zn-Insensitive Pathway for Root-to-Shoot Cadmium Translocation in Rice

Laura Fontanili[1†], Clarissa Lancilli[1,2†], Nobuo Suzui[3], Bianca Dendena[1], Yong-Gen Yin[3], Alessandro Ferri[1], Satomi Ishii[3], Naoki Kawachi[3], Giorgio Lucchini[1], Shu Fujimaki[3], Gian Attilio Sacchi[1] and Fabio Francesco Nocito[1*]

Abstract

Background: Among cereals, rice has a genetic propensity to accumulate high levels of cadmium (Cd) in grains. Xylem-mediated root-to-shoot translocation rather than root uptake has been suggested as the main physiological factor accounting for the genotypic variation observed in Cd accumulation in shoots and grains. Several evidence indicate OsHMA2 – a putative zinc (Zn) transporter – as the main candidate protein that could be involved in mediating Cd- and Zn-xylem loading in rice. However, the specific interactions between Zn and Cd in rice often appear anomalous if compared to those observed in other staple crops, suggesting that root-to-shoot Cd translocation process could be more complex than previously thought. In this study we performed a complete set of competition experiments with Zn and Cd in order to analyze their possible interactions and reciprocal effects at the root-to-shoot translocation level.

Results: The competition analysis revealed the lack of a full reciprocity when considering the effect of Cd on Zn accumulation, and vice versa, since the accumulation of Zn in the shoots was progressively inhibited by Cd increases, whereas that of Cd was only partially impaired by Zn. Such behaviors were probably dependent on Cd-xylem loading mechanisms, as suggested by: i) the analysis of Zn and Cd content in the xylem sap performed in relation to the concentration of the two metals in the mobile fractions of the roots; ii) the analysis of the systemic movement of ^{107}Cd in short term experiments performed using a positron-emitting tracer imaging system (PETIS).

Conclusions: Our results suggest that at least two pathways may mediate root-to-shoot Cd translocation in rice. The former could involve OsHMA2 as Zn^{2+}/Cd^{2+} xylem loader, whereas the latter appears to involve a Zn-insensitive system that still needs to be identified.

Keywords: Rice, Cadmium, Zinc, Translocation, PETIS

Background

Cadmium occurs naturally in the environment as a result of volcanic emissions and the weathering of rocks. Anthropogenic activities can contribute to increase the background levels of Cd in soil, water and living organisms. Important sources of soil contamination are short- or long-range atmospheric depositions from mining activities, phosphate fertilizers and manures, municipal sewage wastes, urban composts and industrial sludges (Alloway and Steinnes 1999; McLaughlin et al. 1999).

Plant-derived foods are the major source of Cd exposure for humans (EFSA 2009). Although Cd is not essential for plant nutrition, it can be easily taken up from the soil by plant roots and accumulated in different plant organs. In this way, Cd contamination of soils can result in moderate Cd accumulation in the edible parts of staple crops, which, in turn, represent the major pathway for Cd entry into the agricultural food chain, as well as the

* Correspondence: fabio.nocito@unimi.it
†Equal contributors
[1]Dipartimento di Scienze Agrarie e Ambientali - Produzione, Territorio, Agroenergia (DISAA), Università degli Studi di Milano, 20133 Milan, Italy
Full list of author information is available at the end of the article

major route for chronic Cd exposure in general populations (Wagner 1993; FAO/WHO 2001). In humans, Cd is known to accumulate in the kidneys and shows a long biological half-life in the whole body, ranging from 10 to 33 years (Ellis et al. 1979).

Among cereals, rice (*Oryza sativa* L.) has a genetic propensity to accumulate high level of Cd in grains (Arao and Ae 2003), sometimes reaching values exceeding the limit of 0.4 mg kg^{-1} established by the Codex Alimentarius Commission of FAO/WHO (CODEX STAN 193–1995 2015). Interestingly, earlier investigations reported finding that rice-eating ethnic groups such as the people from Japan, Thailand, Hong Kong and Taiwan, have the highest renal Cd levels in the world (Perry et al. 1961) and that long-term consumption of rice grown in Cd-contaminated paddy soils can cause a high prevalence of renal proximal tubular dysfunction (Kobayashi 1978; Cai et al. 1990; Kobayashi et al. 2002).

Since rice is the staple crop of more than half of the world's population, most of the Cd exposure through food is likely to come from the consumption of this cereal (Clemens et al. 2013a). For this reason, several efforts have been made to develop a number of agronomic techniques aimed at minimizing the absorption of Cd by rice (Murakami et al. 2007; Makino et al. 2008; Arao et al. 2010) as well as to understand both the biological bases and the genetic variation of Cd accumulation in rice grains, with the special aim of developing new low-Cd varieties (Clemens et al. 2013a; Uraguchi and Fujiwara 2013). This last approach seems to be particularly promising since the existence of a high variability for the trait "Cd accumulation in grain" represents a good starting point for the new challenges of plant breeding (Arao and Ae 2003; Grant et al. 2008; Shi et al. 2009; Uraguchi et al. 2009).

Several processes have been described as involved in Cd accumulation in rice grains. Briefly, Cd is absorbed by roots via transport pathways normally involved in the acquisition of mineral nutrients. OsNramp5 (natural resistance-associated macrophage protein 5) – a Mn transporter – has recently been described as the major Cd uptake system in rice (Ishikawa et al. 2012; Sasaki et al. 2012), although Zn and Fe transporters with limited specificity – such as OsZip1 (Zrt/Irt-like protein 1), OsIRT1 and OsIRT2 (iron regulated transporter 1 and 2), and OsNramp1 – may be implicated in Cd uptake under Zn or Fe deficiency (Ramesh et al. 2003; Nakanishi et al. 2006; Takahashi et al. 2011). Once inside root cells, Cd is translocated to the shoot through the xylem, and then accumulated in leaves and stems. Finally, during the reproductive stage phloem transport seems to be the main process mediating Cd accumulation in the grains (Tanaka et al. 2007; Kato et al. 2010). During this phase Cd remobilization from leaves as well as xylem-to-phloem Cd transfer in the nodes are thought to be essential to determine the final level of Cd into the grains (Rodda et al. 2011; Fujimaki et al. 2010; Uraguchi et al. 2011; Yamaguchi et al. 2012; Kobayashi et al. 2013). Among all these processes xylem-mediated root-to-shoot translocation rather than root uptake has been suggested as the main and most common physiological factor accounting for the genotypic variation observed in Cd accumulation in shoots and grains of rice plants (Uraguchi et al. 2009; Ishikawa et al. 2011). Translocation, in turn, mainly results from the equilibrium between the capability of each rice genotype to retain Cd in the root and the activities involved in xylem loading (Ueno et al. 2010; Miyadate et al. 2011; Nocito et al. 2011). In this equilibrium, a complex root firewall system – involving the vacuolar sequestration of free Cd ions (Ueno et al. 2010; Miyadate et al. 2011) or Cd-phytochelatin complexes (Nocito et al. 2011) – traps the excess of Cd within the root, determining, in this way, the total amount of Cd ions potentially available to be translocated via the xylem in a root-to-shoot direction.

Translocation of Cd ions to the shoots requires active loading into the xylem (Colangelo and Guerinot 2006). Several experimental sources of evidence indicate OsHMA2 – a putative Zn transporter belonging to the P_{1B}-type ATPase family – as the main candidate protein that could be involved in mediating Cd- and Zn-xylem loading in rice (Nocito et al. 2011; Satoh-Nagasawa et al. 2012; Takahashi et al. 2012). On the other hand, Yamaji et al. (2013) proposed OsHMA2 as an influx transporter for both Cd and Zn, involved in the preferential distribution of the two metal ions through the phloem to the developing tissues.

OsHMA2-defective mutants showed a lower translocation ratio for both Zn and Cd and reduced Cd accumulation into the grains as compared to the wild types (Satoh-Nagasawa et al. 2012; Takahashi et al. 2012). Moreover, a careful analysis of mutant phenotypes clearly reveals that root-to-shoot Cd translocation in rice could be a more complex process than previously thought, since that the lack of OsHMA2 activity had more and stronger effects on Zn than on Cd translocation, which indeed was not completely impaired by the mutation (Satoh-Nagasawa et al. 2012). Such a behavior suggests the hypothesis that Cd and Zn may only partially share the same pathways for translocation from root to shoot, underlining at the same time important implication for food safety, especially in the cases where the strategies used for containing Cd accumulation in crops are founded on Zn fertilization (Oliver et al. 1994; Fahad et al. 2015). Considering these aspects, in this paper we present and discuss a set of competition experiments with Zn and Cd which aimed at analyzing their possible interactions and reciprocal effects at the root-to-shoot translocation level.

Results

To better characterize the possible interactions between Zn and Cd translocation pathways we performed two set of experiments. In the first, rice plants were hydroponically grown and exposed for a 10-day period to a range of three increasing Zn external concentrations (0.1, 1, and 10 μM) in the absence or presence of a steady amount of Cd (0.1 μM); in the second, plants were exposed for 10 days to different Cd concentrations (0, 0.01, 0.1, and 1 μM) in the presence of a steady amount of Zn (1 μM). In all the experiments, changes in metal concentrations did not produce either significant effects on the growth of both roots and shoots, or any apparent symptoms of stress (data not shown): at the end of the exposure period, root and shoot dry weights of a single plant were 0.128 ± 0.004 g and 0.588 ± 0.023 g, respectively.

Cd and Zn Partitioning Between Root and Shoot

Results of the first experiment indicated that, in the absence of any source of Cd, Zn concentration in roots and shoots significantly increased as Zn availability in the external medium did. A similar trend was observed in the presence of a steady amount of Cd (0.1 μM); however, in this condition, Zn concentration in the shoots was lowered by the presence of Cd, whilst in the roots it was not significantly affected by the presence of the interfering metal (Fig. 1a and b). Finally, Cd accumulation in shoots and roots was significantly affected changing the external Zn concentration from 0.1 to 1 μM. Interestingly, a further increase in Zn availability in the medium - up to 10 μM - did not produced any additional decrease in Cd accumulation (Fig. 1c and d).

In the second experiment, the concentration of Zn in the shoots significantly decreased as Cd availability in the external medium increased, whilst Zn accumulation in the roots did not seem significantly affected by Cd availability (Fig. 1e and f). Under the same conditions, a dramatic increase in Cd accumulation was observed in both shoots and roots by changing the external Cd concentration from 0.01 to 1 μM (Fig. 1g and h).

Analysis of Root-to-Shoot Translocation of Zn and Cd

Dynamics of root-to-shoot translocation of Zn and Cd were examined by measuring their concentrations in the xylem sap of rice plants exposed to the different combinations of the metals. In these experiments, translocation was estimated as the amount of Zn or Cd ions loaded and transported in the xylem sap for 45 min.

The amount of Zn loaded in the xylem sap linearly increased as Zn concentration in the external medium did, both in the absence and in the presence of 0.1 μM Cd^{2+} (Fig. 2a). The presence of a steady amount of Cd (0.1 μM) slightly decreased Zn xylem loading in all the conditions analyzed (Fig. 2a). In the same experiment, Cd xylem loading was significantly inhibited by changing the external Zn concentration from 0.1 to 1 μM, and then was not affected by a further increase in Zn availability (Fig. 2b). On the other hand, increases in Cd concentration in the external medium progressively reduced Zn xylem loading (Fig. 2c) and resulted in significant increases in the amount of Cd ions loaded in the xylem sap (Fig. 2d). In the latter case, the Cd xylem loading curve started to approach saturation at 0.1 μM.

In both the experimental setups, the ability to load Zn and Cd into the xylem was linearly related to the total amount of Zn and Cd ions accumulated in the shoots over the 10-day period (Additional file 1: Figure S1).

Effect of Zn and Cd Exposure on Thiol Biosynthesis

Since activation of thiol metabolism may potentially allow a greater proportion of Zn and Cd to be retained in roots through vacuolar sequestration, we measured the levels of non-protein thiols (NPTs) in the roots of plants exposed to the different combinations of the two metals. The NPT levels of the roots increased as the Zn concentration in the external medium did, either in the absence or in the presence of 0.1 μM Cd^{2+}. Interestingly, the levels of NPTs measured for each Zn exposure condition significantly increased with the concomitant presence of Cd in the media (Fig. 3a). Finally, the NPT levels in the roots significantly increased as the external Cd concentration did (Fig. 3b).

Fractioning of Zn and Cd in Rice Roots

Fractioning of Zn and Cd retained by roots was carried out using a sequential extraction procedure with buffer and acid (Rauser and Meuwly 1995; Nocito et al. 2011). Additional file 1: Tables S1 and S2 summarize results obtained in representative experiments where the sum of Zn and/or Cd ions recovered in the different fractions accounts for at least 97 % of the total Zn and/or Cd content of the roots. Following extraction, three main metal-ion fractions were obtained: i) buffer soluble (extracts 1–6); ii) acid soluble (extracts 7–9); iii) ash (non-soluble Zn and/or Cd). Extracts 1 to 6 were further resolved into two fractions, named anionic and cationic, by anion-exchange chromatography (Additional file 1: Figure S2).

The procedure which we used for metal ion fractioning allowed us to discriminate Zn and Cd ions: i) retained in complexes with thiol-peptides or other soluble molecules negatively charged in the extraction buffer (anionic fraction); ii) in free cytosolic form and/or bound with ligands (i.e. organic acid and nicotianamine) to form cellular complexes of relatively low thermodynamic stability (cationic fraction); iii) tightly adsorbed to cellular matrices or apoplast components (acid soluble and ash). So, only Zn and Cd ions in the cationic

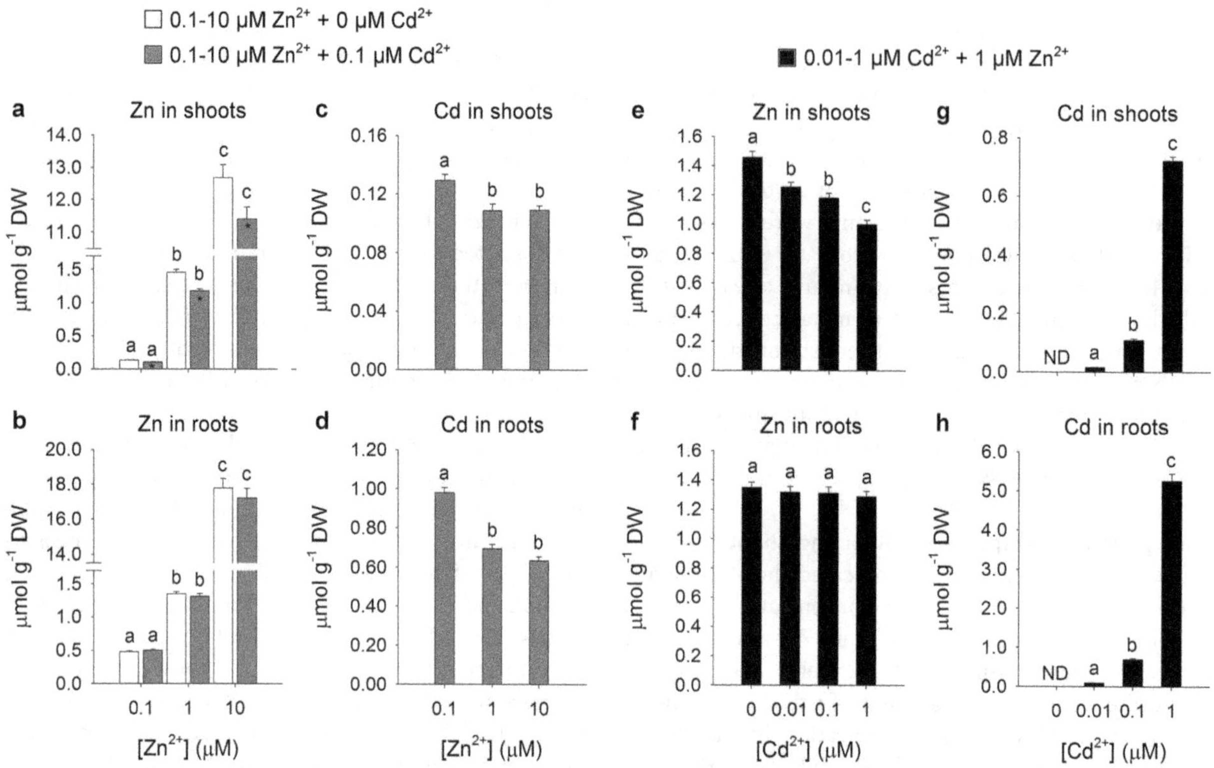

Fig. 1 Zn and Cd accumulation in shoots and roots. Rice plants were hydroponically grown and exposed for a 10-day period to increasing Zn external concentrations (from 0.1 to 10 μM), in the absence or presence of 0.1 μM Cd^{2+} (**a-d**), or to different Cd concentrations (from 0 to 1 μM) in the presence of 1 μM Zn^{2+} (**e-h**). **a** and **b** Zn concentration in shoots and roots in the absence (white bars) and in the presence (grey bars) of a steady amount of Cd. **c** and **d** Cd concentration in shoots and roots in the presence of a steady amount of Cd. E and F, Zn concentration in shoots and roots in the presence of a steady amount of Zn. **g** and **h** Cd concentration in shoots and roots in the presence of a steady amount of Zn. Bars and error bars are means and SE of three experiments run in triplicate ($n = 9$). Different letters indicate significant differences between treatments ($P < 0.05$). Asterisks indicate significant differences between plants exposed or not to 0.1 μM Cd^{2+} ($P \leq 0.001$). ND, not detectable; DW, dry weight

fraction should be considered available for root-to-shoot translocation (Nocito et al. 2011).

An integrated analysis of data in which the total amount of each metal ion retained by roots is divided into two fractions – named mobile (cationic, i.e. potentially available for root-to-shoot translocation) and non-mobile (anionic + acid soluble + ash) – according to Nocito et al. (2011) is reported in Table 1. Results indicated that both mobile and non-mobile fractions of Zn were linearly related to the Zn concentration in the external medium, either in the absence or in the presence of Cd; however, the presence of a steady amount of Cd in the medium significantly enhanced the concentration of Zn ions in the mobile fractions because of the lowering in the amount of the metal in the non-mobile fractions. In these last conditions, increasing the Zn external concentration led to decreased Cd content in the roots (Additional file 1: Tables S2) but did not produce any significant effect on the amount of Cd in the mobile fraction (Table 1). On the other hand, increasing the Cd external concentration significantly increased the amount of Zn and Cd ions measured in the mobile fractions (Table 1).

Finally, the anionic buffer-soluble fractions were further resolved by gel filtration on a Sephadex G-50 column into peaks I, II and III (Additional file 1: Figure S3). Zn/Cd ions into peak I at the void volume of the column (V_0) were ascribed to non-specific adsorption of the metal ions to proteins. Peaks II and III – centered between V_0 and V_t of the column – were designated as the classical high-molecular-weight (HMW) thiol based Cd-binding complexes and low-molecular-weight (LMW) thiol based Zn- and/or Cd-binding complexes, respectively, since the amount of NPTs recovered in these fractions accounted for 22 (1 μM Zn^{2+}, 0 μM Cd^{2+}) to 89 % (1 μM Zn^{2+}, 1 μM Cd^{2+}) of the total GSH equivalents measured in the roots; no thiols were found in peaks II and III of the anionic buffer-soluble fraction obtained from roots of plants grown under 0.1 μM Zn^{2+} in the absence of Cd^{2+} (data not shown). Neither Zn nor Cd ions were found at total volume (V_t) of the column ($K_{av} = 1$). Peaks II (HMW) were centered around $K_{av} = 0.44$ for Cd, whilst peaks III (LMW) were centered around $K_{av} = 0.58$ and $K_{av} = 0.62$, for Zn and Cd, respectively. Data analysis revealed that rice roots sequestered Zn and Cd differently.

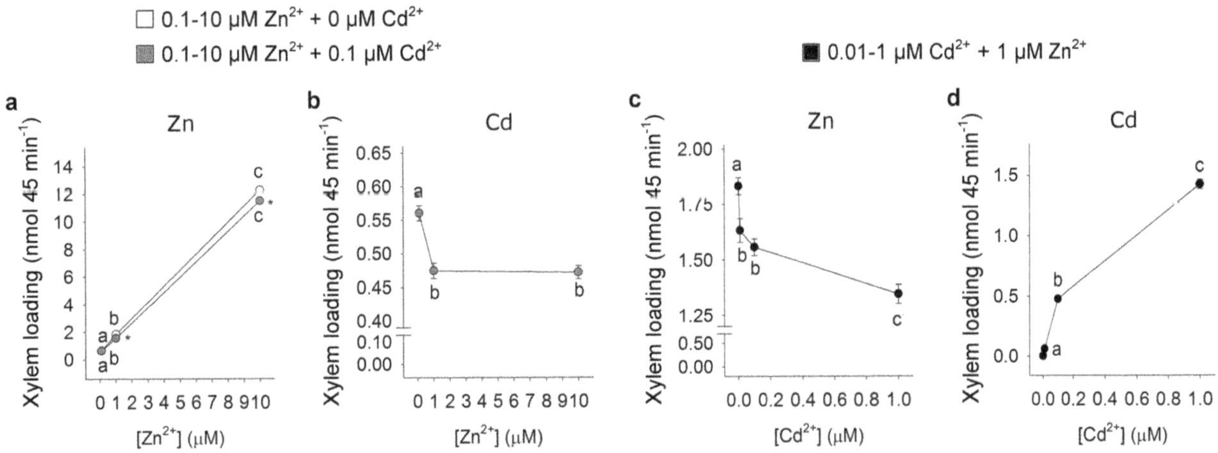

Fig. 2 Zn and Cd xylem loading. Rice plants were hydroponically grown and exposed for a 10-day period to increasing Zn external concentrations (from 0.1 to 10 μM), in the absence or presence of 0.1 μM Cd^{2+} (**a** and **b**), or to different Cd concentrations (from 0 to 1 μM) in the presence of 1 μM Zn^{2+} (**c** and **d**). At the end of the exposure period, shoots were separated from roots and the xylem sap exuded from the cut (root side) surface was collected over a 45 min period. **a** Zn ions loaded and transported in the xylem sap in the absence (white circles) and in the presence (grey circles) of a steady amount of Cd. **b** Cd ions loaded and transported in the xylem sap in the presence of a steady amount of Cd. **c** and **d** Zn and Cd ions loaded and transported in the xylem sap in the presence of a steady amount of Zn. Data are means and SE of three experiments run in triplicate ($n = 9$). Different letters indicate significant differences between treatments ($P < 0.05$). Asterisks indicate significant differences between plants exposed or not to 0.1 μM Cd^{2+} ($P \leq 0.001$)

In all cases Zn appeared in LMW complexes. In the absence of any source of Cd, LMW Zn-binding complexes significantly increased as the Zn external concentration did. Interestingly, the presence of Cd in the external medium significantly enhanced the amount of Zn ions found in LMW complexes. Moreover, the amount of Cd ions in LMW complexes remained constant in all the Zn conditions analyzed, differently from that found in HMW complexes, which instead decreased as Zn external concentration increased. On the other hand, LMW and HMW Cd-binding complexes appeared in a dynamic equilibrium depending on Cd external concentrations, as indicated by the ratio between the amount of Cd ions retained in each complex, which notably changed when moving toward the highest Cd external concentration, in the presence of a steady amount of Zn (Additional file 1: Tables S3 and S4). Finally, in the same conditions, the amount of Zn found in LMW complexes significantly increased as Cd external concentration did.

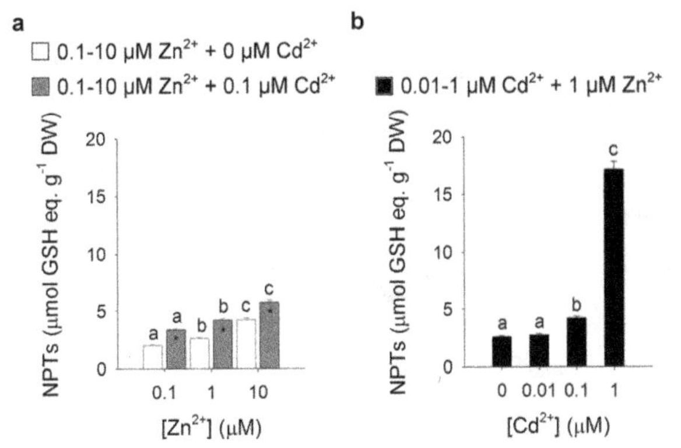

Fig. 3 Total non-protein thiols (NPTs) in the roots. **a** Rice plants were exposed for a 10-day period to increasing Zn external concentrations (from 0.1 to 10 μM), in the absence (white bars) or presence (grey bars) of 0.1 μM Cd^{2+}. **b** Rice plants were exposed for a 10-day period to different Cd concentrations (from 0 to 1 μM) in the presence of 1 μM Zn^{2+}. NPT levels are expressed as GSH equivalents. Bars and error bars are means and SE of three experiments run in triplicate ($n = 9$). Different letters indicate significant differences between treatments ($P < 0.05$). Asterisks indicate significant differences between plants exposed or not to 0.1 μM Cd^{2+} ($P \leq 0.001$). DW, dry weight

Kinetic Analysis of ^{107}Cd Systemic Movement

The systemic movement of Cd in the whole plant was further analyzed using ^{107}Cd in short-term (24 h) PETIS experiments. Figure 4a shows the field of view of a typical experiment and the regions of interest (ROIs) used to estimate the dynamics of Cd in the plants. In particular, six ROIs (background, culture solution, distal roots, proximal roots, shoot base, and proximal shoot) were set for each plant. The experiments were started by applying to the roots fresh marked (^{107}Cd; 0.55 MBq mL^{-1}) culture solutions containing 0.5 mM $CaCl_2$, 0.1 μM $CdCl_2$, and different concentrations of Zn^{2+} (0.1, 1, and 10 μM). ^{107}Cd absorption by roots was immediately observed after injection as clearly shown by the comparison of Fig. 4b and c. In all the conditions analyzed, the amount of Cd in the roots increased over time, reaching a common maximum plateau value at about 11 h both for plants exposed to 0.1 and 1 μM Zn^{2+}, and about 16 h for plants exposed to 10 μM Zn^{2+} (Fig. 4c). Concerning

Table 1 Partitioning of Zn and Cd ions of the roots between potentially mobile and non-mobile fractions

Metal concentration		Mobile fraction		Non-mobile fraction	
Zn	Cd	Zn	Cd	Zn	Cd
μM		*nmol g^−1DW*			
0.1	0	143.03 ± 3.86 (a)	ND	334.78 ± 11.01 (a)	ND
1	0	420.00 ± 14.70 (a)	ND	927.00 ± 28.54 (a)	ND
10	0	3136.82 ± 128.61 (b)	ND	14636.93 ± 584.18 (b)	ND
0.1	0.1	206.02 ± 6.59 (a)	81.66 ± 2.61 (a)	292.83 ± 10.44 (a)	898.72 ± 32.04 (a)
1	0.1	632.25 ± 22.13 (a)	83.00 ± 2.82 (a)	677.22 ± 26.93 (a)	612.57 ± 20.70 (a)
10	0.1	4541.08 ± 177.10 (b)	82.00 ± 3.53 (a)	12669.39 ± 335.86 (b)	552.86 ± 21.10 (b)
1	0	420.00 ± 14.70 (a)	ND	927 ± 28.54 (a)	ND
1	0.01	486.91 ± 17.04 (a)	11.06 ± 0.32 (a)	829.72 ± 27.26 (a)	80.63 ± 2.76 (a)
1	0.1	632.25 ± 22.13 (b)	83.00 ± 2.82 (b)	677.22 ± 26.93 (b)	612.57 ± 20.70 (b)
1	1	696.00 ± 24.36 (b)	349.49 ± 13.28 (c)	591.15 ± 18.18 (b)	4912.21 ± 137.26 (c)

Rice plants were hydroponically grown and exposed for a 10-day period to increasing Zn external concentrations (from 0.1 to 10 μM), in the absence or presence of 0.1 μM Cd^{2+}, or to different Cd concentrations (from 0 to 1 μM) in the presence of 1 μM Zn^{2+}. Mobile (cationic) and non-mobile (anionic + acid soluble + ash) metal ion fractions are derived using data reported in Additional file 1: Table S1 and S2. Data are means and SE of three experiments, each performed with eight plants ($n = 3$). Different letters indicate significant differences between treatments ($P < 0.05$). ND, not detectable; DW, dry weight

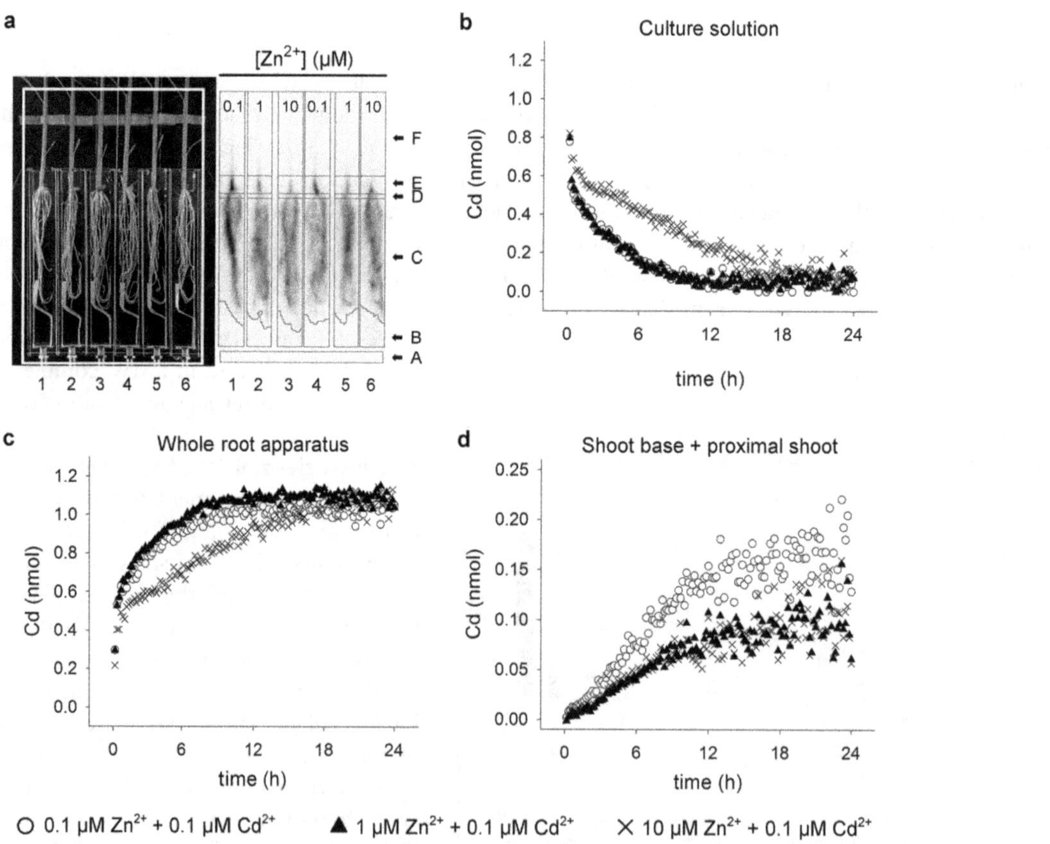

Fig. 4 Time-course analysis of Cd systemic movement. Rice plants were exposed for a 24-h period to different Zn concentrations (from 0.1 to 10 μM), in the presence of 0.1 μM Cd^{2+} enriched with ^{107}Cd. **a** Image of rice plants used in a typical PETIS experiment. The field of view of PETIS is bordered by the white continuous line. ROIs used for time-course analysis are indicated with arrows in the adjacent panel. ROI-A, background; ROI-B, culture solution; ROI-C, distal root; ROI-D, proximal root; ROI-E, shoot base; ROI-F, proximal shoot. **b** and **d** Time-course analysis of Cd dynamics in the culture solution (**b**), whole root apparatus (**c**), and shoot base + proximal shoot (**d**). White circles, black triangles, and thin x refer to the experiments performed in the presence of 0.1 (No. 4), 1 (No. 5) and 10 (No. 6) μM Zn^{2+}, respectively. A representative set of data from two independent experiments performed with two plants for each Zn exposure condition is given

the shoots (Fig. 4d), ^{107}Cd signals appeared in the lower parts of the stems (shoot bases; Additional file 1: Figure S4) within 1 h from the injections and then linearly increased at least up to 10 h. Considering the initial slope of each curve (from 0 to 10 h) we estimated that the rate of Cd translocation was significantly higher in plants exposed to 0.1 μM Zn^{2+} (12.6 ± 0.3 pmol h^{-1}) compared with those exposed to 1 or 10 μM Zn^{2+}, for which the estimated rates of Cd translocation were similar (7.8 ± 0.2 pmol h^{-1} or 7.5 ± 0.4 pmol h^{-1}, respectively). Similar results were obtained in a second independent analysis (Additional file 1: Figure S5).

Discussion

The specific interactions between Zn and Cd in rice – recently reviewed by Chaney (2015) – often appear anomalous if compared to those observed in other staple crops, since the routes used by Cd to enter root cells seem to be closely dependent on Zn nutritional status. Honma and Hirata (1978) described a noticeable effect of Zn starvation on Cd uptake. On the other hand, under Zn sufficiency or excess Cd uptake and accumulation become independent from the levels of Zn in the soils, in strong contrast with other cereal grains and vegetables (Chaney 2015). The peculiar ability of rice to accumulate soil Cd and not Zn in the grains when grown on Zn/Cd co-contaminated soils, in which the level of Zn are up to 100-fold higher than those of Cd, is impressive (Simmons et al. 2003; Chaney 2010). Such a behavior has important implications for human health in subsistence rice consumers, because the level of dietary Zn are known to influence Cd absorption in animal tissues (Fox et al. 1984; Fox 1988; McKenna et al. 1992; Reeves and Chaney 2004). It is now clear that the biological bases of these anomalies are likely due to the presence of multiple transport systems able to mediate Cd uptake (see also the introduction section), whose prevalence, ion specificity and regulation may strongly affect the interaction between Zn and Cd at different levels. Such a relative complexity may make it difficult to study the relationship existing between Zn and Cd at the root-to-shoot translocation level, since possible competitions could be prevented or masked by the relative expression of the transport systems involved in Zn and/or Cd uptake at the root level.

To try to clarify these aspects, we performed a complete set of competition experiments with Zn and Cd using rice plants grown in a soil-free system, aimed at obtaining a wide range of Zn and Cd concentrations at the root level. To these purposes we chose to grow plants under a non-saturating range of Zn external concentrations, in order to allow variable amounts of Cd to be absorbed at least through two main transport systems with different selectivity: i) OsZip1 (i.e. a Zn transporter up-regulated under Zn deficiency and able to move both Zn and Cd; Ramesh et al. 2003); ii) OsNramp5 (i.e. a Mn transporter able to move Cd, but not Zn; Sasaki et al. 2012). In our experimental conditions the final Zn/Cd concentration ratio into the roots ranged from 0.24 (Zn^{2+} 1 μM; 1 μM Cd^{2+}) to 27.11 (Zn^{2+} 10 μM; 0.1 μM Cd^{2+}).

The main results we obtained clearly indicate the lack of a full reciprocity when considering the effect of Cd on Zn accumulation, and vice versa, since the accumulation of Zn in the shoots was significantly inhibited by Cd increases in all the analyzed conditions, whereas that of Cd was only partially impaired by Zn increases (Fig. 1). In fact, Cd accumulation in the shoots was reduced by 16 % by changing the external Zn concentration by one order of magnitude (from 0.1 to 1 μM), then remained unaffected by a further increase in Zn availability, from 1 to 10 μM, indicating that Cd accumulation in this range was a Zn-independent process. However, from these data we cannot conclude that the effects of Zn on shoot Cd accumulation necessarily result from mechanisms involved in root-to-shoot Cd translocation, since Cd absorption by plant roots seemed affected by Zn concentration, as indicated by the total amount of Cd in the whole plant, whose value decreased as Zn concentration in the external medium increased (Fig. 1c and d). A similar observation may be done concerning Zn accumulation in the shoots under different Cd concentrations (Fig. 1e and f). Such behaviors could be reasonably due to the range of Zn concentrations we used for the experiments. However, the ability to load Zn and Cd into the xylem was linearly related to total amounts of Zn and Cd ions accumulated in the shoots over the 10-day period (Fig. 2, Additional file 1: Figure S1). This suggests the hypothesis that the differential effect produced by the two metals on Zn and Cd accumulation in the shoot was reasonably due to the existence of at least two translocation pathways with different metal selectivity. It is also important to consider that the total amounts of Zn and Cd ions in the root tissues are often poor indicators of their actual availability for loading into the xylem and then being translocated in a root-to-shoot direction, since, once inside the cells, Zn and Cd ions may be trapped within the root through selective binding sites or molecules with high affinity for the metals or through transfer across a membrane into an intracellular compartment (Clemens 2006; Ueno et al. 2010; Nocito et al. 2011). Moreover, the selective binding of Zn with low-molecular weight ligand, such as nicotianamine (NA), may promote both symplastic movement and long-distance distribution of the metal ion within the plant (Lee et al. 2011; Deinlein et al. 2012; Haydon et al. 2012). Thus, the concentration of Zn and Cd ions potentially mobile in the plants results from several different biochemical and physiological processes involved in metal chelation, compartmentalization and adsorption (Souza and Rauser 2003; Nocito et al. 2011; Sinclair and Krämer 2012; Clemens et al. 2013b; Olsen and Palmgren 2014).

It has been shown that different photosynthetic organisms respond to excess of Zn and Cd ions by producing phytochelatins (PCs), a class of thiol compounds involved in metal chelation and vacuolar sequestration (Grill et al. 1987; Grill et al. 1988; Zenk 1996; Tennstedt et al. 2009; García-García et al. 2014; Song et al. 2014), which may largely contribute to Zn and Cd retention within the root (Tennstedt et al. 2009; Wong and Cobbett 2009). In our conditions, rising Zn or Cd exposure progressively increased, with different efficiency, the level of NPTs in the roots (Fig. 3). Such a finding may be reasonably ascribable to the activation of PC biosynthesis, as revealed by the gel filtration analysis of the anionic fractions in which most of the Zn and Cd ions were immobilized with thiol compounds in LMW and HMW complexes (Additional file 1: Figure S3, Tables S3 and S4), as previously reported in other papers (Rauser and Meuwly 1995; Souza and Rauser 2003; Nocito et al. 2011). Moreover, fractioning of metals accumulated in the root also revealed that the three main fractions were in a dynamic equilibrium in which the increases in Zn or Cd external concentration, in the presence of a steady amount of Cd or Zn, respectively, resulted in changes in the amounts of the two metal ions in each fraction (Additional file 1: Tables S1 and S2).

Focusing the attention on the cationic fractions (Table 1, Additional file 1: Tables S1 and S2) we can make some educated guesses about changes in the relative mobility of Zn and Cd into the root, since, as discussed above, only the metal ions belonging to these fractions have all the requisites to be considered potentially mobile and then available for root-to-shoot translocation (Nocito et al. 2011; Additional file 1: Figure S2). In fact, the fraction of buffer-soluble Zn and Cd not immobilized by the anion exchanger may have come from Zn and Cd ions in free cytosolic form or bound with organic acids or NA to form cellular complexes of relatively high mobility inside the plant (Krotz et al. 1989; Deinlein et al. 2012).

Increases in Zn external concentration, in the presence of a steady amount of Cd, did not produced significant changes in the amount of Cd ions in the mobile fractions but conversely enhanced the amount of free Zn ions in the same fractions. Such behaviors may be ascribable to competition phenomena between the two ions for both root absorption and negative charges on cellular matrices or apoplast components. In fact, the increases in Zn external concentration not only resulted in a reduced total amount of Cd in the roots, but also displaced non-mobile Cd ions from cellular matrices (Table 1, Additional file 1: Tables S1 and S2). On the other hand, the increases in Cd external concentration, in the presence of a steady amount of Zn, significantly enhanced the amount of both Zn and Cd ions in the mobile fractions (Table 1, Additional file 1: Tables S1 and S2). Also in this case, we can speculate that the gradual saturation of cellular matrices with Cd ions may have displaced Zn ions leading to a transient increase in the activity of the free Zn forms that, in turn, was only partially counterbalanced by a weak increase in the amount of Zn ions immobilized with thiol compounds in the LMW complexes (Additional file 1: Figure S3, Tables S3 and S4). Finally, by plotting the xylem loading data obtained in the two experimental setups as a function of the Zn/Cd or Cd/Zn concentration ratios in the mobile fractions, we can easily deduce that increases in the Zn/Cd ratio did not produce fully reciprocal effects on Zn and Cd translocation, whilst increases in the Cd/Zn ratio resulted in fully reciprocal effects (Fig. 5). Such a finding strongly confirms the hypothesis that Cd ions may use at least two distinct pathways to be translocated from roots to shoots. The first one – shared with Zn – is probably used for Zn translocation in physiological conditions, whilst the second one appears as a Zn-insensitive route that Cd may preferentially use when the first pathway is saturated with Zn. Moreover, the Zn-insensitive pathway we postulate seems to be independent from Cd stress or nutritional status reached by plants under Zn excess (10 μM), since the partial inhibitory effect exerted by Zn on Cd translocation was also observed in short-term PETIS experiments performed with unstressed rice plants (Fig. 4). In fact, time-course analysis of Cd systemic movement in the plants revealed that: i) the rate of Cd translocation was significantly reduced by changing the external concentration of Zn from 0.1 to 1 μM and then remained unaffected following a further increase in Zn availability (Fig. 4d); ii) the rates of Cd translocation measured in the presence of different Zn external concentrations significantly differed just starting from the first hour of exposure (Fig. 4d), suggesting that the hypothetical Zn-insensitive pathway was constitutively expressed and not induced by changes in the nutritional status of the rice plants. It is also noteworthy that rates of Cd translocation measured in the presence of 1 and 10 μM Zn^{2+} did not significantly differ even if the highest Zn concentration we tested strongly reduced the rate of Cd accumulation in the roots (Fig. 4). Such a finding clearly confirms that the gradual saturation of the Zn-dependent pathway with Zn may force Cd ions to move through the second pathway, suggesting that the two translocation routes normally compete for free Cd ions into the root. Moreover, our main conclusion seems to be further supported by the paper of Satoh-Nagasawa et al. (2012) which showed that rice mutants defective for OsHMA2 – the main candidate transport system so far described in rice as involved in xylem-

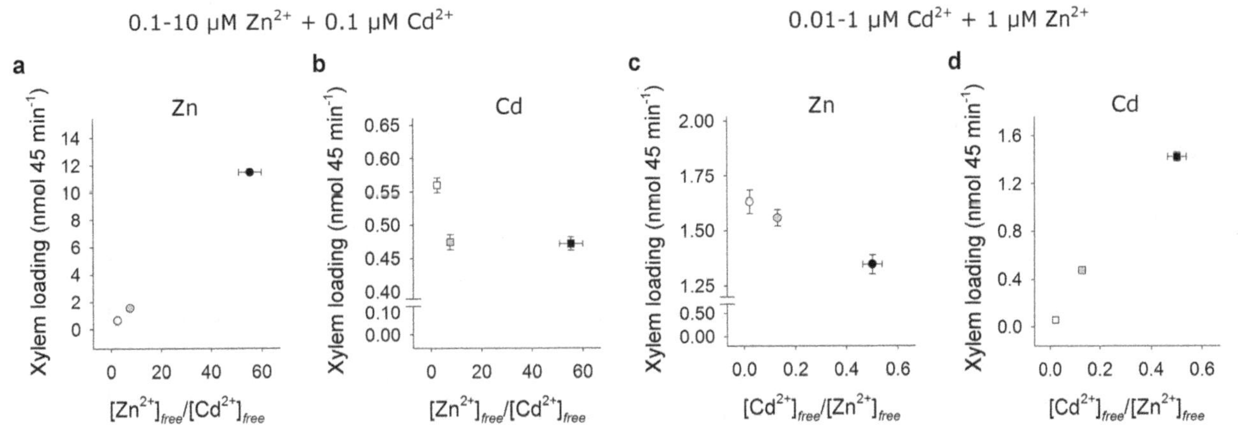

Fig. 5 Analysis of metal xylem loading as a function of Zn and Cd concentration in the mobile fractions. Rice plants were hydroponically grown and exposed for a 10-day period to increasing Zn external concentrations (from 0.1 to 10 μM), in the presence of 0.1 μM Cd^{2+} (**a** and **b**), or to different Cd concentrations (from 0.01 to 1 μM) in the presence of 1 μM Zn^{2+} (**c** and **d**). **a** and **b** Zn (circles) and Cd (square) translocation in plants exposed to 0.1 (*white*), 1 (*grey*) and 10 (*black*) μM Zn^{2+}, in the presence of 0.1 μM Cd^{2+}. **c** and **d** Zn (circles) and Cd (square) translocation in plants exposed to 0.01 (*white*), 0.1 (*grey*) and 1 (*black*) μM Cd^{2+}, in the presence of 1 μM Zn^{2+}. Data reported in figures derive from Fig. 2 and Additional file 1: Tables S1 and S2

mediated Zn^{2+}/Cd^{2+} translocation (Nocito et al. 2011; Satoh-Nagasawa et al. 2012; Takahashi et al. 2012; Satoh-Nagasawa et al. 2013; Yamaji et al. 2013) – had a residual capacity to translocate Cd to the shoots. Finally, the fact that wild-type plants - rather than mutants or genetically manipulated plants - were used in the present study strengthens the notion that long-distance transport of Cd along alternative pathways is a physiological process that generally occurs under Zn excess and does not result from anomalous or compensatory expression of other non-selective transporters.

Conclusions

Our data provide several forms of evidence to support the hypothesis that at least two distinct and competing pathways may take part in mediating root-to-shoot Cd translocation in rice. The first one, prevailing at relatively low Zn concentrations, could involve OsHMA2 as the Zn^{2+}/Cd^{2+} xylem loading system, while the second one appears to involve a Zn-insensitive system that still needs to be identified among transporters showing Cd-efflux activity in the vascular tissues of the roots.

Methods

Plant Material, Growth Conditions and Sampling

Rice (*O. sativa* L. cv. Roma) caryopses were placed on filter paper saturated with distilled water and incubated in the dark at 26 °C. Seven days later, seedlings were transplanted into 5 L plastic tanks (eight seedlings per tank) containing the following complete nutrient solution (pre-growing solution): 1.5 mM KNO_3, 1 mM $Ca(NO_3)_2$, 500 μM $MgSO_4$, 250 μM $NH_4H_2PO_4$, 25 μM Fe-tartrate, 46 μM H_3BO_3, 9 μM $MnCl_2$, 1 μM $ZnCl_2$, 0.3 μM $CuCl_2$, 0.1 μM $(NH_4)_6Mo_7O_{24}$, 30 μM Na_2O_3Si (pH 6.5). Seedlings were kept for a 12-day pre-growing period in a growth chamber maintained at 26 °C and 80 % relative humidity during the 16 h light period and at 22 °C, and 70 % relative humidity during the 8 h dark period. Photosynthetic photon flux density was 400 μmol m^{-2} s^{-1}. At the end of the pre-growing period, plants were differentially exposed for a 10-day period to different concentrations of Zn (0.1, 1 and 10 μM), in the absence or presence of a steady amount of Cd (0.1 μM), or to different concentrations of Cd (0, 0.01, 0.1 and 1 μM), in the presence of a steady amount of Zn (1 μM), by supplementing the pre-growing solution with different amounts of $ZnCl_2$ and/or $CdCl_2$. All hydroponic solutions were renewed daily to minimize nutrient depletion. Plants were harvested and roots were washed for 10 min in ice-cold 5 mM $CaCl_2$ solution to displace extracellular Cd (Rauser 1987), rinsed in distilled water and gently blotted with paper towels. Shoots were separated from roots and the tissues were frozen in liquid N_2 and stored at −80 °C, or analyzed immediately.

For positron-emitting tracer imaging system (PETIS) experiments, rice (*O. sativa* L. cv. Nipponbare) caryopses were germinated on a plastic mesh floating on distilled water, in the dark at 25 °C. Seven days later, seedlings were pre-grown in a one-quarter-strength Kimura B nutrient solution for 7 days and then grown in a full-strength Kimura B nutrient solution for another 7-day period. The Kimura B nutrition solution contained 0.70 mM $(NH_4)_2SO_4$, 0.17 mM Na_2HPO_4, 0.27 mM K_2SO_4, 0.47 mM $MgSO_4 \cdot 7H_2O$, 0.37 mM $CaCl_2$, 11 mg L^{-1} $FeC_6H_5O_7$ (Fe citrate), 0.16 μM $CuSO_4$, 0.15 μM $ZnSO_4 \cdot 7H_2O$, 0.10 μM $Na_2MoO_4 \cdot 2H_2O$,

15 μM H_3BO_3, and 4.6 μM $MnSO_4 \cdot 4H_2O$ (pH 5.5). Plants were kept in a growth chamber maintained at 30 °C and 65 % relative humidity during the 16 h light period and at 25 °C and 65 % relative humidity during the 8 h dark period. Photosynthetic photon flux density was 400 μmol m^{-2} s^{-1}.

Determination of Zn and Cd in Roots and Shoots

Samples of 200 mg fresh weight (FW) were mineralized at 120 °C in 5 mL 14.4 M HNO_3, clarified with 1.5 mL 33 % (w/v) H_2O_2, and finally dried at 80 °C. The mineralized material was dissolved in 5 mL 0.1 M HNO_3 and filtered on a 0.45 μm nylon membrane. Zn and Cd content was measured by inductively coupled plasma mass spectrometry (ICP-MS; Bruker Aurora M90 ICP-MS).

Analysis of Root-to-Shoot Zn and Cd Translocation

At the end of the exposure period, shoots were cut at 2 cm above the roots with a microtome blade. Xylem sap exuded from the lower cut surface was collected for 45 min and stored into 1.5 mL plastic vials. The amount of collected sap was determined by weighing and the concentration of Zn and Cd was measured by ICP-MS.

Determination of Total non-Protein Thiols

Roots were pulverized using mortar and pestle in liquid N_2 and stored frozen in a cryogenic tank. Four hundred milligrams of root powders were extracted in 600 μL of 1 M NaOH and 1 mg mL^{-1} $NaBH_4$, and the homogenate was centrifuged for 15 min at 13 000 *g* and 4 °C. Four hundred microliters of supernatant were collected, 66 μL of 37 % HCl was added and then the mixture was centrifuged again for 10 min at 13 000 *g* and 4 °C. For the quantification, volumes of 200 μL of the supernatant were collected and mixed with 800 μL of 1 M K-Pi buffer (pH 7.5) containing or not 0.6 mM Ellman's reagent {[5,5′-dithiobis(2-nitrobenzoic acid); DTNB]}. The samples' absorbances at 412 nm were then spectrophotometrically measured.

Zn and Cd Fractioning

Metal fractioning was carried out essentially as described by Rauser and Meuwly (1995). Briefly, frozen root tissues (2 g FW) were pulverized in a cold mortar with a pestle and then homogenized with ice-cold N_2-purged 100 mM Tris–HCl (pH 8.6), 1 mM phenylmethanesulfonyl fluoride (PMSF) and 1 % (v/v) Tween 20 at the ratio of 1 mL of buffer to 1 g tissue FW. The homogenate was centrifuged at 4 °C and 48 000 *g* for 6 min, the supernatant (extract 1) was collected and frozen immediately in liquid N_2, and the pellet was resuspended in a volume of N_2-purged 10 mM Tris–HCl (pH 8.6) and 1 % (v/v) Tween 20, previously used to rinse the mortar kept on ice. The suspension was centrifuged again, and the supernatant (extract 2) was collected and added to the extract 1 for freezing. Resuspension and centrifugation of the homogenized tissue debris was repeated four more times to collect extracts 3–6. At the end of this sequence, the pellet was suspended in a volume of ice-cold 100 mM HCl, centrifuged at 4 °C and 48 000 *g* for 6 min and the supernatant (extract 7) was retained. This sequence was repeated two more times to obtain extracts 8 and 9. The exhausted pellet was transferred to a glass tube, mineralized at 120 °C in 10 mL 14.4 M HNO_3, clarified with 3 mL 33 % (w/v) H_2O_2 and finally dried at 80 °C. The mineralized material was dissolved in 5 mL 0.1 M HNO_3 and filtered on a 0.45 μm nylon membrane.

Extracts 1 to 6 were resolved into two fractions, referred to as anionic and cationic, by anion-exchange chromatography. Buffer extract was loaded, at 20 mL h^{-1}, onto a 0.5 × 2 cm column of diethylaminoethyl cellulose (DEAE) Sephadex A-25 (GE Healthcare) equilibrated with 10 mM Tris–HCl (pH 8.6). After loading, the column was washed with 50 mL of equilibrating buffer to remove unadsorbed solutes. All the fluid passing through the anion exchanger was collected for Zn and Cd analysis (cationic fraction). Anionic material was eluted with 6 mL of 10 mM Hepes (pH 8.0) and 1 M KCl. Five milliliters of the anionic fraction so obtained were further resolved by gel filtration on a Sephadex G-50 (Sigma) column (0.8 × 130 cm) equilibrated with 10 mM Hepes (pH 8.0) and 300 mM KCl. The column was developed in equilibrating buffer at 12.5 mL h^{-1} at 4 °C. The absorbance at 254 nm was recorded and fractions of about 5 mL were collected for Zn, Cd, and NPT analysis. The column was calibrated by using 5 mL of 0.25 % (w/v) blue dextran 2000 and 1 % (w/v) $K_3Fe(CN)_6$ to estimate void (V_0) and total volume (V_t), respectively. The partition coefficient, K_{av}, was calculated using the following equation: $K_{av} = (V_e - V_0)/(V_t - V_0)$, where V_e was the elution volume. For total non-protein thiol determination, selected fractions from gel filtration were pooled in a glass tube, lyophilized, and finally analyzed as described above. The amount of Zn and Cd ions in mineralized pellets, extracts and column effluents was measured by ICP-MS.

^{107}Cd Tracer

^{107}Cd was produced as described by Fujimaki et al. (2010). Briefly, a silver foil was bombarded for 2 h with a 17-MeV energetic proton beam, at a current of 5 μA from a cyclotron, at Takasaki Ion Accelerators for Advanced Radiation Application, Japan Atomic Energy Agency. The irradiated material was dissolved in HNO_3, and then diluted in warm water. Silver was precipitated by adding to the solution a 0 to 2 M gradient of HCl. Supernatant, containing ^{107}Cd, was filtered, dried and dissolved in water. Aliquots of ^{107}Cd (6.7 MBq) were added to 12.2 mL of the culture solutions [0.5 mM $CaCl_2$, 0.1 μM nonradioactive $CdCl_2$, different concentrations (0.1, 1, 10 μM) of $ZnSO_4$] used for each plant in the experiments.

PETIS Imaging

Rice plants were transferred into appropriate 120 × 14 × 10 mm plastic vessels, containing 12 mL of full-strength Kimura B nutrient solution. Before starting the experiments, plants were acclimatized for 1 h in aerated 0.5 mM $CaCl_2$ solutions. In a typical experiment, 6 vessels – each containing one plant – were placed in the mid-plane between two opposing detector heads of the PETIS apparatus (a modified type of PPIS-4800; Hamamatsu Photonics, Japan). Detectors were focused on the plants in order to observe the culture solutions, the whole root apparatus, the shoot bases, and the proximal portions of the shoots, in a 12 × 19 cm field of view (FOV). The entire setup was installed in a growth chamber maintained at 30 °C and 65 % relative humidity in continuous light at the density of 400 μmol m^{-2} s^{-1}.

PETIS experiments were started by injecting the ^{107}Cd marked culture solutions in the different plastic vessels. All the solutions were continuously stirred with gentle aeration in order to maintain a uniform composition. The surface level of the solution in the vessels was maintained by supplying fresh 0.5 mM $CaCl_2$ solutions with an appropriate solution supply system. Images of the ^{107}Cd distribution in the FOV were obtained every one minute for 24 h. The data of serial images obtained from the PETIS apparatus were analyzed for ^{107}Cd distribution in specific regions of interest (ROIs; background, culture solution, distal roots, proximal roots, shoot base, and proximal shoot) using NIH ImageJ 1.45 s software (Schneider et al. 2012). ROIs were manually selected on the image data and the time-activity curves (time-courses of signal intensity in the ROIs) were generated and used to estimate the dynamics of Cd in the culture solution, whole root apparatus, shoot base, and proximal shoot, as described by Yoshihara et al. (2014).

Autoradiography

At the end of the PETIS experiments, plants were dissected, fixed on paper sheets with adhesive tape, and then placed in contact with imaging plates (BAS-MS2040, GE Healthcare, Japan) in cassettes for 3 days. The imaging plates were scanned using a Bio Imaging Analyzer (Typhoon FLA 7000, GE Healthcare, Japan) to generate the autoradiographic images of ^{109}Cd in the plants. In fact, ^{109}Cd with a longer half-life (461 days) than ^{107}Cd (6.5 h) was also obtained at a minor ratio (approximately 1:3000) in the production process of ^{107}Cd.

Statistical Analysis

Statistical analysis was carried out using SigmaPlot for Windows version 11.0 (Systat Software, Inc., Chicago, IL, USA). Quantitative values are presented as mean ± standard error of the mean (SE). Significance values were adjusted for multiple comparisons using the Bonferroni correction. Statistical significance was at $P < 0.05$. Student's t-test was used to assess the significance of the observed differences between plants exposed or not to Cd in each Zn exposure condition analyzed. Statistical significance was at $P \leq 0.001$.

Additional File

Additional file 1: Table S1. Fractioning of Zn ions retained in root. **Table S2** Fractioning of Cd ions retained in root. **Table S3** Zn ions retained in the root in complexes with thiols. **Table S4** Cd ions retained in the root in complexes with thiols. **Figure S1** Relationship between Zn and Cd ions loaded in the xylem sap and Zn and Cd concentration in shoots. **Figure S2** Fractioning of Zn and Cd ions in the buffer extract. **Figure S3** Zn- and Cd-binding complexes resolved by gel filtration chromatography. **Figure S4** Autoradiography and time-course analysis of Cd accumulation in the shoot base. **Figure S5** Another set of time-course analysis of Cd systemic movement.

Abbreviations

Cd: cadmium; DEAE: diethylaminoethyl cellulose; DTNB: 5,5′-dithiobis (2-nitrobenzoic acid); FOV: field of view; HMW: high-molecular-weight; ICP-MS: inductively coupled plasma mass spectrometry; LMW: low-molecular-weight; NPTs: non-protein thiols; PCs: phytochelatins; PETIS: positron-emitting tracer imaging system; ROIs: regions of interest; RT-PCR: reverse transcription polymerase chain reaction; Zn: zinc.

Competing Interests

The authors declare that they have no competing interests.

Authors' Contributions

LF, CL, NS, SF, GAS, and FFN conceived and designed the experiments. LF, CL, BD, AF, and FFN carried out the physiological and biochemical analyses. GL performed ICP-MS analysis. LF, NS, Y-GY, SI, NK, and SF performed PETIS experiments. CL and GAS acquired the funds. LF, AF, and FFN wrote the manuscript. All authors discussed, revised and approved the manuscript.

Acknowledgments

This work was supported by grants from the Italian Ministry of Education, University, and Research (FIRB 2008 – PROGRAMMA "FUTURO IN RICERCA": Groundbreaking methods to study the mechanisms of mobilization and acquisition of trace metals in the soil-plant system; RBFR08L2ZT), and the Project AGER Risinnova (grant No. 2010–2369).
We thank Prof Yohey Hashimoto for having had Laura Fontanili as a guest in the framework of the Agreement of Scholarly Exchange and Collaboration between UNIMI (Università degli Studi di Milano) and TUAT (Tokyo University of Agriculture and Technology). We thank Prof Maurizio Cocucci for his valuable contribution during the writing of this paper.
The music of Karlheinz Stockhausen inspired this work.

Author details

[1]Dipartimento di Scienze Agrarie e Ambientali - Produzione, Territorio, Agroenergia (DISAA), Università degli Studi di Milano, 20133 Milan, Italy. [2]Istituto di Biologia e Biotecnologia Agraria (IBBA), Consiglio Nazionale delle Ricerche, 20133 Milan, Italy. [3]Department of Radiation-Applied Biology, Takasaki Advanced Radiation Research Institute, Quantum Beam Science Research Directorate, National Institutes for Quantum and Radiological Science and Technology (QST), 1233 Watanuki, Takasaki, Gunma 370-1292, Japan.

References

Alloway BJ, Steinnes E (1999) Anthropogenic addition of cadmium to soils. In: McLaughlin MJ, Singh BR (eds) Cadmium in Soil and Plants. Kluwer Academic Publishers, Dordrecht, The Netherlands, pp 97–123

Arao T, Ae N (2003) Genotypic variations in cadmium levels of rice grain. Soil Sci Plant Nutr 49:473–479

Arao T, Ishikawa S, Murakami M, Abe K, Maejima Y, Makino T (2010) Heavy metal contamination of agricultural soil and countermeasures in Japan. Paddy Water Environ 8:247–257

Cai S, Yue L, Hu Z, Zhong X, Ye Z, Xu H, Liu YR, Ji RD, Zhang WH, Zhang F (1990) Cadmium exposure and health effects among residents in an irrigation area with ore dressing wastewater. Sci Total Environ 90:67–73

Chaney RL (2010) Cadmium and zinc. In: Hooda PS (ed) Trace Elements in Soils. John Wiley & Sons Ltd, Chichester, UK, pp 409–439

Chaney RL (2015) How does contamination of rice soils with Cd and Zn cause high incidence of human Cd disease in subsistence rice farmers. Curr Pollution Rep 1:13–22

Clemens S (2006) Toxic metal accumulation, responses to exposure and mechanisms of tolerance in plants. Biochimie 88:1707–1719

Clemens S, Aarts MGM, Thomine S, Verbruggen N (2013a) Plant science: the key to preventing slow cadmium poisoning. Trends Plant Sci 18:2

Clemens S, Deinlein U, Ahmadi A, Höreth S, Uraguchi S (2013b) Nicotianamine is a major player in plant Zn homeostasis. Biometals 26:623–632

CODEX STAN 193–1995 (2015) General standard for contaminants and toxins in foods and feed. http://www.fao.org/fao-whocodexalimentarius/download/standards/17/CXS_193e2015.pdf. Accessed 8 Apr 2016.

Colangelo EP, Guerinot ML (2006) Put the metal to the petal: metal uptake and transport throughout plants. Curr Opin Plant Biol 9:322–330

Deinlein U, Weber M, Schmidt H, Rensch S, Trampczynska A, Hansen TH, Husted S, Schjoerring JK, Talke IN, Krämer U, Clemens S (2012) Elevated nicotianamine levels in *Arabidopsis halleri* roots play a key role in zinc hyperaccumulation. Plant Cell 24:708–723

EFSA (2009) Cadmium in food - scientific opinion of the panel on contaminants in the food chain. EFSA J 980:1–139

Ellis KJ, Vartsky D, Zanzi I, Cohn SH, Yasumura S (1979) Cadmium: in vivo measurement in smokers and nonsmokers. Science 205:323–325

Fahad S, Hussain S, Khan F, Wu C, Saud S, Hassan S, Ahmad N, Gang D, Ullah A, Huang J (2015) Effects of tire rubber ash and zinc sulfate on crop productivity and cadmium accumulation in five rice cultivars under field conditions. Environ Sci Pollut Res 22:12424–12434

FAO/WHO (2001) Evaluation of certain food additives and contaminants. http://whqlibdoc.who.int/trs/WHO_TRS_901.pdf.

Fox MRS (1988) Nutritional factors that may influence bioavailability of cadmium. J Environ Qual 17:175–180

Fox MRS, Tao S-H, Stone CL, Fry BE Jr (1984) Effects of zinc, iron, and copper deficiencies on cadmium in tissues of Japanese quail. Environ Health Perspect 54:57–65

Fujimaki S, Suzui N, Ishioka NS, Kawachi N, Ito S, Chino M, Nakamura S (2010) Tracing cadmium from culture to spikelet: noninvasive imaging and quantitative characterization of absorption, transport, and accumulation of cadmium in an intact rice plant. Plant Physiol 152:1796–1806

García-García JD, Girard L, Hernández G, Saavedra E, Pardo JP, Rodríguez-Zavala JS, Encalada R, Reyes-Prieto A, Mendoza-Cózatl DG, Moreno-Sánchez R (2014) Zn-bis-glutathionate is the best co-substrate of the monomeric phytochelatin synthase from the photosynthetic heavy metal-hyperaccumulator *Euglena gracilis*. Metallomics 6:604–616

Grant CA, Clarke JM, Duguid S, Chaney RL (2008) Selection and breeding of plant cultivars to minimize cadmium accumulation. Sci Total Environ 390: 301–310

Grill E, Winnacker EL, Zenk MH (1987) Phytochelatins, a class of heavy-metal-binding peptides from plants, are functionally analogous to metallothioneins. Proc Natl Acad Sci U S A 84:439–443

Grill E, Thumann J, Winnacker EL, Zenk M (1988) Induction of heavy-metal binding phytochelatins by inoculation of cell cultures in standard media. Plant Cell Rep 7:375–378

Haydon MJ, Kawachi M, Wirtz M, Hillmer S, Hell R, Krämer U (2012) Vacuolar nicotianamine has critical and distinct roles under iron deficiency and for zinc sequestration in *Arabidopsis*. Plant Cell 24:724–737

Honma Y, Hirata H (1978) Noticeable increase in cadmium absorption by zinc deficient rice plants. Soil Sci Plant Nutr 14:295–297

Ishikawa S, Suzui N, Ito-Tanabata S, Ishii S, Igura M, Abe T, Kuramat M, Kawachi N, Fujimaki S (2011) Real-time imaging and analysis of differences in cadmium dynamics in rice cultivars (*Oryza sativa*) using positron-emitting ^{107}Cd tracer. BMC Plant Biol 11:172

Ishikawa S, Yasuhiro Ishimaru Y, Igura M, Kuramata M, Abe T, Senoura T, Hased Y, Arao T, Nishizawa NK, Nakanishi H (2012) Ion-beam irradiation, gene identification, and marker-assisted breeding in the development of low-cadmium rice. Proc Natl Acad Sci U S A 109:19166–19171

Kato M, Ishikawa S, Inagaki K, Chiba K, Hayashi H, Yanagisawa S, Yoneyama T (2010) Possible chemical forms of cadmium and varietal differences in cadmium concentrations in the phloem sap of rice plants (*Oryza sativa* L.). Soil Sci Plant Nutr 56:839–847

Kobayashi J (1978) Pollution by cadmium and the *itai-itai* disease in Japan. In: Oehme FW (ed) Toxicity of Heavy Metals in the Environment. Marcel Dekker, Inc, New York, pp 199–260

Kobayashi E, Okubo Y, Suwazono Y, Kido T, Nogawa K (2002) Dose–response relationship between total cadmium intake calculated from cadmium concentration in rice collected from each household of farmers and renal dysfunction in inhabitants of the Jinzu River basin, Japan. J Appl Toxicol 22: 431–436

Kobayashi NI, Tanoi K, Hirose A, Nakanishi TM (2013) Characterization of rapid intervascular transport of cadmium in rice stem by radioisotope imaging. J Exp Bot 64:507–517

Krotz RM, Evangelou BP, Wagner GJ (1989) Relationships between cadmium, zinc, Cd-peptide, and organic acid in tobacco suspension cells. Plant Physiol 91:780–787

Lee S, Persson DP, Hansen TH, Husted S, Schjoerring JK, Kim Y-S, Jeon US, Kim Y-K, Kakei Y, Masuda H, Nishizawa NK, An G (2011) Bio-available zinc in rice seeds is increased by activation tagging of nicotianamine synthase. Plant Biotechnol J 9:865–873

Makino T, Takano H, Kamiya T, Itou T, Sekiya N, Inahara M, Sakurai Y (2008) Restoration of cadmium-contaminated paddy soils by washing with ferric chloride: Cd extraction mechanism and bench-scale verification. Chemosphere 70:1035–1043

McKenna IM, Chaney RL, Tao S, Leach RM Jr, Williams FM (1992) Interactions of plant zinc and plant species on the bioavailability of plant cadmium to Japanese quail fed lettuce and spinach. Environ Res 57:73–87

McLaughlin M, Parker DR, Clarke JM (1999) Metals and micronutrients – food safety issues. Field Crops Res 60:143–163

Miyadate H, Adachi S, Hiraizumi A, Tezuka K, Nakazawa N, Kawamoto T, Katou K, Kodama I, Sakurai K, Takahashi H, Satoh-Nagasawa N, Watanabe A, Fujimura T, Akagi H (2011) OsHMA3, a P_{1B}-type of ATPase affects root-to-shoot cadmium translocation in rice by mediating efflux into vacuoles. New Phytol 189:190–199

Murakami M, Ae N, Ishikawa S (2007) Phytoextraction of cadmium by rice (*Oryza sativa* L.), soybean (*Glycine max* (L.) Merr.), and maize (*Zea mays* L.). Environ Pollut 145:96–103

Nakanishi H, Ogawa I, Ishimaru Y, Mori S, Nishizawa NK (2006) Iron deficiency enhances cadmium uptake and translocation mediated by the Fe^{2+} transporters OsIRT1 and OsIRT2 in rice. Soil Sci Plant Nutr 52:464–469

Nocito FF, Lancilli C, Dendena B, Lucchini G, Sacchi GA (2011) Cadmium retention in rice roots is influenced by cadmium availability, chelation and translocation. Plant Cell Environ 34:994–1008

Oliver DP, Hannam R, Tiller KG, Wilhelm N, Merry RH (1994) The effects of zinc fertilization on cadmium concentration in wheat grain. J Environ Qual 23:705–711

Olsen LI, Palmgren MG (2014) Many rivers to cross: the journey of zinc from soil to seed. Front Plant Sci 5:30

Perry HM, Tipton IH, Schroeder HA, Steiner RL, Cook MJ (1961) Variation in the concentration of cadmium in human kidney as a function of age and geographic origin. J Chron Dis 14:259–271

Ramesh SA, Shin R, Eide DJ, Schachtman DP (2003) Different metal selectivity and gene expression of two zinc transporters from rice. Plant Physiol 133:126–134

Rauser WE (1987) Compartmental efflux analysis and removal of extracellular cadmium from roots. Plant Physiol 85:62–65

Rauser WE, Meuwly P (1995) Retention of cadmium in roots of maize seedlings. Role of complexation by phytochelatins and related thiol peptides. Plant Physiol 109:195–202

Reeves PG, Chaney RL (2004) Marginal nutritional status of zinc, iron, and calcium increases cadmium retention in the duodenum and other organs of rats fed a rice-based diet. Environ Res 96:311–322

Rodda MS, Li G, Reid RJ (2011) The timing of grain Cd accumulation in rice plants: the relative importance of remobilisation within the plant and root Cd uptake post-flowering. Plant Soil 347:105–114

Sasaki A, Yamaji N, Yokosho K, Ma JF (2012) Nramp5 is a major transporter responsible for manganese and cadmium uptake in rice. Plant Cell 24:2155–2167

Satoh-Nagasawa N, Mori M, Nakazawa N, Kawamoto T, Nagato Y, Sakurai K, Takahashi H, Watanabe A, Akagi H (2012) Mutations in rice (*Oryza sativa*) heavy metal ATPase 2 (OsHMA2) restrict the translocation of zinc and cadmium. Plant Cell Physiol 53:213–224

Satoh-Nagasawa N, Mori M, Sakurai K, Takahashi H, Watanabe A, Akagi H (2013) Functional relationship heavy metal P-type ATPases (OsHMA2 and OsHMA3) of rice (*Oryza sativa*) using RNAi. Plant Biotechnol 30:511–515

Schneider CA, Rasband WS, Eliceiri KW (2012) NIH Image to ImageJ: 25 years of image analysis. Nat Methods 9:671–675

Shi J, Li L, Pan G (2009) Variation of grain Cd and Zn concentrations of 110 hybrid rice cultivars grown in a low-Cd paddy soil. J Environ Sci 21:168–172

Simmons RW, Pongsakul P, Chaney RL, Saiyasitpanich D, Klinphoklap S, Nobuntou W (2003) The relative exclusion of zinc and iron from rice grain in relation to rice grain cadmium as compared to soybean: Implications for human health. Plant Soil 257:163–170

Sinclair SA, Krämer U (2012) The zinc homeostasis network of land plants. Biochim Biophys Acta 1823:1553–1567

Song WY, Mendoza-Cózatl DG, Lee Y, Schroeder JI, Ahn SN, Lee HS, Wicker T, Martinoia E (2014) Phytochelatin–metal(loid) transport into vacuoles shows different substrate preferences in barley and *Arabidopsis*. Plant Cell Environ 37:1192–1201

Souza JF, Rauser WE (2003) Maize and radish sequester excess cadmium and zinc in different ways. Plant Sci 165:1009–1022

Takahashi R, Ishimaru Y, Senoura T, Shimo H, Ishikawa S, Arao T, Nakanishi H, Nishizawa NK (2011) The OsNRAMP1 iron transporter is involved in Cd accumulation in rice. J Exp Bot 62:4843–4850

Takahashi R, Ishimaru Y, Shimo H, Ogo Y, Senoura T, Nishizawa NK, Nakanishi H (2012) The OsHMA2 transporter is involved in root-to-shoot translocation of Zn and Cd in rice. Plant Cell Environ 35:1948–1957

Tanaka K, Fujimaki S, Fujiwara T, Yoneyama T, Hayashi H (2007) Quantitative estimation of the contribution of the phloem in cadmium transport to grains in rice plants (*Oryza sativa* L.). Soil Sci Plant Nutr 53:72–77

Tennstedt P, Peisker D, Böttcher C, Trampczynska A, Clemens S (2009) Phytochelatin synthesis is essential for the detoxification of excess zinc and contributes significantly to the accumulation of zinc. Plant Physiol 149:938–948

Ueno D, Yamaji N, Kono I, Huang CF, Ando T, Yano M, Ma JF (2010) Gene limiting cadmium accumulation in rice. Proc Natl Acad Sci U S A 107:16500–16505

Uraguchi S, Fujiwara T (2013) Rice breaks ground for cadmium-free cereals. Curr Opin Plant Biol 16:328–334

Uraguchi S, Mori S, Kuramata M, Kawasaki A, Arao T, Ishikawa S (2009) Root-to-shoot Cd translocation via the xylem is the major process determining shoot and grain cadmium accumulation in rice. J Exp Bot 60:2677–2688

Uraguchi S, Kamiya T, Sakamoto T, Kasai K, Sato Y, Nagamura Y, Yoshida A, Kyozuka J, Ishikawa S, Fujiwara T (2011) Low-affinity cation transporter (*OsLCT1*) regulates cadmium transport into rice grains. Proc Natl Acad Sci U S A 108:20959–20964

Wagner GJ (1993) Accumulation of cadmium in crop plants and its consequences to human health. Adv Agron 51:173–212

Wong CK, Cobbett CS (2009) HMA P-typeATPases are the major mechanism for root-to-shoot Cd translocation in *Arabidopsis thaliana*. New Phytol 181:71–78

Yamaguchi N, Ishikawa S, Abe T, Baba K, Arao T, Terada Y (2012) Role of the node in controlling traffic of cadmium, zinc, and manganese in rice. J Exp Bot 63: 2729–2737

Yamaji N, Xia J, Mitani-Ueno N, Yokosho K, Ma JF (2013) Preferential delivery of zinc to developing tissues in rice is mediated by P-type heavy metal ATPase OsHMA2. Plant Physiol 162:927–939

Yoshihara T, Suzui N, Ishii S, Kitazaki M, Yamazaki H, Kitazaki K, Kawachi N, Yin YG, Ito-Tanabata S, Hashida SN, Shoji K, Shimada H, Goto F, Fujimaki S (2014) A kinetic analysis of cadmium accumulation in a Cd hyper-accumulator fern, *Athyrium yokoscense* and tobacco plants. Plant Cell Environ 37:1086–1096

Zenk MH (1996) Heavy metal detoxification in higher plants – a review. Gene 179:21–30

17

miR2118-triggered Phased siRNAs are Differentially Expressed during the Panicle Development of Wild and Domesticated African Rice Species

K. N. Ta[1,2†], F. Sabot[1†], H. Adam[1], Y. Vigouroux[1], S. De Mita[1,4], A. Ghesquière[1], N. V. Do[2], P. Gantet[2,3] and S. Jouannic[1,2*]

Abstract

Background: Rice exhibits a wide range of panicle structures. To explain these variations, much emphasis has been placed on changes in transcriptional regulation, but no large-scale study has yet reported on changes in small RNA regulation in the various rice species. To evaluate this aspect, we performed deep sequencing and expression profiling of small RNAs from two closely related species with contrasting panicle development: the cultivated African rice *Oryza glaberrima* and its wild relative *Oryza barthii*.

Results: Our RNA-seq analysis revealed a dramatic difference between the two species in the 21 nucleotide small RNA population, corresponding mainly to *miR2118*-triggered phased siRNAs. A detailed expression profiling during the panicle development of *O. glaberrima* and *O. barthii* using qRT-PCRs and *in situ* hybridization, confirmed a delayed expression of the phased siRNAs as well as their lncRNA precursors and regulators (*miR2118* and *MEL1* gene) in *O. glaberrima* compared to *O. barthii*. We provide evidence that the 21-nt phasiRNA pathway in rice is associated with male-gametogenesis but is initiated in spikelet meristems.

Conclusion: Differential expression of the *miR2118*-triggered 21-nt phasiRNA pathway between the two African rice species reflects differential rates of determinate fate acquisition of panicle meristems between the two species.

Keywords: Panicle, Meristem, Small RNAs, phasiRNAs, *miR2118*, Domestication, *Oryza barthii*, *Oryza glaberrima*

Background

A prevailing view in evolutionary developmental biology (Evo-Devo) is that morphological traits evolved mostly by changes in expression patterns of functionally conserved genes rather than through the emergence of new genes (Doebley and Lukens 1998; Carroll 2008). Although much emphasis was placed on changes in transcriptional regulation, gene expression is regulated at many levels. In this context, the regulation of genome expression by small RNAs appears to be an important mechanism in the control of plant development (morphogenesis and phase transition), through post-transcriptional regulation of mRNA abundance (by *miRNAs* and secondary *siRNAs*) and silencing of gene expression (by *siRNAs*) *via* site-specific DNA methylation (Jones-Rhoades et al. 2006; Arikit et al. 2013). Secondary, phased small interfering RNAs (phasiRNAs) are emerging regulators of gene expression in plants (Zheng et al. 2015; Xia et al. 2013). Analyses in large panel of plant species covering plant kingdom revealed that a huge diversity of phasiRNA-generating loci types in plants (Zheng et al. 2015; Xia et al. 2013). An unsuspected diversity of phasiRNAs originating from protein-coding genes was observed. However, most of these loci were clade- or species-specific with distinct expression

* Correspondence: stephane.jouannic@ird.fr
†Equal contributors
[1]IRD, UMR DIADE, 911, avenue Agropolis, BP64501, F-34394 Montpellier, Cedex 5, France
[2]LMI RICE, National Key Laboratory for Plant Cell Biotechnology, Agronomical Genetics Institute, Pham Van Dong road, Hanoi, Vietnam

patterns, suggesting recent and independent evolutionary origins (Xia et al. 2013).

Several types of phasiRNA-generating loci were reported from Asian rice *Oryza sativa* and its wild relative *Oryza rufipogon* from various tissues or organs (Arikit et al. 2013; Zheng et al. 2015; Liu et al. 2013). In *O. sativa*, phasiRNAs triggered by specific miRNAs and originating from non-coding loci, were shown to be involved in panicle development and more specifically in male gametogenesis (Arikit et al. 2013; Johnson et al. 2009; Komiya et al. 2014; Song et al. 2012a, b). The *TAS3*-associated trans-acting siRNAs (or ta-siRNAs), triggered by the microRNA *miR390*, are known to target mRNAs of *Auxin Response Factor* (*ARFs*) involved in various developmental processes including floret and stamen development (Allen et al. 2005; Liu et al. 2007; Nogueira et al. 2007; Song et al. 2012b). Other detected phasiRNAs are panicle-specific 21- or 24-nucleotide (nt) small RNAs. They are produced from numerous polyA-tailed long non-coding RNA (lncRNA) generating loci through an RDR6-dependent pathway triggered by the microRNAs *miR2118* and *miR2275* respectively (Komiya et al. 2014; Song et al. 2012a, b). The 21-nt phasiRNAs may act in a complex with the gamete-specific Argonaute (AGO) protein MEIOSIS ARRESTED AT LEPTOTENE1 (MEL1) (Komiya et al. 2014). In maize, each phasiRNA type exhibits independent spatiotemporal regulation, with 21-nt premeiotic phasiRNAs dependent on stamen epidermal differentiation and 24-nt meiotic phasiRNAs dependent on tapetal cell differentiation (Zhai et al. 2015). However, their function during male gametogenesis remains unclear.

African rice *Oryza glaberrima* was domesticated about 2000 to 3000 years ago from *Oryza barthii*, whereas *O. sativa* was domesticated in Asia almost 10,000 years ago from *Oryza rufipogon* (Second 1982; Linares 2002; Caicedo et al. 2007; Vaughan et al. 2008; Huang et al. 2012; Orjuela et al. 2014; Wang et al. 2014; Zhang et al. 2014a. African rice domestication has a single origin in West Africa (Linares, 2002; Wang et al. 2014) and is associated with a severe genetic bottleneck (Li et al. 2011a, b; Nabholz et al. 2014; Orjuela et al. 2014). It is also associated with major morphological and physiological changes. For example, *O. barthii* and *O. glaberrima* exhibit strikingly different panicle architectures, from a low branching complexity and small grain number in *O. barthii* to a more complex panicle and higher grain number in *O. glaberrima* (Linares 2002). A similar divergence of domestication-related traits was also observed between the Asian wild rice and its cultivated counterpart (Vaughan et al. 2008). It has been suggested that several small RNA loci, such as miRNA-triggered phasiRNAs loci including ta-siRNA locus *TAS3a2*, as well as the microRNA loci *MIR164e*, *MIR390* and *MIR395a/b*, have experienced direct selection during Asian rice domestication (Liu et al. 2013; Wang et al. 2010; Wang et al. 2012). However, a large-scale study of small RNA expression in relation to the early stages of panicle development in African rice is still lacking. To investigate whether the latter varies between the two African species in relation to changes in small RNA expression, we conducted a comparative analysis of small RNA populations accumulating during early panicle development in the two African species. We found conservation of the male gametogenesis-associated *miR2118*-triggered phasiRNA pathway in these African rice species. Our data revealed significant changes in expression of these gamete-specific 21-nt phasiRNAs between *O. barthii* and *O. glaberrima*, associated with differential expression of their regulators, namely *miR2118* and *MEL1*. This is suggestive of a differential rate of determinate meristem fate acquisition during the initial panicle development stages between the two species.

Results

Panicle-associated small RNAs in African rice species

To conduct a comprehensive survey of panicle-derived small RNAs in African rice and characterise qualitative and quantitative differences between the two species, we compared small RNA populations in panicles of *O. glaberrima* and *O. barthii*, using genome-wide small RNA-seq analysis. In order to focus on differences in expression resulting from inter-specific variations and to buffer genotypic variations, we used two RNA bulks extracted from developing panicles at same visual stage (from early rachis elongation to early floret differentiation stages) of 10 genotypes for each species (Additional file 1: Table S1). Over 33.1 and 33.9 million high quality reads were generated from *O. glaberrima* and *O. barthii* libraries respectively (Additional file 1: Table S2). A total of 64 % and 64.5 % of small RNA clusters (distinct small RNAs) ranging from 18 to 28 nucleotides from *O. barthii* and *O. glaberrima*, respectively, were mapped to the reference genome *O. sativa* ssp *japonica* cv Nipponbare MSU v7.0, and were found to be similarly distributed over the *O. sativa* genome (Additional file 1: Table S3, Additional files 2 and 3). As expected, the 21- and 24-nucleotide (nt) small RNAs were the predominant populations of small RNAs in the two species, with a higher number of distinct 24-nt small RNAs (Additional file 2). The high quality small RNA sequences were then categorized, using a BLAST-based filtering annotation pipeline (Additional file 4), into five distinct classes corresponding to the different functional compartments of the rice genome: miRNAs, ncRNAs, repeats, genes (CDS, intron, UTR) and unannotated regions of *O. sativa* reference genome (Additional file 1: Table S4).

The 21-nt small RNA population from the two species, corresponding to 23 798 distinct sequences, exhibited a surprisingly large fraction of small RNAs with a higher expression in *O. barthii* than in *O. glaberrima* (Fig. 1a). This subpopulation corresponds to 29 % of the 21-nt small RNAs mapped, when considering small RNAs that were at least 5-fold more accumulated in *O. barthii*. Other size classes (from 18- to 28-nt, except 21-nt) were not distinguished by such a pattern of distribution between the two species (Additional file 5). The overall 21-nt small RNAs were categorized into five distinct classes corresponding to the different functional compartments of the rice genome: miRNAs (11.3 %), ncRNAs (13.1 %), repeats (12.4 %), genes (CDS, intron, UTR: 25.5 %) and unannotated regions (37.7 %) of *O. sativa* reference genome. However, the distribution of the over-expressed fraction in the wild species showed a strong bias towards unannotated regions (81.4 %; p-value = 0.0) with the remaining classes all severely under-represented: miRNAs (2.5 %; p-value = 8.0×10^{-139}), ncRNAs (0.6 %; p-value = 3.5×10^{-301}), repeats (0.8 %; p-value = 2.3×10^{-261}) and genes (14.8 %; p-value = 1.2×10^{-83}) (Fig. 1a; Additional file 5). This indicated that the over-expressed 21-nt small RNAs detected in *O. barthii* panicles were mainly derived from unannotated regions of the *O. sativa* reference genome (Fig. 1a; Additional file 5).

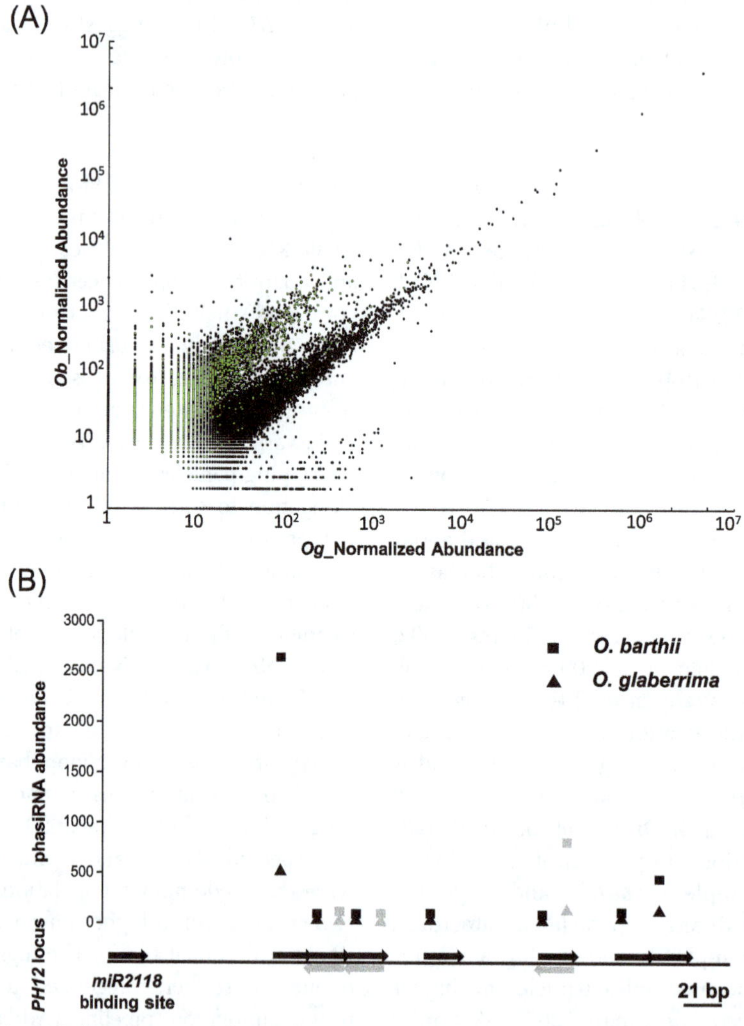

Fig. 1 Relative abundance of panicle-derived 21-nt small RNAs from *O. barthii* and *O. glaberrima*. **a** Relative abundance of 21-nt small RNAs between *O. barthii* (*Ob*) and *O. glaberrima* (*Og*). Normalized abundance of distinct small RNA sequences. Black dots represent global 21-nt small RNAs, red dots unannotated 21-nt small RNAs, and green dots 21-nt small RNAs associated with detected phased loci. **b** Relative abundance of detected phasiRNAs generated from a single locus (*PH-12*) between *O. barthii* and *O. glaberrima*. The lower panel illustrates the position of the *miR2118* binding site and the detected phasiRNAs from the *PH12* locus in the two species (above or below the line if originating from plus or minus strand respectively). The upper panel illustrates the abundance of the corresponding phasiRNAs in *O. glaberrima* (triangles) and *O. barthii* (squares). The abundances of reads from the plus strand are in dark grey, and reads from the minus strand are in light grey

21-nt phasiRNAs are over-represented in African wild rice transcriptomes

Previous studies of *O. sativa* showed that the *miR2118*-triggered 21-nt phasiRNAs specifically expressed in the rice panicle originate from unannotated regions of the genome (Johnson et al. 2009; Song et al. 2012a; Komiya et al. 2014). To determine whether the over-represented 21-nt small RNA fraction in *O. barthii* also corresponds to this class of 21-nt phasiRNAs, we used a dedicated program (see Material and methods), and detected 4 100 distinct phasiRNAs from *O. barthii* and *O. glaberrima*, distributed across 892 loci (denoted "phased loci"), using the *O. sativa* genome as reference (Additional file 6). In addition, 952 distinct 21-nt small RNAs mapped to these 892 loci, but were not detected as phased siRNAs by our procedure. In total, 5 052 distinct 21-nt small RNA sequences associated with these 892 phased loci were found, corresponding to 21.2 % of the panicle-derived 21-nt small RNA population from the two species. Out of these 5 052 small RNAs, 3 694 were detected in both species, 1 352 were detected only in *O. barthii* and six only in *O. glaberrima* (Additional file 6). As previously reported, the detected phased loci were present on all 12 chromosomes but not at the same density, and were organized in clusters or super-clusters, such as on chromosome 12 (Additional file 7). As expected, 86.5 % of these 21-nt phasiRNAs mapped to unannotated regions of the *O. sativa* genome. This fraction of 21-nt phasiRNAs represents 49 % of the unannotated 21-nt small RNA population (4 418 over 8 977 sequences). The 892 phased loci were scanned for the presence of the 22-nt miRNA *miR2118* recognition site in their vicinity (Johnson et al. 2009; Song et al. 2012a; Komiya et al. 2014). 529 phased loci (59.3 %) shared the conserved motif close to one end (from two to 452 bp from the end). Between two and 91 distinct sequences were detected per phased locus, with an average of 6.0 (±4.8, SD) distinct sequences, the majority being 3 or 5 sequences per phased locus (Additional file 5). As expected for double strand RNA processed small RNAs, most of the detected phased loci generated small RNAs from both strands (92.5 %) (Additional file 6).

For 96 % of the phased loci, there was a significant difference in read counts between the two species, and about 71 % of them displayed an abundance ratio higher than 5 (Additional file 6). These 21-nt siRNAs were mainly over-accumulated in *O. barthii* and contributed to 52 % of the *O. barthii* five fold over-accumulated 21-nt small RNAs (Fig. 1a). The abundance of different phasiRNAs from a given locus was variable, with the predominance of one or two phasiRNAs (Fig. 1b; Additional file 8). For the differentially expressed loci, this unequal distribution over the locus was conserved between the two species, with over-accumulation of all the detected phasiRNAs from a single locus (Fig. 1b). Interestingly, the proportion of phased loci sharing the *miR2118* recognition site was significantly higher (p-value = 1.5×10^{-08}) in the differentially expressed group (i.e. higher expression in *O. barthii*) (61 %, $n = 854$) than in the group of non-differentially expressed phased loci (16 %, $n = 38$) (Additional file 6). Therefore, phased loci over-expression in *O. barthii* versus *O. glaberrima* may result from differences in *miR2118* regulation between the two species.

Panicle-associated microRNAs in African rice species

Comprehensive analysis of conserved miRNAs led to the identification of 146 annotated miRNA families expressed in African rice panicles, including 62 canonical families, 20 variant ones and 64 siRNA-like ones, according to the classification of Jeong et al. (2011) (Additional files 6 and 9). Some of the siRNA-like families were previously reported to be associated with transposable elements (TEs) in *O. sativa*, namely *TE-MIRs* (Li et al. 2011b; Yan et al. 2011). In the same way, other annotated *MIR* loci were evidenced in our analysis of 21-nt phased loci as overlapping loci, suggesting that these loci are phasiRNA-generating loci rather than *MIR* ones: *osa-MIR5486, osa-MIR5488, osa-MIR5506, osa-MIR5514*, osa-*miR5516*, osa-*miR5519*, *osa-MIR5517, osa-MIR5527, osa-MIR5530, osa-MIR5791, osa-MIR5796, osa-MIR5800* and *osa-MIR5822* (Additional file 6). Thus, the annotations of these two classes of *MIR* loci (i.e. variant and siRNA-like) should be reconsidered. Similarly to what was reported for *O. sativa* (Jeong et al. 2011), the canonical and variant miRNA families expressed in African rice panicles can be classified according to the length of the highest accumulated small RNA: length ranging from 19- to 25-nt, with 3 main types corresponding to 21-nt (48 %), 24-nt (31 %) and 22-nt (9 %) long mature miRNAs (Additional files 6 and 9).

Most miRNA families were expressed at similar levels in *O. barthii* and *O. glaberrima*. Interestingly, the *miR159/319* families contributed to 64 % of mature miRNA expressed in young panicles of both African species (Additional file 6), in contrast to previous studies of panicle-derived miRNAs in *O. sativa* (Jeong et al. 2011; Peng et al. 2011). This may be related to the specific developmental stages used in our study or may reflect differences between Asian and African rice species. Differential expression affected only a few miRNA families with a trend towards higher expression in *O. barthii* (Fig. 2). These families included the canonical 22-nt *miR2118* and five other ones, namely *miR2275, miR5495, miR5497, miR5516* and *miR5519* (Fig. 2). The latter two were identified in our study as panicle-expressed 21-nt phasiRNAs (Fig. 2; Additional file 6). The canonical 22-nt miRNA *miR2275* was previously reported to trigger 24-nt phasiRNAs (Johnson et al. 2009; Song et al. 2012a). However,

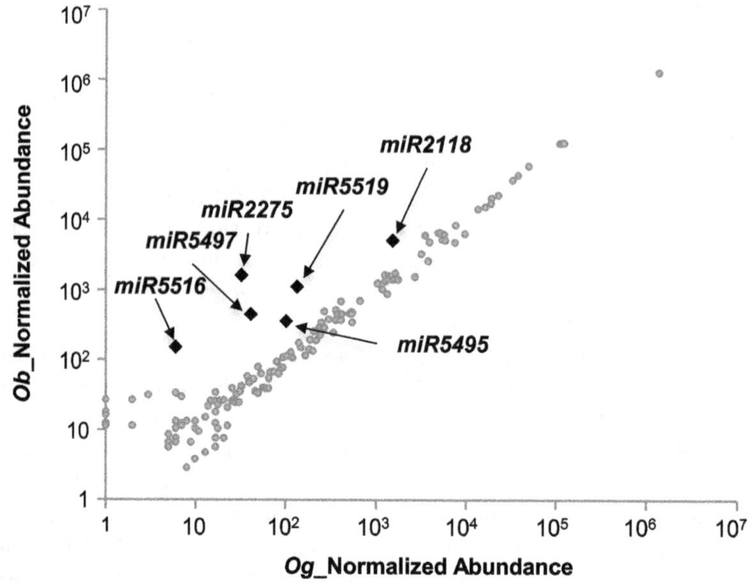

Fig. 2 Relative abundance of miRNA families between *O. barthii* (*Ob*) and *O. glaberrima* (*Og*). Normalized abundance of miRNA reads. Black diamonds represent the differentially expressed miRNA families, and grey points to non-differentially expressed miRNA families between *O. barthii* and *O. glaberrima*

despite over-expression of *miR2275* in *O. barthii,* no major over-expression of a large subset of the 24-nt small RNAs was detected (Additional file 5). The canonical miRNAs *miR5495* and *miR5497* were previously shown in *O. sativa* to be miRNAs specifically expressed in pollen (Wei et al. 2011).

Variations in the expression of *miR2118*, 21-nt phasiRNAs and associated precursors during African rice panicle development

The differential accumulation of specific microRNAs and 21-nt phasiRNAs observed between *O. barthii* and *O. glaberrima* may result from differences in their expression levels, but may also reflect a shift in the timing of their expression during panicle development. To test these hypotheses, expression analysis of *miR2118,* 21-nt phasiRNAs and their RNA precursors was performed on 4 distinct morphological stages of panicles collected from *O. barthii* (accession B88) and *O. glaberrima* (variety CG14) (Fig. 3; see Additional file 10 for histological description). Four different phased loci from distinct clusters and chromosomes were considered for the RNA precursors as well as the phasiRNA with the highest level of accumulation for each locus (Fig. 3; Additional files 8, 11 and 12).

The differential accumulation of *miR2118,* selected phasiRNAs and their corresponding precursors between the two species were validated through Northern-blotting and/or RT-PCRs on RNA bulks similar to the ones used for small RNA deep sequencing (Additional file 8). While neither mature *miR2118* nor 21-nt phasiRNA *phasiPH12-1* were detected at the early morphological stage of panicle development (inflorescence meristem after reproductive transition, stage 1), their expression was initiated earlier in *O. barthii* (early branching stage, stage 2) than in *O. glaberrima* (late branching stage, stage 3). The expression level reached a peak at stage 3 in *O. barthii* and remained at the same level at stage 4 (young flowers with differentiated organs). In *O. glaberrima* the miRNA and phasiRNA were more abundant at stage 4 and displayed a higher expression level than in *O. barthii* at this stage (Fig. 3). Similar patterns of expression were observed during panicle development for 21-nt phasiRNAs from the four phased loci tested (Additional file 12). These findings indicate that the later accumulation of 21-nt phasiRNAs during the panicle developmental time-course in *O. glaberrima* is associated with a later accumulation of *miR2118*. These findings are in agreement with the small RNA-seq data, as the RNA bulks used for sequencing originate from panicles at morphological stages 1 to 3, in which *miR2118* and 21-nt phasiRNAs are accumulated at higher levels in *O. barthii* than in *O. glaberrima*.

Similarly to Komiya et al. (2014), we showed that the 21-nt phasiRNAs from the African species were generated from polyA-tailed long non-coding RNAs (lncRNAs) (Additional file 11). The accumulation patterns of these polyA-tailed lncRNAs were similar to those of the phasiRNAs with a later initiation of expression in *O. glaberrima* than in *O. barthii,* regarding the morphological stages (Fig. 3; Additional file 12). Furthermore, we investigated the accumulation levels of mRNAs corresponding to gene

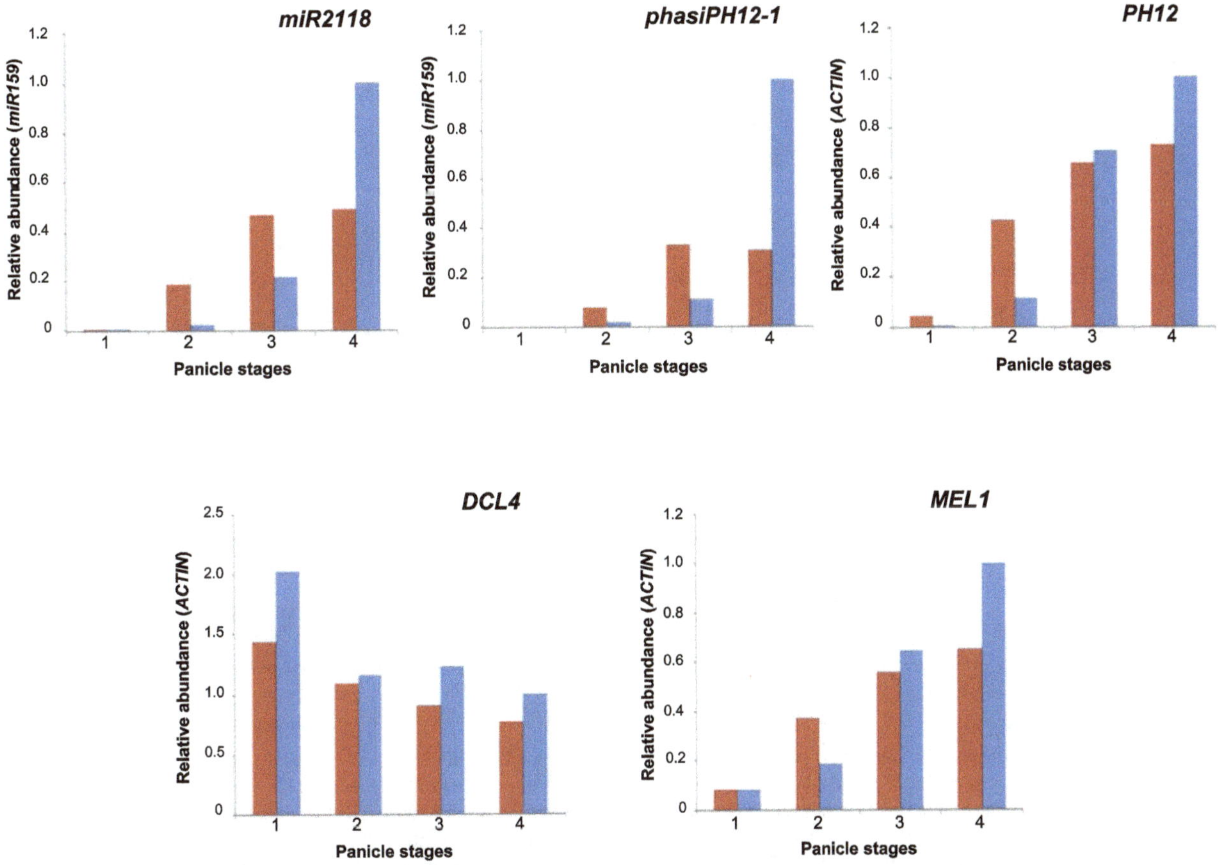

Fig. 3 Time course expression analysis of small RNAs and related-regulators during panicle development. qRT-PCR analysis of *miR2118*, *PH12* precursor, *phasiPH12-1*, *DCL4* and *MEL1* ortholog expression levels during panicle development (from stage 1 to stage 4) in *O. barthii* (red) and *O. glaberrima* (blue). Expression values are relative to *O. glaberrima* stage 4 and the gene or miRNA used as reference is indicated. Stage 1, Inflorescence meristem stage, after initiation of primary branch (i.e. rachis and primary branch meristems); stage 2, early branching stage (panicle with elongated primary and higher order branch development); stage 3, late branching stage (i.e. panicle with elongated primary and secondary branches); stage 4, young flowers with differentiated organs

orthologs encoding other factors involved in the biogenesis of the *miR2118*-triggered 21-nt phased RNAs, such as the OsDCL4 DICER-like protein, and the gamete-specific *Argonaute* protein MEL1 (Song et al. 2012a, b; Komiya et al. 2014). The accumulation level of African rice *OsDCL4* ortholog mRNAs decreases slightly over the 4 stages, while African rice *MEL1* ortholog mRNA accumulation increases from stage 1 to stage 4 in a similar pattern to the small RNAs and lncRNAs (Fig. 3; Additional file 12). Taken together, these data suggest that the accumulation levels of phasiRNAs may depend on *miR2118*, phased loci-associated lncRNAs and African rice *MEL1* ortholog transcript accumulation levels rather than to those of the *OsDCL4* ortholog.

To determine the spatial expression patterns of *miR2118*, *phasiPH12-1* and *PH12* lncRNAs during panicle development, *in situ* hybridization analysis was performed at different developmental stages of young panicles from the two species (Fig. 4). Similar patterns were observed for the two species with variations in spatial patterning of these factors. This analysis revealed that the onset of the *miR2118*-triggered 21-nt phasiRNA pathway is sequential in terms of initiation of expression. *PH12* lncRNAs are detected first in spikelet meristems and are maintained in developing florets, but only in stamens and more specifically in the pollen sac. Subsequently, *miR2118* specifically accumulates in the hypodermis of developing stamens (prior to pollen sac differentiation) at the early floret differentiation stage, then in the epidermis and the pollen sac in differentiated stamens at later stages. Finally, the *phasiPH12-1* phasiRNAs are specifically located in pollen sac of differentiated stamens subsequent to *miR2118* detection (Fig. 4). Eventually, all three signals were observed at the same location at later stages (i.e. pollen sac).

Our spatial expression analysis of *miR2118*, *phasiPH12-1* and *PH12* lncRNA and their differential

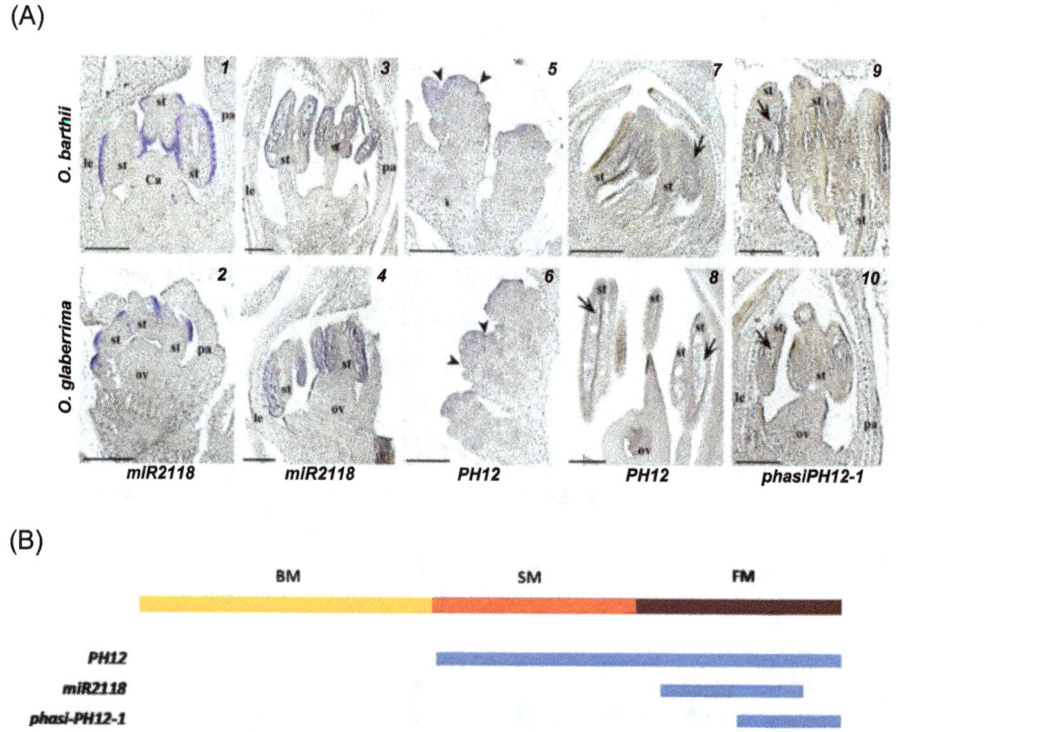

Fig. 4 *In planta* expression analysis of small RNAs and related-regulators during panicle development. **a** *In situ* hybridization of mature *miR2118* (1–4), *PH12* mRNA precursor (5–8) and *phasiPH12-1* (9–10) in spikelet meristem (SMs), floret meristem (FMs) and differentiated floret in *O. barthii* (B88) and *O. glaberrima* (CG14). Mature *miR2118* transcripts were detected in the outer cell layer of stamens (*i.e.* hypodermis) (1–2) and later in pollen sac (3–4). Precursor *PH12* mRNAs were detected in the spikelet meristem (5–6) and in pollen sac of the anthers (arrow) (7–8). The *phasiPH12-1* small RNAs were detected in the pollen sac of the anthers in florets (9–10). This panel of photos illustrates stages for which the transcripts were detected. le, lemma; pa, palea; st, stamen; ca, carpel; ov, ovule. Arrowheads, lemma and palea primordium formation in SMs; arrow, anthers. Scale bars: 100 µm. **b** Schematic view of the expression profiling of *PH12* mRNA precursor, mature *miR2118* and *phasiPH12-1* regarding meristem state and developmental timing based on *in situ* hybridization data. Blue boxes illustrate transcript detection periods. BM: branch meristem; SM: spikelet meristem; FM: floret meristem

accumulation patterns with respect to the morphological stages of panicle development between the two species indicates that at the branching stage, the gamete-associated small RNAs are initiated later and expressed at lower levels in *O. glaberrima* than in *O. barthii*. This suggests that for the same morphological stage of panicle development, stamen differentiation, and therefore floret differentiation, might be delayed in *O. glaberrima* with respect to its wild relative, *O. barthii*.

Variations in expression of a spikelet-related gene during panicle development

To investigate this hypothesis further, we analysed the expression pattern of orthologs of the *O. sativa* E-function MADS-box *LEAFY HULL STERILE1* (*LHS1*)/ *OsMADS1* gene that promotes the determination of meristems (i.e. spikelet and floret meristems) (Jeon et al. 2000; Cui et al. 2010; Khanday et al. 2013). Using *in situ* hybridization, we investigated the expression pattern of the *LHS1* ortholog in *O. barthii* and *O. glaberrima* panicles at the early branching stage. In the two species, *LHS1* mRNA was detected specifically in SMs (Fig. 5). However, while the *LHS1* mRNA signal was observed in some lateral meristems in *O. glaberrima*, it was detected in all lateral meristems in *O. barthii* (Fig. 5). This supports the hypothesis that for the same morphological stage, all terminal and lateral meristems had acquired the spikelet fate in *O. barthii*, whereas in *O. glaberrima* few had done so. These observations are also in agreement with the gamete-associated small RNA profiling data obtained.

Discussion

Conservation of *miR2118*-triggered 21-nt phasiRNAs in Oryza species

Using panicle-derived small RNA transcriptome sequencing in *O. glaberrima* and its wild ancestor, *O. barthii*, we showed that 29 % of the 21-nt small RNA population is drastically repressed (or non activated) in cultivated genotypes during the branching stage of panicle development. This large difference in expression pattern suggests a change in regulation of key regulatory molecules(s). The

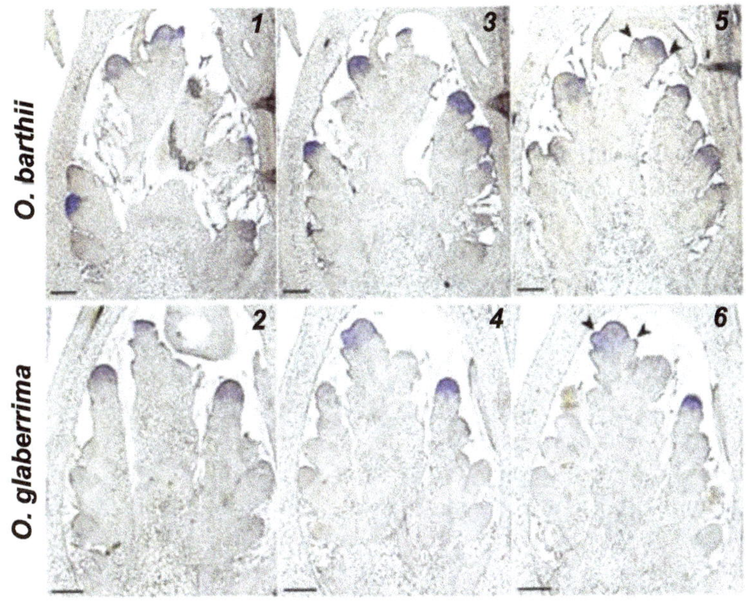

Fig. 5 *In situ* expression analysis of *LHS1* gene in young panicles from African rice. *In situ* hybridization of *LHS1* transcripts in serial sections of spikelet meristems (SMs) of *O. barthii* (1–3) and *O. glaberrima* (4–6); arrowheads, lemma and palea primordium formation in SMs. Scale bars: 100 μm

altered fraction of small RNAs corresponds mainly to 21-nt phasiRNAs generated from nearly a thousand non-coding RNA loci, associated with altered expression of the 22-nt microRNA *miR2118*. This is in agreement with observations in *O. sativa* that panicle-specific 21-nt phasiRNAs are produced from multiple non-coding RNA loci through an RDR6/DCL4-dependent pathway triggered by the microRNA *miR2118* (Johnson et al. 2009; Song et al. 2012a, b). The synthesis of secondary siRNAs (or phasiRNAs) in reproductive organs from non-coding loci has only been reported in rice, maize and *Brachypodium*, suggesting a recent origin of these secondary siRNAs from a common ancestor of grasses (Zheng et al. 2015; Johnson et al. 2009; Song et al. 2012a, b; Zhai et al. 2015, Vogel et al. 2010). However, the low conservation of these loci across the group suggests that they may have species-specific functions (Zheng et al. 2015;, Komiya et al. 2014). Interestingly, *miR2118*-triggered secondary siRNA synthesis is also conserved across distantly related species, such as *Medicago truncatula*, tobacco and tomato. However, in these eudicot species, the *miR2118* family members (including *miR482* family members) were recruited to initiate secondary siRNAs on mRNAs encoding NBS-LRR proteins involved in disease resistance (Zhai et al. 2011; Li et al. 2012; Shivaprasad *et al.* 2012).

Our study revealed conserved features of panicle-associated 21-nt phasiRNAs and associated loci between *Oryza* species. However, this analysis was limited to the fraction of small RNAs corresponding to *O. barthii* and *O. glaberrima* phasiRNAs conserved in *O. sativa* (i.e. mapped to the *O. sativa* genome), and does not provide a complete overview of the complexity of phased loci in African rice species and their evolution during the African domestication process. The genome sequence of *O. glaberrima* accession CG14 was recently published (Wang et al. 2014), however, we obtained better mapping results from our small RNA sequence data using the reference genome *O. sativa* ssp *japonica* cv Nipponbare MSU v7.0 than the *O. glaberrima* CG14 released sequence: 72.1 % of small RNA reads from both African species using Nipponbare MSU v7.0 sequence in contrast to 61.6 % and 56.2 % from *O. barthii* and *O. glaberrima* respectively using the *O. glaberrima* reference genome AGI1.1. For this reason, we continued to perform our small RNA sequence data analysis using the *O. sativa* Nipponbare reference genome. Nevertheless, it might be expected that higher diversification occurred between Asian and African rice since they diverged about 1 million years ago, whereas *O. glaberrima* and *O. barthii* diverged about 3000 years ago. Consequently, our quantitative analysis in African rice will be only marginally biased by evolutionary differences at the sequence level between the two African rice species. However, the number of overlapping phased loci between African rice species and *O. sativa* is quite low. About 1136 phased loci were identified from 4-cm long panicles of *O. sativa* ssp *indica* var 93–11. However, only 416 out of the 892 detected phased loci in African species overlapped phased loci detected in *O. sativa* spp *indica* var 93–11 (Song et al. 2012a). This may be because they were identified in a different way, i.e. based on different stages of panicle development. It should also be noted that more small RNA sequences were characterized in our study. Even if sequences are conserved between *Oryza* species, it cannot be ruled out that species-specific expression patterns of these lncRNA loci may occur. Moreover, 40 % of the detected loci associated with differentially expressed 21-nt phasiRNAs between the two African species had no *miR2118* recognition site motif, suggesting a potentially distinct mechanism of production. Another possible explanation would be that *miR2118* recognition sites are present in the African rice genome at these loci but not in the *O. sativa* genome. In addition, in our comparative analysis, 48 % of the over-expressed 21-nt small RNAs in *O. barthii* mapping to the *O. sativa* genome were not associated with the detected phased loci, and also originated mainly from unannotated regions of the genome. It is possible that these remaining over-expressed 21-nt small RNAs belong to phased loci specific to African rice genomes that were not detectable in our analysis because of sequence divergence from the *O. sativa* reference, or that they are associated with non-coding loci unrelated to phased small RNA generating loci. Recently, Zhang et al. (2014b) showed that a large set of long non-coding RNAs was expressed in anthers and pistils of *O. sativa* rice. Interestingly, only 122 of the 1624 reported lncRNA loci were associated with small RNAs (Zhang et al. 2014b). In agreement with this, only a few of the reported lncRNA loci overlapped with the *miR2118*-associated phasiRNA-generating loci in *O. sativa* and African rice species (data not shown). This suggests that these lncRNAs are mostly not associated with the production of phasiRNAs or other small RNAs and may act through other mechanisms during plant development.

The *miR2118*-triggered 21-nt phasiRNAs are markers of male-gametogenesis in African rice

The function of the 21-nt phasiRNAs during panicle development is still unclear. MicroRNA *miR2118* has been reported to be preferentially expressed in rice and maize stamens, suggesting a role in male gametogenesis (Song et al. 2012a). Recently, it was shown that the function of this class of phased small RNA was dependent on the germ line-specific Argonaute (AGO) protein MEL1 through a direct interaction between MEL1 and phasiRNAs (Komiya et al. 2014). Moreover, as mature *miR2118* was detected in the MEL1-binding small RNA fraction, it was argued that MEL1 might play a role in the first steps of the 21-nt phasiRNA biogenesis pathway as the AGO protein in the *miR2118*-driven RISC triggering to the first cleavage of lncRNA precursors (Komiya et al. 2014). This is in agreement with the expression pattern observed for *MEL1* and *miR2118*. *MEL1* mRNAs were observed to be first detectable by *in situ* hybridization in the hypodermis of developing stamen primordia in a spotty pattern, similarly to that which we observed for *miR2118* microRNA, later becoming restricted to microsporangia and pollen sac (Nonomura et al. 2007). In our study, while *PH12* precursor lncRNAs were detected in the spikelet meristem differentiation stage, *phasiPH12-1* was only detected in the stamens of differentiating florets and co-localized with *PH12* precursor lncRNAs, but only after *miR2118* was observed to accumulate. The latter was first limited to the hypodermis of differentiating stamens and extended to the pollen sac in later stages. Together, these data suggest that the 21-nt phasiRNA regulatory network is initiated early during panicle development from the stage of spikelet meristem establishment, before *miR2118* and *MEL1* expression, through the induction of lncRNA precursor expression. The co-expression of *miR2118* and *MEL1* in the hypodermis of the differentiating stamen would result in the initiation of the 21-nt phasiRNA biogenesis pathway, leading to the accumulation of the phasiRNAs. However, the *trans*-acting factors involved in the regulation of the lncRNA precursors are still unknown. Zhai et al. (2015) have recently shown that maize *miR2118*-triggered 21-nt phasiRNAs correspond to premeiotic siRNAs in anthers, with a similar spatial patterning observed in rice, indicating a conservation of the spatial regulation of this pathway in grasses.

Although direct evidence is still lacking for the involvement in male gametogenesis of *miR2118* and the associated phasiRNAs and lncRNA precursors, their involvement is supported by the phenotypes of *rdr6* and *dcl4* mutants affected in flower and stamen development (Liu et al. 2007; Song et al. 2012b). Moreover the *mel1* mutant phenotype indicates that the MEL1 AGO protein mediates the regulation of germ-line mother cell development and meiosis in both male and female organs, suggesting that the MEL1-phasiRNA complexes may be involved in these processes (Nonomura et al. 2007; Komiya et al. 2014). The remaining question is whether these phasiRNAs can be considered as *ta*-siRNA (secondary siRNAs targeting other mRNAs). The predominant accumulation of MEL1 protein in the cytoplasm would favour a *trans*-acting function of MEL1-phasiRNA complex on mRNAs but without cleavage activity of the putative targets (Komiya et al. 2014; Zhai et al. 2015).

Differential rate of determinate fate acquisition in the panicle meristems of African rice

Our comparative analysis of small RNA expression with respect to the development of the African rice panicle suggests that the spikelet/floret fate acquisition rate differs between the two species: for a similar morphological complexity during the early branching stage, all meristems are converted into spikelets in *O. barthii* whereas only those of the apical part of the panicle branches are converted in *O. glaberrima,* as suggested by the expression pattern of the spikelet-associated MADS-box gene *LHS1/OsMADS1*. This difference was paralleled by the later initiation of expression and lower accumulation during the branching stage of the 21-nt phasiRNA pathway members (*miR2118, MEL1, lncRNAs* and *phasiRNAs*) in *O. glaberrima,* as well as the lncRNA precursors expressed in spikelet meristems. The differential rate of spikelet transition observed between the two species may be explained by a shorter duration of development in *O. barthii* than in *O. glaberrima* but also possibly by the synchronous transition of the apical and lateral meristems from the indeterminate to the determinate state in *O. barthii.*

Theoretical modelling of inflorescence architecture suggests that inflorescence branching complexity and its evolution depends on differences in the timing of floral (i.e. determinate) fate acquisition in apical and/or lateral meristems (Prusinkiewicz et al. 2007). This biological basis is supported by the analysis of various mutants affected in floral meristem identity in different species, notably in *Arabidopsis thaliana, Antirrhinum majus,* petunia and tomato (Koes 2008; Moyroud et al. 2010; Park et al. 2014). A comparative study of the diversity of inflorescence architecture in tomato based on meristem-specific transcriptome analysis provided support for this model (Park et al. 2012). In the case of the grass inflorescence and more specifically the rice panicle, it was also reported that variations in panicle architecture depended on the activity or expression levels of both meristem fate controlling genes and branch-promoting genes (Kyozuka et al. 2014; Zhang and Yuan 2014). In the context of our comparative analysis of African rice species, a differential rate of determinate fate acquisition in panicle meristems would result in a longer or higher rate of branching activity in *O. glaberrima* compared to *O. barthii,* leading to a higher branch complexity in the domesticated species. However, further functional studies are required to define the link between the major regulatory changes of spikelet-related gene expression and the panicle phenotypic variations associated with domestication.

Conclusions

We provide evidence that the male-gametogenesis-specific 21-nt phasiRNA pathway triggered by *miR2118* is conserved in both wild relative and domesticated African rice species. Our study shows that the onset of the *miR2118*-triggered 21-nt phasiRNA pathway is sequential in terms of initiation of expression with variations in spatial patterning of these factors. This pathway is initiated when spikelet meristems are established, through the activation of lncRNA precursor expression. Our study provides evidence that the differential expression of the *miR2118*-triggered 21-nt phasiRNA pathway in *O. barthii* and *O. glaberrima* may be associated with different rates of determinate meristem fate acquisition during branching stage of panicle development. It will be of great interest to determine whether a similar scenario has occurred in the Asian rice species (*O. sativa* vs. *O. rufipogon*), so as to determine whether or not the phenotypic convergence of panicle development observed between the two domestication processes is paralleled by similar changes in the expression of molecular regulatory components associated with spikelet differentiation and panicle architecture complexity.

Methods

Plant materials and panicle sampling

For Illumina sequencing, 10 accessions of *O. glaberrima* and 10 accessions of *O. barthii* (see Additional file 1: Table S1) were grown in the greenhouse at IRD, Montpellier. Around 15 panicles from each accession were collected from 4 to 15 days after induction, to obtain homogeneous representation of early developmental stages corresponding to early rachis elongation to early floret differentiation.

For histological analysis and expression analysis, CG14 and B88 plants were grown in growth chamber at IRD, Montpellier (France) with a 14-10 h day/night cycle at 32 °C/28 °C and humidity at 60 %. Flowering was induced by short day conditions (10-14 h day/night cycle). Panicles were collected at 4 different morphological stages: stage 1, inflorescence meristem stage, after initiation of primary branch (i.e. rachis and primary branch meristems); stage 2, early branching stage (i.e. panicle with elongated primary and higher order branch development); stage 3, late branching stage (i.e. panicle with elongated primary and secondary branches); stage 4, young flowers with differentiated organs (Additional file 10).

Illumina sequencing and data processing

Total RNAs (including small RNAs) were extracted using an RNeasy Plant Mini Kit with RLT and RWT buffers (Qiagen, France). DNase treatments were performed using the RNAeasy-free DNase set (Qiagen, France). Two bulks of total RNAs corresponding to a mix of total RNAs from the 10 accessions of the two species were used for sequencing. Purified small RNA sequencing was performed by Eurofins/MWG Operon (Germany) on an Illumina *Hi-seq 2000* using the TruSeq™ SBS v5 sequencing kit. The complete raw dataset is available in NCBI Gene Expression Omnibus repository [GEO: GSE48346]. The raw data were trimmed by removing adapter sequences and low quality sequences using *CutAdapt* (Martin 2011). All the trimmed reads ranging from 18 to 28 nucleotides were clustered and mapped to *O. sativa ssp japonica* cv Nipponbare genome (MSU release version 7; http://rice.plantbiology.msu.edu/) using *BLAST* (Altschul et al. 1990) (Additional file 4) in order to evaluate library quality. The 18–28 nucleotide reads were them annotated through a filtering process using successive hierarchical *BLAST* (Additional file 4) versus (in order) miRBase v17.0 (Kozomara and Griffiths-Jones 2011), Rfam v7, a home-made repeat database (successive curated concatenation of *RetrOryza*, RepBase, TREP and TIGRRepeats), CDS then gene features (i.e. introns and UTRs) from *Oryza sativa ssp japonica var* Nipponbare MSU v7.0 annotation, and finally the MSU v7.0 rice genome for the sequences not similar to miRNAs, non coding RNAs, Repeat-associated sequences and genes. BLAST filtering was done using *Oryza* genus data form the different databases (i.e. mainly *O. sativa*). Mapping of the sequences to the *O. sativa* MSU v7.0 reference genome was done for the last three steps of our annotation pipeline. This genome was used as a reference because (*i*) of its equal divergence to the two species we analyzed (avoiding favoring one species in mapping analyses rather than the other one), and (*ii*) because of the larger resources in terms of annotation, expression data and genomic information available for this species. The *BLAST* and post-filter parameters used were probability of 85 %, e-value of 10^{-3}, on a size of 85 % of the reads (minimum size of 16). The same *BLAST* parameters were used throughout the analysis. Mapping from *O. glaberrima* and *O. barthii* small RNA sequences was then compared and filtered using a series of homemade *Perl* scripts (available on demand). Comprehensive analysis of the miRNA-associated small RNAs from African rice species was limited to the families present in *O. sativa*, and did not include diverse isoforms. For the same reason, potential African rice specific miRNAs were not investigated. The 21-mers were used in phasing analysis with the *ta-si Prediction* tool from the *UEA sRNA workbench* facilities (http://srna-workbench.cmp.uea.ac.uk/; Stocks et al. 2012). A post-analysis manual filtering was performed for sequences mapping to more than 10 loci in genome, as well as telomere-associated loci, leading to the deletion of 123 loci from the 1015 initial loci reported in *ta-si Prediction* tool outputs. Once the loci were identified, we used the *EMBOSS* software suite v6.5.7.0 (Rice et al. 2000) to extract −500/+500 bases around each locus, and treated them using *MEME* v4.8.1 (Bailey and Elkan 1994). Then, the loci sequences were used as a BLAST database to add other 21-nt sequences that were not identified originally as phased (in order to by-pass the genome divergence), in two pass: first we select only new sequences with 100 % of identity on 85 % of their length (minimum HSP length of 16), then we change the threshold to 85 % of identity on 85 % of their length (min HSP length of 16). Finally, statistical tests of all the processed data were performed using *g-test* and a fixed *p*-value of 10^{-3}. Depending on the experiment, the degree of freedom was adjusted but was generally 1. All the calculations were performed using homemade *Perl* scripts and CPAN statistical modules.

Quantitative RT-PCR analysis

Total RNAs (including small RNAs) were extracted as described before, from the 4 individual morphological stages of young panicles of *O. glaberrima* (CG14) and *O. barthii* (B88). First-stand cDNA was synthesised using SuperScript III cDNA First-strand synthesis system (Invitrogen). Quantitative stem-loop RT-PCR analyses on small RNAs were performed using 100 ng of total RNA according to Varkonyi-Gasic et al. (2007) with reverse transcription at 42 °C in conjunction with small RNA-specific stem-loop primers. Quantitative RT-PCR analyses on mRNAs were performed using 1 μg of total RNA in conjunction with polydT or random hexamer primers according to the manufacturer's instructions. qRT-PCRs were performed using LightCycler 480 thermocycler (Roche, France) in conjunction with SYBR Green I master mix (Roche, France) in 8 μL reaction mix containing 2 μL of diluted RTs and 0.8 μL of forward and reverse primers at 10 μM. The qPCR amplification conditions include 3 stages: pre-incubation (95 °C in 10 min); amplification with 45 cycles (95 °C 15 s and 60 °C 30s); melting curve (95 °C 5 s and 70 °C 1 min). In stem-loop qRT-PCR, the levels of miRNA were normalized by mature *miR159* expression level. In classic qRT-PCR, mRNAs were normalized to the rice *Actin* gene (*LOC_Os03g50885.1*). Each set of experiments was repeated three times, and the relative quantification method with efficiency corrected calculation model (Souaze et al. 1996) was used to evaluate quantitative variations. The primers used are listed in Additional file 1: Table S5.

In situ hybridizations

PCR amplifications were performed with gene-specific antisense primers tailed with a T7 RNA polymerase binding site (see Additional file 1: Table S5 for primer sequences). The resulting DNA fragments were used directly as templates for synthesizing antisense ribo-probes incorporating UTP–digoxigenin (Roche) as the label in conjunction with a T7 Maxi Script kit (Ambion). For *miR2118*, *PH12* precursor and *phasiPH12-1* detection, 0.02 μM of a 5′ digoxigenin–labeled LNA probe complementary to the target (see Additional file 1: Table S3 for primer sequences) was used. *In situ* hybridization experiments were carried out as described by Adam et al.

(2011). Detection was performed using the Vector Blue Alkaline Phosphatase Substrate Kit III (Vector Laboratories). Slides were observed and photographed by Evolution MP5.0 colour Media Cybernetics camera in conjunction with a Leica DMRB microscope and images were processed using Photoshop CS6.

Additional files

Additional file 1: Table S1. List of genotypes used in this study. **Table S2.** Summary statistics of small RNA libraries. **Table S3.** Summary of sequence number (clusters and reads) mapped to *O. sativa* Nipponbare reference genome (MSU7.0) and *O. glaberrima* CG14 reference genome for *O. barthii* and *O. glaberrima* small RNA sequences. **Table S4.** Summary of sequence number (clusters and reads) in each annotation class (i.e. sequence databases) for *O. barthii* and *O. glaberrima* small RNA sequences based on BLAST-filtering. **Table S5.** List of primers used in this study. The bases in brackets represent the ones that were modified LNAs. The sequences in italic correspond to stem-loop region for stem-loop RT-PCRs. The underline sequences correspond to T7 RNA polymerase binding site used for RNA probe synthesis.

Additional file 2: Size distribution, mapping and annotation of *O. barthii* and *O. glaberrima* panicle-derived small RNAs on *O. sativa Nipponbare* genome. (a) Small RNA size distribution of clusters (distinct) and reads (abundance). The size of small RNAs was plotted *versus* frequency (percentage relative to total abundance). (b) Mapping rate (percentage relative to total number) of clusters (i.e. distinct sequences) and total reads on *O. sativa nipponbare* MSU7.0. The percentage values are indicated in the bars. (c) Small RNA annotation frequency (percentage relative to total abundance) of clusters (distinct) and individual reads (abundance). The percentage values are indicated in the bars. (d) Size distribution of the annotated small RNA clusters (i.e. distinct sequences) according the annotation classes. (e) Size distribution of the annotated small RNA sequences (i.e. abundance/reads) according the annotation classes.

Additional file 3: Genomic distribution and abundance of *O. barthii* and *O. glaberrima* panicle-derived small RNAs on *O. sativa nipponbare* genome. (a) Distribution and abundance of *O. barthii* reads vs. *O. sativa nipponbare* genome MSU7.0. (b) Distribution and abundance of *O. glaberrima* reads vs. *O. sativa nipponbare* genome MSU7.0.

Additional file 4: Bio-informatic workflow for small RNA analysis.

Additional file 5: Relative abundance of small RNAs between *O. barthii* and *O. glaberrima*. (a) Relative abundance of small RNAs between *O. barthii* (*Ob*) and *O. glaberrima* (*Og*) according to their size (from 18- to 28-nt small RNAs). LogPlots of normalized abundance of distinct small RNA sequences. (b) Relative abundance of the annotation classes of 21-nt small RNAs between *O. barthii* (*Ob*) and *O. glaberrima* (*Og*). LogPlot of normalized abundance of distinct small RNA sequences. Black dots represent global 21-nt small RNAs, and red dots the class of annotated 21-nt small RNAs. (c) Relative abundance of the annotation classes of 24-nt small RNAs between *O. barthii* (*Ob*) and *O. glaberrima* (*Og*). LogPlot of normalized abundance of distinct small RNA sequences. Black dots represent whole 24-nt small RNAs, red dots represent the class of annotated 24-nt small RNAs.

Additional file 6: 21-nt small RNAs associated to phased loci and mature miRNA families detected in young panicles of *O. glaberrima* and *O. barthii*. (a) *O. sativa* Nipponbare loci associated with 21-nt phased and unphased siRNAs from *O. barthii* and *O. glaberrima* detected using sRNA-workbench facilities. (b) Normalized count of reads (relative abundance for 2 millions of total reads) associated with mature miRNA families in relation to small RNA size.

Additional file 7: Genomic distribution, abundance and complexity of *O. barthii* and *O. glaberrima* 21-nt phased small RNAs on *O. sativa nipponbare* genome (MSU 7.0). (a) Genomic distribution and abundance of *O. barthii* and *O. glaberrima* 21-nt phasiRNAs on *O. sativa nipponbare* genome (MSU 7.0). (b) Distribution of number of detected 21-nt phasiRNAs from African species per phased locus on *O. sativa nipponbare* genome (MSU 7.0).

Additional file 8: Sequences and phasiRNA abundances of the 4 phased loci used for validation. The 4 phased loci reported here were used for stem-loop RT-PCR (phasiRNAs) and classical RT-PCR (lncRNAs) validation. The positions of the phased loci on MSU7.0 chromosomes are indicated. The alignment between *miR2118f* and the phased locus is indicated. The *miR2118* recognition site is in blue. The phasiRNAs detected in the plus strand are highlighted alternately in green and in red. The phasiRNAs detected in the minus strand are underlined alternately in green and in red. The phasiRNA boxed in yellow is the most abundant one and the one used for stem-loop RT-PCR validation and qRT-PCR analysis. Histograms illustrate the relative abundance of the detected phasiRNAs per locus in the two species

Additional file 9: Features of *O. barthii* and *O. glaberrima* panicle-derived microRNAs and abundance of sequences related to microRNA precursor sequences. (a) Size distribution of clusters (distinct) and reads (abundance) related to *O. sativa* microRNAs in *O. barthii* and *O. glaberrima*. (b) Relative abundance of *O. barthii* and *O. glaberrima* reads related to *O. sativa* microRNA precursors. The precursors are classified according to the abundance distribution pattern as defined by Jeong et al. (2011) (i.e. canonical, variant, siRNA-like).

Additional file 10: Histological description of selected developmental stages of African rice panicles. *O. barthii*: 1,3,5,7,9,11; *O. glaberrima*: 2,4,6,8,10,12; stage 1: unbranched stage with elongation of rachis meristem (arrowheads) and formation of primary branch meristems (*) (1,2); stage 2: early branching stage with rachis meristem (arrowheads) and elongating primary branches (3,4). At the end of this stage, secondary branches (white *) are initiated from PBs (*) (5,6); stage 3: late branching stage with elongated secondary branch and spikelet meristem (SM) and floret meristem (FM) differentiation (7,8); stage 4: floret organ differentiation/development (9,10); mature stage: 11 and 12. White arrowhead: vestige of aborted rachis meristem. Scale bar: 100 μm.

Additional file 11: RNA-seq data validation by northern-blotting and semi-quantitative RT-PCR for miRNAs, phasiRNAs and ncRNAs associated with phased loci in panicle-derived small RNA bulks from *O. barthii* (*Ob*) and *O. glaberrima* (*Og*). (a) Northern blot hybridization using *miR2118* as probe. U6 probe was used as control. (b) Stem-loop RT- PCR analysis of various miRNAs and phasiRNAs. *miR159a* probe was used as a loading control. (c) Classic RT-PCR analysis of ncRNAs associated with 21-nt phased small RNA loci using from polydT primer (dT) or random hexamer primers (RH) for the RTs. The *ACTIN* gene (*Os03g50885*) was used as a loading control. (d) Test of specificity of stem-loop RT-PCRs against phasiRNAs on *PH12* and *PH779* loci, in conjunction with phasiRNA specific forward primer (PH12-F and PH779-F primers; phasiRNA label) and long ncRNA forward primer (PH12-F2 and PH779-F2 primers; long ncRNA label). Controls: 1. RT reaction using *O. glaberrima* RNA bulk without stem-loop RT primer; 2. RT reaction using *O. barthii* RNA bulk without stem-loop RT primer; 3. RT-PCR without stem-loop RT matrix; Ø: RT-PCR without RT. See Additional file 1: Table S5 for primer and probe sequences.

Additional file 12: Expression patterns of phasiRNAs and lncRNA precursors associated with 21-nucleotide phased loci in *O. glaberrima* and *O. barthii*. qRT-PCR analysis of 21-nucletiode phasiRNAs expression levels (a) and lncRNAs (b) from 3 distinct phased loci (*PH557*, *PH612* and *PH779*) during panicle development in *O. barthii* (red) and *O. glaberrima* (blue). Expression values are relative to *O. glaberrima* stage 4 and the *ACTIN* gene or *miR159* miRNA used as reference is indicated. Panicle stages: Stage 1, Inflorescence meristem stage, after initiation of primary branch (i.e. rachis and primary branch meristems); stage 2, early branching stage (panicle with elongated primary and higher order branch development); stage 3, late branching stage (i.e. panicle with elongated primary and secondary branches with spikelets and beginning of floret differentiation); stage 4, young flowers with differentiated organs. See Additional file 1: Table S5 for primer sequence.

Competing interests
The authors declare that they have no competing interests.

Authors' contributions

SF established and conducted the bio-informatic pipeline analysis of the small RNA dataset; JS, VY and SF carried out the analysis of bio-informatic out-puts; TKN performed molecular analysis (PCR, RT-PCR, *in situ* hybridization); MS conducted the initial bio-informatic analysis of data set; AH performed the sampling and RNA preparation; JS and VY initiated and organized the project; GP, DNV and GA managed the project. JS, SF, TKN, VY, AH and GP wrote manuscript together. All authors approved the final manuscript.

Acknowledgments

We thank members of the lab for their help with plant care and RNA extractions. We thank members of IRD-Montpellier bioinformatics platform for their technical support. We also thank colleagues from IRD-Montpellier for critical reading of the manuscript. This project was jointly supported by Agropolis Fondation (through the « Investissements d'avenir » programme (ANR-10-LABX-0001-01)) and Fondazione Cariplo under the reference ID EVOREPRICE 1201–004. This work was also supported by the Institut de Recherche pour le Développement (IRD) institutional funding, Agropolis foundation (ARCAD project for post-doctoral fellowship of SDM), Vietnamese government (program 322 to TKN) and the Global Rice Science Partnership (GRiSP) scholarship program (TKN).

Author details

[1]IRD, UMR DIADE, 911, avenue Agropolis, BP64501, F-34394 Montpellier, Cedex 5, France. [2]LMI RICE, National Key Laboratory for Plant Cell Biotechnology, Agronomical Genetics Institute, Pham Van Dong road, Hanoi, Vietnam. [3]Université de Montpellier, UMR DIADE, Place Eugène Bataillon, F-34095 Montpellier, Cedex 5, France. [4]Present address: INRA, Université de Lorraine, UMR 1136 Interactions Arbres/Microorganismes, F-54280 Champenoux, France.

References

Adam H, Marguerettaz M, Qadri R, Adroher B, Richaud F, Collin M, Thuillet AC, Vigouroux Y, Laufs P, Tregear JW, Jouannic S (2011) Divergent expression patterns of *miR164* and *CUP-SHAPED COTYLEDON* genes in palms and other monocots: implication for the evolution of meristem function in angiosperms. Mol Biol Evol 28:1439–1454

Allen E, Xie Z, Gustafson AM, Carrington JC (2005) MicroRNA-directed phasing during trans-acting siRNA biogenesis in plants. Cell 121:207–221

Altschul SF, Gish W, Miller W, Myers EW, Lipman DJ (1990) Basic local alignment search tool. J Mol Biol 215:403–410

Arikit S, Zhai J, Meyers BC (2013) Biogenesis and function of rice small RNAs from non-coding RNA precursors. Curr Opin Plant Biol 16:170–179

Bailey TL, Elkan C (1994) Fitting a mixture model by expectation maximization to discover motifs in biopolymers. In: Altman R, Brutlag D, Karp P, Lathrop R, Searls D (eds) Proceedings of the Second International Conference on Intelligent Systems for Molecular Biology. AAAI Press, Menlo Park

Caicedo AL, Williamson SH, Hernandez RD, Boyko A, Fledel-Alon A, York TL, Polato NR, Olsen KM, Nielsen R, McCouch SR, Bustamante CD, Purugganan MD (2007) Genome-wide patterns of nucleotide polymorphism in domesticated rice. PLoS Genet 3:1745–1756

Carroll S (2008) Evo-Devo and an expanding evolutionary synthesis: a genetic theory of morphological evolution. Cell 134:25–36

Cui R, Han J, Zhao S, Su K, Wu F, Du X, Xu Q, Chong K, Theissen G, Meng Z (2010) Functional conservation and diversification of class E floral homeotic genes in rice (*Oryza sativa*). Plant J 61:767–781

Doebley J, Lukens L (1998) Transcriptional regulators and the evolution of plant form. Plant Cell 10:1075–1082

Huang X, Kurata N, Wei X, Wang Z-X, Wang A, Zhao Q, Zhao Y, Liu K, Lu H, Li W et al (2012) A map of rice genome variation reveals the origin of cultivated rice. Nature 490:497–501

Jeon JS, Jang S, Lee S, Nam J, Kim C, Lee SH, Chung YY, Kim SR, Lee YH, Cho YG et al (2000) *leafy hull sterile1* is a homeotic mutation in a rice MADS box gene affecting rice flower development. Plant cell 12:871–884

Jeong D-H, Park S, Zhai J, Gurazada SGR, De Paoli E, Meyers BC, Green PJ (2011) Massive analysis of rice small RNAs: mechanistic implications of regulated microRNAs and variants for differential target RNA cleavage. Plant cell 23: 4185–4207

Johnson C, Kasprzewska A, Tennessen K, Fernandes J, Nan G-L, Walbot V, Sundaresan V, Vance V, Bowman LH (2009) Clusters and superclusters of phased small RNAs in the developing inflorescence of rice. Genome res 19: 1429–1440

Jones-Rhoades MW, Bartel DP, Bartel B (2006) MicroRNAS and their regulatory roles in plants. Ann Rev Plant Biol 57:19–53

Khanday I, Yadav SR, Vijayraghavan U (2013) Rice *LHS1/OsMADS1* controls floret meristem specification by coordinated regulation of transcription factors and hormone signaling pathways. Plant Physiol 161:1970–1983

Koes R (2008) Evolution and development of virtual inflorescences. Trends Plant Sci 13:1–3

Komiya R, Ohyanagi H, Niihama M, Watanabe T, Nakano M, Kurata N, Nonomura K-I (2014) Rice germline-specific Argonaute MEL1 protein binds to phasiRNAs generated from more than 700 lincRNAs. Plant J 78:385–397

Kozomara A, Griffiths-Jones S (2011) miRBase: integrating microRNA annotation and deep-sequencing data. Nucl Acids Res 39:D152–D157

Kyozuka J, Tokunaga H, Yoshida A (2014) Control of grass inflorescence form by the fine-tuning of meristem phase change. Curr Opin Plant Biol 17: 110–115

Li Y, Li C, Xia J, Jin Y (2011a) Domestication of transposable elements into MicroRNA genes in plants. PLoS ONE 6:e19212

Li Z-M, Zheng X-M, Ge S (2011b) Genetic diversity and domestication history of African rice (*Oryza glaberrima*) as inferred from multiple gene sequences. Theor Appl Genet 123:21–31

Li F, Pignatta D, Bendix C, Brunkard JO, Cohn MM, Tung J, Sun H, Kumar P, Baker B (2012) MicroRNA regulation of plant innate immune receptors. Proc Natl Acad Sci USA 109:1790–1795

Linares OF (2002) African rice (*Oryza glaberrima*): history and future potential. Proc Natl Acad Sci USA 99:16360–16365

Liu B, Chen Z, Song X, Liu C, Cui X, Zhao X, Fang J, Xu W, Zhang H, Wang X et al (2007) Oryza sativa *dicer-like4* reveals a key role for small interfering RNA silencing in plant development. Plant Cell 19:2705–2718

Liu Y, Wang Y, Zhu QH, Fan L (2013) Identification of phasiRNAs in wild rice (*Oryza rufipogon*). Plant Signal Behav 8:e25079

Martin M (2011) *Cutadapt* removes adapter sequences from high-throughput sequencing reads. EMBnetwork J 18:10–12

Moyroud E, Kusters E, Monniaux M, Koes R, Parcy F (2010) LEAFY blossoms. Trends Plant Sci 15:346–352

Nabholz B, Sarah G, Sabot F, Ruiz M, Adam H, Nidelet S, Ghesquière A, Santoni S, David J, Glémin S (2014) Transcriptome population genomics reveals severe bottleneck and domestication cost in the African rice (*Oryza glaberrima*). Mol Ecol 23:2210–2227

Nogueira FTS, Madi S, Chitwood DH, Juarez MT, Timmermans MCP (2007) Two small regulatory RNAs establish opposing fates of a developmental axis. Genes Dev 21:750–755

Nonomura K-I, Morohoshi A, Nakano M, Eiguchi M, Miyao A, Hirochika H, Kurata N (2007) A germ cell specific gene of the *ARGONAUTE* family is essential for the progression of premeiotic mitosis and meiosis during sporogenesis in rice. Plant Cell 19:2583–2594

Orjuela J, Sabot F, Chéron S, Vigouroux Y, Adam H, Chrestin H, Sanni K, Lorieux M, Ghesquière A (2014) An extensive analysis of the African rice genetic diversity through a global genotyping. Theor Appl Genet 127:2211–2223

Park SJ, Jiang K, Schatz MC, Lippman ZB (2012) Rate of meristem maturation determines inflorescence architecture in tomato. Proc Natl Acad Sci USA 109: 639–644

Park SJ, Eshed Y, Lippman ZB (2014) Meristem maturation and inflorescence architecture - lessons from the Solanaceae. Curr Opin Plant Biol 17:70–77

Peng T, Lv Q, Zhang J, Li J, Du Y, Zhao Q (2011) Differential expression of the microRNAs in superior and inferior spikelets in rice (*Oryza sativa*). J Exp Bot 62:4943–4954

Prusinkiewicz P, Erasmus Y, Lane B, Harder LD, Coen E (2007) Evolution and development of inflorescence architectures. Science 316:1452–1456

Rice P, Longden I, Bleasby A (2000) EMBOSS: The European Molecular Biology Open Software Suite. Trends Genet 16:276–277

Second G (1982) Origin of the genic diversity of cultivated rice (*Oryza* spp.): study of the polymorphism scored at 40 isozyme loci. Jap J Genet 57:25–57

Shivaprasad PV, Chen H-M, Patel K, Bond DM, Santos BACM, Baulcombe DC (2012) A microRNA superfamily regulates nucleotide binding site-leucine-rich repeats and other mRNAs. Plant Cell 24:859–874

Song X, Li P, Zhai J, Zhou M, Ma L, Liu B, Jeong D-H, Nakano M, Cao S, Liu C et al (2012a) Roles of DCL4 and DCL3b in rice phased small RNA biogenesis. Plant J 69:462–474

Song X, Wang D, Ma L, Chen Z, Li P, Cui X, Liu C, Cao S, Chu C, Tao Y et al (2012). Rice RNA-dependent RNA polymerase 6 acts in small RNA biogenesis and spikelet development. Plant J 71:378–389

Souaze F, Ntodou-Thome A, Tran CY, Rostene W, Forgez P (1996) Quantitative RT-PCR: limits and accuracy. Biotechniques 21:280–285

Stocks MB, Moxon S, Mapleson D, Woolfenden HC, Mohorianu I, Folkes L, Schwach F, Dalmay T, Moulton V (2012) The UEA sRNA workbench: a suite of tools for analyzing and visualizing next generation sequencing microRNA and small RNA datasets. Bioinformatics 28:2059–2061

Varkonyi-Gasic E, Wu R, Wood M, Walton EF, Hellens RP (2007) Protocol: a highly sensitive RT-PCR method for detection and quantification of microRNAs. Plant Meth 3:12

Vaughan DA, Lu BR, Tomooka N (2008) The evolving story of rice evolution. Plant Sci 174:394–408

Vogel JP, Garvin DF, Mockler TC, Schmutz J, Rokhsar D, Bevan MW, Barry K, Lucas S, Harmon-Smith M, Lail K et al (2010) Genome sequencing and analysis of the model grass Brachypodium distachyon. Nature 463:763–768

Wang Y, Shen D, Bo S, Chen H, Zheng J, Zhu Q-H, Cai D, Helliwell C, Fan L (2010) Sequence variation and selection of small RNAs in domesticated rice. BMC Evol Biol 10:119

Wang Y, Bai X, Yan C, Gui Y, Wei X, Zhu Q-H, Guo L, Fan L (2012) Genomic dissection of small RNAs in wild rice (*Oryza rufipogon*): lessons for rice domestication. New Phytol 196:914–925

Wang M, Yu Y, Haberer G, Marri PR, Fan C, Goicoechea JL, Zuccolo A, Song X, Kudrna D, Ammiraju JSS et al (2014) The genome sequence of African rice (*Oryza glaberrima*) and evidence for independent domestication. Nature Genet 46:982–988

Wei L, Yan L, Wang T (2011) Deep sequencing on genome-wide scale reveals the unique composition and expression patterns of microRNAs in developing pollen of *Oryza sativa*. Genome Biol 12:R53

Xia R, Meyers BC, Liu Z, Beers EP, Ye S, Liu Z (2013) MicroRNA superfamilies descended from *miR390* and their roles in secondary small interfering RNA biogenesis in Eudicots. Plant Cell 25:1555–1572

Yan Y, Zhang Y, Yang K, Sun Z, Fu Y, Chen X et al (2011) Small RNAs from MITE-derived stem-loop precursors regulate abscisic acid signaling and abiotic stress responses in rice. Plant J 65:820–828

Zhai J, Jeong D-H, De Paoli E, Park S, Rosen BD, Li Y, González AJ, Yan Z, Kitto SL, Grusak MA et al (2011) MicroRNAs as master regulators of the plant NB-LRR defense gene family via the production of phased, trans-acting siRNAs. Genes Dev 25:2540–2553

Zhai J, Zhang H, Arikit S, Huang K, Nan G-L, Walbot V, Meyers BC (2015) Spatiotemporally dynamic, cell-type-dependent premeiotic and meiotic phasiRNAs in maize anthers. Proc Natl Acad Sci USA 112:3146–3151

Zhang D, Yuan Z (2014) Molecular control of grass inflorescence development. Ann Rev Plant Biol 65:553–578

Zhang Q-J, Zhu T, Xia E-H, Shi C, Liu Y-L, Zhang Y, Liu Y, Jiang W-K, Zhao Y-J, Mao S-Y et al (2014a) Rapid diversification of five Oryza AA genomes associated with rice adaptation. Proc Natl Acad Sci USA 111:E4954–E4962

Zhang Y-C, Liao J-Y, Li Z-Y, Yu Y, Zhang J-P, Li Q-F, Qu L-H, Shu W-S, Chen Y-Q (2014b) Genome-wide screening and functional analysis identify a large number of long noncoding RNAs involved in the sexual reproduction of rice. Genome Biol 15:512

Zheng Y, Wang Y, Wu J, Ding B, Fei Z (2015) A dynamic evolutionary and functional landscape of plant phased small interfering RNAs. BMC Biol 13:32

18

Molecular Breeding of Thermo-sensitive Genic Male Sterile (TGMS) Lines of Rice for Blast Resistance using *Pi2* Gene

Jiefeng Jiang[1], Tongmin Mou[1*], Huihui Yu[2] and Fasong Zhou[2]

Abstract

Background: Blast disease caused by the fungal pathogen *Magnaporthe oryzae* is one of the big problems in rice production in China, especially for high yield hybrid varieties made from a two-line system in which thermo-sensitive genic male sterile (TGMS) lines are used. In this study, we report the introgression of a rice blast resistance gene *Pi2* from VE6219 into C815S, an elite rice TGMS line, leading to the development of blast resistant TGMS lines through marker assisted selection (MAS) and phenotypic selection approaches.

Results: Four new TGMS lines with blast resistance gene *Pi2* were developed from C815S (an elite TGMS line susceptible to the blast, used as recurrent parent) and VE6219 (a blast resistant line harboring *Pi2*, used as donor parent). The pathogenicity assays inoculated with 53 blast prevalent isolates in glasshouse showed that the blast resistant frequency of the four TGMS lines was 94.3%-98.1% that is equivalent to blast resistant donor parent VE6219. The field evaluation of the new lines and hybrids made from them at a blast epidemic site also showed high resistant levels against the blast. The genetic background of the newly developed TGMS lines were examined using a whole-genome single nucleotide polymorphism (SNP) array (RICE6K) that turned out more than 83% of the genomic markers were derived from the recurrent parent. The critical temperature points of fertility-sterility alteration of the new TGMS lines were between 22°C and 23°C of daily mean temperature, which is similar to that of C815S. The complete male sterility under natural growth conditions at Wuhan last more than 80 days. Their agronomic and grain quality traits meet the requirement for two-line hybrid rice production.

Conclusions: The broad-spectrum and durable rice blast resistant gene *Pi2* was introgressed into the elite TGMS line C815S background. The newly developed TGMS lines can be practically used for two-line hybrid rice breeding and must play an important role in sustainable rice production in China.

Keywords: Rice blast resistance; TGMS line; Two-line hybrid rice; Marker-assisted selection; *Pi2*

Background

Rice blast caused by *Magnaporthe oryzae* is the most destructive diseases widely prevalent in rice fields, leading to significant grain yield and quality reduction. The disease including leaf and panicle blast can cause yield losses from 10% to 30% in severely infected crops (Skamnioti and Gurr 2009). It has been a major problem for hybrid rice production due to the large area monoculture of a crop variety and quickly evolved major race of the pathogene (Bonman et al. 1986; Liu et al. 2010).

So far, 86 blast resistance genes have been documented (Liu et al. 2013). A few of them have already been incorporated into rice varieties, which are now widely cultivated in many countries. The rice blast resistance gene *Pi2* was introduced from a resistant *indica* rice cultivar 5173 into a susceptible cultivar CO39, and the derived isogenic line was named C101A51 (Mackill and Bonman 1992). At least six resistance genes (*Pi2, Pi9, Piz-t, Piz, Pigm* and *Pi50*) were mapped at the same location, close to the centromere of chromosome 6 (Deng et al. 2006; Hayashi et al. 2004; Liu et al. 2002; Zhu et al. 2012). *Pi2*

* Correspondence: tongmin58@mail.hzau.edu.cn
[1]National Key Laboratory of Crop Genetic Improvement, National Center of Plant Gene Research (Wuhan), Huazhong Agricultural University, Wuhan 430070, China
Full list of author information is available at the end of the article

is one of the broad spectrum resistance genes highly effective in rice blast control. Lines carrying *Pi2* were tested with 43 blast isolates collected from 13 countries and showed resistant to 36 of them (Liu et al. 2002). *Pi2* gene has been isolated and characterized, which encodes a NBS-LRR (a nucleotide-binding site and leucine-rich repeat) protein (Zhou et al. 2006). Marker-assisted selection (MAS) has become an important approach to the development of disease resistance cultivars, especially in the utilization of cloned genes, which largely reduces time and cost in breeding programs (Ishihara et al. 2014; Jiang et al. 2012; Narayanan et al. 2002). In this study we employed MAS to develop thermo-sensitive genic male sterile (TGMS) lines by introducing *Pi2* into the TGMS background.

Comparing with three-line rice hybrid production system, two-line system with TGMS has some advantages: (1) The male sterility were controlled by recessive nuclear genes. Any genotype with good combining ability can be used as a male parent. The frequency of obtaining heterotic hybrids in testcrosses is relatively higher than three-line system. (2) It can be used to develop hybrids of *japonica* or basmati in which there are very low frequency of cytoplasmic male sterile (CMS) restorers. (3) It is especially suited for developing sub specific hybrid (*indica/japonica*) because there is no restriction of restoring-maintaining relationship in TGMS system. (4) The cost for hybrid seed production is lower as compared to the three-line system. Two-line hybrid breeding has been widely used in China as an effective approach to the improvement of rice grain yield and quality. However, there is an urgent need to enhance its blast disease resistant. Here we report the line breeding process and main results in the development of TGMS lines with broad spectrum resistance to blast through MAS, and the subsequent check of their genetic background with the aid of a rice genomic SNP array.

Results

Introgression of rice blast resistance gene *Pi2* by MAS

F_1 plants obtained from the cross of C815S (an elite TGMS line) and VE6219 (carrying *Pi2*), were screened for presence of the target resistance gene *Pi2* using the gene linked molecular marker RM527 to identify the true F_1s showing heterozygous and were backcross with C815S in summer season of 2008 at Wuhan. Of 40 BC_1F_1 plants, 16 individual plants were identified to be positive for *Pi2* using the marker RM527. In BC_1F_1, two selection strategies were taken in the subsequent generations (Figure 1). One is to select positive fertile plants and to be used for further backcrossing with C815S. Other is to select male sterile plants that would be ratooned under low temperature for harvesting BC_1F_2 seeds. The advanced backcross progenies of BC_2 and BC_3 were generated from the crosses of selected positive BC_1F_1 (16 plants from 40) and BC_2F_1 (14 plants from 40) by MAS (Figure 1). The same method was repeated for advancing the next generation to reach BC_2F_2 and BC_3F_2. The male sterile plants with homozygous *Pi2* were selected in BC_1F_2, BC_2F_2 and BC_3F_2 populations and ratooned under low temperature to obtain next generation seeds. Finally, we developed four TGMS lines that harbored the homozygous marker linked to *Pi2* and also showed desirable agronomic traits in different backcross generations, which were designated as Hua1034S (BC_1F_6), Hua1032S (BC_2F_6), Hua1033S (BC_2F_6), and Hua1037S (BC_3F_6) (Figure 1 and Figure 2).

Genetic background examination of the newly developed TGMS lines

In this study, RICE6K, a whole-genome single nucleotide polymorphism (SNP) array was used to to analyze the genetic background of the newly developed TGMS lines. The identified high quality markers, 1256 out of 5102, showed polymorphisms between the recurrent parent, C815S, and the donor parent, VE6219. The genetic background recovery of the recurrent parent in Hua1032S, Hua1033S, Hua1034S and Hua1037S were 83.93%, 88.47%, 83.15% and 88.93%, respectively, measured by percentage of the polymorphic marker ratios (Figure 3). All four TGMS lines had large fragments of chromosome 6 because of foreground *Pi2* selection in each generation by MAS. These four TGMS lines must be further self-pollinated and selected since several fragments in each TGMS line were still heterozygous.

Evaluation of blast resistance

Leaf blast resistances of the newly developed TGMS lines and the two parents were assessed using 53 isolates of *M. oryzae* from Zhejiang and Guangdong provinces, China, in greenhouse. The donor parent VE6219, containing *Pi2* gene, showed broad-spectrum resistance to rice blast with a high resistance frequency of 90.6%, and the recurrent parent C815S was found susceptible to the most of isolates with the resistance frequency of 34.0%. The developed TGMS line Hua1037S was resistant against 50 of 53 blast isolates with a resistance frequency of 94.3% slight higher resistance frequency as of VE6219. Other improved TGMS lines viz. Hua1032S, Hua1033S and Hua1034S with 98.1% resistance frequencies showed higher resistance frequency than VE6219 (Table 1).

In order to evaluate the blast resistance of four TGMS lines and their derived hybrids in field, plants of each line were grown at Wangjia village of Yuan-An county, Hubei, China, where the altitude is about 600 meters, a hilly region with frequent blast epidemic in rice growing season, and it is a test site area for evaluation of the blast resistance of the China national rice regional test varieties. The results (Table 2) showed that C815S was

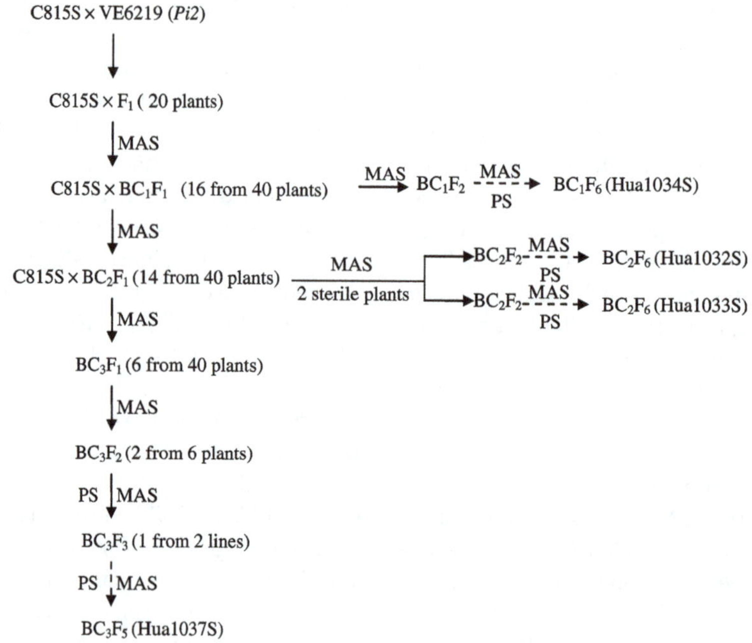

Figure 1 The scheme showing the development of rice blast resistant TGMS lines by introgression of rice blast resistant genes *Pi2* (donor parent: VE6219) into the susceptible popular TGMS recipient parent, C815S, through MAS approaches. MAS: marker assistance selection; PS: phenotypic selection.

susceptible to rice blast with 7 score of leaf blast and 100% of neck blast incidence, respectively. The leaf blast scores and neck blast incidences of the four TGMS lines were less than 1 and 6%, respectively, similar to donor parent VE6219. Two pollen parents (IR24 and R1005) and C815S' derived hybrids were susceptible to both leaf and neck blast. However, the hybrids derived from the four newly developed TGMS lines were resistant against leaf and neck blast (Table 2).

Characterization of fertility-sterility alteration in growth chamber

The critical temperature point (CTP) of fertility-sterility alteration has practical importance especially for a TGMS

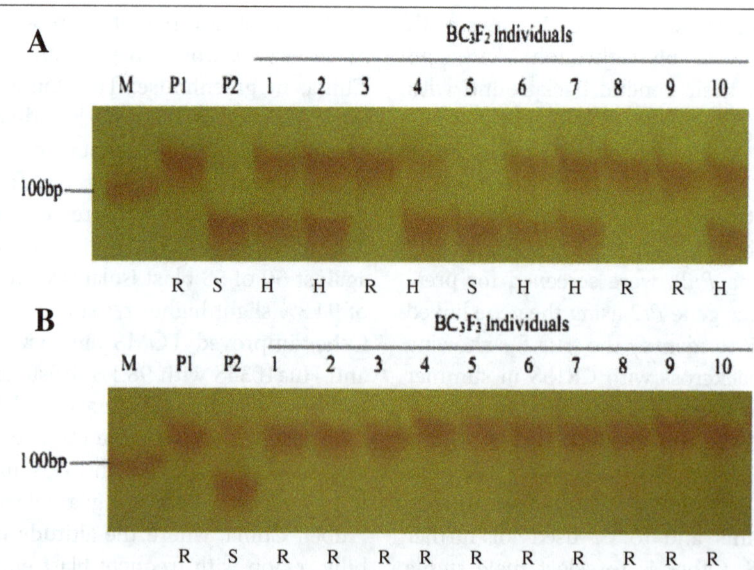

Figure 2 Marker co-dominant and polymorphic for *Pi2* in the populations. (A) marker RM527 for *Pi2* were fractionated on 4% polyacrylamide gel in the BC_3F_2 population, and **(B)** in the BC_3F_3 population. M: 50 bp DNA Marker; P_1: VE6219 (*Pi2* donor parent); P_2: C815S; lanes 1–10: 10 Plants of the BC_3F_2 or BC_3F_3 populations. R, S and H stand for the homozygous resistant, homozygous susceptible and heterozygote genotypes, respectively.

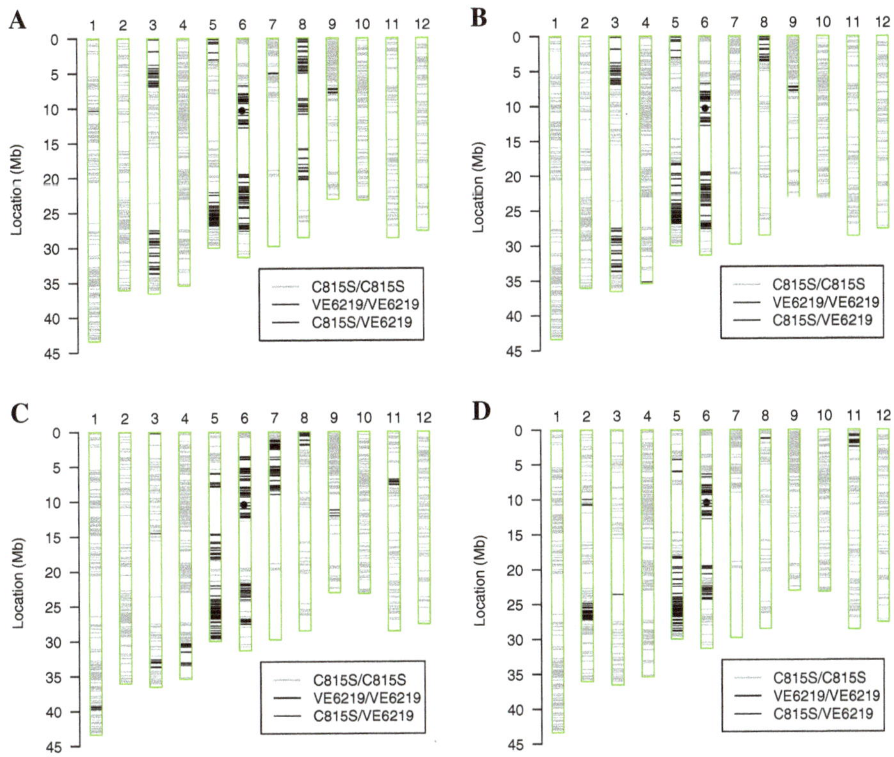

Figure 3 Haplotype maps of genetic background profiling using RICE6K array. (A) Hua1032S, **(B)** Hua1033S, **(C)** Hua1034S and **(D)** Hua1037S. The dots indicate the position of *Pi2* on chromosome 6.

line to safe seed production. In this study, the plants were placed in five growth chambers while the daily mean temperatures (DMT) were set for 21°C, 22°C, 23°C, 24°C and 25°C from 5–16 days after panicle initiation primordial stage for 12 days. The pollen grains collected from top five spikelets from each panicle per plant that headed during 5–16 days after the end of controlled growth chamber environment treatment were observed under microscope. The results showed that the pollen of the recurrent TGMS line i.e. C815S, was completely sterile (pollen sterility more than 99.5%) in the growth chambers with 22°C onwards to 25°C of DMT, but became partial fertile under 21°C, while the new TGMS lines, Hua1037S and Hua1034S had similar CTP of fertility-sterility alteration located in between 21°C and 22°C of DMT (Table 3). Therefore, it will be safe to carry out two-line hybrid seed production under conditions where DMT is higher than 22°C. However, the pollen sterility of Hua1032S and Hua1033S were less than 99.5% under 23°C while completely sterile in growth chambers with 24°C to 25°C DMT (Table 3). So, the natural DMT should be higher than 24°C for seed production safely.

In China, a practical usable TGMS line needs a stable sterility period of longer than 30 days in given region. We sowed TGMS lines at 15-day intervals at the Experimental Farm of Huazhong Agricultural University (HAU). The pollen fertilities in each line from the top five spikelets of primary panicles were investigated dynamically with two-day intervals under microscope from 30 June to 1 October. The pollen sterility data were recorded with average of five panicles from each line. The results showed that the pollen sterile periods of Hua1032S and Hua1033S were 88 days from 30 June to 26 September. Those of Hua1034S and Hua1037S were 80 days from 8 July to 26 September (Figure 4). Analysis of pollen sterility data in relation to temperature weather charts showed that when the DMT during 13 to 16 of September declined to below 22°C, the pollen grains of developed TGMS lines and C815S had shown reversal to partial fertility around 26 September. These phenomena indicated that the sensitive stage to temperature located at 13 days before heading. Although the DMT declined to 22.2°C on 23 August, the pollen of TGMS lines remained completely sterile, indicating their CTPs under natural conditions being similar as in growth chamber (Table 3 and Figure 4).

Evaluation of agronomic and rice quality traits of the newly developed TGMS

Four newly developed TGMS lines were evaluated for their agronomic and grain quality traits as compared to C815S at the Rice Breeding Station of Hainan in the

Table 1 Disease resistance reaction of newly developed TGMS lines (NT), recipient (RP) and donor parents (DP) along with susceptible check (SC) variety to 53 rice blast isolates under artificial inoculation in greenhouse

Blast isolates	C815S	Hua1032S	Hua1033S	Hua1034S	Hua1037S	VE6219	CO39
A13	R	R	R	R	R	R	S
A15	R	R	S	R	R	S	S
B01	S	R	R	R	R	R	S
B05	S	R	R	R	R	R	S
B09	S	R	R	R	R	R	S
B13(1)	R	R	R	R	R	R	S
B13(2)	S	R	R	R	R	R	S
B13(3)	S	R	R	R	R	R	S
B13(4)	S	R	R	R	S	R	S
B13(5)	S	R	R	R	R	R	S
B13(6)	S	R	R	R	R	R	S
B13(7)	S	R	R	R	R	R	S
B13(8)	S	R	R	R	R	R	S
B13(9)	S	R	R	R	R	R	S
B13(10)	R	R	R	R	R	S	S
B13(11)	S	R	R	R	R	R	S
B15(1)	S	R	R	R	R	R	S
B15(2)	S	R	R	R	R	R	S
B15(3)	R	R	R	R	R	R	R
B15(4)	R	R	R	R	R	R	R
B15(6)	R	R	R	R	R	R	R
B15(7)	S	R	R	R	R	R	S
C13(1)	S	R	R	R	R	R	S
C13(2)	S	R	R	R	R	R	S
C13(3)	R	R	R	R	R	R	S
C13(4)	S	R	R	R	R	R	S
C13(5)	R	R	R	R	R	R	S
C13(6)	R	R	R	R	R	S	S
C13(7)	R	R	R	S	S	S	S
C15(1)	S	R	R	R	R	R	S
C15(2)	S	R	R	R	R	R	S
C15(3)	R	R	R	R	R	R	S
G01	R	R	R	R	R	R	S
B9(11–109)	S	R	R	R	R	R	S
C15(12–103)	R	R	R	R	R	R	S
C15(H-1)	S	R	R	R	R	R	S
C15(10–151)	S	S	R	R	R	R	S
B9(09-78-20)	S	R	R	R	R	R	S
C15(12–217)	S	R	R	R	R	R	S
C15(12–211)	S	R	R	R	R	R	S
G1(10–85)	S	R	R	R	R	R	S
C9(06–113)	R	R	R	R	S	R	S
C9(12–180)	S	R	R	R	R	R	S

Table 1 Disease resistance reaction of newly developed TGMS lines (NT), recipient (RP) and donor parents (DP) along with susceptible check (SC) variety to 53 rice blast isolates under artificial inoculation in greenhouse *(Continued)*

B9(06–234)	S	R	R	R	R	S	S
B17(12–105)	S	R	R	R	R	R	S
D5(12–13)	S	R	R	R	R	R	S
G1(09–135)	S	R	R	R	R	R	S
C3(11–216)	R	R	R	R	R	R	S
C3(H-3)	R	R	R	R	R	R	S
D5(H-2)	S	R	R	R	R	R	S
C9(06–263)	S	R	R	R	R	R	S
E3(11–109)	R	R	R	R	R	R	S
B9(10–101)	S	R	R	R	R	R	S
NO. of resistant isolates	18	52	52	52	50	48	3
NO. of susceptible isolates	35	1	1	1	3	5	50
Resistance frequency (%)	34.0	98.1	98.1	98.1	94.3	90.6	5.7

winter season 2011–2012. The results (Table 4) showed non-significant differences for the days from sowing to heading, number of panicles per plant, 1000-grain weight, milling rice percentage, grain length and grain length-width ratio between the newly developed TGMS lines and C815S. The plant height of Hua1034S was higher than C815S by 6.5 cm. The panicle lengths of all four newly developed TGMS lines were longer than that of C815S by 2.1-3.7 cm. The number of grains per panicle of Hua1033S and Hua1034S were more than that of C815S. The spikelet fertility of Hua1032S and Hua1033S, the very important trait in TGMS line multiplication, were 22-31% more than that of C815S. The gel consistency of all the developed TGMS lines were increased 15–29 mm compared with the recurrent parent, C815S. Another desirable grain quality trait was higher amylose content, thus increased taste for customer in China.

Table 2 The performance of leaf and neck blast resistances for the four newly developed TGMS lines, their derived hybrids and control materials under the natural field of blast epidemic area

Entries	Score of leaf blast	Incidence of neck blast (%)
Hua1032S	0	6.0
Hua1033S	0	5.0
Hua1034S	1	0.0
Hua1037S	0	3.0
C815S	7	100.0
CO39	9	100.0
VE6219	0	5.0
Hua1032S/IR24	3	11.0
Hua1033S/IR24	0	11.0
Hua1034S/IR24	0	6.0
Hua1037S/IR24	4	10.0
C815S/IR24	5	29.0
IR24	7	99.0
Hua1037S/R1005	0	0.0
Hua1032S/R1005	0	8.0
Hua1033S/R1005	0	6.0
Hua1034S/R1005	0	0.0
C815S/R1005	4	39.0
R1005	4	54.0

Discussion

The two-line system for hybrid rice production has been widely used in China for increasing rice grain yield. However, the prevalence of rice blast disease is jeopardizing the expansion of the popular hybrids as a result of their susceptibility to the disease. Thus, to enhance the blast resistance of the parental lines is essential for sustainable utilization of the high-yielding hybrids. In this study, we sought to improve rice blast resistance in C815S because several commercial hybrids using C815S as the female parent were widely grown in the *indica*

Table 3 Fertility-sterility alteration behavior of the newly developed TGMS lines and the recurrent parent under five different temperature regimes in growth chambers

Entries	Pollen sterility (%)				
	21°C	22°C	23°C	24°C	25°C
Hua1032S	84.39	96.2	97.1	100.0	100.0
Hua1033S	81.29	95.3	99.4	100.0	100.0
Hua1034S	91.27	99.7	100.0	100.0	100.0
Hua1037S	86.48	100.0	100.0	100.0	100.0
C815S	69.62	99.5	100.0	100.0	100.0

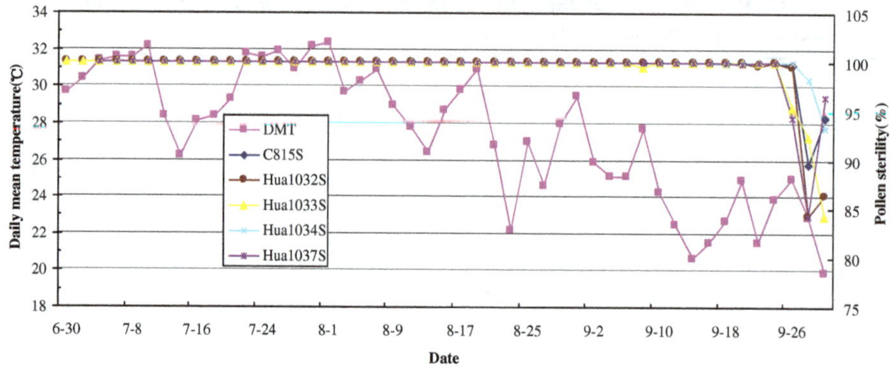

Figure 4 The dynamic pollen fertility expressions of the newly developed TGMS lines and C815S relative to daily mean temperature (DMT) data from 8th July to 1st October in 2012 at Wuhan.

rice growing area of the Central and Southern China. The closely linked and diagnostic marker RM527 was used for tracking *Pi2* in our breeding program, and the resistance gene was incorporated into C815S using backcrossing strategy. The newly developed TGMS lines, Hua1032S, Hua1033S, Hua1034S and Hua1037S, and the hybrids produced thereby, exhibited an elevated blast resistant. Our study clearly indicated that *Pi2* gene has a broad spectrum resistance to rice blast and is valuable for development of blast resistance rice hybrids. MAS-based backcross schemes were shown effective in the improvement of TGMS line for the blast resistance.

The genetic background of the four selected TGMS lines was examined by using a genomic SNP array, RICE6K. The results showed that the genetic background was not recovered to the recurrent parent levels although 3 cycles of backcrosses were performed. To recover a maximum recurrent parent genetic background in a short time, breeders would need a large backcross population and do background selection on top of the foreground check for the target gene (Rajpurohit et al. 2011). Suh et al. (2013) reported more than 92.1% recovery of background of recurrent parent Mangeumbyeo in BC_3F_5 with phenotypic selection without marker assisted background selection (MABS) during introgression of three bacterial blight resistance genes. In this project, we tried to develop new TGMS lines for stronger resistance against blast and with better agronomic traits to increase yields of the hybrid seeds, and our selections on the target traits obviously resulted in genetic drags. In our ongoing breeding program, we are using RICE6K to examine the genetic background of the selected lines and to more precisely remove the negative genetic drags. Interestingly, Hua1034S derived from BC_1F_1 contains 83.15% of genetic loci from C815S, similar to that of Hua1032S (83.93%) and Hua1033S (88.47%), which were derived from BC_2F_1 plants, and Hua1037S (88.93%), which was derived from a BC_3F_1 plant (Figure 1). It might be interpreted as that the recurrent genetic background was recovered faster than theoretically expected in early backcross generations because of selection on morphologic traits of the recurrent parent. In contrary, the recurrent genetic background was recovered slower than theoretically expected in higher backcross generations because of genetic drag of the introgressed gene, *i.e. Pi2*. With the aid of a whole-genome SNP array, the negative genetic drag would be removed more efficiently and the genetic background of a recurrent parent would be recovered quicker in higher backcross generations. The desirable allele *tms* loci of C815S was difficult to track using the linked molecular markers. Up to now, the *tms* gene/s in C815S has not been identified yet although several

Table 4 The performance of main agronomic and grain quality traits of newly developed blast resistant TGMS lines and C815S

TGMS line	DTH[a]	PH (cm)	PN	PL (cm)	NGP	SF (%)	GW (g)	MRP (%)	RGL (mm)	L/W	GC (mm)	AC (%)
Hua1032S	93.0	76.1	13.2	24.4**	145.8	63.6**	25.0	65.4	6.3	2.8	63.5**	12.7*
Hua1033S	93.0	75.9	12.6	23.5**	167.1*	72.6**	25.2	66.2	6.3	2.8	67.5**	12.2*
Hua1034S	97.0	83.7*	13.0	24.5**	160.9*	38.0	25.6	68.2	6.4	2.8	62.0**	15.0*
Hua1037S	96.0	72.6	12.8	23.1*	150.9	47.1	25.9	66.9	6.5	2.7	76.5**	15.4*
C815S	96.0	77.2	11.0	21.0	137.2	41.5	23.7	65.7	6.4	2.7	47.5	10.4

[a]DTH: days to heading, PH: plant height (cm), PN: panicle number, PL: panicle length (cm), NGP: number of grains per panicle, SF: spikelet fertility (%), GW: 1,000-grain weight (g), MRP: milling rice percentage, RGL: rice grain length, L/W: grain length-width ratio, GC: gel consistency, AC: amylose content, *, **Symbols followed after means indicate significant at the 5% or 1% significance level by the *T*-test.

other *tms* genes in TGMS rice have been mapped (Hussain et al. 2012), and the *tms* gene mapping results were influenced by background, even they had same *tms* gene source, e.g. Guangzhan63S and Nongken58S (Ding et al. 2012; Liu et al. 2001; Lu et al. 2005; Xu et al. 2011; Zhang et al. 1994).

Heading date, a critical trait that determines cropping seasons and regional adaptability in rice, depends strongly on photoperiodic responses. More than 16 genes/QTLs have been identified as being involved in the photoperiodic flowering pathway in rice (Matsubara et al. 2008; Yano et al. 2001). *Hd1*, a major photoperiod sensitivity gene on chromosome 6, was tightly linked to a blast resistance gene, *Pi-z^t* (Yano et al. 1997; Yano et al. 2000), the allele at *Pi2*/*Pi9* (Zhou et al. 2006). In our experiments, many combinations from C815S used as female parent were sensitive to photoperiod because they did not heading in normal season growth duration (data not shown). However, the combinations from the above same pollen parents and the newly developed TGMS lines harboring *Pi2* gene could head like other normal varieties. These phenomena implied that no photoperiod sensitive gene replaced the photoperiod sensitive gene in C815S simultaneously when we introgressed the chromosome fragment including *Pi2* gene from nonphotoperiod sensitive variety VE6219 into C815S.

Conventional disease resistance breeding is hard sledding, time-consuming, blindness compared with MAS which offers a powerful strategy to transfer genes from exotic germplasm into cultivated lines. It is important to note that introgression genes in rice should not lead to yield penalty or other drawbacks on important agronomic traits. But MAS only focused on one or several genes, several traits, ignore the comprehensive traits, so we should transvalue the agronomic traits of selected lines. The recently designed rice genomic SNP arrays (Yu et al. 2014; Chen et al. 2014) have offered molecular breeders excellent tools for the genetic background selection which could significantly reduce the recurrent cycles of backcrossing and help to break the negative genetic linkage between a target gene and poor agronomy trait. For TGMS lines, the most important trait is CTP of fertility-sterility alteration which related to safe hybrid rice seed and self seed production. In China, a TGMS line could not be released if its CTP was higher than 24°C of DMT. Seed multiplication will be very difficult if the CTP of a TGMS line was below 21°C of DMT. In our study, the CTPs of Hua1034S and Hua1037S were located at 22°C of DMT similar to the recurrent parent, C815S. However, Hua1032S and Hua1033S were located at 23°C of DMT. So, Hua1034S and Hua1037S can be used for hybrid seed production when the DMT during booting stage is over 22°C and can be used for TGMS line seed multiplication when below 22°C. However, Hua1032S and Hua1033S can be used for hybrid seed production when the DMT during booting stage is over 23°C and can be used for seed multiplication when below 23°C. Finally, Hua1037S was considered as excellent female parent for two-line blast resistant hybrid breeding because of its blast resistance, low CTP and better agronomic traits.

Conclusions

Rice blast disease is one of the most serious worldwide problems for rice production. Recently, it has become more serious and outspread, which urgently requires the development of rice blast resistant varieties. We successfully introgressed a broad-spectrum and durable rice blast resistance gene *Pi2* into an elite TGMS line to develop four blast resistance TGMS lines by using a dual-selection strategy of phenotypic and genotypic selection along with background screening to isolate improved breeding lines. These developed TGMS lines will be practical valuable for developing blast resistance two-line system hybrid rice in China.

Methods

Plant materials

C815S, an elite TGMS line widely used as female parent in two-line system hybrid rice breeding in China, but susceptible to blast was used as TGMS gene donor and recurrent parent. VE6219, a blast resistance breeding line (from the offspring between T1007 and C101A51) possessing *Pi2* gene, was used for blast resistance gene donor parent. The CO39 variety was used as a susceptible control for rice blast evaluation.

Molecular markers and PCR amplification

Adoption of MAS was facilitated by using simple sequence repeat (SSR) marker RM527 for *Pi2* (Chen et al. 2008). Total DNA was extracted according to the procedure of Dellaporta et al. (1983). PCR reactions were performed on a MyCycler™ thermal cycler (BIO-RAD USA). Each 20 µl PCR reaction mixture contained 20 ng genomic DNA, 10 mM Tris–HCl (pH 9.0), 50 mM KCl, 2.5 mM $MgCl_2$, 2 mM dNTPs, 10 µM each of the primer pair and 1 unit Taq DNA polymerase. Template DNA was initially denatured at 94°C for 5 min prior to 35 cycles of denaturation at 94°C (30s), annealing at 55°C (30s), and extension at 72°C (45 s). At the final step, the reaction mixture was incubated at 72°C for 5 min before the completion. The amplified products were then electrophoretically resolved on a 4% polyacrylamide gel in 1 × TBE buffer.

Rice blast evaluation

The newly developed TGMS lines, C815S (TGMS gene donor), VE6219 (blast resistance gene donor) and CO39 (blast susceptible CK) were inoculated in greenhouse by 62 isolates of *M. oryzae* from Zhejiang and Guangdong provinces, China. Leaf and neck blast resistance were identified under natural conditions by planting in the rice blast disease hotspot location, the Wangjia village of Yuan-An county, Hubei, China, where rice blast diseases were epidemic every year. The disease resistance reaction both in greenhouse and under natural fields were carried out according to standard evaluation system for rice (SES) (IRRI 2002).

Genetic background profiling by RICE6K

RICE6K, a whole-genome single nucleotide polymorphism (SNP) array, was used for genetic background profiling of newly developed TGMS line (Yu et al. 2014). RICE6K was developed based on Infinium technology, using representative SNPs selected from more than four million SNPs identified from resequencing data of more than 500 rice landraces. It contains 5102 SNP and InDel markers, about 4500 of which were of high quality in the tested rice lines producing highly repeatable results. For each line, total DNA was extracted from 20 dry seeds. DNA amplification, fragmentation, chip hybridization, single base extension, staining and scanning were conducted by Life Science and Technology Center, China National Seed Group Co., LTD (Wuhan, China), according to Infinium HD Assay Ultra Protocol (http://www.illumina.com/).

Characterization of TGMS lines for fertility-sterility alteration under growth chamber

100 seeds per line were sown at the nursery plots of Experimental Farm of HAU on May 10 of 2012. Uniform and healthy rice seedlings at the 5-leaf stage (about 25 days after sowing) were selected for transplanting of five single plants per hill per plastic pot and each plant was labeled with plastic tags. Care was taken with proper crop management to allow the plants in each plot to grow well and be uniform. Five growth chambers (Model: ZSX1500GS manufactured by The Shanghai Jing Wins and Scientific Equipment Co., Ltd., China) were adjusted for dry runs, one week prior to their actual use for the experiments. Light duration and relative humidity level was set to 14 h and 75% uniformly to all the five growth chambers while the daily mean temperatures were set for 21°C, 22°C, 23°C, 24°C and 25°C, respectively. The temperature and light duration were carefully programmed to follow diurnal patterns at the same time achieve the required daily mean temperature and day light conditions. The plants were placed in growth chambers from 5 to 16 days after panicle initiation primordial stage. The plants were moved out from the growth chambers after 12 days of temperature treatment. The pollen grains collected from top five spikelets from each panicle per plant that headed during 5–16 days after the end of controlled growth chamber environment treatment were observed under microscope. The pollen sterility of each panicle was recorded according to the classification of pollen morphology upon IK-I staining of pollen grains (Virmani et al. 2003). The line with more than 99.5% of pollen sterility on an average was considered as completely sterile.

Dynamic observation of pollen fertility-sterility alteration in the field

120 seeds in each line were sown at 15-day intervals at the Experimental Farm of HAU. Forty uniform and healthy rice seedlings at the 5-leaf stage were transplanted in fields with a spacing of 20 by 25 cm between plants and rows. The pollen fertilities in each line from the top five spikelets of primary panicles were investigated dynamically with two-day intervals under microscope from 30 June to 1 October. The pollen sterility data were recorded with average of five panicles from each line. The daily mean temperature data was provided by the Agricultural Meteorology Department of HAU. Analysis of pollen sterility data in relation to temperature weather charts was carried out to determine the CTP of fertility-sterility alteration under natural conditions.

Evaluation of agronomic and rice grain quality traits

Thirty plants each of the newly developed TGMS lines and its recipient parent C815S were transplanted in the field during the winter season of 2011 at Rice Breeding Station of HAU, Lingshui County of Hainan province, China. Each plot comprised of three rows with 10 plants per row at planting density of 17 cm between plants and 20 cm between rows. The agronomic and rice quality traits were measured according to standard evaluation system for rice (IRRI, 2002). Five individuals in the middle of the second row in each plot were taken for measurements of agronomic traits, including days from sowing to heading, plant height, panicles per plant, panicle length, spikelets per panicle, spikelet fertility and weight of 1,000 grains. *T*-test was to detect statistical differences between developed TGMS line and recurrent parent. Harvested bulk seeds from each plot were used for analyzing rice grain quality traits that included milling rice percentage, rice grain length, grain length-width ratio, gel consistency and amylose content were also measured.

Abbreviations

TGMS: Thermo-sensitive genic male sterile; CMS: Cytoplasmic male sterility; MAS: Marker assisted selection; SNP: Single nucleotide polymorphism; SSR: Simple sequence repeat; Indel: Insertion-deletion; CTP: Critical temperature point; DMT: Daily mean temperature.

Competing interests

The authors declare that they have no competing interests.

Authors' contributions

JFJ carried out the experiments, MAS, characterization of fertility-sterility alteration and evaluation of agronomic and rice grain quality traits. TMT conceptualized the study and did blast evaluation. HHY and FSZ did background profiling by RICE6K. JFJ and TMT prepared manuscript together. All authors read and approved the final manuscript.

Acknowledgments

This research was supported by the Introduction of International Advanced Agricultural Science and Technology Program of China (948 Program, Grant No. 2012-G2), the National High Technology Research and Development Programs of China (863 Programs, "Breeding of New Varieties of Green Super Rice", Grant No. 2014AA10A604), and the Bill & Melinda Gates Foundation "Green Super Rice for the Resources-Poor of Africa and Asia".

Author details

[1]National Key Laboratory of Crop Genetic Improvement, National Center of Plant Gene Research (Wuhan), Huazhong Agricultural University, Wuhan 430070, China. [2]Life Science and Technology Center, China National Seed Group Co., Ltd., Wuhan 430206, China.

References

Bonman JM, Dedios TIV, Khin MM (1986) Physiological Specialization of *Pyricularia-Oryzae* in the Philippines. Plant Dis 70(8):767–769

Chen HD, Xie WB, He H, Yu HH, Chen W, Li J, Yu RB, Yao Y, Zhang WH, He YQ, Tang XY, Zhou FS, Deng XW, Zhang QF (2014) A high-density SNP genotyping array for rice biology and molecular breeding. Mol Plant 7(3):541–553

Chen HQ, Chen ZX, Ni S, Zuo SM, Pan XB, Zhu XD (2008) Pyramiding three genes with resistance to blast by marker assisted selection to improve rice blast resistance of Jin 23B (in Chinese with English summary). Chinese J Rice Sci 22(1):23–27

Dellaporta SL, Wood J, Hicks JB (1983) A plant DNA minipreparation: version II. Plant Mol Biol Rep 1(4):19–21

Deng YW, Zhu XD, Shen Y, He ZH (2006) Genetic characterization and fine mapping of the blast resistance locus *Pigm(t)* tightly linked to *Pi2* and *Pi9* in a broad-spectrum resistant Chinese variety. Theor Appl Genet 113(4):705–713, doi:10.1007/s00122-006-0338-7

Ding JH, Lu Q, Ouyang YD, Mao HL, Zhang PB, Yao JL, Xu CG, Li XH, Xiao JH, Zhang QF (2012) A long noncoding RNA regulates photoperiod-sensitive male sterility, an essential component of hybrid rice. Proc Natl Acad Sci U S A 109(7):2654–2659

Hayashi K, Hashimoto N, Daigen M, Ashikawa I (2004) Development of PCR-based SNP markers for rice blast resistance genes at the *Piz* locus. Theor Appl Genet 108(7):1212–1220, doi:10.1007/s00122-003-1553-0

Hussain AJ, Ali J, Siddiq EA, Gupta VS, Reddy UK, Ranjekar PK (2012) Mapping of *tms8* gene for temperature-sensitive genic male sterility (TGMS) in rice (*Oryza sativa* L.). Plant Breed 131(1):42–47

IRRI (2002) Standard evaluation system for rice (SES). International Rice Research Institute, Los Banos, Laguna, Philippines

Ishihara T, Hayano-Saito Y, Oide S, Ebana K, La NT, Hayashi K, Ashizawa T, Suzuki F, Koizumi S (2014) Quantitative trait locus analysis of resistance to panicle blast in the rice cultivar Miyazakimochi. Rice 7:2, doi:10.1186/s12284-014-0002-9

Jiang HC, Feng YT, Bao L, Li X, Gao GJ, Zhang QL, Xiao JH, Xu CG, He YQ (2012) Improving blast resistance of Jin 23B and its hybrid rice by marker-assisted gene pyramiding. Mol Breed 30(4):1679–1688

Liu G, Lu G, Zeng L, Wang GL (2002) Two broad-spectrum blast resistance genes, *Pi9(t)* and *Pi2(t)*, are physically linked on rice chromosome 6. Mol Genet Genomics 267(4):472–480, doi:10.1007/s00438-002-0677-2

Liu J, Wang X, Mitchell T, Hu Y, Liu X, Dai L, Wang GL (2010) Recent progress and understanding of the molecular mechanisms of the rice-*Magnaporthe oryzae* interaction. Mol Plant Pathol 11(3):419–427

Liu N, Shan Y, Wang FP, Xu CG, Peng KM, Li XH, Zhang Q (2001) Identification of an 85-kb DNA fragment containing *pms1*, a locus for photoperiod-sensitive genic male sterility in rice. Mol Genet Genomics: Mol Genet Genomics 266(2):271–275

Liu Y, Liu B, Zhu XY, Yang JY, Bordeos A, Wang GL, Leach JE, Leung H (2013) Fine-mapping and molecular marker development for *Pi56(t)*, a NBS-LRR gene conferring broad-spectrum resistance to *Magnaporthe oryzae* in rice. Theor Appl Genet 126(4):985–998

Lu Q, Li XH, Guo D, Xu CG, Zhang Q (2005) Localization of *pms3*, a gene for photoperiod-sensitive genic male sterility, to a 28.4-kb DNA fragment. Mol Genet Genomics 273(6):507–511

Mackill DJ, Bonman JM (1992) Inheritance of Blast Resistance in near-Isogenic Lines of Rice. Phytopathology 82(7):746–749

Matsubara K, Kono I, Hori K, Nonoue Y, Ono N, Shomura A, Mizubayashi T, Yamamoto S, Yamanouchi U, Shirasawa K, Nishio T, Yano M (2008) Novel QTLs for photoperiodic flowering revealed by using reciprocal backcross inbred lines from crosses between *japonica* rice cultivars. Theor Appl Genet 117(6):935–945

Narayanan NN, Baisakh N, Cruz CMV, Gnanamanickam SS, Datta K, Datta SK (2002) Molecular breeding for the development of blast and bacterial blight resistance in rice cv. IR50. Crop Sci 42(6):2072–2079

Rajpurohit D, Kumar R, Kumar M, Paul P, Awasthi A, Basha PO, Puri A, Jhang T, Singh K, Dhaliwal HS (2011) Pyramiding of two bacterial blight resistance and a semidwarfing gene in Type 3 Basmati using marker-assisted selection. Euphytica 178(1):111–126

Skamnioti P, Gurr SJ (2009) Against the grain: safeguarding rice from rice blast disease. Trends Biotechnol 27(3):141–150, doi:10.1016/j.tibtech.2008.12.002

Suh JP, Jeung JU, Noh TH, Cho YC, Park SH, Park HS, Shin MS, Kim CK, Jena KK (2013) Development of breeding lines with three pyramided resistance genes that confer broad-spectrum bacterial blight resistance and their molecular analysis in rice. Rice 6:5, doi:10.1186/1939-8433-6-5

Virmani SS, Sun ZX, Mou TM, Ali J, Mao CX (2003) Two-line hybrid rice breeding manual. International Rice Research Institute, Los Banos, Laguna, Philippines

Xu JJ, Wang BH, Wu YH, Du PN, Wang J, Wang M, Yi CD, Gu MH, Liang GH (2011) Fine mapping and candidate gene analysis of *ptgms2-1*, the photoperiod-thermo-sensitive genic male sterile gene in rice (*Oryza sativa* L.). Theor Appl Genet 122(2):365–372

Yano M, Harushima Y, Nagamura Y, Kurata N, Minobe Y, Sasaki T (1997) Identification of quantitative trait loci controlling heading date in rice using a high-density linkage map. Theor Appl Genet 95(7):1025–1032

Yano M, Katayose Y, Ashikari M, Yamanouchi U, Monna L, Fuse T, Baba T, Yamamoto K, Umehara Y, Nagamura Y, Sasaki T (2000) *Hd1*, a major photoperiod sensitivity quantitative trait locus in rice, is closely related to the arabidopsis flowering time gene *CONSTANS*. Plant Cell 12(12):2473–2483

Yano M, Kojima S, Takahashi Y, Lin HX, Sasaki T (2001) Genetic control of flowering time in rice, a short-day plant. Plant Physiol 127(4):1425–1429

Yu HH, Xie WB, Li J, Zhou FS, Zhang QF (2014) A whole-genome SNP array (RICE6K) for genomic breeding in rice. Plant Biotechnol J 12(1):28–37

Zhang QF, Shen BZ, Dai XK, Mei MH, Maroof MAS, Li ZB (1994) Using Bulked Extremes and Recessive Class to Map Genes for Photoperiod-Sensitive Genic Male-Sterility in Rice. Proc Natl Acad Sci U S A 91(18):8675–8679

Zhou B, Qu SH, Liu GF, Dolan M, Sakai H, Lu GD, Bellizzi M, Wang GL (2006) The eight amino-acid differences within three leucine-rich repeats between *Pi2* and *Piz-t* resistance proteins determine the resistance specificity to *Magnaporthe grisea*. Mol Plant-Microbe Interact 19(11):1216–1228

Zhu X, Chen S, Yang J, Zhou S, Zeng L, Han J, Su J, Wang L, Pan Q (2012) The identification of *Pi50(t)*, a new member of the rice blast resistance *Pi2/Pi9* multigene family. Theor Appl Genet 124(7):1295–1304, doi:10.1007/s00122-012-1787-9

PERMISSIONS

All chapters in this book were first published in RICE, by Springer; hereby published with permission under the Creative Commons Attribution License or equivalent. Every chapter published in this book has been scrutinized by our experts. Their significance has been extensively debated. The topics covered herein carry significant findings which will fuel the growth of the discipline. They may even be implemented as practical applications or may be referred to as a beginning point for another development.

The contributors of this book come from diverse backgrounds, making this book a truly international effort. This book will bring forth new frontiers with its revolutionizing research information and detailed analysis of the nascent developments around the world.

We would like to thank all the contributing authors for lending their expertise to make the book truly unique. They have played a crucial role in the development of this book. Without their invaluable contributions this book wouldn't have been possible. They have made vital efforts to compile up to date information on the varied aspects of this subject to make this book a valuable addition to the collection of many professionals and students.

This book was conceptualized with the vision of imparting up-to-date information and advanced data in this field. To ensure the same, a matchless editorial board was set up. Every individual on the board went through rigorous rounds of assessment to prove their worth. After which they invested a large part of their time researching and compiling the most relevant data for our readers.

The editorial board has been involved in producing this book since its inception. They have spent rigorous hours researching and exploring the diverse topics which have resulted in the successful publishing of this book. They have passed on their knowledge of decades through this book. To expedite this challenging task, the publisher supported the team at every step. A small team of assistant editors was also appointed to further simplify the editing procedure and attain best results for the readers.

Apart from the editorial board, the designing team has also invested a significant amount of their time in understanding the subject and creating the most relevant covers. They scrutinized every image to scout for the most suitable representation of the subject and create an appropriate cover for the book.

The publishing team has been an ardent support to the editorial, designing and production team. Their endless efforts to recruit the best for this project, has resulted in the accomplishment of this book. They are a veteran in the field of academics and their pool of knowledge is as vast as their experience in printing. Their expertise and guidance has proved useful at every step. Their uncompromising quality standards have made this book an exceptional effort. Their encouragement from time to time has been an inspiration for everyone.

The publisher and the editorial board hope that this book will prove to be a valuable piece of knowledge for researchers, students, practitioners and scholars across the globe.

LIST OF CONTRIBUTORS

Ulrike Beukert, Zuo Li, Guozheng Liu, Yusheng Zhao and Jochen Christoph Reif
Department of Breeding Research, Leibniz Institute of Plant Genetics and Crop Plant Research (IPK) Gatersleben, Corrensstraße 3, 06466 Stadt Seeland, Germany

Nadhigade Ramachandra
Bayer Bioscience, 500081 Hyderabad, India

Vilson Mirdita
European Wheat Breeding Center, Bayer Crop Science, Am Schwabenplan 8, 06466 Stadt Seeland, Germany

Fabiano Pita
Biometrics and Breeding Research US, Bayer Crop Science, 407 Davis Dr., Morrisville, NC, USA

Klaus Pillen
K. Pillen, Chair of Plant Breeding, Martin-Luther-University Halle-Wittenberg, Betty-Heimann-Str. 3, 06120 Halle/Saale, Germany

Adam H. Price
Institute of Biological and Environmental Sciences, University of Aberdeen, Aberdeen AB24 3UU, UK

Mayuri Munasinghe
Institute of Biological and Environmental Sciences, University of Aberdeen, Aberdeen AB24 3UU, UK
University of Sri Jayewardenepura, Nugegoda, Sri Lanka

Tatsuya Yamamoto, Katsuyuki Oki, Yukimoto Iwasaki and Kotaro Miura
Fukui Prefectural University, Faculty of Biotechnology, 4-1-1 Kenjojima, Matsuoka, Eiheiji-cho, Yoshida-gun, Fukui 910-1195, Japan
Japan Society for the Promotion of Science, Chiyoda-ku, Tokyo 102-8472, Japan

Shuhei Segami
Fukui Prefectural University, Faculty of Biotechnology, 4-1-1 Kenjojima, Matsuoka, Eiheiji-cho, Yoshida-gun, Fukui 910-1195, Japan
Japan Society for the Promotion of Science, Chiyoda-ku, Tokyo 102-8472, Japan

Tomonori Noda, Motoyuki Ashikari and Hidemi Kitano
Bioscience and Biotechnology Center, Nagoya University, Furo-cho, Chikusa, Nagoya, Aichi 464-8601, Japan

Hiroyuki Kanamori, Harumi Sasaki, Satomi Mori and Yuichi Katayose
Agrogenomics Research Center, National Institute of Agrobiological Sciences, 2-1-2 Kannondai, Tsukuba, Ibaraki 305-8602, Japan

Shalabh Dixit, Ramil Mauleon and Hei Leung
Genetics and Biotechnology Division, International Rice Research Institute (IRRI), Los Baños, Laguna, Philippines

Nelzo C. Ereful
Genetics and Biotechnology Division, International Rice Research Institute (IRRI), Los Baños, Laguna, Philippines
The John Bingham Laboratory, National Institute of Agricultural Botany (NIAB), Huntingdon Road, Cambridge CB3 0LE, UK

Katrina Malabanan
Genetics and Biotechnology Division, International Rice Research Institute (IRRI), Los Baños, Laguna, Philippines
Crop Science Cluster, College of Agriculture, University of the Philippines, Los Baños, Laguna 4031, Philippines

Michael Thomson
Genetics and Biotechnology Division, International Rice Research Institute (IRRI), Los Baños, Laguna, Philippines
Texas A &M, Department of Soil and Crop Sciences 2474 TAMU, College Station, TX 77843-2474, USA

Li-Yu Liu, Eric Tsai and Shu-Min Kao
Department of Agronomy, National Taiwan University (NTU), Taipei City 100, Taiwan

Antonio Laurena
Institute of Plant Breeding, University of the Philippines, Los Baños, Laguna, Philippines

David Lee, Ian Mackay and Andy Greenland
The John Bingham Laboratory, National Institute of Agricultural Botany (NIAB), Huntingdon Road, Cambridge CB3 0LE, UK

Wayne Powell
SRUC, Peter Wilson Building, West Mains Road, Edinburgh EH9 3JG, UK

Dan Wang, Jinling Liu, Yue Wang, Minghao Liu, Yufei Deng, Zhilong Wang and Yinghui Xiao
Southern Regional Collaborative Innovation Center for Grain and Oil Crops in China and College of Agronomy, Hunan Agricultural University, Changsha, Hunan 410128, China

Guo-Liang Wang
Southern Regional Collaborative Innovation Center for Grain and Oil Crops in China and College of Agronomy, Hunan Agricultural University, Changsha, Hunan 410128, China
Department of Plant Pathology, Ohio State University, Columbus, OH 43210, USA

Chengang Li, Xinqiu Tan, Yong Liu and Deyong Zhang
Institute of Plant Protection, Hunan Academy of Agricultural Sciences, Changsha, Hunan 410128, China

Houxiang Kang
State Key Laboratory for Biology of Plant Diseases and Insect Pests, Institute of Plant Protection, Chinese Academy of Agricultural Sciences, Beijing 100193, China

Evert Thomas
Bioversity International, Lima, Peru

Eduardo Tovar
The Alexander von Humboldt Biological Resources Research Institute, Laboratory of Conservation Genetics, Bogota, Colombia

Carolina Villafañe
Ministry of Environment and Sustainable Development, Genetic Resources Group, Bogota, Colombia

José Leonardo Bocanegra and Rodrigo Moreno
The Alexander von Humboldt Biological Resources Research Institute, International Affairs, Policy and Cooperation Office, Bogota, Colombia

Anshuman Singh
Rani Lakshmi Bai Central Agricultural University, Jhansi 284003, India
International Rice Research Institute, DAPO 7777, Metro Manila, Philippines

Bertrand C. Y. Collard
Sugar Research Australia, 71378 Bruce Highway, Meringa, Gordonvale, QLD 4865, Australia
International Rice Research Institute, DAPO 7777, Metro Manila, Philippines

Endang M. Septiningsih
Department of Soil and Crop Sciences, Texas A&M University, College Station, TX 77843, USA
International Rice Research Institute, DAPO 7777, Metro Manila, Philippines

Jerome Carandang, Zennia Jean C. Gonzaga and Abdelbagi M. Ismail
International Rice Research Institute, DAPO 7777, Metro Manila, Philippines

Daniel F. Caddell, Patrick E. Canlas and Pamela C. Ronald
Department of Plant Pathology and the Genome Center, University of California, Davis, Davis, CA 95616, USA

Chang-Jin Park
Department of Plant Pathology and the Genome Center, University of California, Davis, Davis, CA 95616, USA
Department of Bioresources Engineering and PERI, Sejong University, Seoul 05006, Republic of Korea

Nicholas C. Thomas
Department of Plant Pathology and the Genome Center, University of California, Davis, Davis, CA 95616, USA
Department of Plant Pathology and Microbiology, University of California, Riverside, Riverside, CA 92521, USA

Sung-Ruyl Kim
Graduate School of Biotechnology & Crop Biotech Institute, Kyung Hee University, Yongin 17104, Republic of Korea
Plant Breeding, Genetics, and Biotechnology Division, International Rice Research Institute, Metro Manila, Philippines

Yun-Shil Gho, Sun-A Park, Anil Kumar Nalini Chandran, Gynheung An and Ki-Hong Jung
Graduate School of Biotechnology & Crop Biotech Institute, Kyung Hee University, Yongin 17104, Republic of Korea

Yunjoo Lee, Gileung Lee, Su Jang, Backki Kim, Mi-Ok Woo and Hee-Jong Koh
Department of Plant Science and Research Institute for Agriculture and Life Sciences, and Plant Genomics and Breeding Institute, Seoul National University, Seoul 08826, South Korea

Min-Seon Choi
Vegetable Crop Division, National Institute of Horticultural and Herbal Science, Rural Development Administration, Muan 534-833, South Korea

Mi-Ra Yoon
Department of Central Area Crop Science, National Institute of Crop Science (NICS), RDA, Suwon 16429, South Korea

Rihua Piao
Rice Research Institute, Jilin Academy of Agricultural Sciences, Gongzhuling, Jilin 136100, China

Joong Hyoun Chin
Graduate School of Integrated Bioindustry, Sejong University, 209, Neungdong-ro Gwangjin-gu, Seoul, South Korea

Ju-Won Kang and Sang-Nag Ahn
College of Agriculture and Life Sciences, Chungnam National University, Daejeon 305-764, South Korea

Hyun-Sook Lee
College of Agriculture and Life Sciences, Chungnam National University, Daejeon 305-764, South Korea
Graduate School of Life Sciences, Tohoku University, 2-1-1 Katahira, Aoba-ku, Sendai 980-8577, Japan

Kazuhiro Sasaki
Graduate School of Life Sciences, Tohoku University, 2-1-1 Katahira, Aoba-ku, Sendai 980-8577, Japan
Institute for Sustainable Agro-ecosystem Services, Graduate School of Agricultural and Life Sciences, The University of Tokyo, 1-1-1 Midoricho, Nishitokyo, Tokyo 188-0002, Japan

Tadashi Sato
Graduate School of Life Sciences, Tohoku University, 2-1-1 Katahira, Aoba-ku, Sendai 980-8577, Japan
Graduate School of Agricultural Science, Tohoku University, 1-1 Amamiya-machi, Tsutsumidori, Aoba-ku, Sendai, Miyagi 980-8555, Japan

Won-Yong Song
POTECH-UZH Cooperative Laboratory, Department of Integrative Bioscience and Biotechnology, Pohang University of Science and Technology, Pohang 790-784, South Korea

Shaohong Zhang, Xiuying He, Junliang Zhao, Yongsheng Cheng, Yuehan Chen, Tifeng Yang, Jingfang Dong, Xiaofei Wang, Qing Liu, Wei Liu, Xingxue Mao, Hua Fu, Zhaoming Chen, Yaoping Liao and Bin Liu
Rice Research Institute, Guangdong Academy of Agricultural Sciences, Guangzhou 510640, China

Zhimei Xie
Rice Research Institute, Hunan Agricultural University, Changsha 410128, China
Yahua Seed Science Research Institute, Longping High-tech, Changsha 410116, China

Shamitha Rao Morey, Ryu Ohsugi, Junko Yamagishi and Naohiro Aoki
Graduate School of Agricultural and Life Sciences, The University of Tokyo, 1-1-1 Yayoi, Bunkyo-ku, Tokyo 113-8657, Japan

Tatsuro Hirose
Graduate School of Agricultural and Life Sciences, The University of Tokyo, 1-1-1 Yayoi, Bunkyo-ku, Tokyo 113-8657, Japan
Central Region Agricultural Research Center, NARO, 1-2-1 Inada, Joetsu, Niigata 943-0193, Japan

Yoichi Hashida
Graduate School of Agricultural and Life Sciences, The University of Tokyo, 1-1-1 Yayoi, Bunkyo-ku, Tokyo 113-8657, Japan
Research Institute for Food and Agriculture, Ryukoku University, 1-5 Yokotani, Seta Oe-cho, Otsu, Shiga 520-2194, Japan

Jie Hu, Xingyuan Chang, Ling Zou, Weiqi Tang and Weiren Wu
Key Laboratory of Genetics, Breeding and Multiple Utilization of Crops, Ministry of Education, Fujian Agriculture and Forestry University, Fuzhou 350002, China
Fujian Key laboratory of Crop Breeding by Design, Fujian Agriculture and Forestry University, Fuzhou 350002, China

Akio Miyao and Hirohiko Hirochika
Advanced Genomics Breeding Section, Institute of Crop Science, NARO, 2-1-2, Kannondai, Tsukuba, Ibaraki 305-8518, Japan

Junliang Zhao, Wu Yang, Shaohong Zhang, Tifeng Yang, Qin Liu, Jingfang Dong, Hua Fu, Xingxue Mao and Bin Liu
Guangdong Key Laboratory of New Technology in Rice Breeding, Guangzhou 510640, China
Rice Research Institute, Guangdong Academy of Agricultural Sciences, Guangzhou 510640, China

Laura Fontanili, Bianca Dendena, Alessandro Ferri, Giorgio Lucchini, Gian Attilio Sacchi and Fabio Francesco Nocito
Dipartimento di Scienze Agrarie e Ambientali - Produzione, Territorio, Agroenergia (DISAA), Università degli Studi di Milano, 20133 Milan, Italy
Istituto di Biologia e Biotecnologia Agraria (IBBA), Consiglio Nazionale delle Ricerche, 20133 Milan, Italy

Clarissa Lancilli
Dipartimento di Scienze Agrarie e Ambientali - Produzione, Territorio, Agroenergia (DISAA), Università degli Studi di Milano, 20133 Milan, Italy
Istituto di Biologia e Biotecnologia Agraria (IBBA), Consiglio Nazionale delle Ricerche, 20133 Milan, Italy

Nobuo Suzui, Yong-Gen Yin, Satomi Ishii, Naoki Kawachi and Shu Fujimaki
Department of Radiation-Applied Biology, Takasaki Advanced Radiation Research Institute, Quantum Beam Science Research Directorate, National Institutes for Quantum and Radiological Science and Technology (QST), 1233 Watanuki, Takasaki, Gunma 370-1292, Japan

F. Sabot, H. Adam, Y. Vigouroux and A. Ghesquière
IRD, UMR DIADE, 911, avenue Agropolis, BP64501, F-34394 Montpellier, Cedex 5, France

K. N. Ta and S. Jouannic
IRD, UMR DIADE, 911, avenue Agropolis, BP64501, F-34394 Montpellier, Cedex 5, France
LMI RICE, National Key Laboratory for Plant Cell Biotechnology, Agronomical Genetics Institute, Pham Van Dong road, Hanoi, Vietnam

S. De Mita
IRD, UMR DIADE, 911, avenue Agropolis, BP64501, F-34394 Montpellier, Cedex 5, France
INRA, Université de Lorraine, UMR 1136 Interactions Arbres/Microorganismes, F 54280 Champenoux, France

N. V. Do
LMI RICE, National Key Laboratory for Plant Cell Biotechnology, Agronomical Genetics Institute, Pham Van Dong road, Hanoi, Vietnam

P. Gantet
LMI RICE, National Key Laboratory for Plant Cell Biotechnology, Agronomical Genetics Institute, Pham Van Dong road, Hanoi, Vietnam
Université de Montpellier, UMR DIADE, Place Eugène Bataillon, F-34095 Montpellier, Cedex 5, France

Jiefeng Jiang and Tongmin Mou
National Key Laboratory of Crop Genetic Improvement, National Center of Plant Gene Research (Wuhan), Huazhong Agricultural University, Wuhan 430070, China

Huihui Yu and Fasong Zhou
Life Science and Technology Center, China National Seed Group Co., Ltd., Wuhan 430206, China

Bin Zhang, Weijun Ye, Deyong Ren, Peng Tian, Youlin Peng, Yang Gao, Banpu Ruan, Li Wang, Guangheng Zhang, Longbiao Guo, Qian Qian and Zhenyu Gao
State Key Laboratory of Rice Biology, China National Rice Research Institute, Chinese Academy of Agricultural Sciences, Tiyuchang Road 359, Hangzhou 310006, China

Amit K. Tripathi, Brijesh K. Gupta and Sneh L. Singla-Pareek
Plant Molecular Biology Group, International Centre for Genetic Engineering & Biotechnology (ICGEB), Aruna Asaf Ali Marg, New Delhi 110067, India

Ashwani Pareek
Stress Physiology and Molecular Biology Laboratory, School of Life Sciences, Jawaharlal Nehru University, New Delhi 110067, India

Anil Kumar Singh
Plant Molecular Biology Group, International Centre for Genetic Engineering & Biotechnology (ICGEB), Aruna Asaf Ali Marg, New Delhi 110067, India
Division of Biotechnology, CSIR-Institute of Himalayan Bioresource Technology, Palampur - 176061, H.P. India

Ritesh Kumar
Plant Molecular Biology Group, International Centre for Genetic Engineering & Biotechnology (ICGEB), Aruna Asaf Ali Marg, New Delhi 110067, India
Division of Applied Life Science, Plant Molecular Biology and Biotechnology Research Center, Gyeongsang National University, Jinju 660-701, Korea

Index

www.ingramcontent.com/pod-product-compliance
Lightning Source LLC
LaVergne TN
LVHW080905210926
848485LV00004B/596